D.G.T. Thomas and D.I. Graham (Eds)

Malignant Brain Tumours

With 135 Figures
including 40 Colour Plates

Springer-Verlag
London Berlin Heidelberg New York
Paris Tokyo Hong Kong
Barcelona Budapest

David G.T. Thomas, MA, FRCP(G), FRCS Ed.
Professor, Institute of Neurology, Gough-Cooper Department of
Neurological Surgery, The National Hospital, Queen Square,
London WC1N 3BG

David I. Graham, MBBCh, PhD, FRCPath, FRS(Ed)
Professor, Department of Neuropathology, Institute of
Neurological Sciences, Southern General Hospital, Glasgow
G51 4TF

ISBN-13: 978-1-4471-1879-4   e-ISBN-13: 978-1-4471-1877-0
DOI: 10.1007/ 978-1-4471-1877-0

British Library Cataloguing in Publication Data
Malignant Brain Tumours
   I. Thomas, D.G.T.   II. Graham, David I.
   616.99

Library of Congress Cataloging-in-Publication Data
A catalog record for this book is available from the Library of Congress

© Springer-Verlag London Limited 1995
Softcover reprint of the hardcover 1st edition 1995

Typeset by EXPO Holdings, Malaysia.
28/3830-543210 Printed on acid-free paper

Contents

**3   Current Concepts of the Histopathological
     Classification of Tumours of the Central Nervous
     System**
     S.R. VandenBerg, R.B. Hessler and M.B.S. Lopes 29

Colour plate section appears between pages 42 and 43

4   General Introduction to the Clinical Features of Malignant Brain Tumours

5   Complications of Primary Malignant Brain Tumours

10   Diagnostic Imaging of Brain Tumours

Preface

Malignant brain tumours remain a not uncommon cause of disability and premature death, particularly in children and middle-aged adults. This is only too well known by those concerned with the clinical management of such patients. Thus, not only are malignant cerebral gliomas very aggressive tumours, usually associated with a rapidly fatal outcome, but also it has been recognised more recently that low grade gliomas often have a poor prognosis. Furthermore, the increased application of non-invasive neuroradiological imaging to patients with other types of tumour has disclosed an even larger number of metastatic brain tumours than has been recognised previously. In the last decade, the spectrum of malignant brain tumours has been further widened by the onset of tumours related to acquired immunodeficiency syndrome (AIDS). Nevertheless, all is not gloom, and within the community of life-science professionals in the neuro-oncological specialities, both scientific and clinical, there is some optimism that matters, albeit slowly, will change for the better. This hope is based largely on the realisation that large amounts of pertinent information exist about the biology of these tumours at molecular and cellular levels. It is also based on the ability to image them by CT and MR methods, often at an early stage in the disease, as well as on the capacity to achieve accurate histopathological correlation with the diagnostic imaging. Thus, the necessary prerequisites for the optimum treatment of these diseases by surgery, by radiation, and by drug treatment exist.

Perhaps, as never before, it is appreciated that a multidisciplinary approach to these difficult tumours is required and that advances will only be made by bridging between basic scientific and clinical disciplines. We believe that this indeed is happening and are confident that such an approach will be seen in improved patient outcome in the not too distant future.

D.G.T. Thomas
London 1994

D.I. Graham
Glasgow 1994

Acknowledgements

Special thanks are due to Dr. H.A.C. Cockburn, for assistance in proof reading and Miss H. Gossling for preparation of the manuscript.

Contributors

Junaid Ashraf, MBBS, FRCS Ed, FRCS Ed (SN)
Registrar, Neurosurgery
The Royal Free Hospital
Pond Street
London NW3 2QG
UK

Jay H. Beckstead, MD
Professor of Pathology
Department of Pathology
L113
Oregon Health Sciences University
Portland
Oregon 97201
USA

K. Bevan, BSc
Ludwig Institute for Cancer Research
Middlesex Hospital Branch
Courtauld Building
Riding House Street
London W1
UK

Michael Brada, BSc, MRCP, FRCR
Senior Lecturer and Consultant in Radiotherapy and Oncology
The Institute of Cancer Research
The Royal Marsden Hospital
Downs Road
Sutton
SURREY SM2 5PT
UK

R. Bradford, MD, FRCS
Consultant Neurosurgeon and Senior Lecturer in Neurosurgery
Department of Neurosurgery
The Royal Free Hospital
Pond Street
London NW3 2QG
UK

Richard L. Davis, MD
Professor of Pathology
Department of Pathology
School of Medicine, HSW 501
University of California, San Francisco
San Francisco
CA 94143-0506
USA

N. de Tribolet, MD
Service de Neurochirugie
Centre Hopital
Universitaire Vaudois
CH-1011 Lausanne
Switzerland

D.I. Graham, MBBCh, PhD. FRCPath
Professor of Neuropathology
Department of Neuropathology
Institute of Neurological Sciences
Southern General Hospital
Glasgow G51 4TF
UK

Anita E. Harding, MD, FRCP
Institute of Neurology
Queen Square
London WC1N 3BG

R.B. Hessler, MD
Assistant Professor of Pathology (Neuropathology)
Department of Pathology
The George Washington University Medical Center
Washington DC 20037
USA

J.V. Hunter, MRCP, FRCR
Department of Radiology
Massachusetts General Hospital
Boston
Massachusetts
USA

C.E. Jones, BSc
University Department of Clinical Neurology
Institute of Neurology
Queen Square
London WC1 3BG

B.E. Kendall, FRCP, FRCR, FRCS
Lysholm Radiological Department
The National Hospital for Neurology and Neurosurgery
Queen Square
London WC1 3BG
UK

J. Koivukangas, MD, PhD
Department of Neurosurgery
University of Oulu
Oulu
Finland

P. Krauseneck, MD
Nervenklinik Bamberg
Neurologische Klinik
Bamberg
Germany

M.B.S. Lopes, MD
Assistant Professor of Pathology (Neuropathology)
Department of Pathology
Box 214
University of Virginia School of Medicine
Charlottesville
VA 22908
USA

B. Müller, MD
Nervenklinik Bamberg
Neurologische Klinik
Bamberg
Germany

Mark Noble, PhD
Ludwig Institute for Cancer Research
Middlesex Hospital Branch
Courtauld Building
Riding House Street
London W1
UK

M.F. Pell, FRACS
Consultant Neurosurgeon
St. Vincent's Hospital
Sydney
Australia

J. Punt, FRCS
Consultant Neurosurgeon
Queens Medical Centre
Clifton Boulevard
Nottingham NG7 2UH
UK

Y. Sawamura, MD
Department of Neurosurgery
Hokkaido University School of Medicine
Nishi 7-chome Kita 15-jo
Kita-ku
060 Sapporo
Hokkaido
Japan

Yuen T. So, MD, PhD
Assistant Professor of Neurology
Department of Neurology
University of San Francisco
San Francisco General Hospital
1001 Potrero Avenue
San Francisco
CA 94110-0870
USA

Kintomo T. Takakura, MD
Department of Neurosurgery
University of Tokyo
7-3-1 Hongo
Bunkyo-ku
Tokyo 113
Japan

David G.T. Thomas, MA, FRCP(G), FRCS Ed
Consultant Neurosurgeon
Gough-Cooper Department of Neurological Surgery
Institute of Neurology
The National Hospital
Queen Square
London WC1N 3BG
UK

N.V. Todd, MD, FRCS
Consultant Neurosurgeon
Newcastle General Hospital
Westgate Road
Newcastle Upon Tyne NE4 6BE
UK

Scott R. VandenBerg, MD, PhD
Associate Professor of Pathology
Director of Neuropathology
Division of Neuropathology
Department of Pathology
University of Virginia School of Medicine
Charlottesville
Virginia 22908
USA

L. Vanuytsel, MD
Department of Radiotherapy
University Hospital St. Rafael
Capucijnevvoer 33
3000 Leuven
Belgium

D. Venter, MD
Department of Pathology
University of Melbourne
Parkville
Victoria 3052
Australia

1 Cellular and Molecular Approaches to the Study of Gliomas and Glial Development

Mark Noble, Karen Bevan and Deon Venter

It is striking to realize that the methods used for classification of gliomas have undergone relatively little change since the original studies of Bailey and Cushing in 1926. This conservative approach to glioma classification is all the more surprising when compared with the drastic changes which have occurred in the fields of both cancer research and glial biology, changes which have dramatically altered our thinking in those very same areas of research which underlie any attempt to develop a maximally useful classification system for glial tumours.

This chapter first reviews current knowledge about the development of the CNS glia and then discusses the discrepancies which occur when attempts are made to apply this body of knowledge to an understanding of glioma biology. Finally, approaches to the study of gliomas are suggested which might improve on our currently poor understanding of this important group of tumours.

Glial Cell Development in the Rat Optic Nerve

The starting point for a cellular biological analysis of glial development has been the optic nerve, the simplest part of the CNS. This tissue has been particularly useful for developmental studies because of the limited number of cell-types it contains. All of the neurons which send their axons through the optic nerve have their cell bodies located elsewhere, either in the retina or in the brain. Thus, the only neuroectodermally derived cells in this tissue are the astrocytes and oligodendrocytes, and their progenitor cells. Until 1983, the major question related to development of these optic nerve glia was whether they came from a single progenitor cell or whether astroblasts and oligodendroblasts represented two distinct glial lineages. However, since this time, it has become clear that the optic nerve is a substantially more complex tissue that was previously appreciated.

Current studies indicate that the optic nerve contains three differentiated glial cell-types, of which the first to develop is the *type-1 astrocyte*. This population of astrocytes appears to be specifically derived from the optic stalk, the outpouching of the neural tube which represents the embryonic anlage of the optic nerve, and is first seen at embryonic day 16 (E16; Small *et al.*, 1987). Type-1 astrocytes, and their precursors, seem to contribute to the morphogenetic development of the optic nerve by offering a preferred substrate for growing axons (Silver and Sapiro, 1981; Noble *et al.*, 1984). These cells also apparently interact with endothelial cells to induce formation of the blood – brain barrier (Janzer and Raff, 1987), and (as discussed below) are a source of mitogen for other cells in the nerve (Noble and Murray, 1984; Noble *et al.*, 1988; Raff *et al.*, 1988).

The next two glial cell-types to appear during development of the rat optic nerve are both derived from a single progenitor cell, the *oligodendro-cyte-type-2 astrocyte (O-2A) progenitor cell* (Raff *et al.*, 1983a,b), which migrates into the optic nerve from the optic chiasm during embryogenesis beginning on E16 (Small *et al.*, 1987). In vitro studies suggest strongly that division of the O-2A progenitor cells is promoted by the type-1 astrocytes derived from the optic stalk, through the secretion of platelet-derived growth factor (PDGF; Noble and Murray, 1984; Noble *et al.*, 1988; Raff *et al.*, 1988). Beginning at the time of birth (= E21), dividing O-2A progenitors begin to generate *oligodendrocytes* (Miller *et al.*, 1985) and these cells go on to enwrap large axons of the CNS with myelin sheaths. After another 7–10 days, the first *type-2 astrocytes* appear in the optic nerve (Miller *et al.*, 1985); these cells are thought to extend processes which are associated with axons at the nodes of Ranvier (ffrench-Constant and Raff, 1986; Miller *et al.*, 1989), the regions between consecutive myelin sheaths where ion fluxes occur during transmission of impulses along the myelinated axon. Thus, the O-2A lineage appears to be specialized to create the anatomical special-izations which characterize the myelinated tracts of the CNS (ffrench-Constant and Raff, 1986).

The cellular mechanisms involved in controlling the differentiation of O-2A progenitors into oligodendrocytes and type-2 astrocytes are quite different. Oligodendrocytic differentiation of progenitors induced to divide by type-1 astrocytes (or PDGF) is controlled by an internal clock which causes clonally related progenitors to synchronously differentiate (Temple and Raff, 1986; Raff *et al.*, 1988). In the presence of PDGF, the induction of differentiation is associated with a cessation of division (Noble *et al.*, 1988; Raff *et al.*, 1988). Similarly, if O-2A progenitors are grown in chemically-defined medium, in the absence of mitogens, cells differentiate rapidly into oligodendrocytes without undergoing any cell division. Thus, under these conditions there is a strict association between the onset of oligodendro-cytic differentiation and the cessation of cell division. In contrast to this internally controlled differentiation pathway, the differentiation of O-2A progenitors into type-2 astrocytes requires the presence of appropriate inducing factors (Hughes *et al.*, 1988; Lillien *et al.*, 1988).

A further member of the O-2A lineage is the O-2A*adult* progenitor cell, which differs from its perinatal counterpart in antigen expression, mor-phology, cell-cycle length, motility, time-course of differentiation, in the manner in which it generates oligodendrocytes and seemingly also in its capacity for extended self-renewal (Wolswijk and Noble, 1989; Noble *et al.*, 1992; Wren *et al.*, 1992). Two properties of O-2A*adult* progenitors are particularly noteworthy. First, they express several properties which suggest they are specialized to be more in keeping with the physiological requirements of the adult nervous system than O-2A^{perinatal} progenitors (Wolswijk and Noble, 1989; Wren *et al.*, 1992). Among these properties is included the ability to undergo asymmetric division and differentiation, such that a single clone of cells can produce both differentiated end-stage cells and more progenitors (Wren *et al.*, 1992). This pattern of division and differentiation is commonly seen in stem cell populations, and several lines of evidence indicate that the O-2A ^{adult} progenitor is the first stem cell to be identified in the adult nervous system. A second striking feature of O-2A^{adult} progenitors is that they appear to be derived from a subpopulation of O-2A^{perinatal} progenitors (Wren *et al.*, 1992). This generation of O-2A^{adult} progenitors from O-2A^{perinatal} progenitors suggests that progenitor populations are developmentally nested (Wren *et al.*,), such that cells with properties appropriate for early development give rise both to terminally committed end-stage cells and also to a new group of precursor cells with properties more appropriate for later developmental periods. For more detailed reviews on the O-2A lineage, see Anderson (1989), Noble *et al.* (1989) and Raff (1989).

Control of Self-Renewal in the O-2A Lineage

One of the recent findings to emerge from studies on the O-2A lineage has been to demonstrate a previously unsuspected linkage between the effects of growth factors on the differentiation of precursor cells and the effects of nuclear oncogenes on this same process. It has been found that O-2A^{perinatal} progenitors induced to divide by either PDGF or by basic fibroblast growth factor will generate oligodendrocytes, although the specific programs of differentiation activated by either individual mitogen are very different. In contrast, the simultaneous application of both mitogens to O-2A^{perinatal} progenitors inhibits oligodendrocytic differentiation and causes these cells to undergo continuous self-renewal (Bögler *et al.*, 1990). A similar inhibition of differentiation is seen when SV40 large T antigen, an oncogene of the nuclear immortalizing family (Jat and Sharp, 1989), is introduced into O-2A^{perinatal} progenitors (M. Noble *et al.*, unpublished observations). These results raise the possibility that specific combinations of growth factors might promote tumour cell growth not only by causing cell division, but also by contributing to the inhibition of differentiation which prevents tumour cells from following a normal progression from a dividing precursor cell into a non-dividing end-stage cell.

There are certain striking differences between O-2A^{adult} progenitors and O-2A^{perinatal} progenitors with respect to the control of self-renewal, and these differences may also have relevance for consideration of glioma biology. O-2A^{perinatal} progenitors induced to divide by type-1 astrocytes seem to represent a self-extinguishing population. As mentioned earlier, families of clonally related O-2A^{perinatal} progenitors generally undergo a limited number of cell divisions before all members of the family synchronously differentiate into oligodendrocytes (Raff *et al.*, 1985; Temple and Raff, 1986; Raff *et al.*, 1988). In contrast to the symmetric division and differentiation of O-2A^{perinatal} progenitors, when O-2A^{adult} progenitors are grown in identical conditions they are not self-extinguishing. Instead, through a process of asymmetric division and differentiation, these cells both produce oligodendrocytes and continuously replenish the progenitor pool (Wren *et al.*, 1990). In this population, which appears to be intrinsically immortal by virtue of its asymmetric behaviour, it is easy to see that processes which increase the probability of self-renewal could turn a cell from a harmless stem cell to a slowly-growing tumour. Strikingly, studies to date indicate that this probability of self-renewal can be dramatically increased simply by exposing the cell to basic fibroblast growth factor (FGF) instead of to the PDGF secreted by type-1 astrocytes (G. Wolswijk *et al.*, unpublished observations). Thus, a simple environmental change can, at least in vitro, alter the behaviour of this cell in a way which could initiate a focal and inappropriate cell growth by causing cells to divide without differentiating into non-dividing end-stage cells.

Glial and Neuronal Development in the Retina and Cortex

The analysis of cellular relationships in the retina and the cortex has been pursued rather differently than in the optic nerve. Although cellular biology has played some role in these studies, the most critical contributions have come from the use of retroviruses (i.e. RNA viruses) to label clones of related cells (for review, see Price, 1987). The first step is genetically to engineer the genome of a retrovirus to encode bacterial β-galactosidase, an enzyme which can be visualized histochemically. A small number of retroviral particles are then injected into a region of the developing nervous system, where occasional retroviruses will successfully infect single cells and have their viral genome incorporated into the cellular DNA. Following a productive integration of viral genes, all progeny resulting from an infected cell will express the β-galactosidase gene, and can subsequently be recognized. By initially injecting small numbers of viral particles, the frequency of successful infection is brought low enough to allow the subsequent identification of groups of clonally related cells.

The lineage relationships between the cells of the retina, as revealed by retroviral labelling, have been surprising. All possible relationships between the different cell-types of the retina have thus far been demonstrated, and clones have been visualized which consist only of neurons, only of retinal Müller cells (the retinal macroglial cell), and of both Müller cells and neurons (Turner and Cepko, 1987). Although these results are consistent with the view that the earliest retinal progenitors are all pluripotent cells, capable of giving rise to all retinal populations, it is also possible that even the earliest retinal progenitor cells will turn out to be heterogeneous in their developmental potential.

An additional surprising insight into retinal development has come with the observations that the astrocytes which are found in the retinas of many species are not even derived from the retinal progenitor cells. These astrocytes instead appear to be derived from astrocyte precursors of the optic nerve, and migrate into the retina during embryonic development (Watanabe and Raff, 1988). Thus, although the cells which form the retina begin development as just a farther portion of the same neural tube outpouching which forms the embryonic optic stalk, the retina develops into a tissue remarkably different from the optic nerve. The retinal progenitors develop into many different kinds of neurons and into retinal Müller cells, while the progenitor cells of the embryonic optic nerve give rise only to one population of astrocytes.

Developmental studies on the cortex have been as surprising as studies on the retina. Injection of β-galactosidase expressing retroviruses into the cortex of embryonic day 16 rats has led to the identification of several different kinds of clones (Price and Thurlow, 1988). Some clones appear to consist only of astrocytes, while other clones appear to consist only of neurons. However, the neuronal clones are not confined to one type of neuron. Even in clones with two identifiable members, these two β-galactosidase expressing neurons may be very different types of cells (e.g. pyramidal and non-pyramidal neurons). In addition, some clones consist of both obvious neurons and cells which are organized horizontally in the white matter tracts. As the only cells thus far known to be horizontally organized in the white matter are cells of the O-2A lineage, it may well be that at least some progenitor cells give rise to neurons, oligodendrocytes and type-2 astrocytes.

Ramifications of Current Developmental Studies for Neuropathology

Three concepts emerge from contemporary developmental neurobiological studies: (a) the lineage relationships between the cell types of the CNS are different from those predicted by earlier investigators; (b) the range of cell-types present in the CNS is different than envisaged by earlier investigators; and (c) the rules which govern development in one region of the CNS may not necessarily be valid in other regions.

Two additional important concepts have also specifically emerged from studies on the O-2A lineage. First, our studies on this lineage indicate that the very criteria used in assessing cells morphologically is not supported when tested against cellular biological analysis. Specifically, morphologists have used the presence of 10 nm (intermediate) filaments as a significant identifying element for "astroblasts" and the absence of such filaments as an identifying element for "oligoblasts". Yet, the O-2A^{perinatal} progenitor cell is rich in vimentin intermediate filaments (Raff *et al.*, 1984), whereas the O-2A^{adult} progenitor cell has no detectable intermediate filaments (Wolswijk and Noble, 1989). Thus, these cells would have been classified by ultrastructural criteria as "astroblasts" or "oligoblasts", respectively, even though both of the progenitors can develop into either oligodendrocytes or type-2 astrocytes. If morphology cannot accurately be used to classify normal glial precursors, then one must question the usefulness of such techniques for the correct classification of the gliomas, a population of transformed cells which are often suggested to be derived from glial precursors. Secondly, the demonstration that division of O-2A progenitors is promoted by cells of a different glial lineage (i.e. type-1 astrocytes) demonstrates that control of cell division in the CNS can be mediated by mitogens secreted by normal glial cells.

A Serological Analysis of Gliomas

The first technique the authors applied to glioma analysis was immunostaining. The antigenic phenotype of glioma cells in vitro and in situ was examined with antibodies which have been useful in studying normal glial populations in rodents and in humans (for details of these studies, see Kennedy *et al.*, 1987). Of these antibodies, four were particularly useful: anti-fibronectin, anti-glial fibrillary acidic protein (GFAP) and the A4 and A2B5 monoclonal antibodies. Fibronectin is a glycoprotein of the extracellular matrix and is associated with many cell types, such as fibroblasts and endothelial cells, but is usually not found on normal CNS cells except during early embryonic development (Schachner *et al.*, 1978; Stewart and Pearlman, 1987). GFAP, an intermediate filament protein, is specifically expressed by astrocytes within the CNS (Bignami *et al.*, 1972; Dahl and Bignami, 1975; Raff *et al.*, 1983a). Monoclonal antibody A2B5 (Eisenbarth *et al.*, 1979) has been used in studies of the rat optic nerve to distinguish between type-1 astrocytes (and their precursors) and cells of the O-2A lineage. Type-1 astrocytes are large flat cells which are A2B5$^-$, while type-2 astrocytes are A2B5$^+$ process-bearing cells with small cell bodies (Raff *et al.*, 1983a). O-2A progenitors and immature oligodendrocytes are also A2B5$^+$ (Raff *et al.*, 1983a; Raff *et al.*, 1984) and so are some cells not of the O-2A lineage; for example, many neurons are A2B5$^+$ (Eisenbarth *et al.*, 1979; Schnitzer and Schachner, 1982). Monoclonal antibody A4 is a more generally useful reagent which labels all CNS cells derived from the neural tube, but not cell types with other embryological origins (Cohen and Selvendran, 1981; Miller *et al.*, 1984). Thus astrocytes, oligodendrocytes, neurons and ependymal cells are all A4$^+$, whereas microglia, meningeal cells, brain endothelial cells, and cells of all other tissues are A4$^-$.

Glioma-Derived Cells can be Divided into Two Superfamilies on the Basis of Antigen Expression

Monoclonal antibody A4 could be used to subdivide the gliomas into two main antigenically-defined families. Surprisingly, the largest group of tumours was A4$^-$. Of the glioma-derived cultures, examined at passages 1 and 3, 129/149 labelled only with antibodies to fibronectin. The fibronectin$^+$ (FN$^+$) cells were uniformly A4$^-$GFAP$^-$A2B5$^-$. In contrast, a smaller number of tumours (14/149) were GFAP$^+$ and either A2B5$^+$ or A2B5$^-$; a still smaller number (2/149) were A2B5$^+$ and GFAP$^-$. All of the cells which were labelled with anti-GFAP or A2B5 antibodies were also A4$^+$ (four gliomas not included).

Confirmation that the FN$^+$ cells seen were truly tumour cells, and not contaminating fibroblast-like cells of non-tumour origin, came with several observations (Kennedy *et al.*, 1987). First the FN$^+$ glioma-derived cells were aneuploid, unlike normal human meningeal cells grown in culture. In addition, 22 of 25 tumours examined in sectioned biopsies contained FN$^+$ areas with no apparent morphological or antigenic relationship to blood vessels. Tumours which were clearly either FN$^+$ or A4$^+$ in section consistently gave rise to tissue culture populations with antigenic phenotypes like that of the tumour of origin. Similar sorts of observations have led other investigators also to conclude that the FN$^+$ cells derived from gliomas are correctly identified as tumour cells (McKeever *et al.*, 1987).

The most convincing demonstration that the FN$^+$ cells isolated from many gliomas are not only tumour cells, but are representative of the population found within the tumour mass in situ, has come with molecular biological studies. In these studies the authors have used the recent observations that a high proportion of gliomas have lost genetic material from chromosome 10 as a basis for characterizing FN$^+$ cells (James *et al.*, 1988). Patients were first identified whose lymphocyte-derived DNA showed them to be heterozygous at the position defined by the PLAU chromosome 10 probe (Pearson *et al.*, 1987), whose tumour biopsy specimens showed that one of the two alleles for this chromosome had been lost during neoplastic development and whose gliomas had given rise to FN$^+$ tissue culture populations. In all cases evaluated thus far, the genetic aberration seen within the tumour in situ is accurately reflected in tissue culture, thus demonstrating unequivocally that the FN$^+$ cells are bona fide tumour cells (Authors' unpublished observations).

Antigenic and Morphological Analyses of Gliomas Yield Conflicting Classification Systems

As summarized in Table 1.1, there was a remarkable lack of correlation between antigenic and morphological classification of individual tumours. By far the largest group of gliomas, consisting of the FN$^+$ gliomas, was not labelled by antibodies which have been used in the study of normal glial cells. These fibronectin$^+$ tumours were seen in every morphologically-defined category of glioma. Moreover, all gliomas which did express normal glial antigens (e.g. GFAP) had morphologically identical counterparts which were fibronectin$^+$.

Along with the major difference between the results of our antigenic studies and the predicted phenotypes expected from traditional neuropathological analysis, other discrepancies were also seen. For example, neuropathological analysis has been interpreted to indicate that cells derived from benign astrocytomas should express more astrocytic characteristics than those derived from malignant astrocytomas. In contrast, it

Table 1.1 Antigen expression in human glioma-derived cell cultures, analysed by original tumour histology*

Histological classification	Antigenic Phenotype	
	FN⁺	A4⁺
Astrocytoma grade I	6/6	0/6
Astrocytoma grade II	12/13	1/13
Astrocytoma grade III	22/24	2/24
Astrocytoma grade IV	56/68	9/68
Oligodendroglioma	12/12	0/12
Ependymoma	3/3	0/3
Medulloblastoma	8/10	2/10
Other (2 chordoma, 4 mixed o/a, 1 PNET, 1 rhabdomyosarcoma, 4 gliomas not otherwise specified)	10/13	2/13
Total of all gliomas	128/149	17/149

* Does not include four cultures, three of which contained both A4⁺ and FN⁺ populations at passage 3, and one of which did not stain with any of the antibodies used.

was found that almost all of the GFAP⁺ cultures were derived from the most malignant gliomas (9 GFAP⁺ cultures derived from 68 grade IV tumours) and almost no GFAP⁺ cultures were derived from the most benign tumours (1 GFAP⁺ culture from 6 grade I and 13 grade II astrocytomas). In addition, none of the 12 oligodendrogliomas examined yielded cells which expressed oligodendrocyte-specific antigens.

Potential for Positive Feedback Loops as Contributors to Growth of Gliomas

A striking paradox for those studying human neoplasms has been the knowledge that these tumours grow effectively in situ, yet often grow poorly, or not at all, in tissue culture medium lacking serum or defined mitogens. Thus, human tumours are by and large not wholly growth factor independent. We have made similar observations in our glioma cultures. The interpretation we have made of such observations is that glioma cells are supplied in vivo with factors which they themselves are incapable of producing, but which are required to induce the glioma cells to divide. Our studies on normal glial development raised the possibility that these additional factors might be supplied from normal cells contained within the tumour mass.

It is clear that tumours growing in situ contain both tumour cells and endothelial cells. Our observations that most astrocytomas give rise to FN⁺GFAP⁻ cells in culture suggested to us that many of the GFAP⁺ cells seen within gliomas in situ might be normal astrocytes, growing within the tumour as a consequence of secretion of astrocytic mitogens by tumour cells and/or endothelial cells. Thus, we found it a reasonable hypothesis that many gliomas contain at least endothelial cells, normal astrocytes and tumour cells. Hence, we determined whether endothelial cells or astrocytes were able to promote the division of cells derived from gliomas.

Our studies have indicated that each of the three cell types which are found in tumours in situ are capable of stimulating DNA synthesis in the two other cell types (Noble *et al.*, work in progress). Thus, astrocytes stimulate DNA synthesis in endothelial cells and in many tumour populations, endothelial cells stimulate astrocytes and many tumour populations, and many tumour cells secrete growth factors which stimulate DNA synthesis in endothelial cells and/or astrocytes. These results suggest to us that a major factor controlling the growth of gliomas in situ might be the interplay between the tumour cells and the normal cells of the immediate microenvironment. Tumour cells could stimulate the growth of nearby normal cells, and these cells could in turn release mitogens which promote division of the tumour cells. Similar ideas have also been proposed by Westermark, Heldin and their colleagues (Westermark *et al.*, 1985). It will now be of interest to identify specific mitogens which contribute to such positive-feedback loops, and to determine whether it is possible to interrupt tumour growth with appropriate antibodies against externally-supplied mitogens. Interference of this nature has been used in vitro to show that antibodies against platelet-derived growth factor inhibit the ability of type-1 astrocytes to promote division of O-2A progenitor cells (Noble *et al.*, 1988; Raff *et al.*, 1988).

Conclusion

By applying some of the tools of cellular biology to a study of gliomas it has been found that the families of tumours recognized by serological analysis are substantially different from those recognized by the analysis of cytology. These findings are of particular interest when considered in the light of the value of serological classification of the leukaemias, where the use of biological classification systems has been associated with improvements in the ability to predict clinical prognosis and to assign patients to appropriate therapeutic regimens (Greaves, 1981; Greaves *et al.*, 1981; van Eys *et al.*, 1986). Moreover, extensive analysis of lymphoid tumour cells has led to the view that the pattern of antigens expressed by these cells is consistent with the antigenic phenotypes of normal lymphoid cells (Greaves, 1981). By analogy with the experience of those who have studied antigen expression by leukaemic cells, it will be of interest to determine whether the expression of particular antigenic phenotypes by glioma-derived cells is of any prognostic value, either in respect to prediction of longevity or prediction of response to different therapeutic regimens. It will also be of interest to determine whether the FN[+] cells derived from gliomas correspond with any cell-types found in the developing or mature CNS, or whether tumours of this tissue can evolve into cells with no detectable relationship with their tissue of origin. If the gliomas accurately reflect the lineage history of at least a subset of the CNS glia, then the indications that FN[+] cells can occasionally be derived from GFAP[+] gliomas (Westphal and

Hermann, 1989) raise the interesting possibility that there exists in the CNS an astrocyte-like cell which can be induced to differentiate along apparently non-CNS pathways. The identification of such a cell would not only drastically alter our views about the development potential of glia and glial precursor cells, but also would indicate strongly that the antigenic approach to glioma classification is potentially of value in our attempts to understand the biology of this poorly understood tumour group and in our attempts to understand the lineage history of the CNS.

As our knowledge about the different cell-types which make up the CNS continues to grow, and the number of useful molecular and serological markers increases, it will be of continued interest to apply these new tools to the study of gliomas. Although cytological analysis, as currently practised, is of obvious value to the clinician and the patient, it is likely that modifications to current neuropathological practice will enhance our ability to help afflicted individuals. Current experience in related fields suggests that useful modifications will evolve around detailed analysis of the expression of lineage-specific antigens and of the genetic aberrations associated with tumour progression.

References

Anderson D (1989) New roles for PDGF and CNTF in development of the nervous system. Trends in Neurosci 12: 83–85

Bignami A, Eng LF, Dahl D, Uyeda CT (1972) Localisation of the glial fibrillary acidic protein in astrocytes by immunofluorescence. Brain Res 43: 429–435

Bögler O, Wren D, Barnett S, Land H, Noble M (1990) Cooperation between two growth factors promotes extended self-renewal, and inhibits differentiation, of O-2A progenitor cells. Proc Natl Acad Sci USA 87: 6368–6372

Cohen J, Selvendran SY (1981) A neuronal cell-surface antigen is found in the CNS but not in peripheral neurons. Nature 291: 421–423

Dahl D, Bignami A (1975) GFAP from normal and gliosed human brain. Demonstration of multiple and related polypeptides. Biochem Biophys Acta 386: 41–51

Eisenbarth GS, Walsh FS, Nirenberg M (1979) Monoclonal antibody to a plasma membrane antigen of neurons. Proc Natl Acad Sci USA 76: 4913–4917

ffrench-Constant C, Raff MC (1986) The oligodendrocyte-type-2 astrocyte cell lineage is specialized for myelination. Nature 323: 335–338

Greaves MF (1981) Analysis of the clinical and biological significance of lymphoid phenotypes in acute leukemia. Cancer Res 41: 4752–4766

Greaves MF, Janossy G, Peto J, Kay H (1981) Immunologically defined subclasses of acute lymphoblastic leukaemia in children: their relationship to presentation features and prognosis. Br J Haem 48: 179–197

Hughes S, Lillien LE, Raff MC, Rohrer H, Sendtner M (1988) Ciliary neurotrophic factor induces type-2 astrocyte differentiation in culture. Nature 335: 70–73

James CD, Carlbom E, Dumanski JP *et al.* (1988) Clonal genomic alterations in glioma malignancy stages. Cancer Res 48: 5546–5561

Jat PS, Sharp PA (1989) Cell lines established by a temperature-sensitive simian virus 40 large-T antigen gene are growth restricted at the nonpermissive temperature. Molec Cell Biol 9: 1672–1681

Janzer R, Raff MC (1987) Astrocytes induce blood-brain barrier properties in endothelial cells. Nature 325: 253–257

Kennedy PGE, Watkins BA, Thomas DGT, Noble M (1987) Antigenic expression by cells derived from gliomas does not correlate with morphological classification. Neuropath Appl Neurobiol 13: 327–347

Lillien LE, Sendtner M, Rohrer H, Hughes, SM, Raff MC (1988) Type-2 astrocyte development in rat brain cultures is initiated by a CNTF-like protein produced by type-1 astrocytes. Neuron 1, 485–494

McKeever PE, Smith BH, Taren JA et al. (1987) Products of cells cultured from gliomas. IV; Immunofluorescent, morphometric and ultrastructural characterisation of two different cell types growing from explants of human gliomas. Am J Pathol 127: 358–372

Miller RH, Williams BP, Cohen J, Raff MC (1984) A4: An antigenic marker of neural tube-derived cells. J Neurocytol 13: 329–338

Miller RH, David S, Patel ER, Raff MC (1985) A quantitative immunohistochemical study of macroglial cell development in the rat optic nerve: in vivo evidence for two distinct astrocyte lineages. Devel Biol 111: 35–43

Miller RH, Fulton B, Raff MC (1989) A novel type of glial cell associated with nodes of Ranvier in rat optic nerve. Eur J Neurosci 1: 172–180

Noble M, Fok-Seang J, Cohen J (1984) Glia are a unique substrate for the in-vitro growth of central nervous system neurons. J Neurosci 4: 1892–1903

Noble M, Murray K (1984) Purified astrocytes promote the in vitro division of a bipotential glial progenitor cell. EMBO J 3: 2243–2247

Noble M, Murray K, Stroobant P, Waterfield M, Riddle, P (1988) Platelet-derived growth factor promotes division and motility and inhibits premature differentiation of the oligodendrocyte-type-2 astrocyte progenitor cell. Nature 333: 560–562

Noble M, Wolswijk G, Wren D (1989) The complex relationship between cell division and the control of differentiation in oligodendrocyte-type-2 astrocyte progenitor cells isolated from perinatal and adult rat optic nerves. Prog Growth Factor Res 1: 179–194

Noble M, Wren D, Wolswijk G (1992) The O-2A^{adult} progenitor cell: A glial stem cell of the adult central nervous system. Semin Cell Biol 3: 413–422

Pearson PL, Kidd KK, Willard HF (1987) Human gene mapping by recombinant DNA techniques. Cytogen Cell Genet 46: 390–566

Price J (1987) Retroviruses and the study of cell lineage. Development 101: 409–419

Price J, Thurlow L (1988) Cell lineage in the rat cerebral cortex: a study using retroviral-mediated gene transfer. Development 104: 473–482

Raff MC, Abney ER, Cohen J, Lindsay R, Noble M (1983a) Two types of astrocytes in cultures of developing rat white matter: differences in morphology, surface gangliosides and growth characteristics. J Neurosci 3: 1289–1300

Raff MC, Miller RH, Noble M (1983b) A glial progenitor cell that develops into an astrocyte or an oligodendrocyte depending on the culture medium. Nature 303: 390–396

Raff MC, Williams BP, Miller R (1984) The in vitro differentiation of a bipotential glial progenitor cell. EMBO J 3: 1857–1864

Raff MC, Abney ER, Fok-Seang J (1985) Reconstitution of a developmental clock *in vitro*: a critical role for astrocytes in the timing of oligodendrocyte differentiation. Cell 42: 61–19

Raff MC, Lillien LE, Richardson WD, Burne JF, Noble, M (1988) Platelet-derived growth factor from astrocytes drives the clock that times oligodendrocyte development in culture. Nature 333: 562–565

Raff MC (1989) Glial cell diversification in the rat optic nerve. Science 243: 1450–1455

Schachner M, Schoonmaker G, Hynes RO (1978) Cellular and subcellular localisation of LETS protein in the nervous system. Brain Res 158: 149–158

Schnitzer J, Schachner M (1982) Cell type-specificity of neural cell surface antigen recognized by monoclonal antibody A2B5. Cell Tiss Res 224: 625–636

Silver J, Sapiro J (1981) Axonal guidance during development of the optic nerve: the role of pigmented epithelia and other intrinsic factors. J Comp Neurol 202: 521–538

Small RK, Riddle P, Noble M (1987) Evidence for migration of oligodendrocyte-type-2 astrocyte progenitor cells into the developing rat optic nerve. Nature 328: 155–157

Stewart GR, Pearlman AL (1987) Fibronectin-like immunoreactivity in the developing cerebral cortex. J Neurosci 7: 3325–3333

Temple S, Raff MC (1986) Clonal analysis of oligodendrocyte development in culture: evidence for a developmental clock that counts cell divisions. Cell 44: 773–779

Turner DL, Cepko CL (1987) A common progenitor for neurons and glia persists late in development. Nature 328: 131–136

van Eys J, Pullen J, Head D et al. (1986) The French-American-British (FAB) classification of leukaemia – the Pediatric Oncology Group experience with lymphocytic leukaemia. Cancer 57: 1046 1051

Watanabe T, Raff MC (1988) Retinal astrocytes are immigrants from the optic nerve. Nature 332: 834–837

Westphal M, Herrmann H-D (1989) Growth factor biology and oncogene activation in human gliomas and their implication for specific therapeutic concepts. Neurosurg 25: 681–694

Westermark B, Nister M, Heldin C-H (1985) Growth factors and oncogenes in malignant glioma. Neurolog Clin 3: 785–799

Wolswijk G, Noble M (1989) Identification of an adult-specific glial progenitor cell. Development 105: 387–400

Wren D, Wolswijk G, Noble M (1992) In vitro analysis of origin and maintenance of O-2A[adult] progenitor cells. J Cell Biol 116: 167–176

2 Molecular Genetics of Brain Tumours

C.E. Jones and A.E. Harding

Genetic Techniques in Studies of Brain Tumours

Through the application of molecular biological techniques to the investigation of tumour biology, an understanding of some of the mechanisms involved in neoplasia is beginning to emerge. In studying tumours, it is important to determine which of the gross or submicroscopic alterations to the genome found in tumour cells are causative, and which are a consequence of transformation. In order to do this it is necessary to look for consistent changes, or changes specifically associated with the type or stage of tumour growth.

Genetic studies of tumours initially concentrated on cytogenetic analyses, which will detect gross chromosomal abnormalities. These can be numerical or structural in nature. The normal chromosomal complement is diploid; numerical alterations involve the loss (monosomy) or gain (trisomy or polysomy) of parts of chromosomes or whole chromosomes, or the gain of an entire complement of chromosomes (polyploidy). Structural alterations include rearrangements within chromosomes, such as deletions, duplications and inversions, or rearrangements between chromosomes such as translocations. These changes can result in cells having a balanced or unbalanced genome, depending on whether there is any net loss or gain of genetic material. Cytogenetics can also detect the gain of additional chromosomal fragments, termed double minutes (DMs), which are frequently found in tumour cells and are an indication of gene amplification.

Karyotype analysis requires that cells are actively dividing, so it may not be possible in solid tumours as many have a very low mitotic index and there may not be enough suitable cells for study. This is often overcome by culturing tumour cells, but this in itself can give rise to further problems. Cultured tumour cells may not necessarily represent the major cell component in vivo, particularly if the tumour has a heterogeneous cellular constitution such as is found in gliomas. Additional alterations to the genome may arise in culture that were not present in the original tumour, resulting in new cell clones which could take over as the dominant cell line in vitro. Conditions in cell culture obviously differ from those in vivo and thus there will be selection for characteristics that give rise to an in vitro growth advantage. Cells often accumulate cytogenetic aberrations, such as polysomy and polyploidy, with increasing time in culture. Such difficulties can be partly overcome by the use of short-term (up to 72 h) cultures. There are also problems of infection of cultures and of the possibility of inadvertently culturing normal cells present in the tumour.

Cytogenetic techniques can only detect changes in the genome if the size of the region involved is greater than about 3000–5000 kb. Thus a loss or gain of genetic material detected cytogenetically may affect many genes simultaneously. In the case of a balanced translocation, it may be that only the genes which lie at the position of the breakpoints are affected, but the precise location of these cannot be determined by karyotype analysis.

Cytogenetic findings can give clues as to where to look for changes in the genome in more detail using molecular genetic techniques (Davies and Read, 1988). These also have the advantage of not requiring actively dividing cells for study. The use of Southern blotting analysis, separating DNA fragments and identifying specific parts of the genome, has now been widely applied to the study of tumours. This technique (Fig. 2.1) involves the isolation of DNA from tissue and its digestion into small fragments by the use of restriction enzymes. These enzymes, which occur naturally in bacteria, are capable of recognizing specific DNA sequences and cleave double stranded DNA symmetrically at these sites, but, equally importantly, nowhere else. Following digestion of DNA, the resulting fragments can be separated on the basis of size by applying an electric current to the DNA and running it through an agarose gel. The separated DNA fragments can then be transferred from the gel onto a nitrocellulose or nylon filter by Southern blotting. The DNA bound to the filter is then denatured to single stranded form and incubated with a DNA probe. This probe consists of a labelled (usually radioactive) fragment of single stranded DNA which is homologous to a sequence or sequences found in the genome. After hybridization, the excess unbound probe is washed off the filter and the position of the bound probe can be visualized, in the case of a radioactive probe by exposing the filter to an X-ray film.

Because of naturally occurring variations in restriction enzyme recognition sequences it is possible to generate fragments of different sizes from each half of a chromosome pair if an individual bears different sites on each one, that is they are heterozygous. Such variations are known as restriction fragment length polymorphisms (RFLPs). Southern blotting and hybridization can thus be used to detect the presence or absence of a specific sequence in tumours from individuals whose normal cells are found to be heterozygous but have a deletion in their tumour cell DNA (Fig. 2.2). This can also be determined by a reduction in signal intensity of around 50% in a fragment from a tumour sample in a homozygous individual. In addition, this technique can detect amplification of a sequence, as in these circumstances the hybridization signal will be increased, or in some cases a rearrangement in the DNA if this either involves a restriction site or creates a restriction fragment of differing size when compared to that found in normal cells from the same individual.

Another technique similar to Southern blotting frequently used in molecular biology is that of Northern blotting. This involves the isolation of RNA from tissue followed by agarose gel electrophoresis, blotting onto nylon membrane and hybridizing with a probe in a similar way to Southern blotting. Unlike DNA, RNA transcripts are already single stranded and of varying lengths, and represent only the parts of a gene (called exons) that are translated into protein. Northern blotting can be used to demonstrate whether or not a gene is being expressed, and to what extent, and whether the gene transcript differs from that found in normal cells.

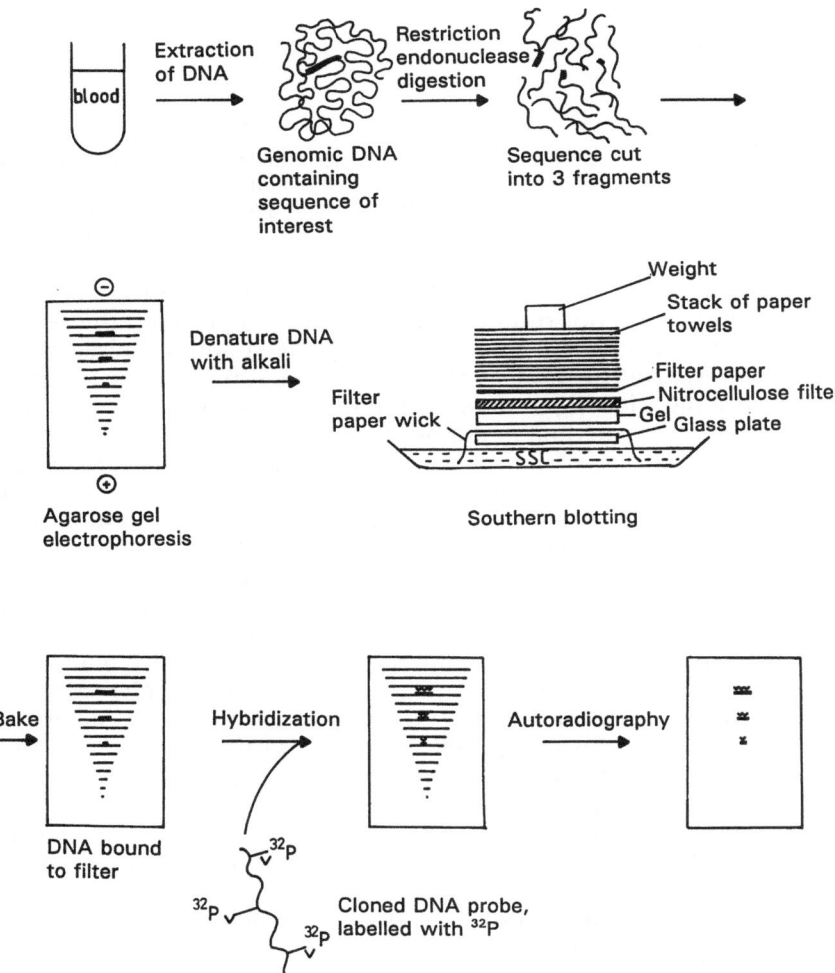

Fig. 2.1 Diagrammatic summary of the techniques used for analysing restriction fragment length polymorphisms (RFLPs). DNA extracted from leucocytes (or tumour tissue) is digested with a given restriction endonuclease. In this example, the genomic DNA sequence of interest (thick line) contains two restriction sites and is thus cut into three fragments of different lengths. The DNA fragments (about 10^6) are separated by agarose gel electrophoresis, shorter fragments migrating faster than longer ones. They are then denatured with alkali and transferred to a nitrocellulose (or nylon) filter by Southern blotting. After baking (or exposure to ultraviolet light), the DNA is permanently bound to the filter. A DNA probe complementary to the sequence of interest is labelled with radioactive phosphorus and hybridized to the DNA on the filter. After autoradiography, the position of three fragments can be visualized (From Harding, 1988, with permission from Churchill Livingstone)

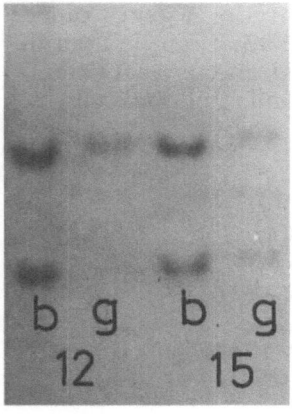

Fig. 2.2 DNA samples from blood (b) and glioma tissue (g) showing loss (12) and maintenance (15) of heterozygosity for RFLPs detected by the probe D17S58 which maps to a locus on chromosome 17

Southern blotting can separate DNA molecules ranging from about 100 bp to 30–40 kb, so until recently a gap has existed between the resolving powers of cytogenetic and molecular genetic techniques. Pulsed field gel electrophoresis (PFGE) bridges this gap to some extent. PFGE separates large DNA molecules (from 50 to 2000 kb). At present, technical problems limit the use of this method in tumour DNA analysis. It is necessary to generate a single cell suspension in preparing samples suitable for PFGE, and difficulties in interpretation may occur due to the different methylation patterns found in different tissues (and possibly in different tumours) which lead to DNA fragments of variable size.

Labelled DNA probes can also be used directly on tissue sections or chromosome spreads. This technique is known as in situ hybridization and involves first denaturing and then incubating slides with a labelled probe (radioactive or biotinylated). The slides are developed using a photographic emulsion or labelled antibody and examined under a microscope. The chromosomal location of specific DNA fragments can thus be determined, or expressed sequences can be detected in fixed tissue sections. This can demonstrate, for example, overexpression of a specific mRNA in tumours when compared to normal tissues.

The technique of molecular cloning will not be discussed in detail here, but it should be noted that this is a very powerful tool, potentially allowing the isolation, production and manipulation of any DNA or RNA sequence in the genome. Once isolated, the exact nucleotide sequence can be determined and, from this, information on the amino acid sequence of the protein and structure of the gene can be deduced. Cloning techniques are employed in the production of DNA probes, and an example of the use of such methods can be found later in this chapter when the isolation and characterization of a gene found to be amplified in a glioma, GLI, will be discussed.

Molecular Basis of Tumorigenesis

There is increasing evidence that some tumours are produced by two genetic events, predicted by the "two hit" hypothesis of Knudson (1971), in a single cell. This particularly applies to tumours which are clearly inherited in some cases, for example retinoblastoma and acoustic neuroma. In familial cases, one mutation is inherited and the other is in the tumour stem cell (e.g. the retinoblast, a somatic cell mutation). In sporadic cases both "hits" occur in the stem cell. These observations were first made in retinoblastomas, following the identification of deletions in the long arm of chromosome 13 in some individuals with these tumours. This disorder can be familial (40% of cases) or sporadic in occurrence. The growth of tumours of the embryonic retina is due to a loss in function of both copies of the Rb gene, which has recently been cloned, and maps to chromosome 13. In familial cases, one allele of the Rb gene is defective or deleted in the germ line of the individual. This is inherited as an autosomal dominant trait. All that is required for tumour formation is a mutation in the other Rb homologue, hence the bilateral and multifocal occurrence of inherited retinoblastomas. Sporadic cases require that mutations occur in both Rb genes in a single cell, which, being a statistically less likely event, leads to unilateral, unifocal lesions (Cavenee et al., 1983).

There is evidence to suggest that the function of the Rb gene in normal cells is in the control of stem cell division and differentiation, that is, it acts as a potential tumour suppressor gene. Retinoblastoma cells in culture express both neuronal and glial markers, suggesting that they have arisen from bipotent stem cells, and the fact that lesions in familial cases of retinoblastoma can occur in sites other than the retina, including the brain and bone (osteosarcoma), indicates that this mechanism for malignant transformation can occur in other types of stem cells. Structural rearrangements of the Rb gene have also been reported in some melanomas, breast cancers and small cell lung carcinomas (Vile, 1989).

The two hit hypothesis explains the mechanism of tumorigenesis in some inherited tumour syndromes, but a "multistep" model for tumour development is increasingly emerging, and is probably more common. In this, tumorigenesis is dependent on several molecular events, and oncogenes appear to be frequently involved in this process (Gilbert, 1988). DNA sequences called proto-oncogenes are widespread throughout the human genome; these are homologous to viral sequences capable of producing tumours in experimental animals. They appear to encode products which are important in normal cellular growth and differentiation but do not usually cause neoplastic growth. This probably requires activation of proto-oncogenes by mutagenic agents such as radiation, hydrocarbons or viral infection. Activation may involve the deletion of nearby suppressor genes, or translocation of the proto-oncogene to a different part of the genome (e.g. another chromosome) where an adjacent sequence may act as promoter, as occurs in chronic myeloid leukaemia.

Over 30 human oncogenes have been characterized (Gilbert, 1988), and the mechanisms by which they cause neoplastic growth include the production of abnormal types or amounts of growth factors or growth factor receptors, or of cellular kinase enzymes. For example, a gene on the short arm of chromosome 7 codes for the epidermal growth factor receptor (EGFR). The c-erb B oncogene product is homologous to EGFR, but lacks its external EGF binding portion, perhaps removing the mechanism for homeostatic control. The action of oncogenes is not specific to particular tumours; for example the N-myc gene product is transcribed at unusually high levels in retinoblastoma and several other tumours including neuroblastoma, and it may be that the myc gene product comes under the control of the Rb gene. Expression of the myc gene correlates with cell proliferation, and the level of myc expression is repressed following cell differentiation.

The multistep model can be divided into three stages (Farber, 1984): initiation, promotion and progression. Initiation, which is the primary event, involves alteration in the DNA of the target cell, for example by the action of a virus, a chemical carcinogen or through a mistake in DNA replication or repair. The cell must undergo division to establish this as a heritable change. Thus the origin of the tumour is of a clonal nature, arising from a single cell. It should be noted that the primary genetic event may be a mutation in a gene necessary for genetic stability, defects of which can give rise to mutations in other genes that may bring about transformation.

Promotion involves the action of an agent which does not in itself cause cell transformation, but can increase the proliferation of an initiated cell and accelerate tumour growth. For example, promoting agents can take the form of hormones, viruses, immunosuppressants or angiogenic substances. Tumour progression is the process by which a neoplasm becomes more aggressive in its clinical and biological characteristics over time. This tendency is often accompanied by an accumulation of genetic aberrations, which may be either a secondary consequence of the increased state of malignancy or be directly involved in bringing about this state. At the genetic level it is thought that tumour progression results from mutations in the genome of the original transformed clone giving rise to subpopulations of cells with an increased growth potential.

Thus at a molecular level transformation can be thought of as arising from alterations to genes controlling cell growth and differentiation. These can manifest themselves as quantitative or qualitative changes in the expression of such genes, either by loss, reduction or overexpression of a normal gene product or by alteration of the function of a gene product.

Molecular Genetics of Benign Brain Tumours

Molecular genetic analyses of meningiomas and acoustic neuromas are, to date, more advanced than those of gliomas, and the results have been important in identifying candidate chromosomal regions for inherited tumour syndromes, particularly the central form of neurofibromatosis (NF 2). Acoustic neuromas are the hallmark of NF 2, and are usually bilateral in this condition. In contrast, sporadic acoustic neuromas are unilateral. The difference in presentation between sporadic and familial acoustic tumours parallels the situation in retinoblastoma referred to above, in which loss of a specific chromosomal region is implicated in tumour development.

Early cytogenetic studies showed that loss of chromosome 22 was frequent in meningiomas; these tumours also occur in NF 2. These observations led Seizinger et al. (1986) to look for heterozygosity in acoustic neuromas using RFLP analysis. In seven of 16 informative cases (one bilateral, six unilateral acoustic neuromas), the tumours showed loss of genes from chromosome 22 which were present in the patients' leucocyte DNA. Studies with DNA markers mapping to other chromosomes were normal. Further analyses of one meningioma and of cervical root neurofibromas from NF 2 patients showed selective loss of all or part of chromosome 22, and the pattern of loss of alleles in NF 2 acoustic neuromas suggested that the long arm of 22 was most consistently involved (Seizinger et al., 1987a). Genetic linkage studies subsequently established that the gene for NF 2, which is presumed to be a tumour suppressor gene, is near the centre of the long arm of chromosome 22 (Wertelecki et al., 1988).

Loss of heterozygosity for chromosome 22 RFLPs has also been demonstrated in 17 of 40 informative meningiomas. Nine of these 17 were studied cytogenetically, and seven had lost one complete copy of chromosome 22; the other two had normal karyotypes (Seizinger et al., 1987b). Both DNA and cytogenetic analyses showed changes in other chromosomes, but these were less consistent than the chromosome 22 data. One explanation for this observation, which contrasts with the findings in acoustic neuromas, is that meningioma cells may be less genomically stable than those in acoustic neuromas, demonstrating aggressive histological behaviour in some cases (Martuza et al., 1988). Some support for this comes from the study of Dumanski et al. (1987), who found loss of alleles from four chromosomes in an anaplastic meningioma, although two other anaplastic tumours retained heterozygosity for all markers tested. Overall, these authors found loss of chromosome 22 alleles in 14 of 30 meningiomas. Five of these tumours had lost heterozygosity for some loci but not others, and it was possible to define a "meningioma locus" between 22q12.3 and the telomere.

The gene locus for von Hippel–Lindau disease (VHLD) has been mapped to the short arm of chromosome 3 and is linked to that for the human homologue of the raf 1 oncogene (Seizinger et al., 1988). This is of interest given that renal cell carcinomas had previously been shown to exhibit loss of heterozygosity for chromosome 3p markers. More recent studies have suggested that the VHLD gene acts as a tumour suppressor gene, as loss of 3p alleles has been demonstrated in 11 renal cell cancers, one phaeochromocytoma, and three haemangioblastomas from cases of VHLD (Tory et al., 1989).

The observations in acoustic neuromas, meningiomas, and tumours from VHLD tumours described above suggest that the two hit hypothesis is applicable to their pathogenesis. This may not apply in the majority of other nervous system tumours, including the neurofibromas which are an integral component of von Recklinghausen disease (NF 1). The gene locus for NF 1 has been mapped to the long arm of chromosome 17 (Barker et al., 1987; Seizinger et al., 1987c). The latter paper reported linkage of the NF 1 locus to that for the nerve growth factor receptor (NGFR) gene, a candidate gene for NF 1. However, crossovers between the two loci indicated that a defect of NGFR was unlikely to be the basis of NF 1, and the disease locus has since been mapped to the region between the NGFR locus and the centromere (Fain et al., 1989). Loss of heterozygosity for chromosome 17 markers has not been observed in more than 40 neurofibromas from NF 1 patients studied to date. Some neurofibrosarcomas show loss of sequences from chromosome 17, but in most cases these have been derived from the short arm and have not included the NF 1 locus (Chung et al., 1990). Losses of 17p are also observed in all grades of glioma, which occur in some patients with NF 1. The existence of a tumour suppressor gene on 17p (p53) has been proposed on the basis of studies in colon cancers (Baker et al., 1989). It seems unlikely that this is related to the NF 1 locus on 17q in terms of tumorigenesis in this disease, and more complex mechanisms than the two hit system are probably operative in this context.

Cytogenetic and Molecular Genetic Studies of Malignant Brain Tumours: Gliomas

Several cytogenetic investigations into the chromosome content of glioma cells, from both fresh biopsy material and cultured cells, have been performed, the most extensive of which is that of Bigner et al. (1988a). In a study of 54 malignant human gliomas which were karyotyped from direct chromosome preparations or after short term culture, 12 were found to have normal stemline karyotypes or had lost only one sex chromosome. Further analysis of 38 of the 42 tumours with abnormal autosomal karyotypes showed that six had near triploid or tetraploid stemlines and the rest had near diploid stemlines with chromosomal losses, gains or structural abnormalities. These included gains of chromosome 7 (81%), losses of chromosome 10 (59%) and the presence of double minutes (56%). The incidence of structural abnormalities in specific chromosomes was compared to what might be expected if these were random on the basis of chromosome arm length; structural abnormalities involving chromosomes 9p and 19q were found to occur at a significantly high frequency.

The six polyploid tumours in this study showed a tendency for loss of two copies of chromosome 22 and a lower incidence of chromosome 7 gains and the presence of double minutes. It is possible that these mainly polyploid tumours are evolved by a mechanism different from that of the near-diploids. A comparison of different ploidy groups (normal or loss of one sex chromosome vs. near-diploid vs. polyploid) for their clinical implications among glioblastoma multiforme and gliosarcoma patients showed no significant correlation with patient survival. Extreme heterogeneity of karyotype has also been reported, with regional variation within a tumour. Rebiopsy of recurrent malignant tumours may show "resistant" populations of cells with new, more uniform, karyotypes (Schmidek, 1987).

Initial investigations into the DNA content of malignant human gliomas focused on overall measurements of cellular DNA content using flow cytometry. This method, although not revealing the precise nature of genetic abnormalities, is a very rapid process, enabling many thousands of cells to be screened accurately. The results indicated that most glioma cells are near-diploid, but that a greater proportion of near-triploid and near-tetraploid cells occurs in gliomas of higher malignancy grade (Ahyai, 1988). This is in keeping with results from other types of tumours, where more malignant neoplasms have a higher proportion of cells with a polyploid DNA content.

Recently, much work has been done to investigate genomic abnormalities in gliomas using DNA probes. The results from this have confirmed the findings of cytogenetic studies and have brought to light additional changes not detected by karyotyping. Studies of gliomas have so far shown that loss of sequences on chromosomes 10, 13, 17 (see Fig. 2.2) and 22 is common (James et al., 1988). There is an indication that losses of chromosome 10 sequences are restricted to tumours of high malignancy grade, whereas those on other chromosomes occur in tumours of all grades. James and colleagues used a total of 51 polymorphic DNA probes, including at least one from each chromosome, to look for loss of heterozygosity in 53 gliomas. Loss of sequences on chromosome 10 was found in 28/29 glioblastomas. Two glioblastomas were not informative at the chromosome 10 loci tested, but showed a reduction in signal intensity of around 50%. No losses on chromosome 10 were detected in gliomas of lower malignancy grades. The frequency of losses on other chromosomes was found to be significantly above that expected at random for chromosomes 13, 17 and 22. Chromosome 13 losses occurred in 5 cases (14%), losses on 17 occurred in 8 cases (22%) and those on 22 occurred in 7 cases (19%). These losses were found in tumours of each malignancy grade, but those on chromosome 17 were found only in tumours of astrocytic differentiation.

These results suggest that the loss of a tumour suppressor gene on chromosome 10 could be the genetic mechanism by which a lower grade tumour progresses to become a glioblastoma. Fujimoto et al. (1989a) determined the common region of loss on chromosome 10 to be between the middle of the short arm and the band 10q23. Thirteen cases of glioblastoma multiforme were screened with ten polymorphic DNA markers mapped to chromosome 10; ten tumours showed loss of heterozygosity. One tumour was found to retain sequences both proximal and distal to the sequence that was lost on chromosome 10, indicating that there had been an interstitial deletion. Four anaplastic astrocytomas were also tested but no losses were found.

Further investigations into the loss of chromosome 17 sequences in astrocytic tumours were carried out by James et al. (1989). Thirty-five gliomas were included in this study, of which 24 displayed astrocytic differentiation. Chromosome 17 losses were detected in eight tumours, all of which showed astrocytic differentiation, using a total of seven RFLP markers. These included two grade II tumours, two grade III tumours and four grade IV tumours. In addition, one primitive neuroectodermal tumour (PNET) was found to have lost sequences on chromosome 17. In one grade IV tumour it was found that there had been complete loss of one chromosome 17 with reduplication of the remaining homologue. Four other gliomas, and the PNET, showed a loss of sequences on one homologue coupled with reduplication of the corresponding region on the other homologue, presumably arising from mitotic recombination.

The region of loss and reduplication in these tumours varied, but commonly involved the short arm from p11.2-pter. One grade IV tumour had a deletion of this region without reduplication of the remaining homologue. Two other tumours, grade III and grade IV, had undergone multiple mitotic recombination. This resulted in two regions of loss of alleles on one homologue with reduplication of the corresponding regions on the remaining homologue. The regions involved were the tip of the long arm and the same region involving the short arm mentioned above. The fact that, in most cases, loss of genetic material on one chromosome was accompanied by reduplication of the same region on the other chromosome explains why losses on chromosome 17 have not been observed at a high frequency in cytogenetic studies of gliomas, since there is no net loss of material. Loss of heterozygosity for chromosome 17 sequences was also observed in 40% of anaplastic astrocytomas and glioblastomas by Fults *et al.* (1989). These results imply the existence of a tumour suppressor gene located within the common region of loss (p11.2-pter). Thus the mechanism of inactivation of this gene may be analogous to that of the retinoblastoma gene, whereby a mutation in one copy of the gene and loss of the other (with or without reduplication of the defective copy) can promote tumorigenesis.

It is of interest that the p53 gene maps to chromosome 17p13.1. The function of this gene is not yet clear, and it has been suggested that p53 could either be an oncogene or that it acts as a tumour suppressor gene. Recent work supports the latter hypothesis. Only mutant forms of the p53 gene are able to act in cell transformation in vitro. The p53 gene product, like that of the well-characterized Rb tumour suppressor gene, can interact with the large T-antigen of simian virus 40 (SV40) and it is possible that the mechanism of transformation of SV40 involves inactivation of the p53 or Rb protein and thus loss of the tumour suppressor function of these genes. The deletion of sequences on chromosome 17 and the recent findings of p53 gene mutations in colon tumours prompted a search for p53 mutations in other tumours which show losses of 17p sequences, including gliomas (Nigro *et al.*, 1989). Among five glioblastomas which showed loss of heterozygosity for 17p alleles, four were found to have point mutations in the p53 gene. The one tumour in which no mutation was detected was found to express p53 mRNA, but no protein product could be detected, implying that a mutation elsewhere may have affected p53 mRNA translation or stability. These results, and those from other tumours, support the hypothesis that p53 is a tumour suppressor gene, the loss of function of which (by deletion or mutation) may play a role in the development of many cancers.

Oncogenes and Gliomas

Molecular genetic studies have also demonstrated amplification of known oncogenes in over 40% of gliomas, and recently led to the identification of a new probable oncogene, gli, discovered by virtue of its amplification in a glioma (Kinzler *et al.*, 1987). Oncogenes can be grouped according to the different biochemical mechanisms by which they stimulate the growth of a cell. Oncogenes encoding growth factors include sis (encoding platelet derived growth factor, PDGF). Those encoding growth factor receptors include erb B (epidermal growth factor receptor, EGFR) and fms (colony stimulating factor, CSF-1) and are often found to have tyrosine kinase activity. A third group includes genes encoding membrane proteins with tyrosine kinase activity; it is possible that some or all of these may prove to be associated with growth factor receptors on further investigation. Such oncogenes include src, fps, ros, fgr and abl. The ras oncogenes make up a separate group. Ras genes code for GTP-binding proteins and may be functionally related to G proteins which have a role in mediating the response of a cell to hormone receptors. Lastly, oncogenes that encode nuclear proteins include myc, myb and fos. The mRNA levels of these are regulated by serum growth factors and they may be mediators of the mitogenic signal response to these growth factors. The myc gene product is known to bind to DNA and possibly has a role as a regulator of the expression of other genes. The four genetic mechanisms by which a proto-oncogene can be activated are by gene amplification, gene loss, alterations in gene structure or changes in the regulation of gene expression. The use of cytogenetic and molecular genetic techniques can identify such changes.

The EGFR gene is amplified in about 38% of gliomas (Libermann et al., 1985; Wong et al., 1987). The EGFR receptor molecule has been well studied, and consists of an extracellular EGF-binding regulatory domain, a transmembrane portion and an intracellular effector domain which catalyses the phosphorylation of tyrosine residues. The EGF molecule is a potent mitogen for many cells, including those of neuroectodermal origin. Bigner et al. (1988b) examined 64 gliomas for EGFR gene amplification and mRNA expression to determine whether there is any correlation to tumour histology or patient prognosis. Overall, 44% were found to have EGFR gene amplification, the majority of these positive cases (24/48 cases) being glioblastoma multiforme. In addition, 2/6 anaplastic astrocytomas and 2/5 gliosarcomas were found to have amplification. There was no statistically significant difference in survival among those patients with amplification and those without and no significant difference in tumour histology between these two groups of tumours could be determined.

In situ hybridization of EGFR in tumours with amplification showed that EGFR mRNA expression was not limited to cells of a particular morphological type within a tumour, although one glioblastoma had increased EGFR expression in areas of more malignant morphology and low levels of expression in areas with lower grade morphology. The higher prevalence of EGFR amplification among glioblastomas, which has also been observed in other studies, implies that amplification may be a late event in tumour progression. There is also evidence for a rearrangement of the EGFR gene occurring in some tumours with amplification. Humphrey et al. (1988), investigating EGFR amplification and expression in human glioma xenografts (whereby the tumour is grown subcutaneously in athymic mice), demonstrated that five of six tumours with amplification had rearrangements of the EGFR gene with expression of a low molecular weight protein. In one case there appeared to be a loss of the external (EGF) binding domain of the protein, resulting in a protein that is structurally similar to the v-erb B product.

A combination of cytogenetic and molecular genetic techniques has been used to identify amplified sequences in gliomas. Trent et al. (1986) demonstrated the amplification, rearrangement and overexpression of c-myc in a human glioblastoma cell line, and Kinzler et al. (1987) identified a new gene, gli, which was amplified and highly expressed in a human glioma.

The first step is to undertake a cytogenetic analysis of the tumour for evidence of gene amplification in the form of homogeneously staining regions or DMs. Once a tumour containing one of these has been identified, molecular genetic techniques can be employed. First, gel denaturation–renaturation is used to detect sequences amplified more than 20 times. The next step is to test the tumour DNA for amplification of known oncogenes using Southern blot analysis and hybridization to labelled oncogene probes. This led Trent and colleagues to find amplification and rearrangement of the c-myc gene in a human glioma cell line, and c-myc was also found to be highly expressed. The study by Bigner et al. (1988b) of 64 gliomas identified 2 glioblastomas (out of a total of 48) with N-myc amplification.

Kinzler et al. (1987) found that the amplified sequences identified in a malignant human glioma, using the above technique, did not hybridize to 20 known oncogene probes. This implied that the amplified sequence had not been previously described. DNA enriched for amplified sequences by the gel denaturation-renaturation technique was used to construct a tumour DNA library. Out of 31 clones tested, three were found to contain sequences that were specifically amplified in the tumour DNA, unique sequences free of repeats which detected single copy sequences in normal DNA by Southern blot analysis. The new probes generated were used to determine the level of amplification in the tumour which was found to be approximately 75 times per haploid genome. The normal chromosomal location of the amplified sequences was found to be 12cen-q14.3. The newly discovered putative oncogene was named gli. Although any trans- forming ability of gli has not yet been demonstrated, it has been shown to be amplified in a small proportion of gliomas (2/48 glioblastomas, Bigner et al., 1988b), as well as other types of tumours. Further investigations of the gli gene have shown that it bears homology to the Kruppel family of zinc finger proteins, and that the product may bind to DNA and may have a role in development or tissue specific differentiation (and perhaps neoplasia).

Other oncogenes reported to be expressed, amplified and/or rearranged in gliomas include N-ras, N-myc, v-fos, Rosl and v-sis. Gerosa et al. (1989) found overexpression of N-ras in 5/5 glioblastomas, both in cell lines and biopsy specimens from the same patient. All these glioblastomas also showed increased EGFR expression. Fujimoto et al. (1989b), using tissue collected from ten primary tumours of neuroectodermal origin, found increased expression of N-myc and v-sis, together with gene amplification of around 80-fold for N-myc and 3–4-fold for v-sis in a recurrent glioblas- toma multiforme. The authors suggested that myc expression may be cor- related with the degree of malignancy of neuroectodermal tumours. Further support for this comes from a study of the amount of c-myc onco- protein detected in glioma tumour tissue specimens and cell lines using immunofluorescent monoclonal antibodies (Engelhard et al., 1989). High levels of c-myc oncoprotein expression were detected in glioblastoma tumour specimens and glioma-derived neoplastic cell lines, but no expres- sion could be detected in a benign brain tissue specimen or in a non- neoplastic glial cell line.

Birchmeier *et al.* (1987) examined 45 different human cell lines for expression of Rosl transcripts but only found overexpression in lines originating from glioblastomas. Little is known about Rosl, but structural similarities to other oncogenes including v-erb B and v-fms indicate a possible role as a hormone or growth-factor receptor gene.

The sis oncogene maps to chromosome 22 (deleted in some gliomas) and is related to the PDGF gene and its receptor PDGFR (Schmidek, 1987). PDGF is secreted by platelets into serum and stimulates the growth of cells derived from connective tissue and glia. The PDGF receptor molecule, like EGFR, has tyrosine kinase activity. It is known that the viral oncogene v-erb B bears homology to the EGFR gene, and that transformation by this virus involves the expression of a truncated EGF receptor-like molecule which fails to respond to normal cell control mechanisms. The finding of alterations to the EGFR gene, revealed by restriction enzyme digests, and of abnormal protein products, some with properties such as increased EGF binding, has led to the speculation that increased production of an altered EGFR gene product may be acting in a similar way to the v-erb B gene in bringing about transformation. The PDGF B chain also has a viral counterpart, v-sis, which induces gliomas at a very high frequency in experimental animals. Studies using Northern blotting and in situ hybridization have shown a high level of expression of the PDGF A chain along with lower levels of expression of PDGF B chain and PDGF receptor in glioma cells, while the neighbouring cells of the proliferating vascular endothelium showed high levels of PDGF B chain and PDGF receptor expression than of the PDGF A chain. It has been suggested that this represents autocrine growth stimulation of endothelial cells in response to interaction with glioma cells (Hermansson *et al.*, 1988).

Molecular Genetics of Neuroblastomas

The most commonly detected genetic alterations found in neuroblastomas are loss of sequences on chromosome 1p and amplification of the N-myc oncogene, often indicated by the presence of DMs. DNA content is also found to vary. Using flow cytometry and measurements of N-myc amplification, Dominici *et al.* (1989) found that among 17 neuroblastomas studied there emerged two different subsets of tumour. Those that were aneuploid (65% of tumours) were without amplification and were usually found to be from patients under 2 years of age with localized or stage IV S tumours. The near-diploid tumours, however, often had N-myc amplification (in 3/6 cases) and were usually seen in patients of over 2 years of age with advanced clinical stages.

Hayashi *et al.* (1989) found similar results in a study of 51 children with neuroblastomas. Tumours were classified into 4 groups according to karyotype: near-diploid (42–47 chromosomes) 11 cases; hyperdiploid (50–56 chromosomes) 4 cases; near-triploid (60–77 chromosomes) 33 cases; and hypotetraploid (80–83 chromosomes) 3 cases. Patients with near-diploid or hypotetraploid karyotypes were found to have several structural abnormalities including loss of material on chromosome 1 with or without the presence of DMs or HSRs. Most of these patients were aged one year or older and had advanced tumours. Patients with hyperdiploid and near-triploid karyotypes had fewer structural abnormalities and all but one were under one year of age, with localized tumours and long-term survival.

Brodeur (1989) reported that among 646 neuroblastomas and 16 ganglioneuromas, N-myc amplification was found in none of the ganglioneuromas, 10% of stage IVS tumours, 4% of low stage tumours and 32% of advanced stage tumours. N-myc amplification was found to be highly correlated with rapid tumour progression even in patients with early disease. In a further study on the same group of tumours, Fong *et al.* (1989), using chromosome 1p RFLPs, found that 28% (13/47) of tumours examined showed partial monosomy for 1p. The common region of loss was from 1p36.1 to 1p36.3. Loss of heterozygosity for 1p showed a significant correlation with N-myc amplification. Fong speculates that these two genetic events may be related, characterizing a subset of more aggressive tumours. The frequency of loss of heterozygosity as determined by molecular genetic techniques (28%) is substantially lower than that previously reported from cytogenetic studies (around 70%). This may reflect a bias in sampling, since cytogenetic analysis is frequently performed on tumours from patients with advanced stages of disease and near-diploid karyotypes, or from cultured cell lines. Also, due to difficulties in performing cytogenetic techniques on primary tissue, only about 20% of primary tumours can be karyotyped.

Ritke *et al.* (1989) analysed seven neuroblastoma cell lines in an attempt to define the structural alterations on chromosome 1p more precisely. Their results indicate that the region of loss is located closer to 1p32 than previously suggested. They also found evidence for the translocation of a region including the mycL gene and a deletion of sequences distal to mycL on 1p. Other tumours for which loss of heterozygosity for 1p has been described include phaeochromocytoma, medullary thyroid carcinoma and melanoma, which are also of neural crest origin. N-myc expression is also found in retinoblastoma and astrocytoma.

Conclusion

It is likely that increased understanding of the molecular mechanisms of malignant brain tumours will have important clinical implications. This has already proved to be the case in neuroblastomas; data on gliomas are currently incomplete but certain consistent observations are emerging. Genetic analysis may potentially be useful in diagnosis, probably leading to new classification among groups of tumours. As such, molecular genetics could affect the choice of treatment made by the clinician, but it is also possible that delineating the precise mechanisms involved in tumorigenesis will direct new avenues for further therapy.

References

Ahyai A (1988) Flow cytometric analysis of cellular DNA content in human astrocytomas and oligodendrogliomas. Neurosurg Rev 11: 177–187

Baker SJ, Fearon ER, Nigro JM et al. (1989) Chromosome 17 deletions and p53 gene mutations in colorectal carcinomas. Science 244: 217–220

Barker D, Wright E, Nguyen K et al. (1987) Gene for von Recklinghausen neurofibromatosis is in the pericentromeric region of chromosome 17. Science 236: 1100–1102

Bigner SH, Mark J, Burger PC et al. (1988a) Specific chromosomal abnormalities in malignant human gliomas. Cancer Res 88: 405–411

Bigner SH, Burger PC, Wong AJ et al. (1988b) Gene amplification in malignant human gliomas: clinical and histopathologic aspects. J Neuropathol Exp Neurol 47: 191–205

Birchmeier C, Sharma S, Wigler M (1987) Expression and rearrangement of the Rosl gene in human glioblastoma cells. Proc Natl Acad Sci USA 84: 9270–9274

Brodeur GM (1989) Clinical significance of genetic rearrangements in human neuroblastomas. Clin Chem 35 (Suppl 7): B38–42

Cavenee WK, Dryja TP, Phillips RA et al. (1983) Expression of recessive alleles by chromosomal mechanisms in retinoblastoma. Nature 305: 779–784

Chung R, Farmer G, Anderson K et al. (1989) Distinct deletion on the short arm of chromosome 17 in tumorigenesis of neurofibrosarcomas and astrocytomas. Am J Hum Genet 45: A17

Davies KE, Read AP (1988) Molecular Basis of Inherited Diseases. IRL Press, Oxford

Dominici C, Negroni A, Romeo A et al. (1989) Association of near-diploid DNA content and N-myc amplification in neuroblastomas. Clin Exp Metastasis 7: 202–211

Dumanski JP, Carlbom E, Collins VP, Nordenskjold M (1987) Deletion mapping of a locus on human chromosome 22 involved in the oncogenesis of meningioma. Proc Natl Acad Sci USA 84: 9275–9279

Engelhard HH, Butler AB, Bauer KD (1989) Quantification of the c-myc oncoprotein in human glioblastoma cells and tumour tissue. J Neurosurg 71: 224–232

Fain PR, Goldgar DE, Wallace MR et al. (1989) Refined physical and genetic mapping of the NF1 region on chromosome 17. Am J Hum Genet 45: 721–728

Farber E (1984) The multistep nature of cancer development. Cancer Res 44: 4217–4223

Fong CT, Dracopoli NC, White PS et al. (1989) Loss of heterozygosity for the short arm of chromosome 1 in human neuroblastomas: correlation with N-myc amplification. Proc Natl Acad Sci USA 86: 3753–3757

Fujimoto M, Fults DW, Thomas GA et al. (1989a) Loss of heterozygosity on chromosome 10 in human glioblastoma multiforme. Genomics 4: 210–214

Fujimoto M, Sheridan PJ, Sharp ZD et al. (1989b) Proto-oncogene analyses in brain tumours. J Neurosurg 70: 910–915

Fults D, Tippets RH, Thomas GA, Nakamura Y, White R (1989) Loss of heterozygosity for loci on chromosome 17p in human malignant astrocytoma. Cancer Res 49: 6572–6576

Gerosa MA, Talarico D, Fognani C et al. (1989) Overexpression of N-ras oncogene and epidermal growth factor receptor gene in human glioblastomas. J Natl Cancer Inst 81: 63–67

Gilbert F (1988) Neuro-oncogenesis: recessive genes, activated oncogenes, and chromosome abnormalities in the development of neuro-ectodermal cancers. In: Rosenberg RN, Harding AE (eds) The Molecular Biology of Neurological Disease. Butterworths, London, pp 109–124

Harding, AE (1988) Molecular genetics and neurological disease. In: Kennard C (ed) Recent Advances in Clinical Neurology 5. Churchill Livingstone, Edinburgh

Hayashi Y, Kanda N, Inaba T et al. (1989) Cytogenetic findings and prognosis in neuroblastoma with emphasis on marker chromosome 1. Cancer 63: 126–132

Hermansson M, Nister M, Betsholtz C et al. (1988) Endothelial cell hyperplasia in human glioblastoma: coexpression of mRNA for platelet-derived growth factor (PDGF) B chain and PDGF receptor suggests autocrine growth stimulation. Proc Natl Acad Sci USA 85: 7748–7752

Humphrey PA, Wong AJ, Vogelstein B et al. (1988) Amplification and expression of the epidermal growth factor receptor gene in human glioma xenografts. Cancer 48: 2231–2238

James CD, Carlbom E, Dumanski JP et al. (1988) Clonal genomic alterations in glioma malignancy stages. Cancer Res 48: 5546–5551

James CD, Carlbom E, Nordenskjold M, Collins VP, Cavenee WK (1989) Mitotic recombination of chromosome 17 in astrocytomas. Proc Natl Acad Sci USA 86: 2858–2862

Kinzler KW, Bigner SH, Bigner DD et al. (1987) Identification of an amplified, highly expressed gene in a human glioma. Science 236: 70–73

Knudson AG (1971) Mutation and cancer: statistical study of retinoblastoma. Proc Natl Acad Sci USA 68: 820–823

Libermann TA, Nusbaum HR, Razon N et al. (1985) Amplification, enhanced expression and possible rearrangement of EGF receptor gene in primary human brain tumours of glial origin. Nature 313: 144–147

Martuza RL, Seizinger BR, Jacoby LB, Rouleau GA, Gusella JF (1988) The molecular biology of human glial tumours. Trends Neurosci 11: 22–27

Nigro JM, Baker SJ, Preisinger AC et al. (1989) Mutations in the p53 gene occur in diverse human tumour types. Nature 342: 705–708

Ritke MK, Shah R, Valentine M, Douglass EC, Tereba E (1989) Molecular analysis of chromosome 1 abnormalities in neuroblastoma. Cytogenet Cell Genet 50: 84–90

Schmidek HH (1987) The molecular genetics of nervous system tumors. J Neurosurg 67: 1–16

Seizinger BR, Martuza RL, Gusella JF (1986) Loss of genes on chromosome 22 in tumorigenesis of human acoustic neuroma. Nature 322: 644–647

Seizinger BR, Rouleau GA, Ozelius LJ et al. (1987a) Common pathogenetic mechanism for three tumor types in bilateral acoustic neurofibromatosis. Science 236: 317–319

Seizinger BR, de la Monte S, Atkins L et al. (1987b) A molecular genetic approach to human meningioma: loss of genes on chromosome 22. Proc Natl Acad Sci USA 84: 5419–5423

Seizinger BR, Rouleau GA, Ozelius LJ et al. (1987c) Genetic linkage of von Recklinghausen neurofibromatosis to the nerve growth factor receptor gene. Cell 49: 589–594

Seizinger BR, Rouleau GA, Ozelius LJ et al. (1988) Von Hippel–Lindau disease maps to the region of chromosome 3 associated with renal cell carcinoma. Nature 32: 268–269

Tory K, Brauch H, Linehan M et al. (1989) Allele deletion analysis of tumors associated with von Hippel–Lindau disease (VHL) identifies VHL as a tumor suppressor gene. Cytogenet Cell Genet 51: 1092

Trent J, Meltzer P, Rosenblum M et al. (1986) Evidence for rearrangement, amplification, and expression of c-myc in a human glioblastoma. Proc Natl Acad Sci USA 83: 470–473

Vile R (1989) Tumour suppressor genes. Br Med J 298: 1335–1336

Wertelecki W, Rouleau GA, Superneau DW et al. (1988) Neurofibromatosis 2: clinical and DNA linkage studies of a large kindred. New Engl J Med 319: 278–283

Wong AJ, Bigner SH, Kinzler KW, Hamilton SR, Vogelstein B (1987) Increased expression of the EGF receptor gene in malignant gliomas is invariably associated with gene amplification. Proc Natl Acad Sci USA 84: 6899–6903

3 Current Concepts of the Histopathological Classification of Tumours of the Central Nervous System

S.R. VandenBerg, R.B. Hessler and M.B.S. Lopes

Introduction

Surgical neuropathology of CNS tumours has been markedly influenced by increasing numbers of special morphological procedures, especially immunohistochemistry and ultrastructural immunocytochemistry. These techniques have provided a greater understanding of tumour histogenesis with respect to cytoskeletal proteins, membrane proteins, growth factors, oncogenes and growth kinetics. In this context, a revised classification system for tumours of the nervous system was formulated by a WHO Meeting on the Histological Typing of the Tumors of the Central Nervous System in Zurich, Switzerland (28–31 March 1990).

Clinical neuro-oncology has been dramatically affected by advances in neuroimaging techniques, increased numbers of neurosurgical procedures at critical sites and the expanded use of stereotaxic biopsies. Accordingly, neuropathologists are called upon to perform diagnostic evaluations on diminutive tissue fragments, often from heterogeneous neoplasms. In these circumstances, defining key histopathological features for the classification of brain tumours is crucial. The purpose of this review is to provide an interpretation of the recently revised WHO Brain Tumor Classification (Tables 3.1–3.4) from the perspective of the distinctive cytological, histopathological and neuroimaging features of the tumours arising from the central neuroaxis and the meninges.

Four major categories of tumours are significantly revised: (a) astrocytic tumours (Table 3.1), (b) neuronal and mixed neuronal–glial tumours (Table 3.2), (c) embryonal tumours (Table 3.3), and (d) tumours of meningothelial cells (Table 3.4). First, the range of neoplastic astrocytic phenotypes has been broadened with recognition of the pleomorphic xanthoastrocytoma which not only has distinctive clinicopathological features but also has therapeutic implications. Likewise, the neuronal and mixed neuronal–glial group is expanded by the addition of one neuronal (central neurocytoma) and two mixed (dysembryoplastic neuroepithelial tumour, desmoplastic infantile ganglioglioma) neoplasms. The recognition of these tumour designations represents an important advance due to the distinctive clinical biology of these entities.

The most significant revision with regard to embryonal tumours is the elimination of the glioblastoma multiforme and the poorly understood gliomatosis cerebri from this group. The remaining tumours commonly arise in the developing and immature brain in contrast to the anaplastic tumours of the adult brain. The neuroblastoma, ependymoblastoma and the primitive neuroectodermal tumour (PNET) designations have been

Table 3.1   Tumours of neuroepithelial tissue

WHO 1990	WHO 1979

Astrocytic tumours
  1. Astrocytoma
    Variants: fibrillary, protoplasmic
       gemistocytic, or mixed
  2. Anaplastic (malignant) astrocytoma
  3. Glioblastomas
    Variants: giant cell glioblastoma
       gliosarcoma
  4. Pilocytic astrocytoma
  5. Pleomorphic xanthoastrocytoma
  6. Subependymal giant cell astrocytoma
    (usually in association with tuberous
    sclerosis)
	Astrocytic tumours
	  1. Astrocytoma
	    Variants: fibrillary, protoplasmic, or
	       gemistocytic
	  2. Pilocytic astrocytoma
	  3. Subependymal giant cell
	    astrocytoma (ventricular tumour of
	    tuberous sclerosis)
	  4. Astroblastoma
	  5. Anaplastic [malignant] astrocytoma

Oligodendroglial tumours
  1. Oligodendroglioma
  2. Anaplastic (malignant) oligodendroglioma
	Oligodendroglial tumours
	  1. Oligodendroglioma
	  2. Mixed oligo-astrocytoma
	  3. Anaplastic (malignant)
	    oligodendroglioma

Ependymal tumours
  1. Ependymoma
    Variants: cellular, papillary, epithelial
       clear cell, mixed
  2. Anaplastic (malignant) ependymoma
  3. Myxopapillary ependymoma
  4. Subependymoma
	Ependymal and choroid plexus tumours
	  1. Ependymoma
	    Variants: myxopapillary, papillary,
	       or subependymoma
	  2. Anaplastic (malignant)
	    ependymoma
	  3. Choroid plexus papilloma
	  4. Anaplastic (malignant) choroid
	    plexus papilloma

Mixed gliomas
  1. Mixed oligo-astrocytoma
  2. Anaplastic (malignant) oligo-astrocytoma
  3. Others	

Choroid plexus tumours
  1. Choroid plexus papilloma
  2. Choroid plexus carcinoma	

Neuroepithelial tumours of uncertain origin
  1. Astroblastoma
  2. Polar spongioblastoma
  3. Gliomatosis cerebri	

Table 3.2 Tumours of neuroepithelial tissue

WHO 1990	WHO 1979
Neuronal and Mixed Neuronal–Glial Tumours 1. Gangliocytoma 2. Dysplastic gangliocytoma of cerebellum (Lhermitte–Duclos) 3. Desmoplastic infantile ganglioglioma 4. Dysembryoplastic neuroepithelial tumour 5. Ganglioglioma 6. Anaplastic (malignant) ganglioglioma 7. Central neurocytoma 8. Olfactory neuroblastoma (Esthesioneuroblastoma) Variant: olfactory neuroepithelioma	Neuronal tumours 1. Gangliocytoma 2. Ganglioglioma 3. Ganglioneuroblastoma 4. Anaplastic (malignant) gangliocytoma and ganglioglioma 5. Neuroblastoma
Pineal tumours 1. Pineocytoma 2. Pineoblastoma 3. Mixed pineocytoma/pineoblastoma	Pineal cell tumours 1. Pineocytoma (pinealocytoma) 2. Pineoblastoma (pinealoblastoma)

Table 3.3 Tumours of neuroepithelial tissue

WHO 1990	WHO 1979
Embryonal tumours 1. Medulloepithelioma 2. Neuroblastoma Variant: ganglioneuroblastoma 3. Ependymoblastoma 4. Retinoblastoma 5. Primitive neuroectodermal tumours (PNETs) with multipotent differentiation: neuronal, astrocytic, ependymal, muscle, melanotic, etc. (a) Medulloblastoma Variants: desmoplastic medulloblastoma medullomyoblastoma melanocytic medulloblastoma (b) Cerebral or spinal PNETs	Poorly differentiated and embryonal tumours 1. Glioblastoma Variants: glioblastoma with sarcomatous component [mixed glioblastoma and sarcoma giant cell glioblastoma 2. Medulloblastoma Variants: desmoplastic medulloblastoma medullomyoblastoma 3. Medulloepithelioma 4. Primitive polar spongioblastoma 5. Gliomatosis cerebri

added to this category. The first two tumours identify a more restricted histogenesis while the PNET designation recognizes an unrestricted, heterogeneous differentiation.

Finally, the variety of epithelial and metaplastic phenotypes associated with tumours of meningothelial cells has become clearer with the recognition of the microcystic, secretory, clear cell, chordoid and "metaplastic" variants. In addition, the new designation of atypical meningioma reflects a rudimentary attempt to define the histopathological features which may predict a clinically aggressive behaviour. Immunohistochemical and

Table 3.4 Tumours of the meninges

WHO 1990	WHO 1979
Tumours of meningothelial cells	Meningioma
1. Meningioma	1. Meningotheliomatous (endotheliomatous,
Histological types:	syncytial, arachnotheliomatous)
(a) meningothelial (syncytial)	2. Fibrous (fibroblastic)
(b) transitional/mixed	3. Transitional (mixed)
(c) fibrous (fibroblastic)	4. Psammomatous
(d) psammomatous	5. Angiomatous
(e) angiomatous	6. Haemangioblastic
(f) microcystic	7. Haemangioblastic
(g) secretory	8. Papillary
(h) clear cell	9. Anaplastic (malignant) meningioma
(i) chordoid	
(j) lymphoplasmacyte-rich	
(k) metaplastic variants	
(xanthomatous, myxoid,	
osseous, cartilagenous, etc.)	
2. Atypical meningioma	
3. Anaplastic (malignant) meningioma	
(a) variants of 1(a)–(k) (see above)	
(b) papillary	

histochemical silver techniques that permit more sensitive quantitation of the mitotic index will most likely play an important role in the diagnostic interpretation of these tumours and permit a better resolution of other applicable histopathological features.

Tumours of Neuroepithelial Tissue

Astrocytic Tumours

Of the three major changes that have occurred in the designation of astrocytic tumours, two have been made due to the impact of the immunohistochemical localization of glial fibrillary acidic protein (GFAP) which has permitted the identification of pleomorphic astroglial populations. First, glioblastomas are now included in the astrocytic tumours and secondly, the astrocytic origin of the pleomorphic xanthoastrocytoma is well established (Table 3.1). Recognition of this astrocytic tumour with an unusual stromal induction is significant from the perspective of astroglial–stromal interactions. The capacity of astrocytic tumour cells to secrete, modify and interact with a wide variety of extracellular matrices has been recognized with increasing frequency (Rubinstein, 1991) and is currently an area of intense basic investigation.

The third revision involves elimination of the rare astroblastoma (Bonnin and Rubinstein, 1989) due to the uncertainty of its histogenesis. Culture studies of two tumours (Rubinstein and Herman, 1989) suggested a tanycytic origin, but further investigation is necessary for a more definitive classification.

The astrocytic tumours are now divided into six categories (Table 3.1). The first three groups include the diffusely infiltrating tumours which are separated on the basis of histopathological features of anaplasia. Although the current WHO classification does not endorse one specific "grading" system for correlation to clinical behaviour, the categories of astrocytoma, anaplastic astrocytoma and glioblastoma tacitly designate the slower growing (astrocytomas) and the malignant tumours (anaplastic astrocytomas and glioblastomas). The principal caveat for histopathological classification and grading of these three groups with limited biopsy tissue is the degree of intratumoral heterogeneity (Bigner *et al.*, 1981; Shapiro *et al.*, 1981; Shapiro and Shapiro, 1984). Thus, the assessment of tumour "grade" should include precise neuroimaging data of the neoplasm with specific attention to the biopsy target areas in addition to histopathological examination.

The last three groups of astrocytic tumours (pilocytic astrocytoma, pleomorphic xanthoastrocytoma and subependymal giant cell astrocytoma) have distinctive clinicopathological features that justify separate classification. Despite a wide range of histopathological features, all have macroscopically circumscribed borders compared to the preceding astrocytomas and usually exhibit a slower growth with a relatively indolent biological behaviour and a lower capacity for anaplastic progression. The clinical prognosis is usually favourable following surgical resection. These distinctive histopathological features and preferential locations suggest a more selective, including regional, astroglial origin.

Astrocytomas

The reported incidence of astrocytomas varies due to a number of factors, including different classification systems, surgical versus autopsy data, and the differences between referral and primary treatment centres (Zimmerman, 1969; Percy *et al.*, 1972; Poirier *et al.*, 1985; Codd and Kurland, 1985; Zülch, 1986). In the University of Virginia series of approximately 6500 referral cases, astrocytomas, anaplastic astrocytomas and glioblastomas multiforme comprise nearly 75% of cerebral astrocytic tumours (Russell and Rubinstein, 1989). The astrocytomas commonly occur in the third and fourth decades whereas the anaplastic tumours tend to arise in the fourth and fifth decades. In general, males are more often affected than females (Russell and Rubinstein, 1989).

Astrocytomas can be characterized further by similarities to the basic morphological types of astrocytes in normal and reactive brain, i.e. fibrillary, protoplasmic and gemistocytic. In general, protoplasmic astrocytes are more abundant in the regions of the cerebral cortex and deep grey nuclei, such as the striatum, while the fibrillary astrocytes are more abundant at the borders of the cortex and white matter. The gemistocytic astrocytoma is composed of cells which superficially resemble reactive, metabolically active astrocytes. Recent data (see below), however, suggest that the resemblance to benign, reactive astrocytes is spurious and that gemistocytic astrocytomas have a more aggressive biological behaviour than previously believed.

Subclassifying astrocytomas by only general morphological traits has significant limitations because these do not consider the astroglial physiological heterogeneity which may be regionally specific (Denis-Donini *et al.*, 1984; Hansson, 1988). Recent evidence emphasizes that this heterogeneity may also be modified by local cellular interactions (Chamak *et al.*, 1987; Beyer *et al.*, 1990; Cockram, 1990). Therefore, the complex functional diversity of neoplastic astrocytic populations further confounds the correlation of specific histological features to biological activity. A dilemma for contemporary neuropathology is the determination of specific neurochemical properties, in addition to morphology, which could identify tumour cell populations as more likely to act aggressively or to undergo anaplastic progression.

Fibrillary Astrocytomas. Some authors have proposed the subdivision of fibrillary astrocytomas into diffuse and circumscribed forms (Russell and Rubinstein, 1989). However, these "circumscribed" astrocytomas are more appropriately classified with the pilocytic astrocytomas (see below). Fibrillary astrocytomas, as diffusely infiltrating astrocytomas, most often occur in the cerebral hemispheres of adults, and the brainstem of children and adolescents. In the cerebellum, these tumours are more common in young and middle-aged adults. Macroscopically, fibrillary astrocytomas are often difficult to distinguish from adjacent brain, but are typically more firm due to the abundance of glial fibrils. This texture can readily be appreciated during the preparation of tumour smears.

The typical cytoplasmic morphology of fibrillary astrocytes ranges from a scant perinuclear rim with ill-defined borders to a more stellate shape. Cells commonly have multiple cytoplasmic processes which form the fibrillary matrix observed in both the smear and histological preparations (Figs 3.1 and 3.2). This matrix is often accentuated around the vascular elements. In comparison to normal fibrous astrocytes, the chromatin is hyperchromatic and more coarsely stippled. Special histochemical and immunohistochemical studies are rarely needed to discriminate fibrillary astrocytomas from other astrocytic tumours. However, the affinity of glial fibrils for phosphotungstic acid haematoxylin (PTAH) staining combined with the conspicuous GFAP immunoreactivity is a property of these tumours that is useful for diagnostic applications (Perentes and Rubinstein, 1987; Cosgrove *et al.*, 1989). In addition, fibrillary astrocytomas also demonstrate immunoreactivity for glutamine synthetase, aldolase C, S-100 protein, vimentin, neuron specific enolase and the Leu-7 epitope (Pilkington and Lantos, 1982; Perentes and Rubinstein, 1987).

Protoplasmic Astrocytomas. This histological variant as a pure tumour is rare (less than 1%) (Russell and Rubinstein, 1989). Protoplasmic astrocytes may be a component of pilocytic astrocytomas (see below), and inadequately sampled pilocytic astrocytomas may be misinterpreted as protoplasmic astrocytomas for this reason.

Protoplasmic astrocytomas are located most often in the cerebrum of children and young adults. There appears to be a predilection for the temporal lobes, although the tumours may occur in any cortical region. The

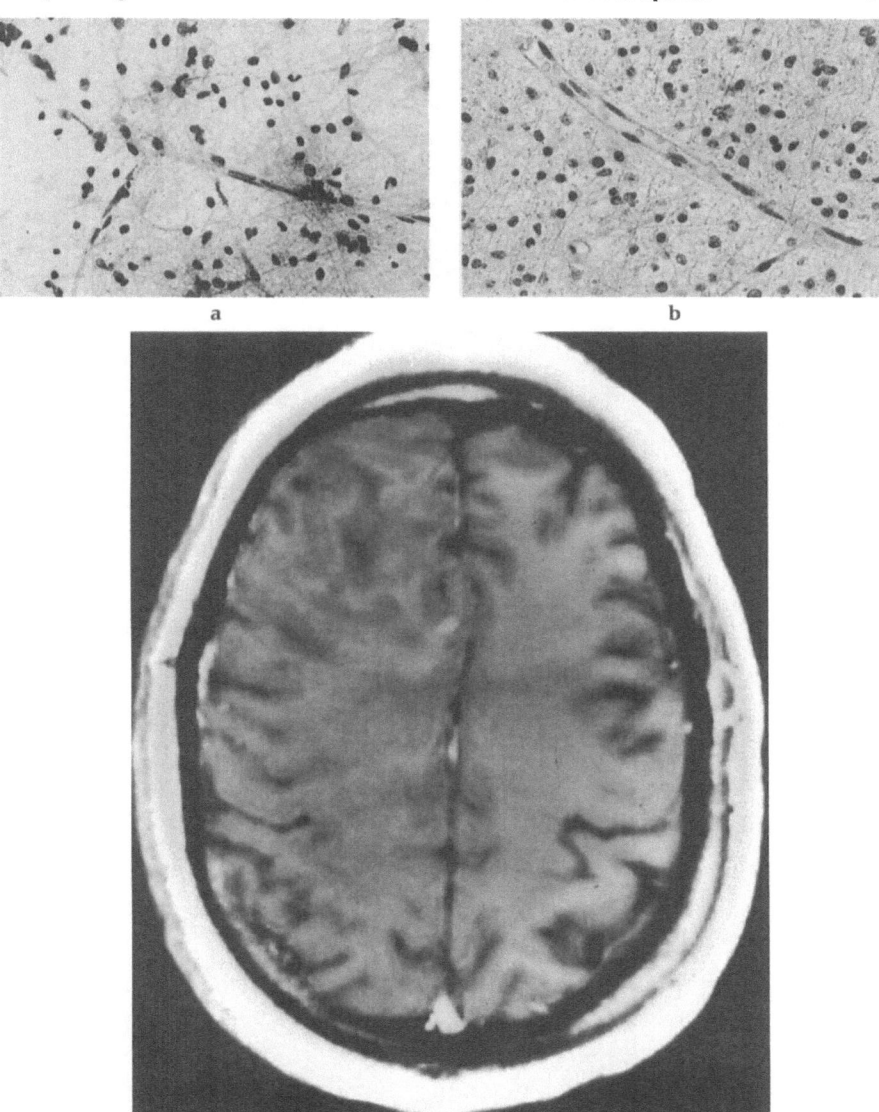

c

Fig. 3.1   Fibrillary astrocytoma. **a** A smear preparation demonstrates slightly atypical astro-cytes in a moderately fibrillary matrix. Note the delicate blood vessels (Morris stain, original magnification ×400). **b** Similar features are present on the paraffin-embedded section (Haematoxylin and eosin, original magnification ×400). **c** The same tumour appears as a large, ill-defined mass with minimal contrast enhancement on a T_1 weighted magnetic res-onance image following the administration of intravenous gadolinium–DTPA [Axial scan: SE 0.5 s/20 ms (1.5 T)]

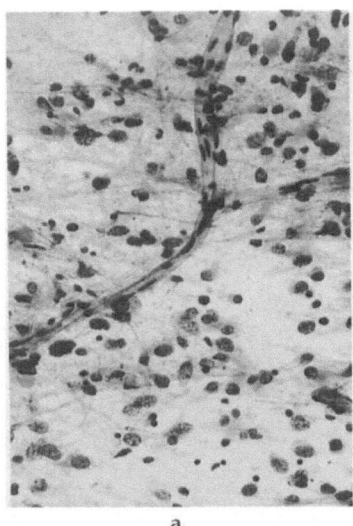

a

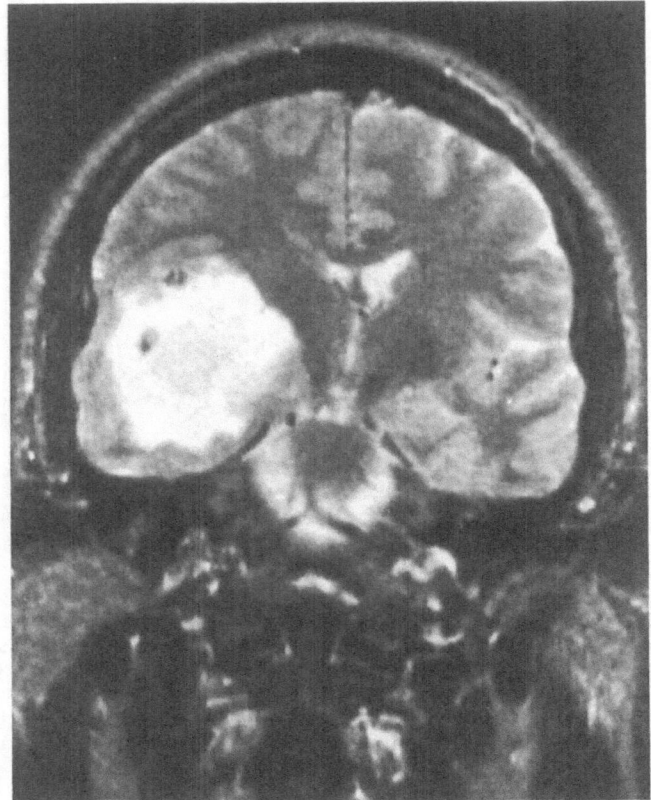

b

tumours are usually superficial and expand the cortex as gelatinous, homogeneous masses. Variable cysts are common. The tumour is composed of a relatively homogeneous cell population in an abundant eosinophilic matrix with extensive microcystic degeneration (Fig. 3.3). The cells typically have a stellate geometry with short, delicate processes without PTAH positive glial fibrils. GFAP immunoreactivity is like wise sparse (Russell and Rubinstein, 1989).

Gemistocytic Astrocytomas. Gemistocytic astrocytomas are rare tumours, usually accounting for no more than 9%–19% of astrocytomas in most studies (see Krouwer *et al.*, 1991). As a pure histological variant (i.e. greater than 60% gemistocytes/high power field) the tumours are limited to the cerebral hemispheres. However, variable numbers of cells with a gemistocytic character may be encountered in fibrillary astrocytomas and anaplastic astrocytomas. Recent data suggest that astrocytic tumours composed of a gemistocytic component greater than 20% have a poorer prognosis than other astrocytomas. These tumours have a clinical behaviour more similar to anaplastic astrocytomas (Krouwer *et al.*, 1991). One autopsy series found that 80% of gemistocytic astrocytomas progress to glioblastoma (Russell and Rubinstein, 1989).

Macroscopically, gemistocytic astrocytomas are soft, grey and homogeneous in appearance. The borders can appear to be well circumscribed, and a cystic component can at times be appreciated. Smears of gemistocytic astrocytomas are characterized by cells with abundant eosinophilic, round to slightly angulated cytoplasm, and eccentric nuclei (Fig. 3.4). The matrix is not conspicuously fibrillated, but varies with abundance of fibrillary astrocytes in the tumour. The tumour cells may produce only short, inconspicuous processes, a feature which distinguishes neoplastic from reactive gemistocytic astrocytes. Multinucleated cells are not uncommon, and the nuclei demonstrate varying degrees of morphological atypia, with occasionally prominent nucleoli. The chromatin pattern is, in general, more coarse than in other astrocytomas. This pattern is better demonstrated in permanent sections, where the chromatin often coalesces into small clumps. The tumour cells are strongly immunoreactive for GFAP, which is consistent with the ultrastructural finding of bundles of intracytoplasmic filaments (Russell and Rubinstein, 1989).

Neuroimaging Correlations. Astrocytomas, usually regardless of histological type, show similar features on MR and CT images (Figs 3.1C and 3.2B) (see Chap. 10). With MR imaging, there is an increase in T_2 and decrease in T_1 signal intensities compared to normal brain. The tumours are often well demarcated, with only minimal to moderate oedema, which is often difficult to distinguish from the actual tumour mass. One intriguing report

---

◀ **Fig. 3.2**   Fibrillary astrocytoma. **a** A smear preparation comparable to Fig. 3.1 shows moderate pleomorphism as the only atypical feature. Mitoses, necrosis or endothelial proliferation are not found in smears of the tumour (Morris stain, original magnification ×350). **b** The tumour has areas of poor demarcation from the adjacent cortex and moderate mass effect on a T_2 weighted magnetic resonance image [B, Coronal scan: SE 2.5 s/90 ms (1.0 T)]

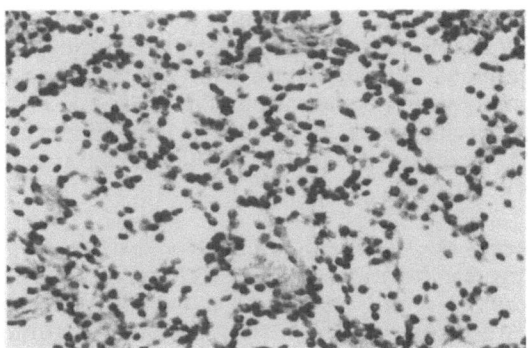

Fig. 3.3 Protoplasmic astrocytoma. These tumours have uniform cytoarchitecture with short, delicate cytoplasmic process and often indistinct borders. The matrix is often indistinct and rarely fibrillary with frequent microcystic spaces (Haematoxylin and eosin, original magnification ×250)

has recently described a gemistocytic astrocytoma demonstrating both relatively high T_1 and T_2 signal intensities compared to the usual MR images of astrocytomas. Shortening of the T_1 relaxation time was attributed to protein hydration layering due to the increased cytoplasmic protein of the tumour cells (Abe *et al.*, 1990). Combined with the histopathological features, MRI data would lead to more accurate diagnostic interpretations, especially considering the morphological heterogeneity of the more anaplastic astrocytomas. In one series, an overall accuracy for MRI diagnosis was 94.4% for attempting to classify astrocytic tumours as astrocytomas, anaplastic astrocytomas or glioblastomas based on 7 MRI characteristics, i.e. crossing of the midline, oedema, heterogeneity, haemorrhage, border definition, cyst/necrosis, and mass effect (Dean *et al.*, 1990).

Anaplastic Astrocytoma

All types of astrocytic tumours appear to be variably capable of progression to anaplasia. The incidence of progression to anaplastic astrocytomas within the group of astrocytic gliomas may be as high as 80% (Scherer 1940; Russell and Rubinstein, 1989). The course of anaplastic change can vary widely, adding to the difficulty of prediction in individual cases. In one series of 79 recurrent low-grade astrocytomas, progression occurred in 50% of the cases (Laws *et al.*, 1984). Other data suggest that progression can be expected in two-thirds of recurrent astrocytomas (Russell and Rubinstein, 1989).

The histopathological criteria to specifically classify astrocytic tumours along the spectrum from well-differentiated astrocytomas to glioblastomas have varied among laboratories (Davis, 1989). These features include the magnitude of cellularity, degree of cellular pleomorphism (cytoplasmic and/or nuclear), number of mitoses, prominence of endothelial/pericytic proliferation and necrosis.

Clinical survival data generally support the separation of three histopathological groups of astrocytic tumours. The largest cooperative

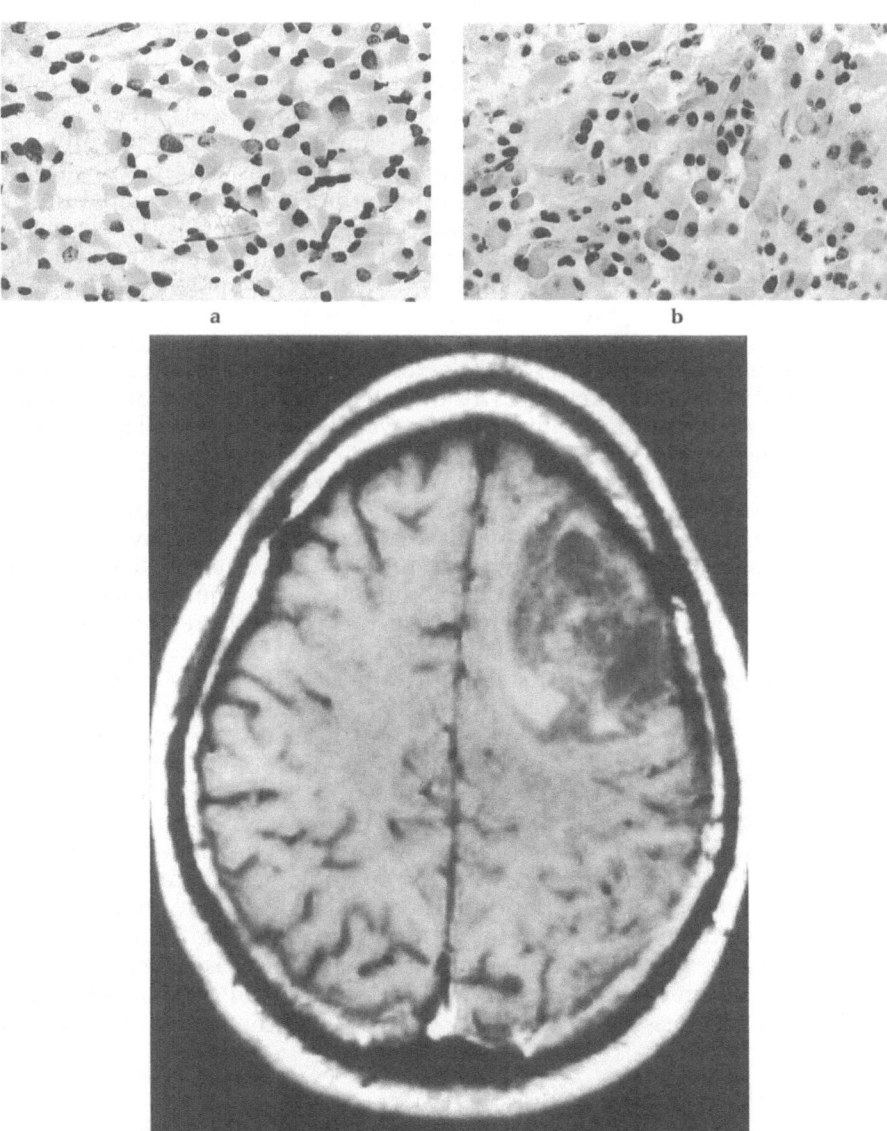

Fig. 3.4 Gemistocytic astrocytoma. **a** Smear preparations typically demonstrate cells with an abundant, intensely eosinophilic cytoplasm, eccentric nuclei, and well-defined cytoplasmic margins (Morris stain, original magnification ×400). **b** Similar features are demonstrated in a tissue section of the same tumour. Tumours in which gemistocytic astrocytes predominate should be considered biologically aggressive (Haematoxylin and eosin, original magnification ×400). **c** This tumour demonstrates a mixed signal intensity and patchy enhancement on a T_1 weighted magnetic resonance image following administration of intravenous gadolinium–DTPA [Axial scan: SE 0.5 s/20 ms (1.5T)].

studies using this three-tiered approach (Nelson *et al.*, 1983; Burger *et al.*, 1985) have used the designation of astrocytomas, anaplastic astrocytomas and glioblastomas. The two systems primarily differed on the use of vascular proliferation and mitotic figures as discriminating features for increasing anaplasia. However, both systems used necrosis as an important characteristic to separate anaplastic astrocytomas from glioblastomas. In contrast, another approach (Davis, 1989) differs by not emphasizing necrosis as an important single feature. Vascular proliferation and the combination of high cellularity associated with intense nuclear and cytoplasmic pleomorphism are used in this grading system as the most suitable criteria for the identification of glioblastoma multiforme. Daumas-Duport *et al.* (1988a) proposed a numerical grading system for diffusely infiltrating supratentorial astrocytomas based on the presence or absence of cytological atypia, vascular proliferation, mitoses and necrosis. The presence of each feature added another grade increment without weighing of particular combinations. The application of this system was tested in two large series (Daumas-Duport *et al.*, 1988a; Kim *et al.*, 1991) and showed strong correlation of this histopathological ranking with survival times.

Recent studies for developing a more quantitative assessment of mitotic activity in astroglial tumours have used several types of innovative morphological approaches, including (a) immunohistochemical detection of DNA synthesis as indicated by BUdR (bromodeoxyuridine) uptake in vivo or in vitro, (b) immunohistochemistry of the cell-cycle related antigens, Ki-67 and proliferating cell nuclear antigen (PCNA), and (c) silver nucleolar organizer region (AgNOR) staining (Burger *et al.*, 1986; Giangaspero *et al.*, 1987; Hoshino *et al.*, 1989; Kajiwara *et al.*, 1990; Louis *et al.*, 1991; Plate *et al.*, 1991). In general, labelling indices of the cell-cycle markers corresponded to the classification of tumour anaplasia and clinically aggressive behaviour; however, a strict quantitative correlation between labelling indices and the biological behaviour for individual cases awaits more data. Nevertheless, these techniques promise to provide a valuable complementary parameter in combination with histopathological features to distinguish astrocytoma differentiation and anaplasia. The term anaplastic astrocytoma encompasses a heterogeneous group of tumours with the common features associated with anaplasia, and thus it is difficult to make generalizations on their smear appearance (Fig. 3.5). In general, anaplastic tumours have a less dense fibrillary matrix than the well-differentiated

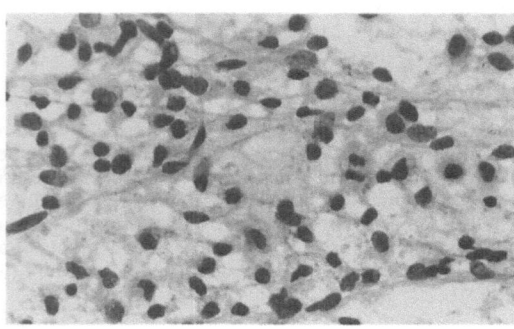

Fig. 3.5 Anaplastic astrocytoma. A smear preparation of this astrocytic tumour typically shows moderate nuclear and cytoplasmic pleomorphism and mitotic activity (Morris stain, original magnification ×400)

astrocytomas, and thus the tissue is softer during smear preparation. Magnetic resonance data complement the histopathological indicators of increased tumour anaplasia with signal heterogeneity, less well-defined margins, increased mass effect and vasogenic oedema. There is focal contrast enhancement, but unlike glioblastomas, this is not a constant feature.

Glioblastoma Multiforme

Glioblastoma multiforme has a peak incidence in the fifth and sixth decades, occurring predominantly in the cerebral hemispheres, and is the most common intracranial tumour of this age group. Overall, it represents approximately 15%–20% of all intracranial tumours and 50% of gliomas. Astrocytomas and, particularly, anaplastic astrocytomas do progress to glioblastoma in a significant number of cases (Muller et al., 1977; Laws et al., 1984; Dropcho et al., 1987), although these tumours may also arise de novo. There is a slight male predominance (3 : 2) (Russell and Rubinstein, 1989 Burger et al., 1991). Paediatric cases also occur and must be separated from the embryonal tumours (see below).

Glioblastomas most often present as an expansive type of lesion, with ring enhancement on CT and MR images (Fig. 3.6C). However, this tumour invariably infiltrates the adjacent brain in an aggressive manner. Tumour cell permeation along fibre tracts is common, such as the characteristic "butterfly" neoplasms by involvement of the corpus callosum. The malignant potential is also reflected by the highest incidence of metastasis for any glial neoplasm beyond the nervous system (Liwnicz and Rubinstein, 1979).

Glioblastomas exhibit extensive but variable cytoplasmic and nuclear pleomorphism on both smear and permanent preparations (Fig. 3.6A and 3.6B). The cellular forms range from scant cytoplasm with round to oval, densely hyperchromatic nuclei to bizarre, multinucleated giant cells. Cytoplasmic changes include lipidization and heteroplastic cytokeratin expression (Kepes and Rubinstein, 1981; Mørk et al., 1988). Mitoses, including atypical forms, are frequent but may significantly vary within an individual tumour. Exuberant endothelial/pericytic proliferation ("endothelial proliferation") combined with necrosis is commonly regarded as definitive for glioblastoma. These two features, which can be readily appreciated on smears, can distinguish glioblastoma from anaplastic astrocytomas (Fig. 3.6A; Plate 3.1).

The mechanisms of endothelial proliferation in astroglial tumours are not well defined, but probably represent perturbation of the normal interactions between the microvascular elements and astroglia which appear to mutually modulate mitogenic potential and differentiation (Goldstein, 1988). These interactions may, in part, be mediated by a number of growth factors including endothelial cell growth factor, transforming growth factor β, platelet-derived growth factor and epidermal growth factors (Libermann et al., 1987; Mauro and Bulfone, 1990; Jones et al., 1990; Cockram, 1990). Additionally, the presence of renin in some glioblastoma cells has implicated the renin–angiotensin II cascade in neovascular angiogenesis of glioblastomas (Ariza et al., 1988).

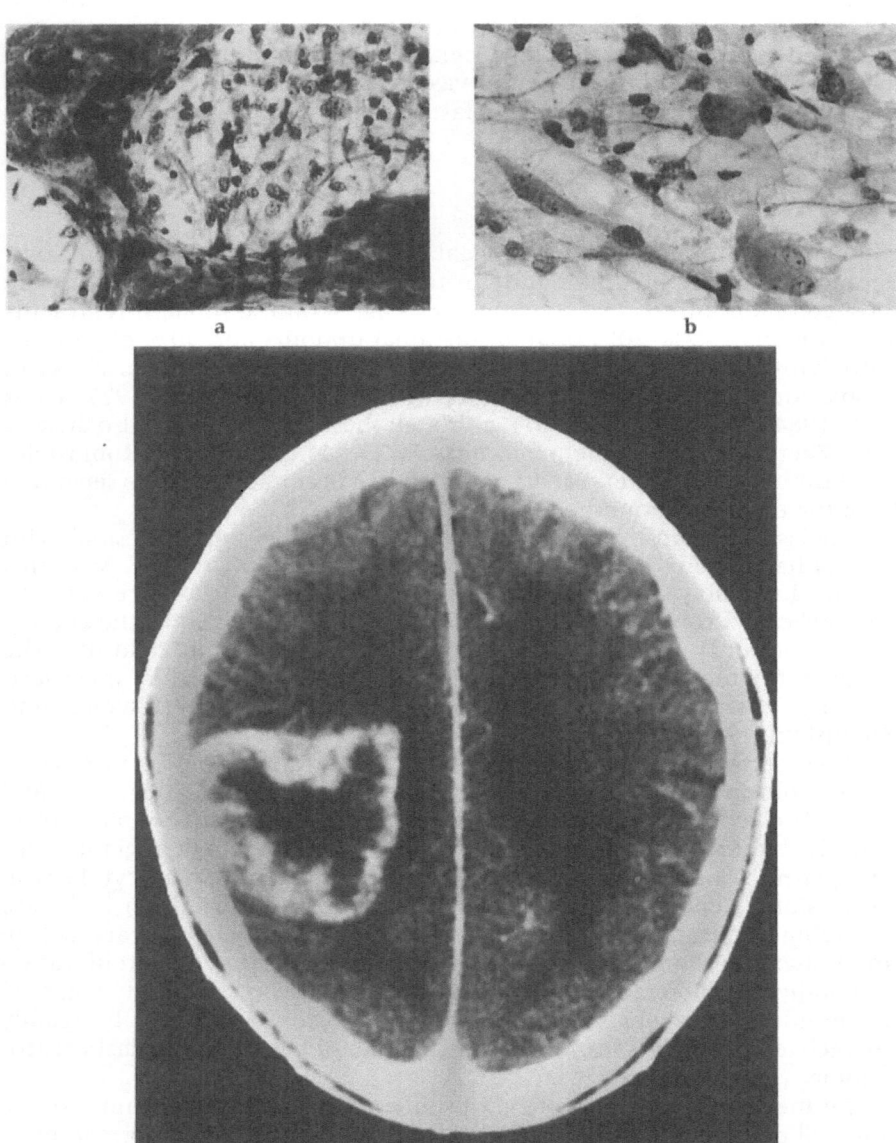

a

b

c

Fig. 3.6 Glioblastoma multiforme. **a** An intraoperative smear contains markedly pleomorphic astrocytes adjacent to an area of endothelial proliferation (Morris stain, original magnification ×200). **b** At higher magnification, the pleomorphic aspect of the tumour cells is easily appreciated in a less cellular area of the smear (Morris stain, original magnification ×400). **c** An enhanced CT scan of a glioblastoma demonstrates the characteristic ring of enhancement (See Plate 3.1)

Colour Plate Section

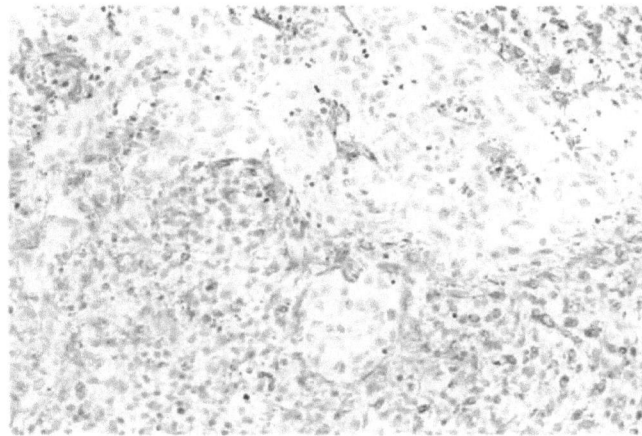

Plate 3.1 Immuno-histochemical staining for GFAP clearly distinguishes the neoplastic glial populations from areas of endothelial/pericytic proliferation in glioblastomas (GFAP–avidin–biotin complex immunostaining with haematoxylin counterstain, original magnification ×200)

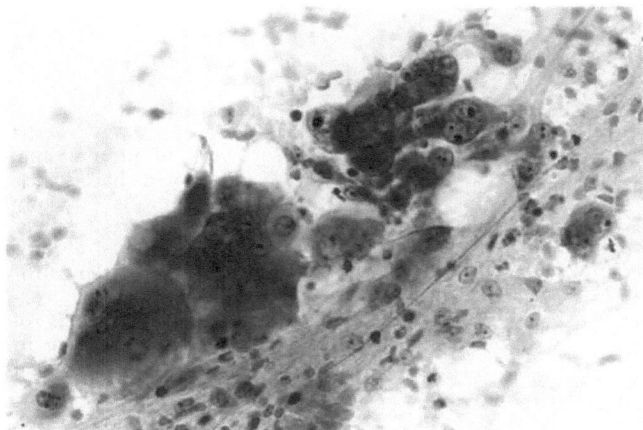

Plate 3.2 An intra-operative smear of a giant cell glioblastoma shows clusters of bizarre, multinucleated giant cells adjacent to smaller neoplastic astrocytes. The matrix usually has a poorly defined fibrillary quality. These tumours are macroscopically circumscribed (Morris stain, original magnification ×400)

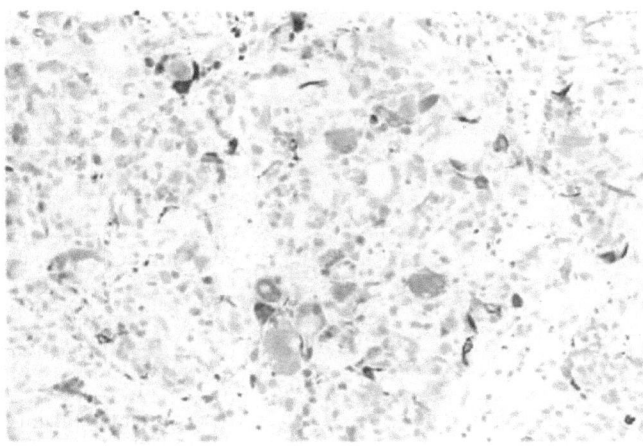

Plate 3.3 GFAP immunoreactivity in the giant cell glioblastoma demonstrates the glial nature of both the giant cells and the more typical astrocytic tumour cells (GFAP–avidin–biotin complex immunostaining with haematoxylin counterstain, original magnification ×200)

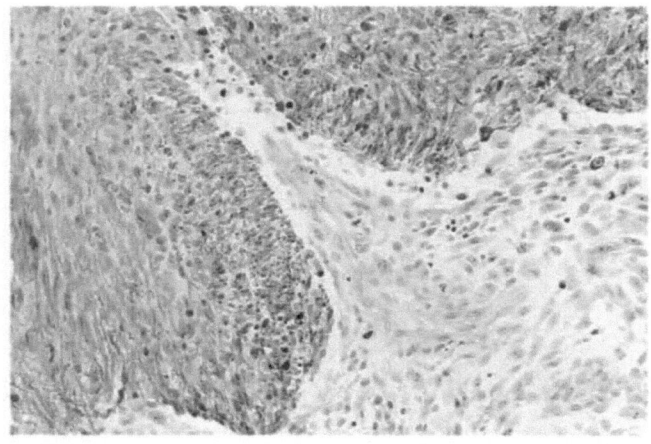

Plate 3.4 The astrocytic component of a gliosarcoma is easily distinguished from the sarcomatous tumour cells by positive immunohistochemical staining for GFAP (GFAP–peroxidase–anti peroxidase immunostaining with haematoxylin counterstain, original magnification ×200)

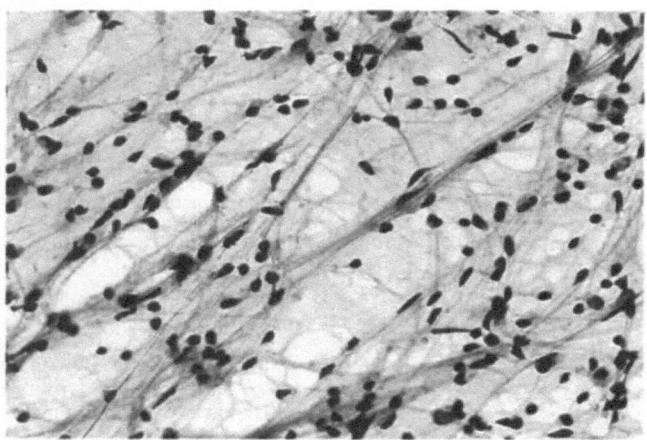

Plate 3.5 A meshwork of protoplasmic elongated fibrous astrocytes is readily observed in a smear preparation of a cerebellar (pilocytic) astrocytoma. This corresponds to areas of microcystic change in the paraffin-embedded tissue (Morris stain, original magnification ×400)

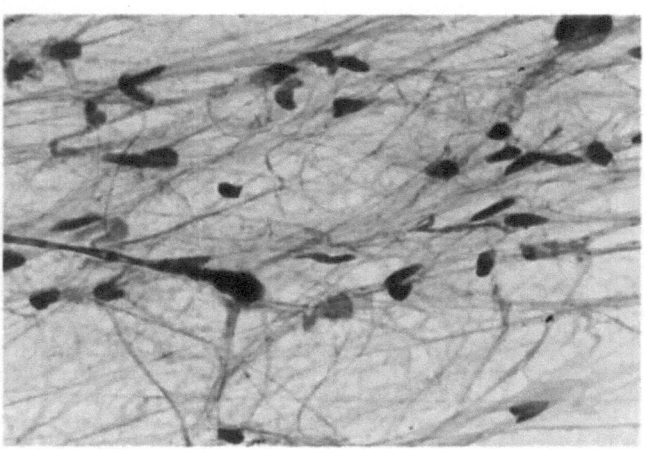

Plate 3.6 Rosenthal fibres in a smear preparation of a thalamic pilocytic astrocytoma appear as irregularly elongated beads of eosinophilic material, often in continuity with cellular processes. Dense wire-like processes form the typical intercellular matrix of these tumours (Morris stain, original magnification ×600)

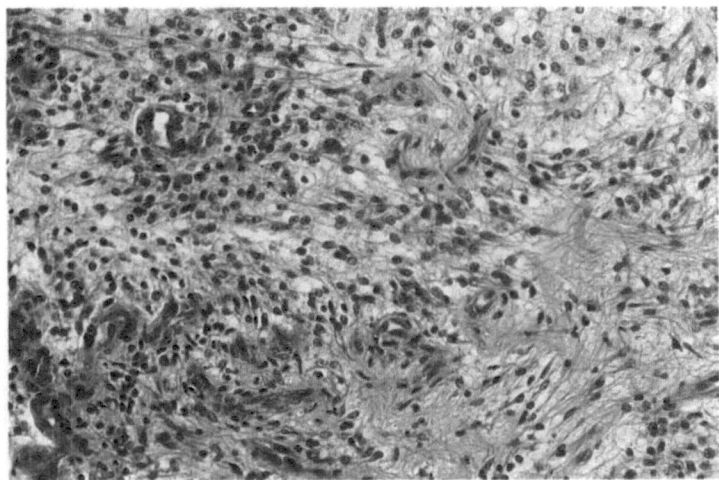

Plate 3.7 A tissue section of the tumour shown in Plate 3.5 shows the typical biphasic arrangement of stellate and more piloid astrocytes in a densely fibrillary matrix. Endothelial/pericytic proliferation is a common feature without ominous prognostic implications (Haematoxylin and eosin, original magnification ×400)

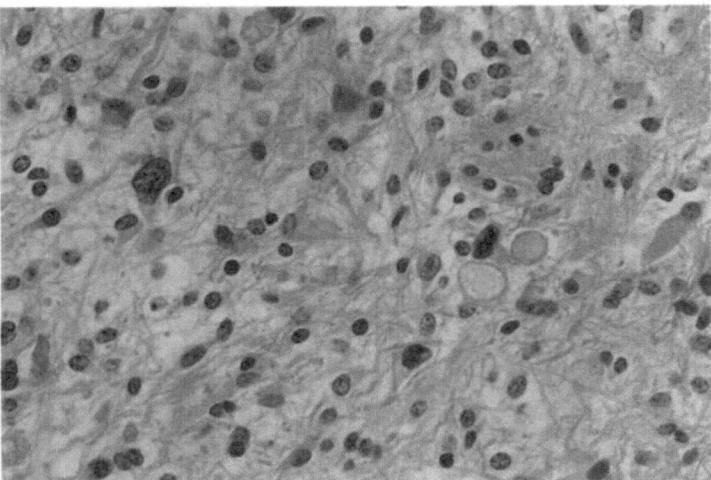

Plate 3.8 Granular eosinophilic bodies are common in pilocytic astrocytomas and appear as circular structures with eosinophilic rims and granular, hyalinized cores. Note the Rosenthal fibres and cytoplasmic pleomorphism with prominently thickened processes. A dense mesh of delicate cellular processes is typically a significant component of the intercellular matrix. Nuclear pleomorphism is not considered an anaplastic feature in these tumours (Haematoxylin and eosin, original magnification ×400)

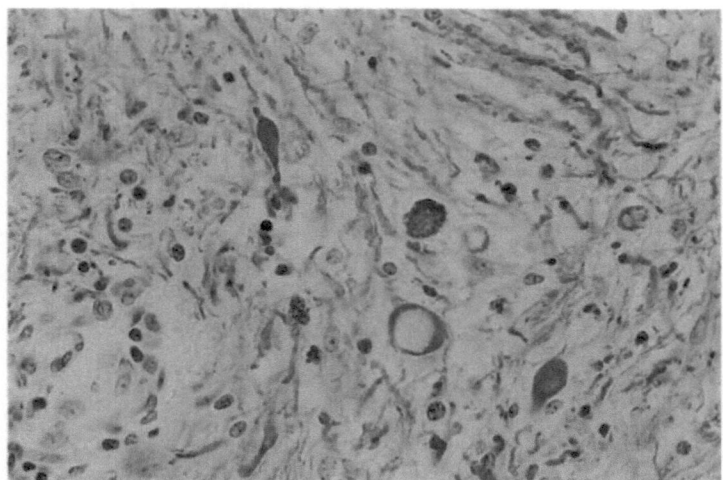

Plate 3.9 Granular eosinophilic bodies have heterogenous GFAP immunoreactivity. Note the bodies with diffuse, granular staining and with positivity limited to a peripheral ring (arrowheads). Rosenthal fibres have variable GFAP immunoreactivity but are invariably highlighted by PTAH staining. In this field, the intensely positive Rosenthal fibre is admixed with dilated tumour cell processes (GFAP–avidin–biotin complex immunostaining, original magnification ×400)

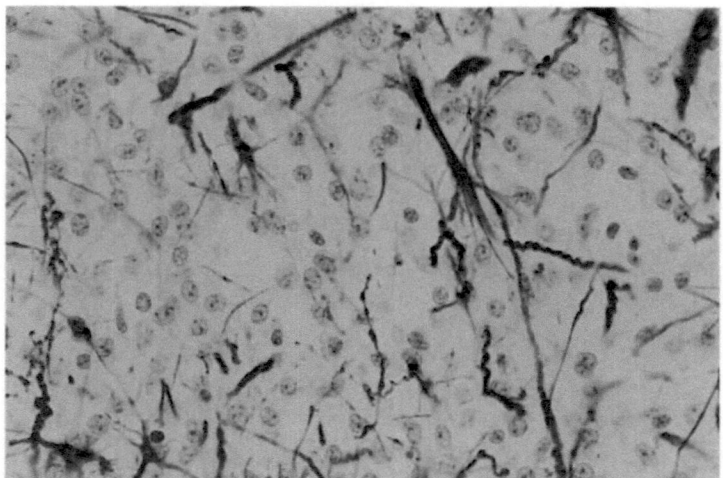

Plate 3.10 GFAP immunohistochemistry in typical cerebellar (pilocytic) astrocytomas reveals heterogeneous glial cell populations. Both fibrillated astrocytes and unreactive, more stellate glial cells with poorly defined processes are present (GFAP–avidin–biotin complex immunostaining, original magnification ×400)

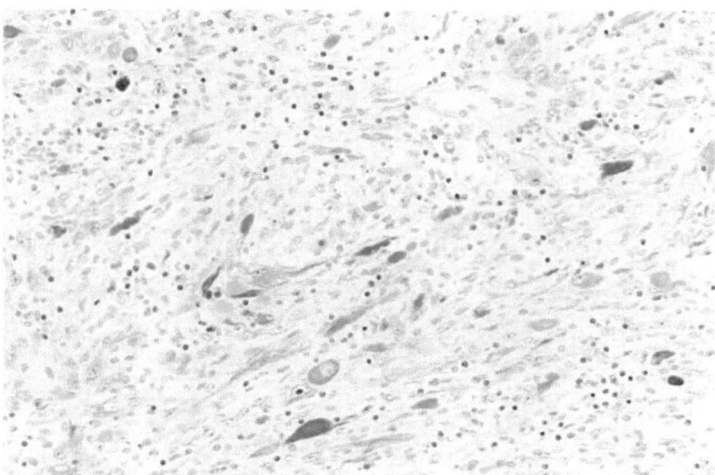

Plate 3.11 GFAP immunohistochemistry in pleomorphic xanthoastrocytoma unequivocally demonstrates the astrocytic, as opposed to histiocytic, nature of the tumour with the characteristic bundles of elongated tumour cells admixed with larger, pleomorphic cells. Note the multinucleated cell containing GFAP (arrow) (GFAP–peroxidase–antiperoxidase immunostaining, original magnification ×200)

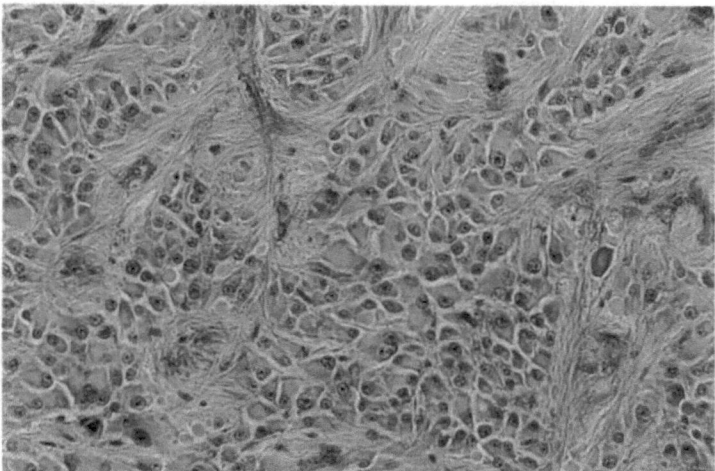

Plate 3.12 The neoplastic astrocytes in the subependymal giant cell astrocytoma are heterogenous cell populations composed of large pyramidal cells admixed with strap-like and smaller fibrillated astrocytes. The acellular areas of the matrix can be densely fibrillary, reminiscent of the subependymoma. Note the characteristic coarse chromatin pattern and conspicuous nucleoli of the larger polygonal cells (PTAH, original magnification ×200)

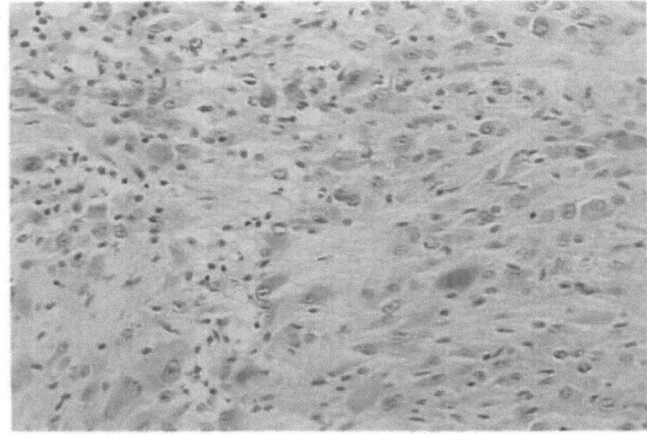

Plate 3.13 Immuno-histochemistry for GFAP also demonstrates heterogeneous glial cells in the subependymal giant cell astrocytoma (GFAP–peroxidase–antiperoxidase immunostaining, original magnification ×400)

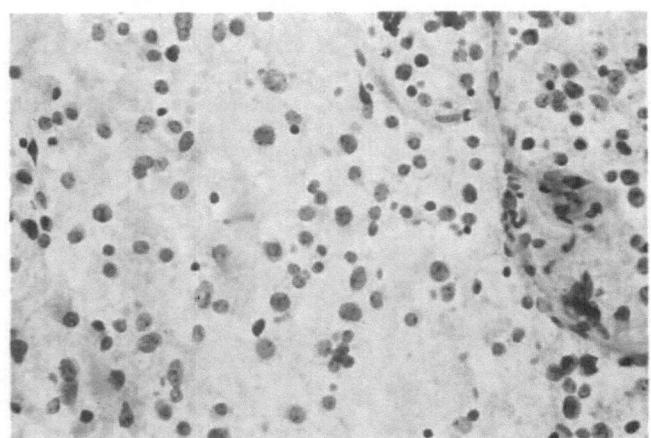

Plate 3.14 Smear preparations of oligodendrogliomas invariably reveal homogeneous populations of cells with inconspicuous cytoplasmic margins in a poorly defined matrix (Morris stain, original magnification ×400)

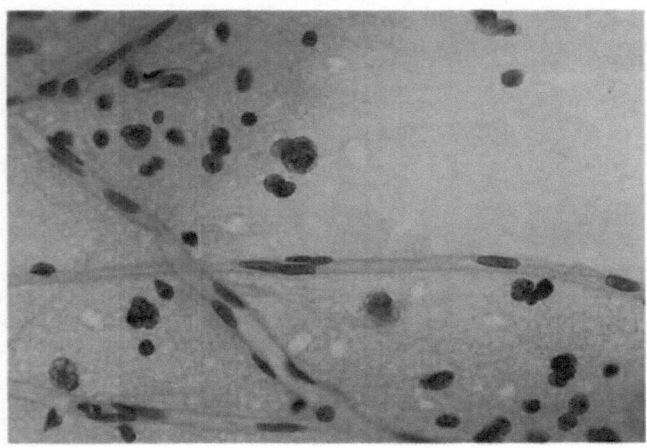

Plate 3.15 In contrast to astrocytomas, the nuclei in an oligodendroglioma are usually more pleomorphic, often moderately lobulated and contain coarse chromatin with chromatin nodes. Delicate, interlacing blood vessels are usually abundant (Haematoxylin and eosin, original magnification ×600)

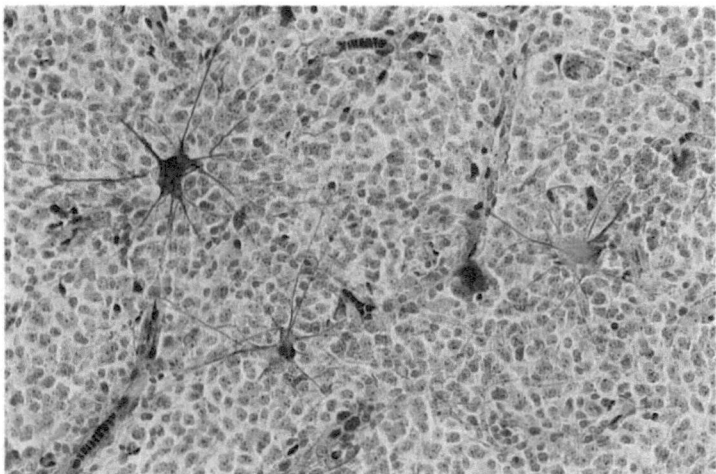

Plate 3.16 Although typical oligodendrogliomas do not have a neoplastic astrocytic component, reactive astrocytes can be readily visualized by special stains (PTAH, original magnification ×200)

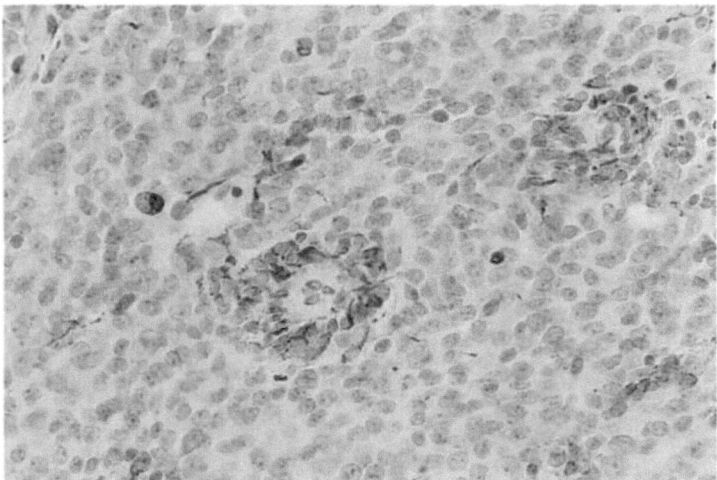

Plate 3.17 GFAP immunoreactivity is not uncommon in the stubby perivascular processes of the tumour cells in oligodendrogliomas. Cells without a precise relationship to blood vessels may also have GFAP immunoreactivity with the characteristic globular geometry of the "gliofibrillary" oligodendrocyte. These features are not interpreted as evidence of a neoplastic astrocytic population in the typical oligodendroglioma (GFAP–avidin–biotin complex immunostaining with haematoxylin counterstain, original magnification ×400)

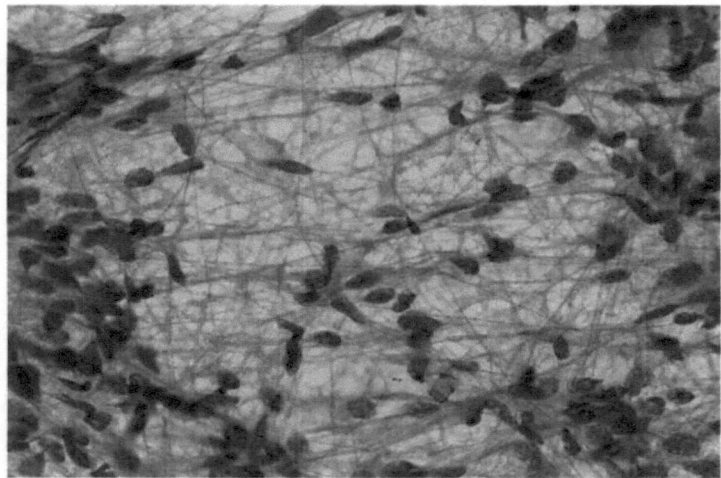

Plate 3.18   Smear preparations of ependymomas accentuate the abundant fibrillary intercellular matrix as well as the unevenly distributed, punctate ("open") chromatin of the tumour cells (Haematoxylin and eosin, original magnification ×400)

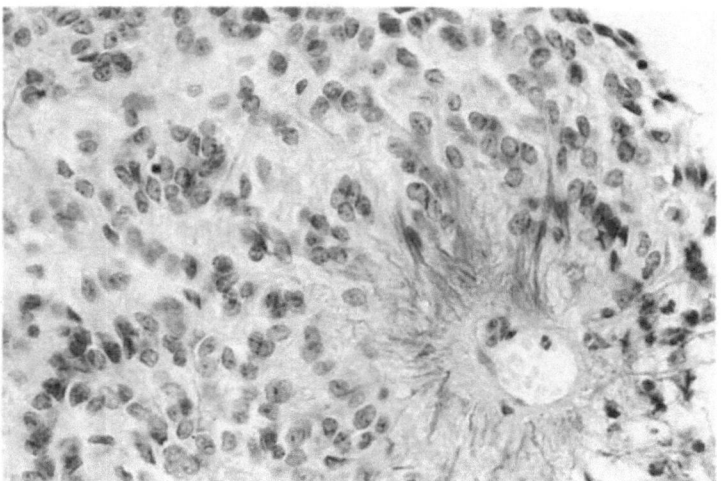

Plate 3.19   GFAP immunohistochemistry in ependymomas highlights the fibrillary processes which are usually more prominent in the perivascular rosettes. GFAP immunoreactivity is extremely stable in paraffin-embedded tissue, and was present in this tissue after more than 30 years of storage (GFAP–peroxidase–antiperoxidase immunostaining with haematoxylin counterstain, original magnification ×400)

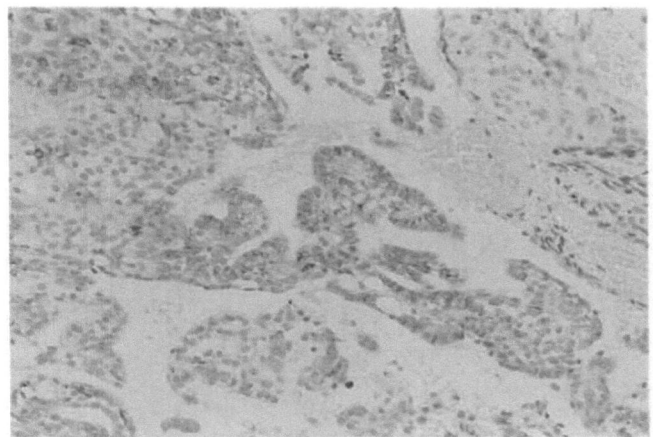

Plate 3.20 In the papillary variant of the ependymoma, both the epithelial cells lining the papillary structures and the fibrillary cells in the more compactly-arranged areas show immunoreactivity for GFAP (GFAP–avidin–biotin complex immunostaining with haematoxylin counterstain, original magnification ×200)

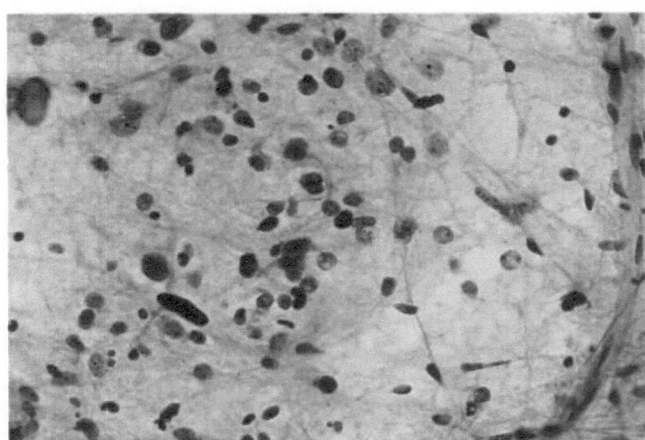

Plate 3.21 The intraoperative smear of a mixed oligoastrocytoma can clearly discriminate between the two cell populations in these gliomas. The oligodendroglial cells have the typical nuclear morphology as described in Plate 3.15. These cells are admixed with neoplastic astrocytes with fibrillary processes (Morris stain, original magnification ×400)

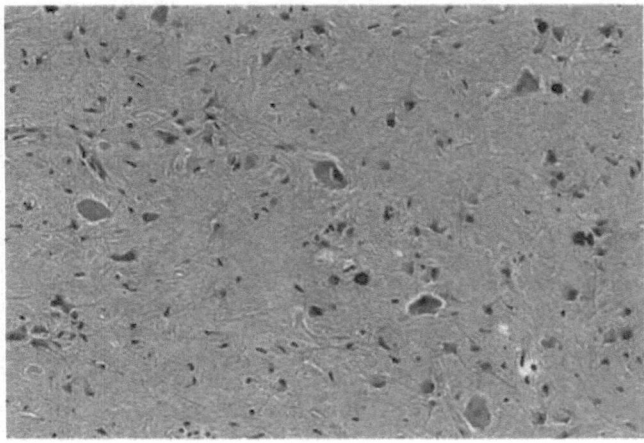

Plate 3.22 In this gangliocytoma, abnormal ganglion cells are enmeshed in a dense, acellular fibrillary matrix. Note the small foci of microcalcifications (Haematoxylin and eosin, original magnification ×200)

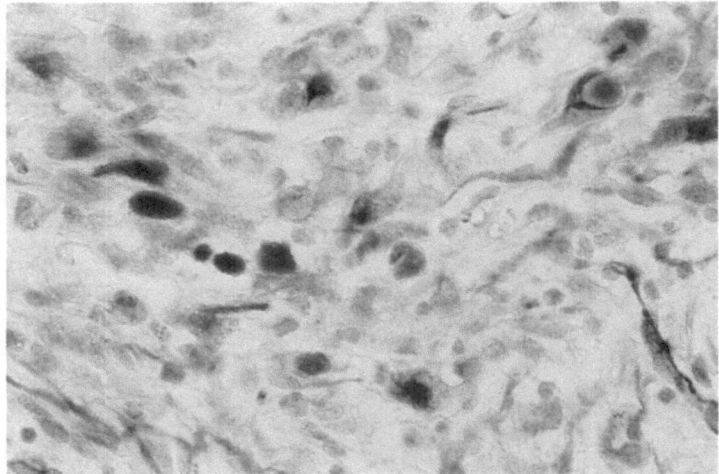

Plate 3.23   Dual labelling immunohistochemistry for GFAP (purple) and a phosphorylation-associated neurofilament (NF-H) epitope (red-brown) in a desmoplastic infantile ganglioglioma delineates the astrocytic and neuronal cell populations which are usually admixed within a desmoplastic stroma (GFAP and NFP1A3–peroxidase–antiperoxidase immunostaining, original magnification ×400)

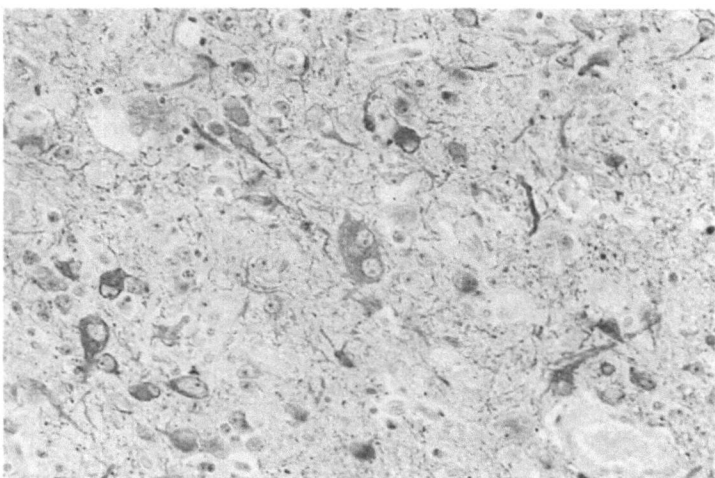

Plate 3.24   The neoplastic ganglion cells of a ganglioglioma demonstrate immunoreactivity for a phosphorylation independent neurofilament (NF-H/M) epitope. Note the binucleate cell in the centre of the field (SM133–avidin–biotin complex immunostaining with haematoxylin counterstain, original magnification ×400)

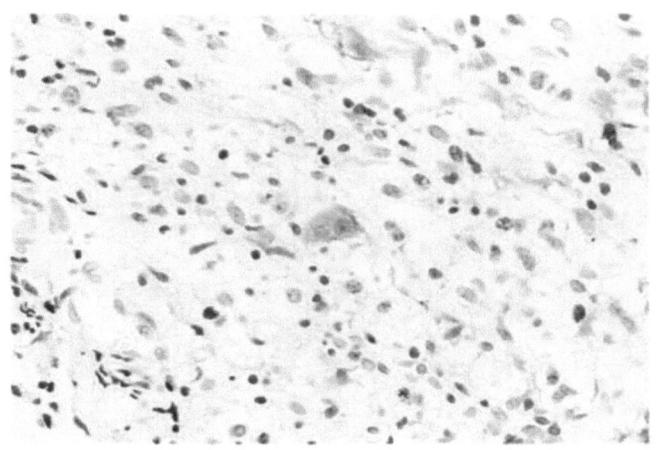

Plate 3.25 The ganglion cells in this cerebellar ganglioglioma showed immunoreactivity for calbindin-28kD, a calcium binding protein selectively present in Purkinje cells (Calbindin–D28K–avidin–biotin complex immunostaining with haematoxylin counterstain, original magnification ×200)

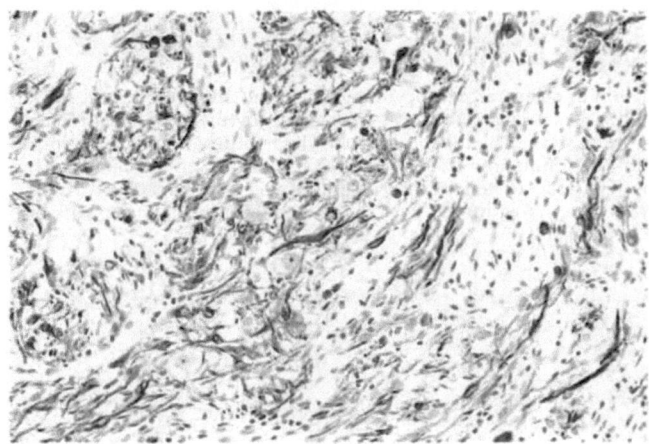

Plate 3.26 GFAP immunohistochemistry highlights the neoplastic astrocytic component of this ganglioglioma. Note the large ganglion cells in the field (GFAP–peroxidase–antiperoxidase immunostaining with haematoxylin counterstain, original magnification ×200)

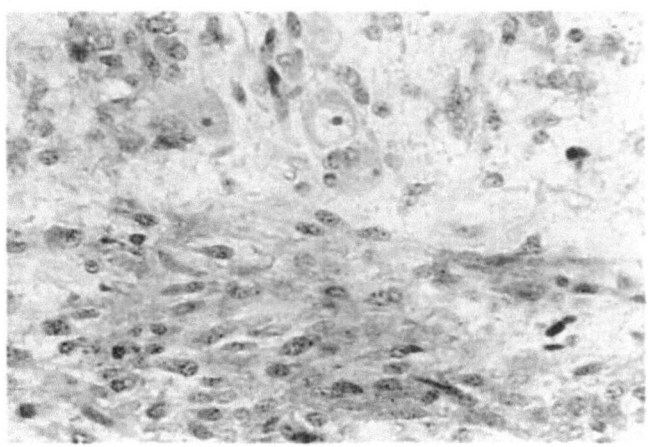

Plate 3.27 GFAP immunoreactivity marks the cellular, anaplastic astrocytic component of this anaplastic (malignant) ganglioglioma. Note the engulfed neoplastic ganglion cells (GFAP–peroxidase–antiperoxidase immunostaining, original magnification ×400)

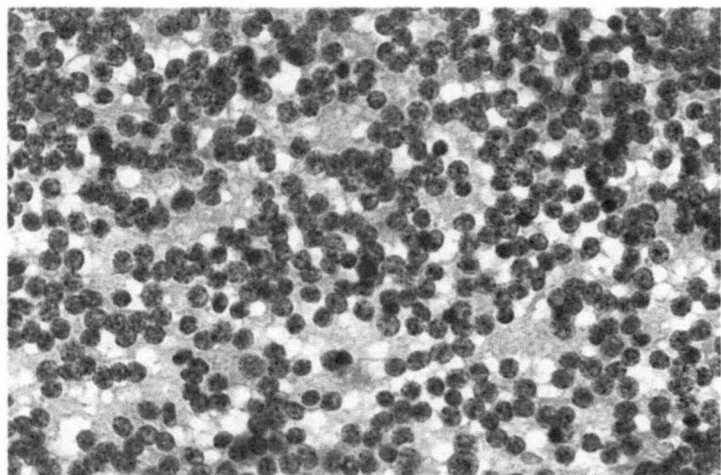

Plate 3.28   A smear preparation of a central neurocytoma shows the relatively round nuclei with evenly dispersed chromatin which are characteristic of this tumour. In smear preparations, the nuclei of central neurocytomas, in contrast to oligodendrogliomas, are more smoothly contoured with minimal lobulation and more delicately dispersed chromatin. The eosinophilic matrix is composed of delicate neuritic processes and recalls smears of intrinsic or local circuit neurons, such as the internal granular layer of the cerebellum (Haematoxylin and eosin, original magnification ×400)

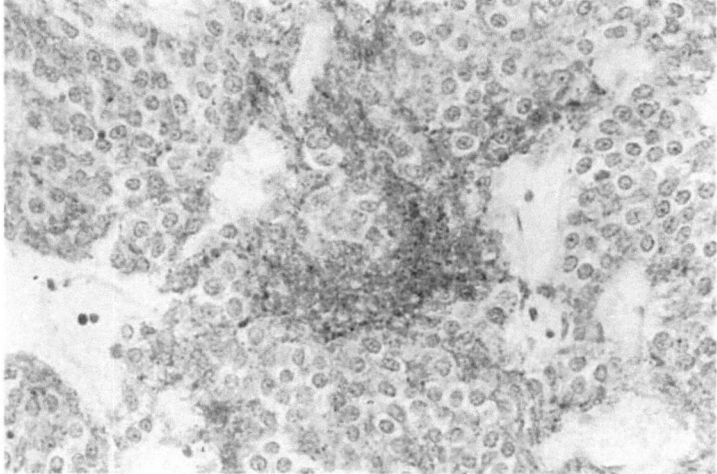

Plate 3.29   Although synaptophysin immunoreactivity is distributed throughout central neurocytomas, it is more conspicuous in the more acellular zones. Ultrastructurally, these areas appear as a meshwork of neuritic processes (Synaptophysin–avidin–biotin complex immunostaining with haematoxylin counterstain, original magnification ×400)

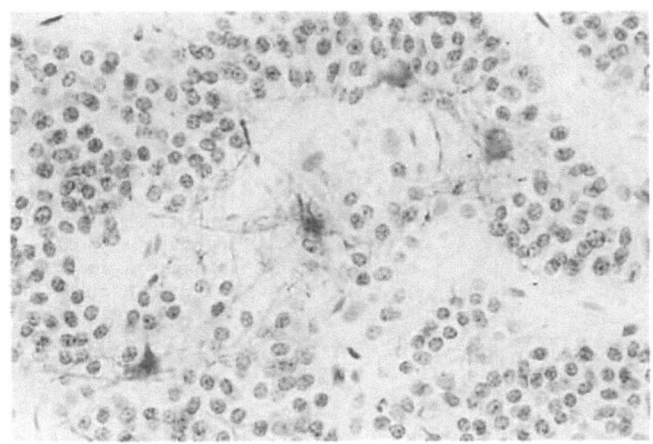

Plate 3.30 GFAP immunohistochemistry usually demonstrates only a reactive astrocytic component, frequently near blood vessels. This pattern recalls the reactive astrocytic population in oligodendrogliomas (GFAP – avidin – biotin complex immunostaining with haematoxylin counterstain, original magnification ×400)

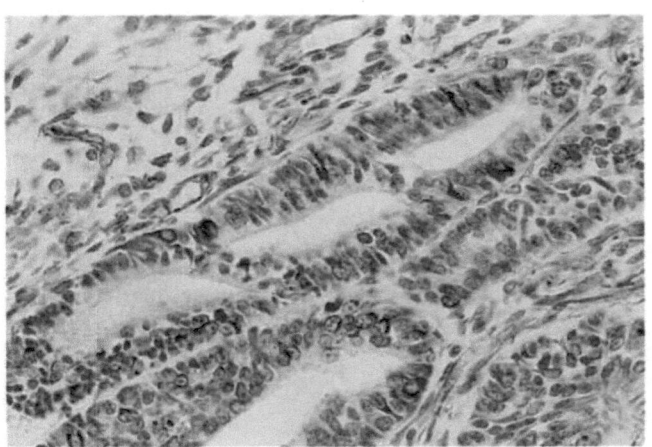

Plates 3.31 The primitive neuroepithelium of medulloepithelial tubules and rosettes displays immunoreactivity for vimentin (vimentin–peroxidase–antiperoxidase immunostaining with haematoxylin counterstain, original magnification ×400)

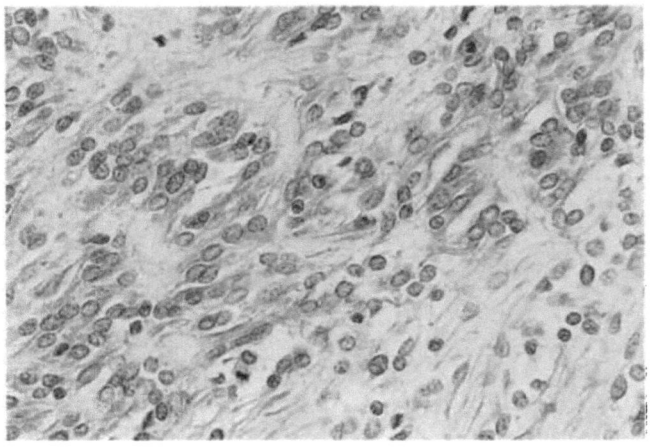

Plate 3.32 Immunoreactivity for class III β-tubulin may be the earliest evidence for neuronal differentiation in neuroblastomas (TUJ1–peroxidase–anti peroxidase immunostaining with haematoxylin counterstain, original magnification ×400)

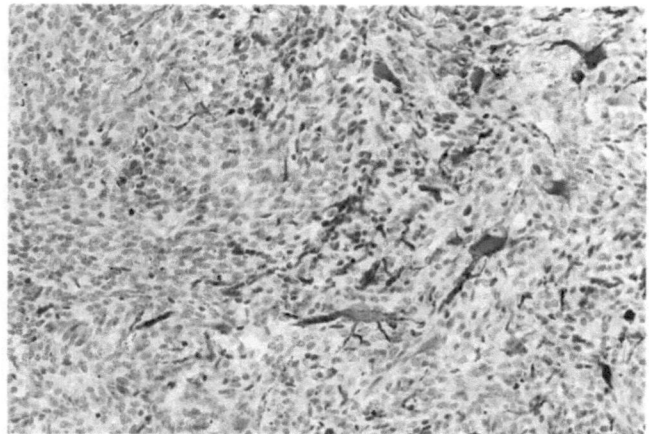

Plate 3.33 GFAP immunohistochemistry highlights the reactive "stromal" astrocytes in this medulloblastoma (GFAP–peroxidase–anti peroxidase immunostaining with haematoxylin counterstain, original magnification ×200)

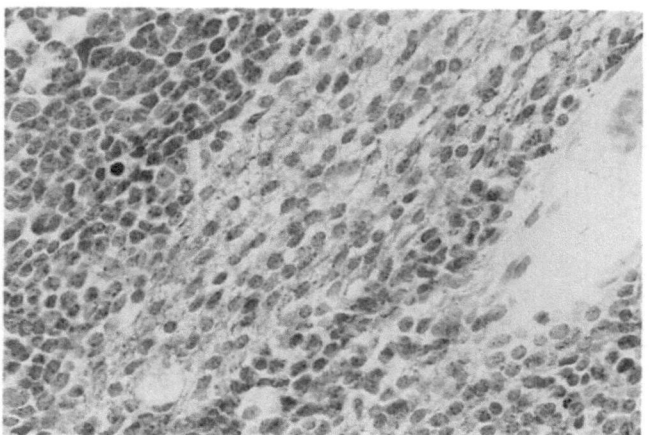

Plate 3.34 Immunohistochemistry for a neurofilament epitope (NF-H) demonstrates the fine neuritic processes in an area of neuronal differentiation (NFP1A3–avidin–biotin complex immunostaining with haematoxylin counterstain, original magnification ×400). This field corresponds to the more fibrillary cellular component in Fig. 3.28c (see page 82)

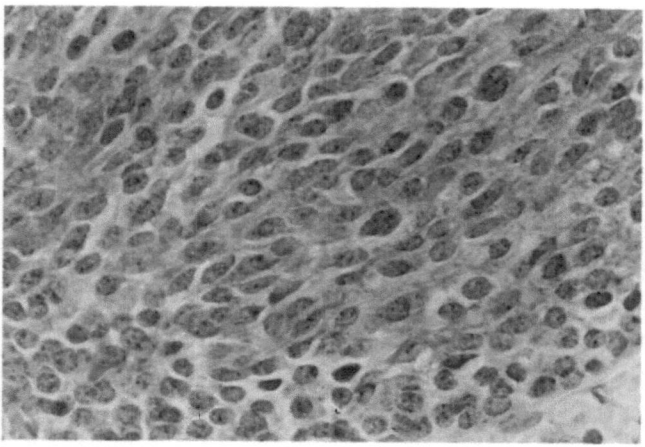

Plate 3.35 Immunoreactivity for class III β-tubulin in the same area as in Plate 3.34 shows a similar distribution as the neurofilament labelling (TUJ1–avidin–biotin complex immuno-staining with haematoxylin counterstain, original magnification ×600)

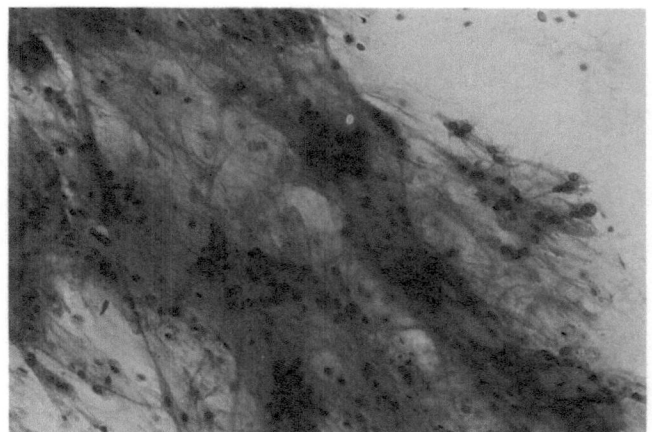

Plate 3.36 In this smear preparation of a chordoid meningioma, the abundant muco-polysaccharide matrix may suggest another diagnosis to the un-wary. The correct inter-pretation depends on the identification of the characteristic meningo-thelial nuclear features (Morris, original magni-fication ×200)

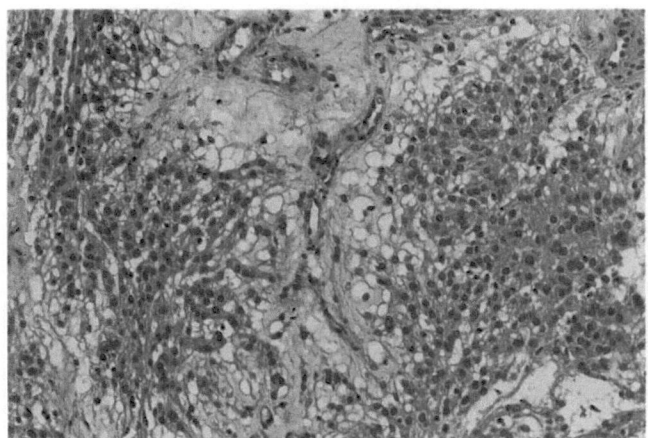

Plate 3.37 A tissue section of the same tumour illustrated in Plate 3.36 reveals populations of typical meningothelial cells clustered between the lakes of extracellular matrix (Haematoxylin and eosin, original magnification ×200)

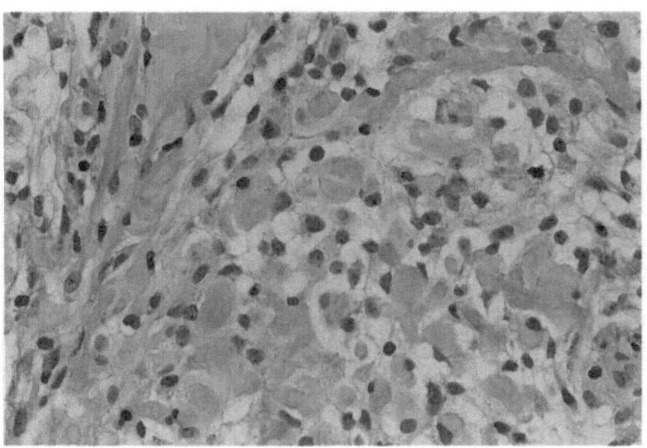

Plate 3.38 The secre-tory ("wet") menin-gioma contains "pseudohyaline", circu-lar inclusions with-in epithelioid-appearing cells. These inclusions are characteristic of this variant and are notably PAS-positive. Vascular hyalinization is com-mon and may be related to the secretory activity of the tumour cells (Haematoxylin and eosin, original magnification ×400)

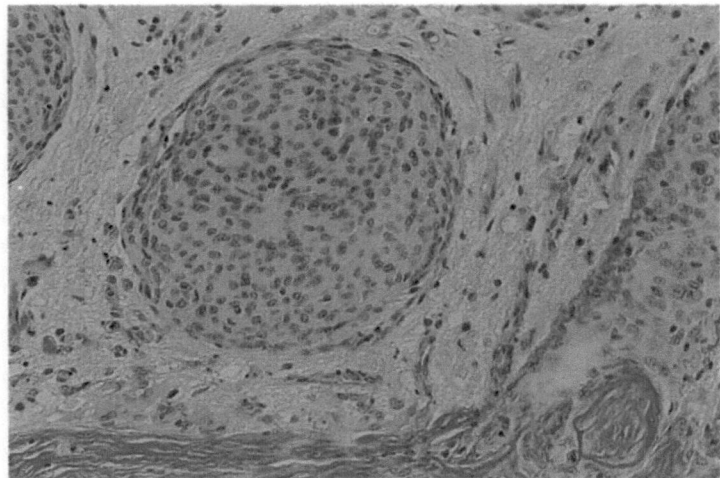

Plate 3.39 For meningiomas of any histopathological variant, brain invasion is a strong predictor of "malignant" behaviour. In this field, large clusters and individual tumour cells have breached the pial membrane and are situated within the neuropil. True invasion significantly differs from an uneven interfaces of the "expanding" tumour with the adjacent brain and an intact pial border (Haematoxylin–Van Gieson, original magnification ×200)

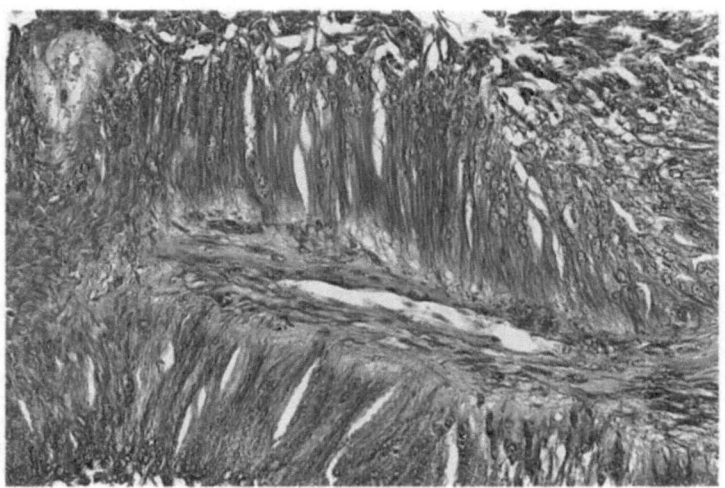

Plate 3.40 Papillary meningiomas have an epithelial architecture which ranges from low cuboidal cells with well-defined borders to pseudostratified tall columnar cells. The tall columnar cells with relatively delicate processes may demonstrate prominent staining with PTAH. In contrast to similar cytoarchitectural arrangement in ependymomas, there is invariably no GFAP immunoreactivity (PATH, original magnification ×200)

Allelotypic analysis by determination of restriction fragment length polymorphism provides an intriguing complement to the morphological classification of glioblastomas. Although there appears to be extensive loss of heterozygosity in many autosomes (Fults *et al.*, 1990), losses of heterozygosity for loci on chromosomes 10 and 17 (17p) are frequent. Changes on chromosome 10 appear to be more selective for glioblastomas compared to less anaplastic astrocytomas (James *et al.*, 1988; Bigner *et al.*, 1990; Wanatabe *et al.*, 1990; Fults *et al.*, 1990). In contrast, loss of heterozygosity for loci on chromosome 17 appears to occur in astrocytomas regardless of the degree of anaplasia (Bigner *et al.*, 1988a; James *et al.*, 1989).

In addition to allelotypic differences, abnormal expression of growth factor receptors may extend the histopathological classification of glioblastomas and anaplastic astrocytomas. Glioblastomas commonly have epidermal growth factor receptor (EGFR) gene amplification with frequent expression of aberrant gene/transcripts in both primary tumours and in tumour lines (Bigner *et al.*, 1987; James *et al.*, 1988; Jones *et al.*, 1990; Sugawa *et al.*, 1990; Humphrey *et al.*, 1990; Burgart *et al.*, 1991). Heteroplastic expression of endothelin receptors has also been demonstrated in astrocytic tumours with the highest density in glioblastomas (Kurihara *et al.*, 1990). Abnormal expression of growth factor ligands, including transforming growth factor α, platelet-derived growth factor and basic fibroblast growth factor may also be more frequently associated with anaplastic astrocytic tumours (Betsholtz *et al.*, 1989; Takahashi *et al.*, 1990; Ekstrand *et al.*, 1991; Mapstone *et al.*, 1991).

Quantitative image processing combined with either in situ hybridization or immunohistochemistry of abnormal growth factor receptor (Humphrey *et al.*, 1990) or oncogene (Salgaller *et al.*, 1990) proteins will be an important adjunct to routine morphology for characterizing anaplastic glial tumours. In one intriguing report of multiple cerebral anaplastic gliomas (Leifer *et al.*, 1989), metastasis beyond the nervous system was selectively accompanied by immunohistochemical detection of a *ras* protein epitope.

Giant Cell Glioblastoma. Giant cell glioblastoma multiforme appears, on the basis of clinicopathological data, to be a distinct entity in which the patients may frequently exceed the median survival time reported for glioblastomas (Russell and Rubinstein, 1989; Margetts and Kalyan-Raman, 1989). Although there is no predominant location, one series suggested a slight predilection for the temporal lobe (Margetts and Kalyan-Raman, 1989). The tumours are typically well circumscribed and neuroimaging also demonstrates a heterogeneous enhancement pattern without peripheral prominence. Despite the clear radiographic and macroscopic demarcation, the adjacent parenchyma and leptomeninges are frequently infiltrated. Necrosis is commonly a significant element and the tumours can partially be cystic. The typical giant cells, often in the context of extensive necrosis, are the most impressive cellular component in intraoperative smears (Plate 3.2). Careful examination for fibrillary cells confirms the astrocytic nature of these tumours.

In tissue sections, the tumour is characterized by large, bizarre, multinucleated giant cells containing large vesicular nuclei with prominent nucleoli. These tumour cells tend to surround blood vessels, often forming pseudorosettes. These islands of viable tumour are commonly separated by extensive areas of necrosis. Endothelial proliferation is exceptional, and perivascular lymphoplasmocytic infiltrates are common. The majority of the remaining tumour cells are plump or pyramidal, with coarse cytoplasmic processes with relatively abundant reticulin deposition. Smaller, more fusiform astrocytic tumour cells may also be present. The giant cells can be immunoreactive for GFAP, although GFAP is usually more prominent in the fusiform cells (Plate 3.3). Atypical mitoses are common and heavily lipidized tumour cells have been described (Gherardi *et al.*, 1986).

Gliosarcoma. Although vascular proliferation is a significant feature of all glioblastomas, gliosarcomas are tumours in which the cellular elements derived from the vasculature become a major sarcomatous component. The reported frequency for sarcomatous change in glioblastomas ranges from approximately 2% to 8% (Morantz *et al.*, 1976; Meis *et al.*, 1991). Histologically, the relative proportions of glial and sarcomatous components vary, and the sarcomatous elements may overwhelm the primary glial neoplasm to complicate diagnostic interpretation. Intraoperative smears can discriminate the two cell populations (Fig. 3.7), but special stains should be used with adequate tissue sampling to confirm the diagnosis. The sarcomatous elements can readily be highlighted by reticulin deposition and GFAP immunohistochemistry should be used to confirm the glial component (Plate 3.4).

The histopathological appearance of the sarcomatous component can vary from resembling a malignant fibrous histiocytoma to fibrosarcoma. Osteoid and chondroid metaplasia and rhabdomyosarcomatous elements are occasionally noted within the sarcoma (Barnard *et al.*, 1986; Russell and Rubinstein, 1989). Immunohistochemical and ultrastructural studies suggest that the histogenesis of the sarcomatous component reflects an origin from an undifferentiated mesenchymal cell associated with the vas-

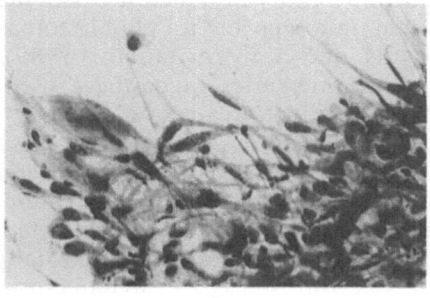 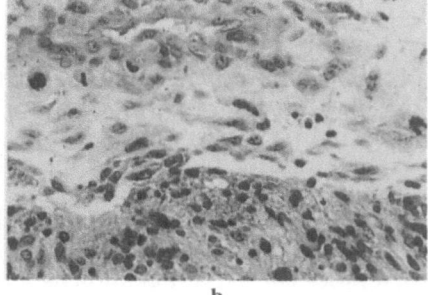

a   b

Fig. 3.7 Gliosarcoma. **a** Bizarre, pleomorphic and spindle-shaped cells are seen in a smear of the sarcomatous area (Morris stain, original magnification ×300). **b** Section of the paraffin embedded tumour demonstrates a more dispersed sarcomatous component that is clearly delineated from the compactly arranged neoplastic astroglial cells. Note the numerous mitoses in the sarcomatous area (Haematoxylin and eosin, original magnification ×400) (See Plate 3.4)

cular adventitia which can differentiate along endothelial, smooth muscle and pericytic cell types (Grant *et al.*, 1989; Ho, 1990; Ng and Poon, 1990; Miller *et al.*, 1991; Haddad *et al.*, 1991).

Pilocytic Astrocytoma

Pilocytic astrocytomas commonly occur in children and young adults and most often arise in the cerebellum, the region of the third ventricle, the lower brainstem, the optic chiasm and nerves and, less frequently, in the cerebral cortex (Burger *et al.*, 1991; Russell and Rubinstein, 1989). Magnetic resonance images reveal well-demarcated, often cystic lesions with increased relaxation times on T_1 and T_2 images (Fig. 3.8). Contrast enhancement is often evident and due to the typical vascular proliferation in both the solid component of cystic tumours and in the cyst wall.

Histologically, these tumours are composed of a biphasic population of stellate and elongate bipolar astrocytes (Plate 3.7) with only moderate cellularity. The relative abundance of these constituents varies, but cytoarchitectural heterogeneity is always present. The piloid processes may be arranged in tight bundles, particularly prominent around the abundant vascular component. Intraoperative smears readily demonstrate the admixtures of both stellate and bipolar astrocytes in a matrix of stout and finely fibrillated glial processes that give an impression of jumbled piano wires (Plate 3.5). Vascular endothelial proliferation, as well as nuclear pleomorphism, including multinucleated cells, are common features which do not indicate anaplasia or sinister clinical behaviour. In contrast, solid tumours composed of entirely piloid-type astrocytes without the other typical cytoarchitectural features of pilocytic astrocytomas should be considered more appropriately as a variant of a well-differentiated, diffusely infiltrating astrocytoma rather than a pilocytic astrocytoma.

Pilocytic astrocytomas typically display degenerative features that include Rosenthal fibres, granular eosinophilic or hyaline bodies (GEB), microcystic degeneration and vascular hyalinization (Plates 3.5, 3.6, 3.8, 3.9). GEB appear as either diffusely hyalinized spherical droplets, or as structures with a granular core and an eosinophilic rim. Immunohistochemistry for GFAP or vimentin demonstrates diffuse reactivity in the hyalinized droplets, and peripheral reactivity in the granular bodies. Ultrastructurally, these structures appear as many small autophagic vacuoles incorporated within larger membrane-bound bodies. The granular cores contain abundant immunoreactivity for the serpins α-1 antitrypsin and α-1 antichymotrypsin (Friedberg *et al.*, 1991). Extensive nuclear atypia, necrosis and increased mitotic activity are not regular features and may indicate rare anaplastic progression. Despite the macroscopically well-circumscribed character, the tumour cells may extend into the leptomeninges (particularly prominent in the posterior fossa) or track along the optic nerve. Leptomeningeal spread or infiltration along nerve fibres does not indicate anaplastic or malignant progression of the tumour.

The distinctive histopathological features of pilocytic astrocytomas suggest an origin from a distinct glial cell population related to the radial glia or progenitor cell population (O-2A) which most likely persists in the paediatric and young adult nervous system (ffrench-Constant and Raff, 1986; Beahr *et al.*, 1987; Norton and Farooq, 1989). The commonly biphasic nature of these tumours with respect to GFAP immunoreactivity (Plate 10) and the regional predilection would be consistent with this hypothesis. Normal O-2A derived astrocytes lack the EGF responsiveness and have a low mitotic activity in monolayer culture (Raff *et al.*, 1983), analogous to the usual indolent growth of the pilocytic astrocytomas. The fibrillated processes of pilocytic astrocytomas are strongly immunoreactive for both GFAP and vimentin, a characteristic shared by the fetal radial glia (Pixley and DeVellis, 1984; Schiffer *et al.*, 1986; Wilkinson *et al.*, 1990).

Pleomorphic Xanthoastrocytoma

The pleomorphic xanthoastrocytoma (PXA) is characteristically a superficial cerebral tumour that involves the temporal lobe within the second decade. The reported age at first craniotomy ranges from 6 to 42 years but the average age is less than 20 years and there is usually a history

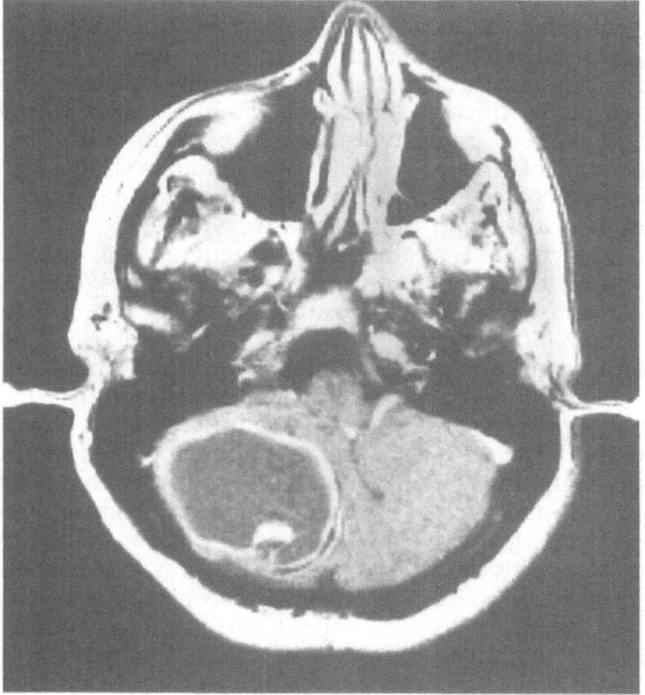

Fig. 3.8   Pilocytic astrocytoma.
a Both the cyst wall and the nodular component of a cerebellar pilocytic astrocytoma in a 17-year-old demonstrates enhancement in a T_1 weighted magnetic resonance image following administration of intravenous gadolinium DTPA [Axial scan: SE 0.5 s/20 ms (1.0 T)].

of epilepsy preceding the clinical signs of tumour (Pasquier *et al.*, 1985 for review; Kawano, 1991). The neoplasms are commonly cystic with mural nodules and involve the overlying leptomeninges. Dural involvement is, however, exceptional (Kepes *et al.*, 1979; Strom and Skullerud, 1983; Kawano, 1991). The tumours are only moderately firmer than the adjacent brain but their marked cohesiveness is readily appreciated when making intraoperative smears. Although there is usually a well-defined macroscopic border with the subjacent brain, there is always focal microscopic infiltration.

The heterogeneous astroglial population, readily demonstrated by GFAP immunohistochemistry (Plate 3.11), varies from plump polygonal cells without processes to more fusiform cells arranged in interlacing bundles within a variably fibrillated matrix. Cellularity is usually moderate but may be focally increased. The pleomorphic nuclei are commonly hyperchromatic and multinucleated giant cells are invariably present (Fig. 3.9).

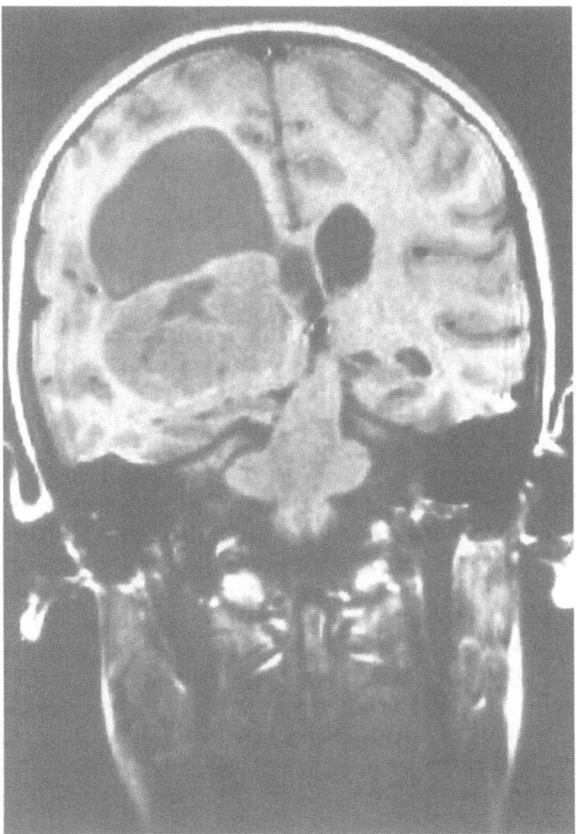

Fig. 3.8b   The cystic and solid areas of a thalamic pilocytic astrocytoma are evident on a T_1 weighted magnetic resonance image without contrast [Coronal scan: SE 0.68 s/20 ms (1.5T)]. Both neuroimages emphasize the macroscopically distinct borders of these tumours which are commonly appreciated with neuroimaging (See Plates 3.5–3.10)

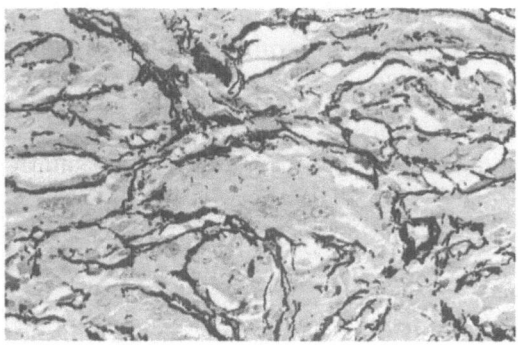

Fig. 3.9 Pleomorphic xanthoastrocytoma. Reticulin is present surrounding both individual and groups of cells, which corresponds to a basal lamina secreted by the neoplastic astrocytes. In contrast to mesenchymal tumours, collagen deposition is rare (Reticulin stain, original magnification ×400) (See Plate 3.11)

Mitoses may be present but are never abundant. The extremely low percentage of S-phase cells by cytofluorometric analysis (0.25%) confirms the insignificant growth fraction of the tumour (Hosokawa *et al.*, 1991). Cytoplasmic lipid, especially in the polygonal and giant cells, may be prominent but can be highly variable among tumours.

The abundance of reticulin stroma is a striking feature that delineates fascicles of cells and is variably distributed between single tumour cells. Although this stroma may be present in any portion of the tumour, it is most prominent in association with leptomeningeal involvement and the prominent vascular elements. However, a number of cases which have all other typical histopathological features except the prominent reticulin stroma have been reported (Kawano, 1991). Ultrastructural studies have confirmed the presence of a basal lamina between cells (Weldon-Linne *et al.*, 1983; Kepes *et al.*, 1989), a feature which the neoplastic astrocytes share with subpial astrocytes, suggesting a histogenic relationship between pleomorphic xanthoastrocytomas and this astrocytic phenotype (Russell and Rubinstein, 1989; Whittle *et al.*, 1989).

The number of reports describing pleomorphic xanthoastrocytomas with anaplastic progression are too limited to determine a meaningful incidence, but the presence of abundant mitoses, necrosis, and vascular endothelial/ pericytic proliferation should be considered significant evidence of malignant transformation. Although rarer than for the diffusely infiltrating astrocytomas, anaplastic progression and malignant change appears to occur more frequently than with pilocytic and subependymal giant cell astrocytomas. The histopathological character of the infiltrating margin may be important with respect to the tumour recurrence and anaplastic progression which may occur (Weldon-Linne *et al.*, 1983; Kepes *et al.*, 1989; Daita *et al.*, 1991). Variable perivascular lymphocytic infiltration is not uncommon in pleomorphic xanthoastrocytomas but has no specific histopathological significance.

Subependymal Giant Cell Astrocytoma

Subependymal giant cell astrocytoma (SEGA) is a tumour typically occurring in the first to second decades in association with tuberous sclerosis. In this context, it may be the first clinical manifestation of this disease. However, similar tumours may also occur without any association to pha- comatoses. The tumours are well circumscribed, frequently nodular, multi- cystic and calcified (Russell and Rubinstein, 1989). They most commonly arise from the wall of the lateral ventricles, over the basal ganglia or, less commonly, from the third ventricle.

The SEGAs display a heterogeneous glial cell population ranging from polygonal cells with abundant glassy eosinophilic cytoplasm to randomly oriented, more elongated forms in a variable fibrillated matrix (Plate 3.12). The nuclei have a finely granular chromatin pattern with distinct nucleoli. Pyramidal giant cells are not uncommon. Considerable nuclear pleomor- phism, variable mitoses and occasional multinucleated cells are present but are not indicative of anaplasia. In a large, recently reported series of 20 SEGAs, histological features associated with anaplasia in the diffusely infiltrating astrocytomas, such as endothelial proliferation, necrosis, mitoses and marked cellular pleomorphism did not correlate with an adverse clinical course or survival time (Shepherd et al., 1991). Despite the few examples of recurrent tumours, even after 47 years (Halmagyi et al., 1979) no cases with malignant transformation have been described.

The histogenesis of SEGAs is not well understood, but astroglial, mixed glial-neuronal and ependymal cell origins have been hypothesized. The astrocytic character of the tumour cells is readily demonstrated by both GFAP immunoreactivity (Plate 3.13) and ultrastructural studies (Sima and Robertson, 1979; De Chadarévian and Hollenberg, 1979; Trombley and Mirra, 1981; Nakamura and Becker, 1983; Russell and Rubinstein, 1989). However, a significant population of non-GFAP immunoreactive cells may be present in some tumours, mainly in patients with tuberous sclerosis. Ultrastructural evidence for neuronal differentiation has been reported in a small number of cases (Nakamura and Becker, 1983). Evidence for an ependymal origin includes tumours with 9 + 2 cilia associated with cyto- plasmic lumina (Bancel et al., 1990) and extensive perivascular rosettes with blepharoplasts (Halmagyi et al., 1979).

Oligodendroglial Tumours

Oligodendroglioma

Tumours composed predominantly of oligodendrocytic glia represent 5%–15% of intracranial gliomas and have a peak incidence in the fourth and fifth decades. There is a slight male predominance to that generally associated with gliomas (2 : 1–3 : 2) (Mørk et al., 1985). These tumours are more common in the frontotemporal region but arise throughout the neuraxis, often in relative proportion to the volume of white matter.

Oligodendrogliomas and astrocytomas have a similar appearance with neuroimaging, except for the increased calcification which is helpful for identification (Fig. 3.10). The tumours usually have a macroscopic appear- ance of a grey and often gelatinous mass from which smear preparations are easily made. Microcalcification, present in approximately 90% of oligo- dendrogliomas, is often revealed in intraoperative smears (Burger

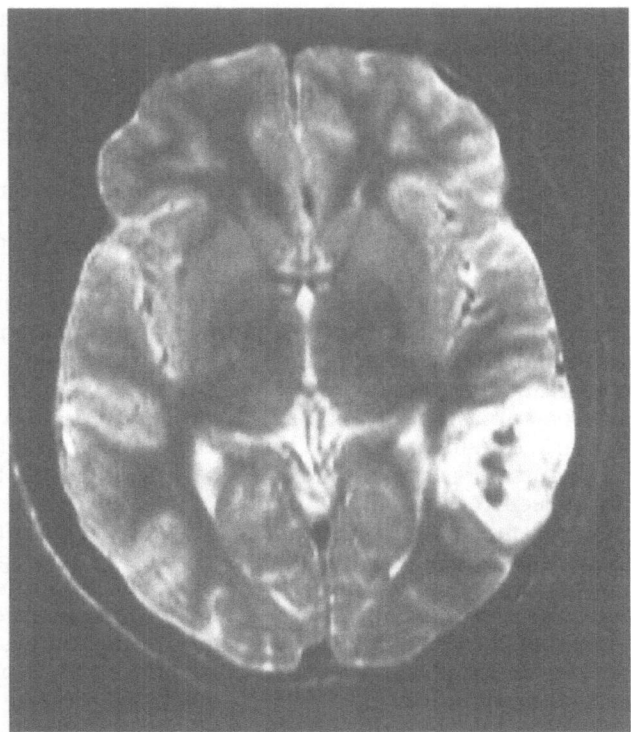

Fig. 3.10 Oligodendroglioma. Areas of low signal intensity represent macroscopic calcification on a T_2 weighted magnetic resonance image. Minimal mass effect or oedema is associated with the tumour [Axial scan: SE 2.5 s/90 ms (1.0 T)]

et al., 1991). The principal cytological features which distinguish oligodendrogliomas from astrocytomas are the nuclear morphology and a delicate matrix composed of indistinct cellular processes (Plate 3.14). Nuclear lobulation with delicate chromatin and commonly distinct nucleoli are particularly prominent in smear preparations. The intercellular matrix lacks the conspicuously fibrillated background of astrocytomas. Other features readily identified on smears are the characteristic delicate network of capillaries and focal microcystic degeneration. A majority of oligodendrogliomas have been demonstrated to be Leu-7 antigen immunoreactive (Perentes and Rubinstein, 1986); however this surface antigen is not particularly selective in the diagnosis of oligodendrogliomas, since it may also be detected in central neurocytomas, astrocytomas, ependymomas and choroid plexus papillomas. Immunohistochemical studies for galactocerebroside (GC) have indicated a high percentage of oligodendrogliomas with GC positive cells. In one series, the majority of tumours contained abundant populations of GC positive cells resembling typical oligodendroglia (de la Monte, 1989).

In contrast to smear preparations, the histopathological features of oligo-dendrogliomas in paraffin-embedded sections are more variable and present a wide variety of cytoarchitectural patterns (Plate 3.15). The most constant feature is the conspicuous network of delicate capillaries distrib-uted in a `pseudolobular' pattern. Dystrophic changes within the vessels are common, a feature which may be related to the incidence of sponta-neous, often fatal, haemorrhage reported in oligodendrogliomas (Liwnicz et al., 1987).

Reactive astrocytes are often found scattered throughout the tumour, adjacent to or associated with the vascular stroma. These astroglia can be readily highlighted by PTAH staining (Plate 3.16) or GFAP immunohisto-chemistry. GFAP immunoreactivity has been well documented in neoplas-tic oligodendrocytes that usually display prominent cytoplasm with distinct borders. These have been designated "gliofibrillary oligodendrocytes" (Plate 3.17) (Herpers and Budka, 1984) and may compose approximately 50% of the neoplastic cells (Herpers and Budka, 1984; Nakagawa, et al., 1986). GFAP immunoreactivity in neoplastic oligodendroglia recapitulates the normal ontogeny of oligodendrocytes in which GFAP is transiently expressed prior to myelinogenesis (Choi and Kim, 1985). Experimental studies in rodent development have suggested that type-2 astrocytes and oligodendroglia share a progenitor cell (O-2A) that can be identified by immunohistochemistry (A2B5+) (Raff et al., 1983). A study of 28 oligoden-drogliomas has demonstrated that a high proportion contain "A2B5+" cells, most with and some without the coexpression of the oligodendroglial-gly-colipid (GC) (de la Monte, 1989). Therefore, a proportion of oligoden-droglial tumours may contain variable numbers of the "A2B5+" progenitor cells and the GFAP immunoreactive phenotypes in these tumours could potentially occur with astrocytic (type-2) differentiation from these pro-genitor cells. This may be conspicuous in the more poorly differentiated oligodendrogliomas, since the proportion of GFAP-negative immunoreac-tive cells which more closely resemble gemistocytic astrocytes than the typical "gliofibrillary oligodendrocytes" tends to increase with anaplastic progression (Kros et al., 1990). However, the histogenesis of these GFAP immunoreactive cells remains unclear and they may arise from alterations in the neoplastic "gliofibrillary oligodendrocytic" phenotype or by astro-cytic (type-2) evolution from the O-2A-like progenitor populations ele-ments that may be variably present within the oligodendrogliomas (Kros et al., 1992).

Anaplastic oligodendroglioma

The recognition of anaplastic oligodendrogliomas is generally based on similar criteria for anaplasia in the diffusely infiltrating astrocytomas, including necrosis, vascular proliferation, high cellularity with a high mitotic activity and increased nuclear pleomorphism (Fig. 3.11) (Smith et al., 1983; Mørk et al., 1986; Burger et al., 1987a; Burger, 1989). However, more precise "grading" of these features as prognostic factors is problem-atic. Using univariate analysis, it has been found that necrosis, increased mitotic rate, nuclear atypia and vascular proliferation are statistically

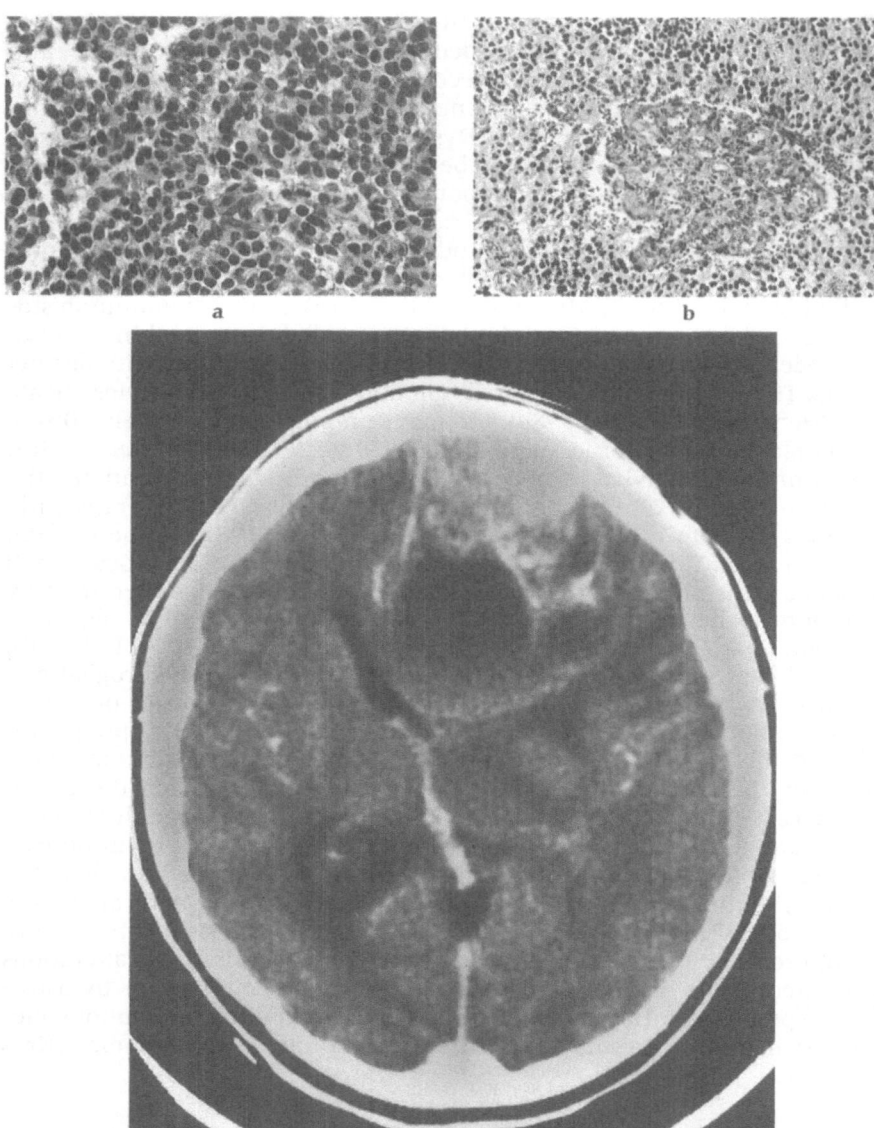

Fig. 3.11 Anaplastic oligodendroglioma. **a** Increased cellularity and mitoses, although invariably present in anaplastic oligodendrogliomas, may not be as significant as endothelial proliferation as a hallmark feature of anaplasia of an oligodendroglioma (Haematoxylin and eosin, original magnification ×400). **b** Marked "glomeruloid" endothelial proliferation is a significant feature of this anaplastic tumour (Haematoxylin and eosin, original magnification ×200). **c** Mixed signal intensity and non-uniform contrast enhancement, significant oedema, mass effect and a possible cystic component are demonstrated in an enhanced CT scan of the tumour in A and B. This image was interpreted preoperatively as malignant progression in a glioma. The tumour rapidly recurred despite a bulk resection

relevant factors. Multivariate assessment, however, appears to demonstrate necrosis as the only independent prognostic feature (Burger, 1989). The development of a reliable grading system based on these data has yet to be fully developed and verified. Progression to glioblastoma must be regarded as a rare yet recognized event in anaplastic oligodendrogliomas. This transformation may be heralded by an increase of "neoplastic astrocytes" in these tumours.

Ependymal Tumours

Ependymoma

Ependymomas represent between 2% and 6% of gliomas. Multiple tumours may be associated with von Recklinghausen's disease (Russell and Rubinstein, 1989). Unlike the diffusely infiltrating astrocytomas and oligodendrogliomas, ependymomas arise predominantly in children and adolescents where they may comprise as high as 10% of the brain tumours (Jellinger and Seitelberger, 1970). In this group, the tumours commonly occur in the ventricles and the fourth ventricle is the most frequent site. In contrast, the tumours that arise in adults are usually located in the spinal cord and filum terminale (Slooff et al., 1964).

In contrast to the astrocytomas, the macroscopic appearance is highly variable and dependent on location. The intraventricular tumours tend to be soft and papillary in contrast to the more homogeneous, somewhat granular tumours which arise in the parenchyma. In all locations, the tumours are relatively well defined which corresponds to the macroscopic "pushing tumour borders". Computed tomographic and MR imaging commonly demonstrate similar features for the exophytic intraventricular and the intramedullary lesions. The heavily vascularized nature of these tumours accounts for variable enhancement with intravenous contrast. On MR imaging, these tumours are usually hypointense on T_1 and slightly hyperintense on T_2 images (Fig. 3.12D). Cystic degeneration and calcification are common.

The appearance of ependymomas on intraoperative smears reflects the neuroepithelial nature of these tumours (Fig. 3.12A and B; Plate 3.18). Nuclear chromatin is usually distributed irregularly into delicate nodes, contrasting to the more coarse quality of staining in astrocytomas and oligodendrogliomas. Nucleoli are relatively inconspicuous. A diagnostic feature of these tumours is the proclivity of the tumour cells to orientate in a polarized fashion around blood vessels to form "pseudorosettes" and around small extracellular spaces to form ependymal rosettes, epithelial-lined clefts and tubules. In the case of perivascular "pseudorosettes", ependymoma cells may often be distinguished from the perivascular arrangements of other glial tumours by their long-tapering, fibrillated processes extending to the vessel wall. These processes may be highlighted by PTAH staining and GFAP immunohistochemistry (Plate 3.19) and may also permeate the more cellular populations remote from the rosettes. Although the tumours usually have discrete microscopic interfaces with

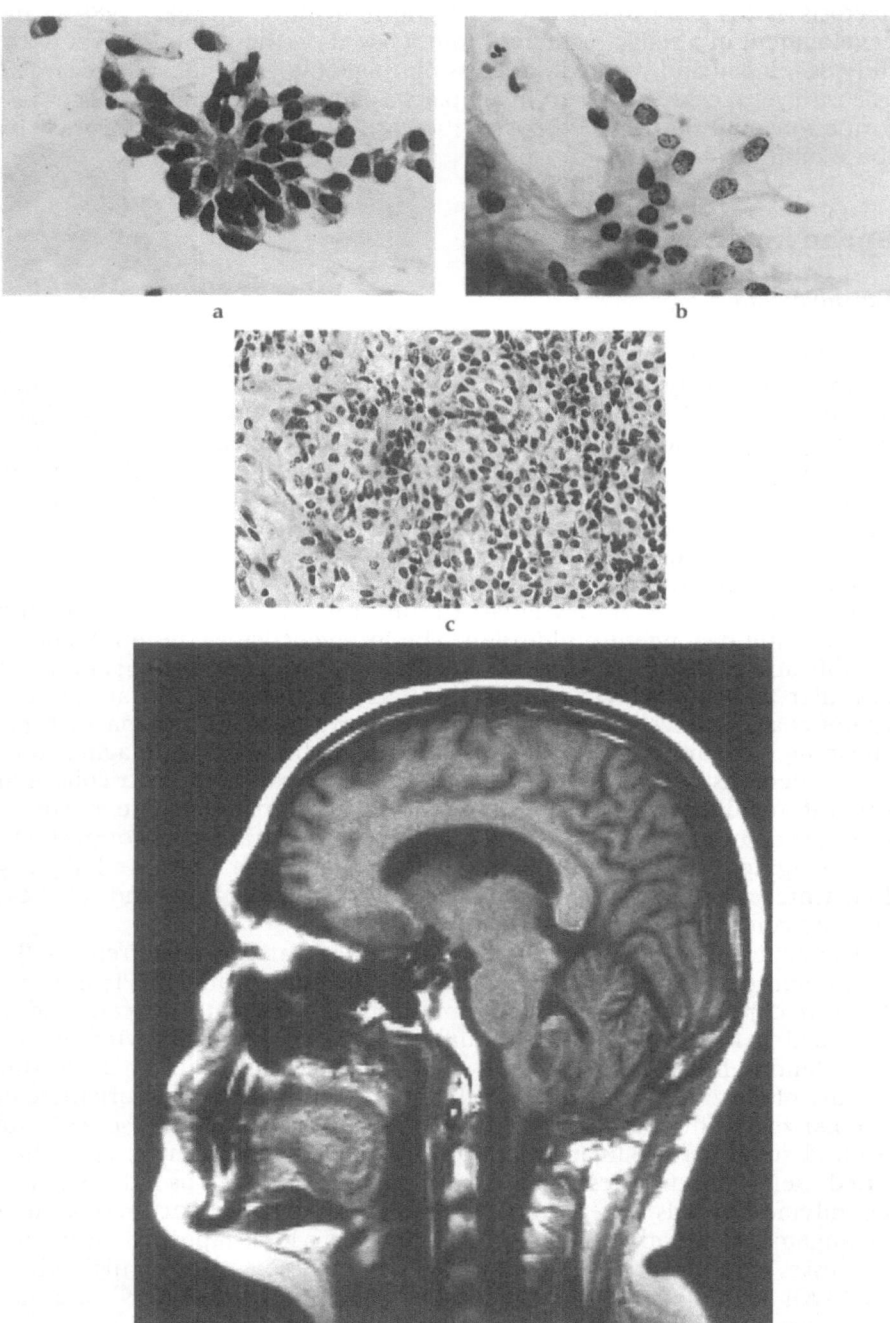

the surrounding parenchyma, the occurrence of CSF and extramedullary spread has been documented (Russell and Rubinstein, 1989).

The current classification recognizes three histological variants: cellular, papillary and clear cell. Although the presence of these subtypes has no prognostic importance, their recognition is important when considering a differential diagnosis. The "cellular" ependymoma usually has relatively few rosetting structures and thereby is also more conspicuously cellular with more amorphous arrangements that are not interrupted by anuclear zones of cellular processes. Increased numbers of mitoses restricted to foci of this histological variant are not particularly indicative of anaplastic transformation. The papillary variant is rare, and more often encountered in the cauda equina. However, when occupying the fourth ventricle these tumours may be distinguished from choroid plexus papillomas by the presence of more typical cytology in other fields and the characteristic neuroglial stroma (Plate 3.20). Similarly, the clear cell variant (Fig. 3.12C) may be distinguished from an oligodendroglioma by the presence of glial fibrils, well demonstrated by the PTAH stain. Most often, tumours will be a mixture of different proportions of the numerous ependymal variants.

Anaplastic Ependymoma

For ependymal tumours, there are no histopathological features that appear to precisely designate grades of anaplasia or serve as prognostic indicators for recurrence or clinical malignant progression (Ross and Rubinstein, 1989). This situation has been complicated, in part, from tumour locations that preclude accurate evaluation of complete removal and, also in part, by imprecise designations of ependymomas, anaplastic ependymomas and ependymoblastomas (Rubinstein, 1989). Given this caution, anaplastic ependymomas are those tumours which are recognized by a histological character which is dominated by nuclear and cytoplasmic pleomorphism, increased mitotic activity, necrosis and endothelial proliferation (Fig. 3.13). In contrast to ependymomas, the anaplastic tumours tend to invade microscopically adjacent structures. (See also Ependymoblastomas, p. 78.)

Myxopapillary Ependymoma

The myxopapillary ependymoma is a distinct clinicopathological entity which is restricted to the cauda equina, where it arises from the conus medullaris and filum terminale. On rare occasions these tumours occur as primary extradural lesions in the pre- and postsacral regions, where they

---

◀ **Fig. 3.12** Ependymoma. **a** A smear preparation demonstrates tumour cells with tapering processes in a typical rosette-like arrangement (Morris stain, original magnification ×600). **b** The "open" chromatin pattern and the finely fibrillated intercellular matrix are typical features (Haematoxylin and eosin, original magnification ×600). **c** The clear cell variant of ependymoma may be difficult to distinguish from an oligodendroglioma (Haematoxylin and eosin, original magnification ×400). Careful examination should reveal the more characteristic nuclear features of an ependymoma. PTAH staining and GFAP immunohistochemistry readily distinguish between the two types of tumours (See Plate 3.19). **d** In a T_1 weighted magnetic resonance image, the tumour appears as a mass with heterogeneous signal intensity within the fourth ventricle [Sagittal scan: SE 0.5 s/17 ms (1.0 T)] (See Plate 3.18)

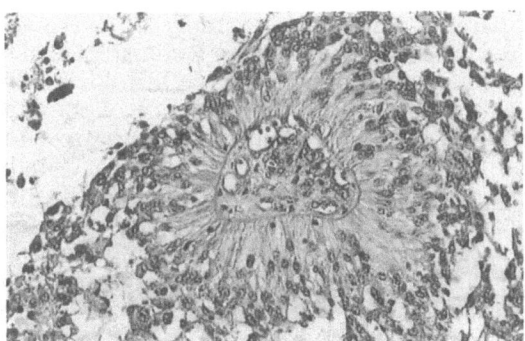

Fig. 3.13 Anaplastic ependymoma. Endothelial proliferation is present in the centre of the perivascular rosette of this anaplastic ependymoma. Note the increased cellular pleomorphism (Haematoxylin and eosin, original magnification ×200)

presumably arise from ependymal rests (Morantz *et al.*, 1979; Pulitzer *et al.*, 1988). These tumours are macroscopically discrete, sausage-shaped growths typically attached to the filum terminale or spinal nerve root.

Histologically, myxopapillary ependymomas contain an admixture of highly fibrillated and epithelioid cells with an abundant supportive connective tissue stroma (Fig. 3.14). The arrangement of elongated fibrillary cells extending fine processes to hyalinized vessels gives these tumours the characteristic papillary appearance. The hyalinized matrix stains positively for mucin. Another characteristic feature are the so-called "balloons"–spherical, PAS-positive fibrillated structures. Ultrastructural studies have confirmed the ependymal nature of this neoplasm (Rawlinson *et al.*, 1973; Specht *et al.*, 1986). However, certain features, such as an abundant basal lamina, the relatively paucity of cilia, and numerous and often dense cellular interdigitations, are restricted to the myxopapillary tumours. The presence of glial fibrils can be confirmed by immunohistochemistry for GFAP, a feature which distinguishes this tumour from others which are usually

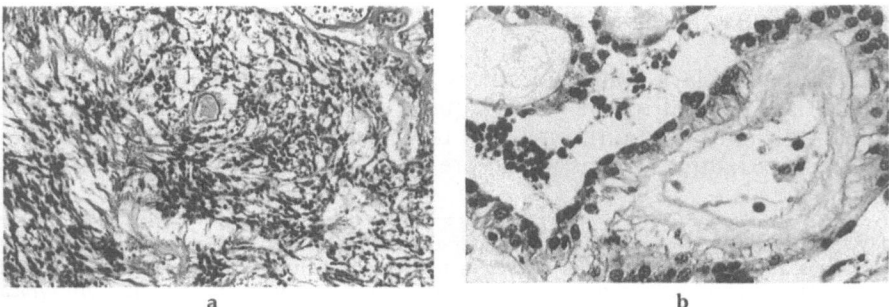

a   b

Fig. 3.14 Myxopapillary ependymoma. **a** The cells may be palisaded in an extracellular matrix with mucinous degeneration (Haematoxylin and eosin, original magnification ×300). **b** Commonly, there is a prominent epithelial cytoarchitecture. Note the characteristically thickened, hyalinized vessel (Haematoxylin and eosin, original magnification ×400)

considered in the differential diagnosis of tumours in this region, i.e. paragangliomas and schwannomas.

Subependymoma

Subependymomas are usually incidental findings at autopsy where they present as nodular, discrete masses at the wall of a ventricular cavity. Less frequently, these tumours may become clinically apparent due to size or location that causes an obstruction of the CSF flow and increased intracranial pressure. The tumours are well circumscribed and do not commonly undergo anaplastic transformation. However, symptomatic haemorrhage with necrosis from these lesions has also been reported (Changaris *et al.*, 1981; Lombardi *et al.*, 1991). The fourth ventricle is the most common site, accounting for 66%–70% in two published series (Scheithauer, 1978; Lombardi *et al.*, 1991); however, these tumours may arise in other locations such as the lateral ventricles, septum pellucidum, foramen of Monro or spinal cord.

These gliomas have a characteristic histological appearance with a distinctive cytoarchitecture reflecting a composite of ependymal and astrocytic features. Areas with definitive features of ependymomas, however, are usually more sparse and poorly developed in comparison to areas resembling fibrillary astrocytomas. The tumour cells are typically arranged as clusters that alternate with densely fibrillary and poorly cellular zones. The fibrillary quality of the matrix is evident even in the cellular areas (Fig. 3.15). The transitional/composite nature of the astrocytic and ependymal differentiation was confirmed by electron microscopic and tissue culture observations (Fu *et al.*, 1974; Azzarelli *et al.*, 1977). Despite the occurrence of histopathological features associated with increased atypia and mitotic activity in these tumours, tumour location and successful resection are more important prognostic factors (Lombardi *et al.*, 1991). The histogenesis of this lesion remains to be determined, but their discrimination from ependymomas is important to avoid needless adjuvant therapy.

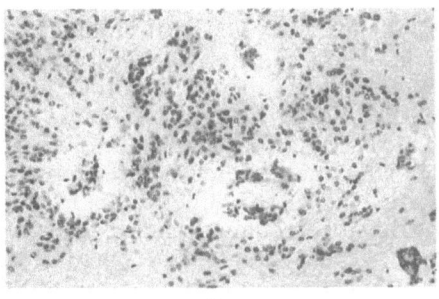 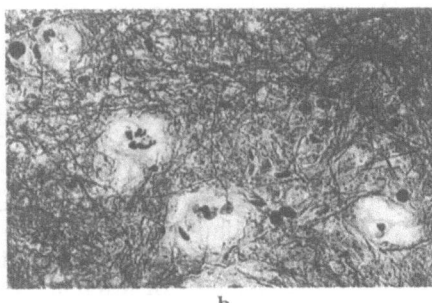

a   b

Fig. 3.15 Subependymoma. **a** Conspicuous cellular palisading and a dense fibrillary matrix are characteristic features (Haematoxylin and eosin, original magnification ×150). **b** The elaborate network of glial processes contain abundant PTAH-positive fibrils (PTAH, original magnification ×400)

Mixed Gliomas

Tumours with mixtures of various glial cell types, including astrocytes, oligodendroglia and, less frequently, ependymal cells, may be identified when adequate tissue is available for examination. These tumours pose a number of important issues. First, when the different glial cell components have markedly different proportions, the proper designation of the tumour relative to cellular anaplasia may be problematic. This dilemma is especially present in oligo-astrocytomas where features of anaplasia may not be equivalent in the two glial populations. Secondly, the diverse glial cell components commonly are not admixed in equivalent fashion throughout the tumours. The different cell populations may be intimately admixed or may be relatively discrete in adjacent regions, so that thorough examination of large mixed gliomas is important to classify appropriately the neoplasm. Thirdly, tumour growth and progression may be associated with a significant change in the character of the mixed glial cell population and certain glial cell components may be more likely to undergo anaplastic transformation. Fourthly, the origin and histogenesis of these tumours remain completely unknown, except for the possible implications of the studies which demonstrate bipotential cells associated with central gliogenesis in the rodent (Lillien and Raff, 1990).

Mixed Oligo-astrocytoma

These mixed tumours are composed of oligodendrocytes and a significant proportion of neoplastic astrocytes which may be either focally or diffusely distributed. Frequently, the prevalent cell component remains oligodendroglial (Rubinstein, 1972), but the mixed tumours are explicitly different from approximately 60% of the oligodendrogliomas which contain reactive astrocytes (Russell and Rubinstein, 1989). Analogously, GFAP expression in "gliofibrillary" oligodendrocytes is not evidence *per se* for an astrocytic component. Smear preparations in these cases are very helpful for identifying the distinct cytomorphological features of both cell populations (Plate 3.21).

The histogenesis of these tumours remains unclear, but the tumours may arise from cells which are "transitional" between mature oligodendrocytes and astrocytes (Herpers and Budka, 1984) or from the A2B5[+] progenitor cells discussed above (see Oligodendrogliomas p. 51). A common glial progenitor cell (O-2A) with a bipotential capacity for differentiation into oligodendrocytes and stellate type-2 astrocytes has been demonstrated in the developing and adult rat optic nerve (Raff *et al.*, 1983; Miller *et al.*, 1989). Although these data are principally restricted to rodent tissue in culture and solid human tumours (de la Monte, 1989; Bishop and de la Monte, 1989), the existence of such a progenitor cell population in the adult nervous system may provide a key to understanding the histogenesis of oliodendrogliomas and mixed oligo-astrocytomas (see Chap. 1, Noble and Venter). Nevertheless there are currently no data to exclude the possibility that one glial component induced neoplastic transformation in the second cell population.

Malignant Oligo-astrocytoma

The precise proportion of oligo-astrocytomas that progress to anaplastic tumours is not known. The astrocytic component of the mixed gliomas appears more susceptible to anaplastic progression (Muller *et al.*, 1977; Russell and Rubinstein, 1989). The histopathological features are thus similar to those previously discussed for anaplastic astrocytomas and oligodendrogliomas.

Choroid Plexus Tumours

Choroid Plexus Papilloma

Choroid plexus papillomas are relatively rare tumours in adults and comprise approximately 2%–5% of intracranial tumours in the paediatric population (Laurence, 1979). Congenital tumours have been reported (Matson, 1953; Tomita and Naidich, 1987). The most common sites are the lateral, fourth and third ventricles with preferential involvement of the lateral ventricles in young children and the fourth ventricle in adults (Russell and Rubinstein, 1989).

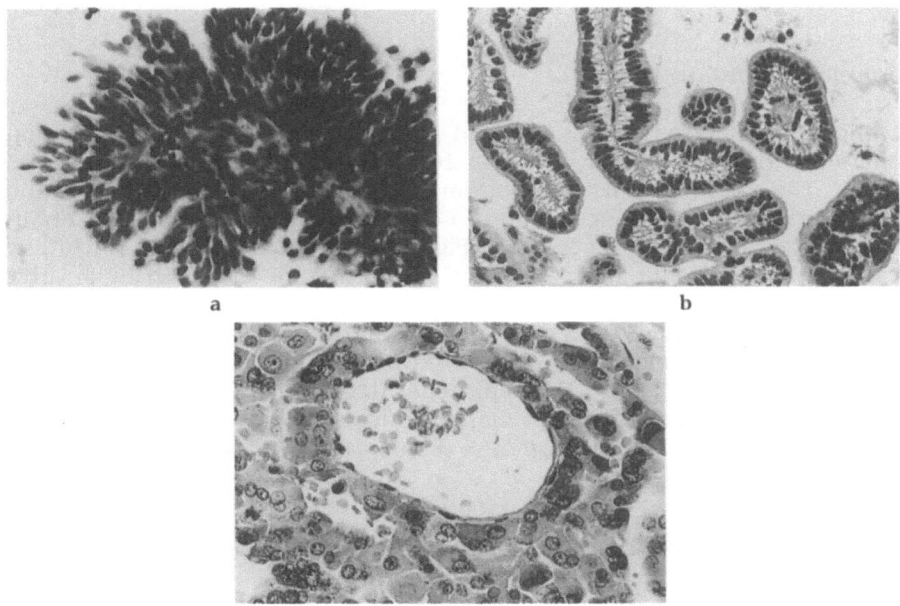

Fig. 3.16 Choroid plexus neoplasms. **a** Smear preparations of choroid plexus papillomas typically demonstrate tumour cells in a papillary arrangement around a vascular core (Morris stain, original magnification ×400). **b** This pattern is the hallmark histopathological feature in paraffin embedded tissue. Note the calcification, which is common in this tumour (Haematoxylin and eosin, original magnification ×400). **c** The typical architecture becomes markedly distorted with increased cellular pleomorphism in a choroid plexus carcinoma (Haematoxylin and eosin, original magnification ×400)

These tumours are macroscopically pink or grey lobulated masses which are commonly calcified. Choroid plexus papillomas mimic the histological features of choroid plexus tissue with a cuboidal to low columnar epithelium enclosing a fibrovascular core. The papillomas, however, may be more cellular and exhibit a greater cellular pleomorphism than normal choroid plexus (Fig. 3.16B). These features are readily appreciated on intraoperative smear preparations (Fig. 3.16A). Additional histological features include xanthomatous (Kepes, 1970) and oncocytic change (Kepes, 1983; Bonnin *et al.*, 1987), bone (Cardozo *et al.*, 1985) and cartilage formation (Salazar *et al.*, 1986), or pigmentation (Reimund *et al.*, 1990).

Mitotic figures and microscopic parenchymal invasion are rare. The prognostic significance of these features, when present, is uncertain. Some authors state that when present together, mitotic activity and brain invasion suggest an increased potential for recurrence and anaplastic progression (Russell and Rubinstein, 1989; Paulus and Janisch, 1990), whereas others have found no such correlation (McGirr *et al.*, 1988). Recurrence of histologically benign tumours and subarachnoid seeding have been reported (Matson, 1953; Wolfson and Brown, 1977; McGirr *et al.*, 1988). Ultrastructural studies of these tumours reveal a cuboidal epithelium with typical apical/basal polarity, apical microvilli and a basal lamina (Carter *et al.*, 1972). Apical junctional complexes, in contrast to ependymomas, are limited to tight junctions.

Choroid Plexus Carcinoma

Malignant transformation in choroid plexus papillomas is an uncommon phenomenon, occurring in less than 20% of cases (Lewis, 1967; Laurence, 1979). The carcinomas usually demonstrate marked cytological atypia and increased mitotic rates, in addition to disorganization of the regular architectural features of the papilloma (Fig. 3.16C). Other tumours, including embryonal carcinoma and choriocarcinoma, may be excluded by positive immunohistochemical staining for α-fetoprotein and β-human chorionic gonadotropin (β-HCG). The value of transthyretin (prealbumin) immunohistochemistry (Matsushima *et al.*, 1988) to discriminate choroid plexus neoplasms reliably from other epithelial neoplasms is limited (Albrecht *et al.*, 1991). The absence of this marker, however, may correlate with poor prognosis in choroid plexus tumours (Paulus and Jänisch, 1990).

Neuronal and Mixed Neuronal–Glial Tumours

Gangliocytoma

Gangliocytomas are rare tumours composed of mature neuronal populations. The major issue in classifying these lesions is the true nature of the neuronal cells with respect to a neoplastic versus hamartomatous origin. In contrast to gangliogliomas, there is no neoplastic glial component and therefore no potential for anaplastic progression. In regard to the preferential site of origin, clinical setting and macroscopic features, these lesions closely resemble gangliogliomas (see Gangliogliomas, p. 65).

Histologically, these tumours are composed of abnormal ganglion cells with a fibrous, acellular fibrillary matrix, presumably arising from reactive astrocytes (Plate 3.22). Microcalcification is common. As in gangliogliomas, the abnormal cytoarchitecture and orientation of the ganglion cells may be readily demonstrated by silver stains for neuritic processes and by cytoskeletal immunohistochemistry.

Dysplastic Gangliocytoma of the Cerebellum

The dysplastic gangliocytoma of the cerebellum (Lhermitte–Duclos disease) is a lesion occurring primarily in young adults; however, a congenital case has been reported (Roessmann and Wongmongkolrit, 1984). Magnetic resonance imaging has recently proved to be quite sensitive in identifying this lesion preoperatively based on the characteristic thickening of the cerebellar folia (Fig. 3.17 and Plate 3.22) (Reeder *et al.*, 1988; Milbouw *et al.*, 1988; Ashley *et al.*, 1990). Although the tumours are associated with a favourable postoperative course, recurrences may follow subtotal resection (Banerjee and Gleathill, 1979; Marano *et al.*, 1988).

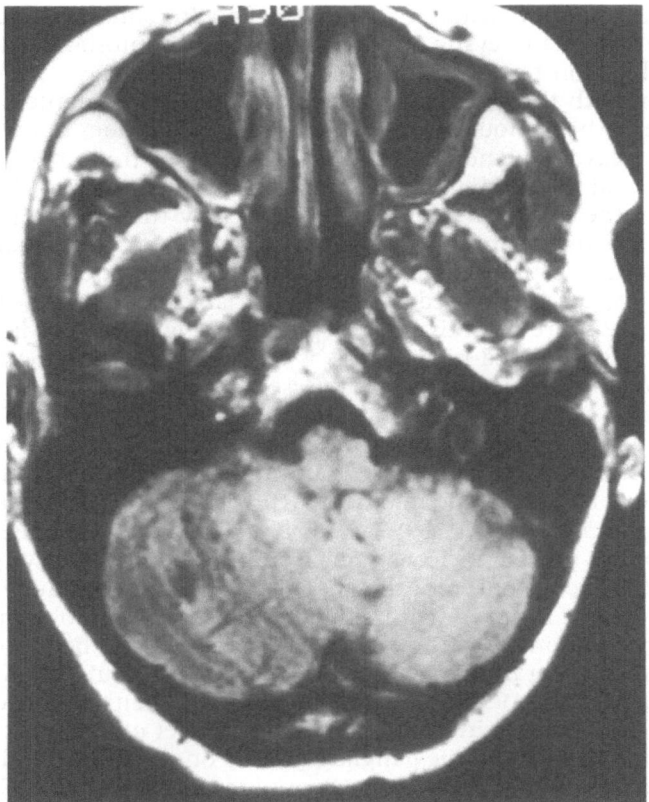

Fig. 3.17 Dysplastic gangliocytoma of the cerebellum. The typical enlarged and thickened cerebellar folia are apparent on this T_1 weighted magnetic resonance image. The majority of these lesions can be accurately diagnosed by MRI [axial scan: SE 0.8 s/20 ms (1.5 T)]

Histologically, this lesion is characterized by a loss of the normal cortical architecture and thickening of the cerebellar folia. The normal cerebellar cortical architecture is altered and usually consists of an outer layer of abnormally myelinated parallel fibres overlying an inner layer of abnormal, hypertrophied ganglion cells which resemble pyramidal-type neurons. The underlying white matter is greatly attenuated.

Two views on the pathogenesis of this lesion predominate in the literature. Some believe that this lesion results from the progressive hypertrophy of granular layer neurons, possibly due to hyperexpression of neurofilaments (Reznik and Schoenen, 1983; Yachnis *et al.*, 1988). Immunohistochemical detection of Purkinje cell synaptic proteins and surface membrane proteins suggests that this lesion is a cytogenetically complex hamartoma related to Purkinje cell lineages (Shiruba *et al.*, 1988; Russell and Rubinstein, 1989; Faillot *et al.*, 1990).

Desmoplastic Infantile Ganglioglioma

The desmoplastic infantile ganglioglioma is a rare variant of the mixed neuronal–glial neoplasms which invariably become clinically apparent during early infancy, usually within the first four months. The massive tumours are partially cystic and occur in a supratentorial location with a slight predilection for the frontal and parietal lobes (Fig. 3.18d). No communication with the ventricular system has been documented. The interface between the tumour and the surrounding brain does not usually have a definite cleavage plane. The hallmark feature of these neoplasms is an abundant and often dense desmoplasia. This confers a characteristic firmness to the neoplasms and usually arises by tumour interaction with the leptomeninges. Both neoplastic astrocytes and the usually globoid neurons compose the neoplastic neuroepithelial component (Fig. 3.18a; Plate 3.23) which is admixed within the variable desmoplasia (Fig. 3.18c). In addition, a more primitive cell population (Fig. 3.18b) which does not yet have a well-defined histogenesis is frequently observed (VandenBerg, 1991). Mitoses and micronecrosis, if present, occur only in association with this primitive cellular component. These features, however, do not appear to predict a poor outcome. A favourable prognosis, following complete or near-complete surgical resection, is an important clinical feature of these tumours and has been documented in several reports (Gambarelli *et al.*, 1982; VandenBerg *et al.*, 1987a; Ng *et al.*, 1990; VandenBerg, 1991). Current follow-up data in 13 patients from the University of Virginia series indicate no evidence of tumour recurrence, with a postoperative survival of 0.75–14.5 years. Tumours with similar clinical and pathological features, except a restriction to astrocytic differentiation, have been described as desmoplastic cerebral astrocytomas of infancy (Taratuto *et al.*, 1984; De Chadarévian *et al.*, 1990). The relationship between the histogenesis of these two desmoplastic infantile tumours is not yet understood.

The desmoplastic infantile ganglioglioma is one of the several distinctive CNS tumours in which the astrocytes produce a basal lamina. Collectively, these tumours implicate neoplastic neuroepithelial cells in an important role in forming and modifying the extracellular matrix. As previously described, several other CNS neoplasms also have the salient feature of a reticulin-positive stroma associated with the neuroepithelial component. This feature may be conspicuous without a true fibrous desmoplasia or collagen deposition, as in the pleomorphic xanthoastrocytomas (Kepes *et al.*, 1979, 1989). Alternatively, this may also be combined with conspicuous collagen deposition and fibrovascular proliferation, as in the desmoplastic infantile ganglioglioma and desmoplastic cerebral astrocytomas of infancy (Taratuto *et al.*, 1984; De Chadarévian *et al.*, 1990) or to a lesser degree in mixed neuronal–glial cell neoplasms (Probst *et al.*, 1979; Tang *et al.*, 1985) and in the rare gliofibroma (Fried, 1978; Iglesias *et al.*, 1984). In some tumours, the typically superficial location implicates an interaction with the leptomeninges as an important factor for stromal induction. However, this may not be an essential feature in all CNS tumours which have distinctive stromal characteristics. There is accumulating evidence for the secretion of extracellular matrix components by astrocytic tumour cells (see Russell and Rubinstein, 1989, for review), including possible complex chondromucoid matrices (Kepes *et al.*, 1984) and specific types of collagen (McComb *et al.*, 1987; Paulus and Peiffer, 1988; Sage and Iruela-Arispe, 1990). Neoplastic neuronal cells may also secrete collagens (DeClerck and Lee, 1985; Tsokos *et al.*, 1985, 1987). The ability of some CNS tumours to secrete serine proteinases and serpins (see Rao *et al.*, 1990a; Sawaya, 1990) as a function of cellular differentiation (Keohane *et al.*, 1990) presents another mechanism by which these neoplasms may create a distinctive stroma. The desmoplastic infantile ganglioglioma may be a tumour in which the capacity to secrete, induce and/or modify a distinctive extracellular matrix may occur with differentiation. This property may correlate with its limited brain invasion and characteristic lack of metastatic potential.

Dysembryoplastic Neuroepithelial Tumour

The dysembryoplastic neuroepithelial tumour (DNT) was described in a detailed report of 39 cases (Daumas-Duport *et al.*, 1988b). The tumours usually arise during the first two decades in the temporal cortex of patients with long-standing complex partial seizures and have an excellent prognosis after surgical resection. Consistent with a long, indolent course, radiological evidence of an overlying skull defect is common. With CT imaging, the lesions are hypodense, often cystic and show a modest and heterogeneous enhancement. Focal calcifications may be identified and peritumoral oedema is not usually present. The MRI characteristics are indistinguishable from low-grade gliomas.

The DNT are usually macroscopically apparent from the cortical surface. All cases are associated with microscopic cortical dysplasia, often with poor delineation of the tumour from the adjacent brain. The tumours characteristically have a multinodular histological architecture composed of either compact or "alveolar" patterns of mixed astrocytic, oligodendroglial and neuronal cell populations (Fig. 3.19). The "alveolar" pattern is marked by accumulation of extracellular mucoid material where the oligodendroglial component is usually predominant. No cellular anaplasia,

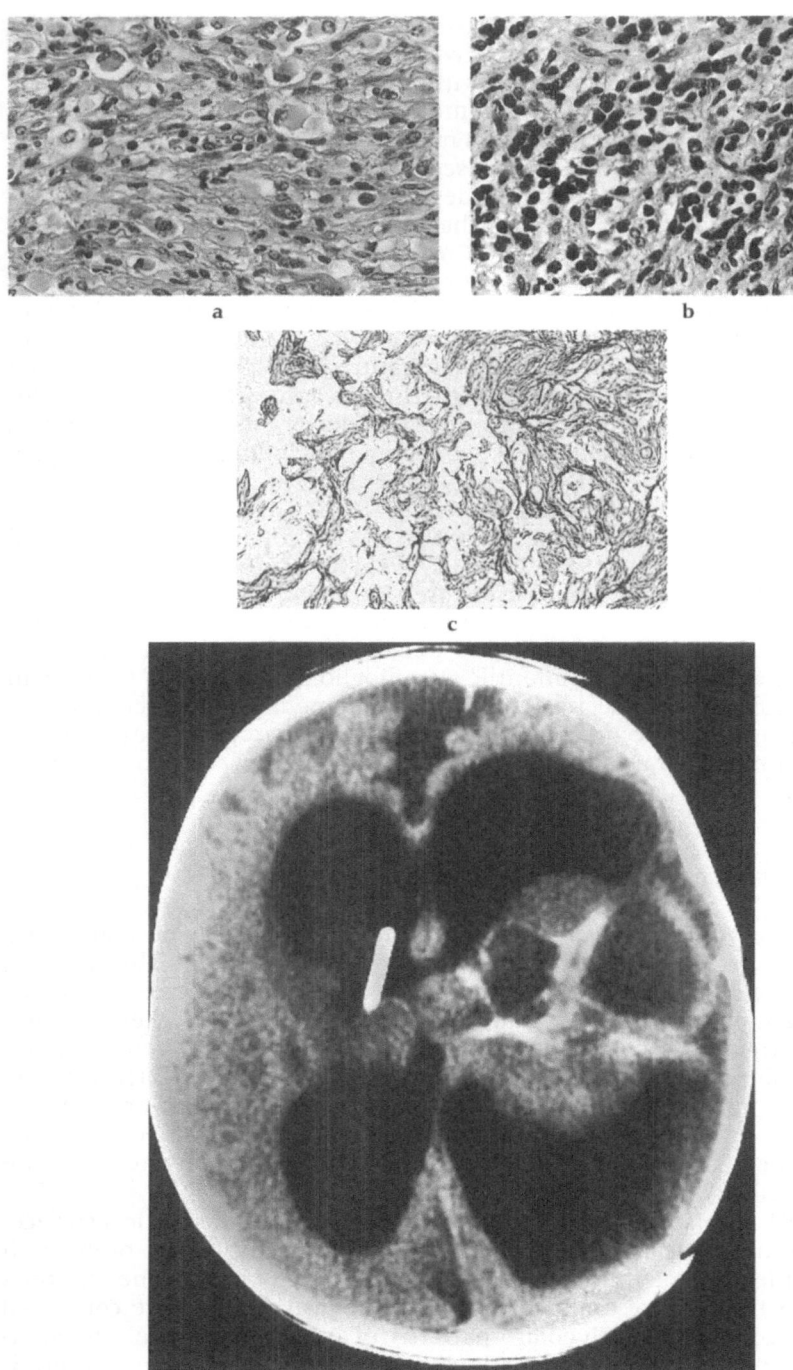

endothelial proliferation or mitoses are present. The heterogeneous nature of these neoplasms necessitates thorough examination of the lesion for diagnosis.

Ganglioglioma

Gangliogliomas, tumours with varying proportions of neoplastic neuronal and glial cells, are not common, representing only about 1% of all brain tumours (Kalyan-Raman and Olivero, 1987). The tumours are slightly more common in the paediatric group, comprising 4%–5% of CNS tumours (Sutton et al., 1983). The category of mixed neurogliogenic tumours is well established (Courville, 1930, 1931; Wolf and Morton, 1937; Courville and Anderson, 1941; Bailey and Beiser, 1947) and the largest review in the literature reports 120 cases (Russell and Rubinstein, 1989). In this series, 80% of the tumours occurred in patients under the age of 30, with 10% presenting in infants.

The clinical history typically involves a seizure disorder, often preceding surgical treatment by years. Gangliogliomas demonstrate a distinct preference for the temporal lobes, although they may occur anywhere within the central neuraxis. Computed tomographic imaging typically demonstrates low density, only mild contrast enhancement and foci of calcification. The tumours are often cystic, and may actually appear as a large cyst with a mural nodule (Kalyan-Raman and Olivero, 1987). Studies by MRI show well-delineated tumours, with low signal on T_1 and hyperdensity on T_2 weighted images (Fig. 3.20c).

These tumours are macroscopically well demarcated and relatively firm, a feature which is readily perceived during preparation of intraoperative smears. Abnormal neuronal cells and astroglia, often enmeshed in either a fibrovascular stroma or a fibrillary matrix, are easily appreciated in smear preparations (Fig. 3.20a). Two aspects of the histopathological features of these mixed neuronal–glial neoplasms provide important variables for consideration. First, the neoplastic cell populations often display phenotypic traits that reflect selected cell lineages of the region (Kawai et al., 1987; Takahashi et al., 1989). In addition to neurofilament protein (Trojanowski et al., 1984) (Plate 3.24), neoplastic ganglionic cells of a cerebellar ganglioglioma may express calbindin-28kD, a membrane-associated calcium binding protein which is selectively expressed in the cerebellum by Purkinje cells (Plate 3.25). Likewise, the glial elements of these cerebellar neoplasms may display features typical for the most frequent glioma of

---

◄ **Fig. 3.18** Desmoplastic infantile ganglioglioma. **a** The typically heterogeneous neoplastic populations composed of globoid neurons admixed with astrocytes are enmeshed in a variably desmoplastic stroma (Haematoxylin and eosin, original magnification ×400). **b** The primitive populations are highly cellular and commonly form a delicate, fibrillary matrix (Haematoxylin and eosin, original magnification ×400). **c** The reticulin stain highlights the typical desmoplastic stroma (Reticulin stain, original magnification ×200). **d** ACT scan of a $3\frac{1}{2}$-month-old child demonstrates a massive, multicystic irregular frontal-parietal tumour which does not communicate with the ventricles. Note the extensive hydrocephalus and the apparent involvement of the cortical surface (See Plate 3.23)

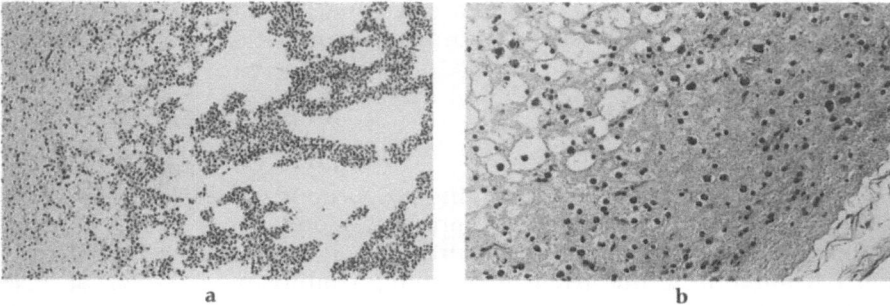

a   b

Fig. 3.19 Dysembryoplastic neuroepithelial tumour. **a** The characteristic alveolar pattern of the dysembryoplastic neuroepithelial tumour is demonstrated in this field. Oligodendroglial-appearing cells predominate in these areas (Haematoxylin and eosin, original magnification ×80). **b** The cortical architecture adjacent to the areas of microcystic change is typically disrupted by the heterogeneous neoplastic cells (Haematoxylin and eosin, original magnification ×200)

the region, a pilocytic astrocytoma. The specific glial elements, usually astrocytic (Plate 3.26), contribute to the potential for recurrence and anaplastic progression. Oligodendroglial elements are significantly less common in these tumours (Allegranza *et al.*, 1990). In addition to playing a role in the exceptional anaplastic progression of these tumours, the glial component may also modulate the maturation, survival and phenotypic character of the neuronal population by secretion of growth factors such as NGF (Cockram, 1990).

The stromal component of gangliogliomas comprises the second set of variable histopathological features that may range from delicate fibrovascular elements (Fig. 3.20b) admixed with an abundant glial population to a relatively intense fibrous reaction that nearly supplants the glial matrix. A variable angiomatosis may accompany the fibrovascular proliferation, but perivascular lymphocytic infiltration is a common feature regardless of the specific character of the stroma.

Malignant Ganglioglioma

In Rubinstein's series of 120 cases, 10% contained an anaplastic component (Russell and Rubinstein, 1989). Magnetic resonance imaging can usually distinguish these tumours from gangliogliomas by increased T_2 signal intensity and contrast enhancement. The anaplastic progression invariably involves the glial component and may culminate in a histological picture indistinguishable from glioblastoma multiforme (Plate 3.27). Similarly this malignant transformation has also occurred in oligodendroglial elements of these mixed tumours (Allegranza *et al.*, 1990). Immunohistochemistry for neurofilaments and GFAP is useful in recognizing the different cell populations where the highly cellular glial component is predominant (Plate 3.27). GFAP immunohistochemistry is invaluable to distinguish between malignant ganglioglioma and a ganglioneuroblastoma in limited biopsy material.

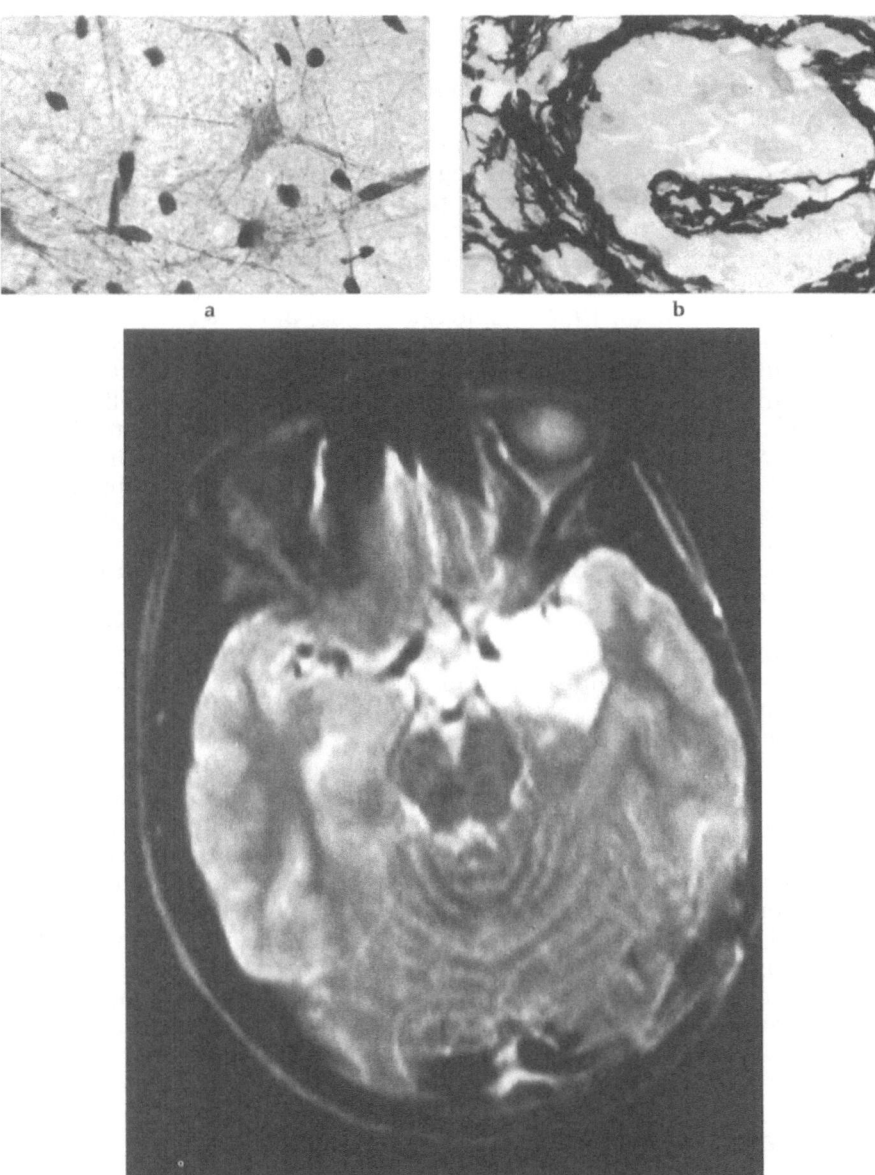

Fig. 3.20 Ganglioglioma. **a** An intraoperative smear of a ganglioglioma shows an abnormal ganglion cell with a large nucleus, prominent nucleolus and abundant cytoplasm admixed with fibrillary astrocytes (Morris stain, original magnification ×600). **b** The reticulin stain highlights the fibrous stroma which is most prominently associated with the ganglionic cells of the tumour (Reticulin stain, original magnification ×800). **c** A well demarcated ganglioglioma in the medial temporal lobe has a lobulated appearance and relatively uniform hyperintense signal on a T_2 weighted magnetic resonance image. Note the negligible oedema and minimal mass effort [Axial scan: SE 2.5 s/90 ms (1.0 T)] (See Plates 3.24 and 3.25)

Central Neurocytoma

Central neurocytomas are rare tumours occurring in young adults as discrete, partially calcified masses which project into the ventricles, preferentially near the midline adjacent to the foramen of Monro (Fig. 3.21c). The tumours may be massive, involving both lateral and third ventricles. Regardless of size, these tumours appear to have a favourable prognosis following successful surgical resection. Central neurocytomas must be distinguished from the gliomas and neuroblastomas which may arise in similar locations (Hassoun *et al.*, 1982; Pearl *et al.*, 1985; Wilson *et al.*, 1985; Townsend and Seaman, 1986; Nishio *et al.*, 1988; Barbosa *et al.*, 1990; Patil *et al.*, 1990; Kubota *et al.*, 1991). The significance of a single case in the clinical setting of neurofibromatosis is uncertain (Pearl *et al.*, 1981).

The central neurocytomas are composed of homogeneous cell populations with a characteristic morphology. The nuclei are round to slightly lobulated and contain delicate, diffusely distributed chromatin. Poorly defined cytoplasmic processes form a typically fibrillated intercellular matrix (Fig. 3.21a, b). These features are readily appreciated on smear preparations (Plate 3.28). Acellular zones of the intercellular matrix are not uncommon. The tumour vasculature is usually abundant in the form of delicate capillaries forming a branching pattern reminiscent of oligodendrogliomas. Focal micronecrosis may be present without any apparent relationship to cellular atypia or other anaplastic features. In our experience (Hessler *et al.*, 1991), mitoses and endothelial/pericytic proliferation are uncommon, although these features have been reported in one case (von Deimling *et al.*, 1990).

Ultrastructural analysis of central neurocytomas has been essential to demonstrate the hallmark features of a relatively well-differentiated neuronal tumour and remains the definitive diagnostic technique. The cellular processes contain parallel arrays of microtubules in addition to commonly abundant clear and dense-core vesicles. Variably developed synapses indicate neuronal maturation (Hassoun *et al.*, 1982; Wilson *et al.*, 1985; Pearl *et al.*, 1985; Townsend and Seaman, 1986; Nishio *et al.*, 1988; von Deimling *et al.*, 1990; Hessler *et al.*, 1991). The neurocytic nature of the tumour also can be demonstrated by immunohistochemical staining for synaptophysin (von Deimling *et al.*, 1990; Hessler *et al.*, 1991) and neuron-associated cytoskeletal proteins (Plates 3.29 and 3.30) (Hessler *et al.*, 1991).

Pineal Parenchymal Tumours

Classification of pineal parenchymal neoplasms is impeded by their rarity, with an estimated overall incidence of <0.1% (Scheithauer, 1985) of intracranial neoplasms. They comprise about 11%–28% of the pineal region tumours in children (D'Andrea *et al.*, 1987; Edwards *et al.*, 1988). The largest collective clinicopathological experience, representing 41 cases, was described by Russell and Rubinstein (1989). The essence of this classification of pineal-derived tumours includes three groups. Although the histopathological features of these neoplasms demonstrate a wide spec-

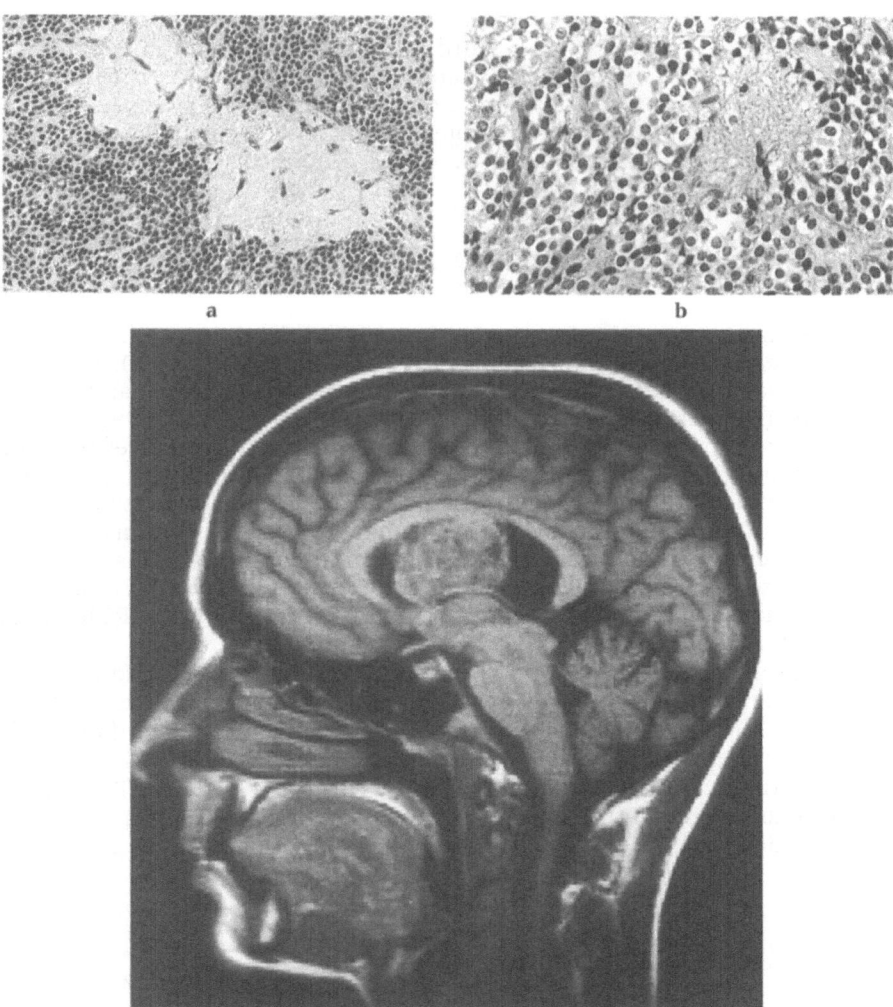

a   b

c

Fig. 3.21   Central neurocytoma. **a** The relatively dense fibrillary matrix is conspicuous in acellular areas (Haematoxylin and eosin, original magnification ×200). **b** The tumour cells can have a perinuclear halo of cytoplasmic retraction which is similar to oligodendrogliomas. The typically more extensive acellular areas of neurocytomas are useful in distinguishing between the two types of tumours (Haematoxylin and eosin, original magnification ×400). **c** A well-circumscribed intraventricular neurocytoma demonstrates a heterogeneous pattern of signal intensity in a T_1 magnetic resonance image [Sagittal scan: SE 0.5 s/17 ms (1.0T)] (See Plates 3.28–3.30)

trum of differentiation, both sides appear quite distinct. Highly cellular populations of cells with primitive pineal phenotypes compose the pineoblastomas at one end. At the other extreme, pineocytomas display moderately cellular and well-differentiated cell populations with mature pineocytic characteristics. The third group, the mixed pineocytoma/ pineoblastoma, encompasses the neoplasms which are intermediate in cytoarchitectural features and cellular patterns. This category is suitable for the tumours which have been previously designated as malignant pineocytomas or "pineocytomas without neuronal differentiation" (Herrick and Rubinstein, 1979).

Pineocytoma

Although pineocytomas are more common in adults than children (mean 42 ± 19 years) (Herrick and Rubinstein, 1979; Borit *et al.*, 1980; Vaquero *et al.*, 1990a), the age at initial diagnosis ranges from approximately 11 to 78 years (Herrick and Rubinstein, 1979; Vaquero *et al.*, 1990a). The tumours are typically well delineated from the adjacent structures without evidence of infiltration. The frequently cystic tumour is soft with a variably granular texture. Although focal necrosis may be present, extensive necrosis is uncommon. Computed tomographic imaging usually demonstrates an isodense neoplasm with variable calcification and contrast enhancement. Hypointense T_1 weighted signals and hyperintense T_2 weighted signals with no abnormal signal intensity of the adjacent brain are typical MRI features (Disclafani *et al.*, 1989; Vaquero *et al.*, 1990a).

Microscopically, pineocytomas represent the most well-differentiated pineal parenchymal neoplasm with moderate cellularity and no cellular atypia (Fig. 3.22). A hallmark feature is the arrangement of tumour cells surrounding delicately fibrillated, often interconnected acellular zones, previously termed "pineocytomatous rosettes" (Borit *et al.*, 1980). These may vary in size and regularity, but never should be confused with the smaller, less well-developed neuroblastic rosettes of medulloblastomas or neuro-

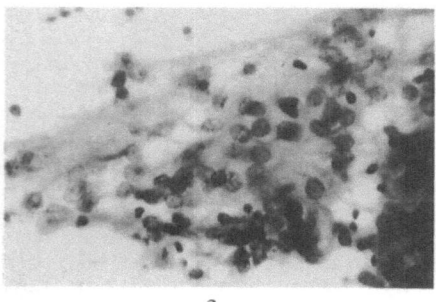
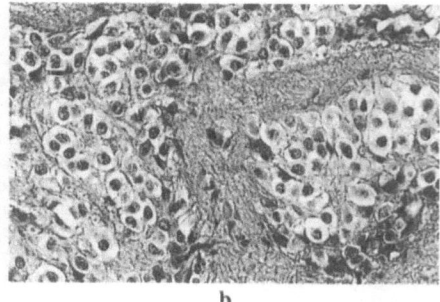

a   b

Fig. 3.22   Pineocytoma. **a** A smear preparation shows uniform, round nuclei with a finely dispersed chromatin pattern in a poorly defined, partially fibrillated matrix (Morris stain, original magnification ×400). **b** Sheets of tumour cells demarcated by a variably vascular stroma are characteristic (Haematoxylin and eosin, original magnification ×200)

blastomas. Silver impregnation stains demonstrate the characteristically delicate argyrophilic process with a terminal expansion which fill the acellular matrices. Most cellular processes may also be demonstrated by the silver carbonate impregnation which labels pineocytes (Herrick and Rubinstein, 1979). Although these rosettes were previously described as evidence for "neuronal" differentiation (Herrick and Rubinstein, 1979), they should be more appropriately considered as a manifestation of neoplastic pineocytic maturation. They are an important hallmark of the well-differentiated pineocytomas and appear to be associated with a more benign clinical course (see Russell and Rubinstein, 1989, for review).

Both neuron and photosensory cell-associated proteins have been demonstrated immunohistochemically in pineocytomas, including neurofilament epitopes and synaptophysin, and retinal S-antigen, respectively (Korf et al., 1986; Perentes et al., 1986; Collins, 1987; Rushing et al., 1991). Ganglion cell differentiation can occur in pineocytomas and may be accompanied by neoplastic astrocytes (Herrick and Rubinstein, 1979). In this context, these rare tumours may represent gangliocytomas and gangliogliomas of the pineal as forms of heteroplastic differentiation rather than particular steps in pineocytic cellular maturation.

The significance of neoplastic astrocytic differentiation in pineal parenchymal tumours is problematic. Glial histogenesis, either from the stromal astrocytes which are intimately associated with the pineal vasculature or from pineocytic progenitors, is as yet unknown. An origin from pineal parenchymal cells may be hypothesized on the basis of the transient and early photosensory epithelial differentiation which occurs during ontogeny (Reiter, 1981) and the partial photosensory differentiation which may occur in the primitive pineoblastomas as Flexner rosettes and photosensory fleurettes. A sporadic expression of an intrinsic glial phenotype, analogous to Müller cell differentiation in retinoblastomas, may occur and give rise to an intrinsic astroglial population. However, such an early phenotypic expression of photosensory lineages is only associated with the pineoblastomas and mixed pineocytoma/pineoblastomas and not the more mature pineocytomas. Therefore, the presence of astrocytic differentiation in a pineocytic neoplasm would not necessarily correlate with an indolent clinical behaviour (Herrick and Rubinstein, 1979). Likewise, proliferation of stromal astrocytes in the absence of pineocytomatous rosettes would not be a reliable indicator of tumour maturation.

The general lack of cellular pleomorphism in pineocytomas can occasionally be punctuated by giant cells without any diagnostic relevance. Mitotic activity may be present but does not appear to correlate directly with clinical behaviour. Previous reviews (Russell and Rubinstein, 1989) also emphasized the hallmark feature of a lobulated grouping of unipolar tumour cells demarcated by, and often aligned along, a delicate vascular stroma. However, this pattern alone is often associated with tumours which tend to be clinically aggressive and disseminate along CSF pathways similar to the pineoblastoma. Therefore the lobulated architecture alone is not particularly useful in distinguishing more immature pineal parenchymal tumours from pineocytomas. The lobular architecture appears to be a property of the early pineocytic differentiation in transitional tumours, the mixed pineocytoma/pineoblastomas. The papillary pattern previously described in both clinically aggressive and indolent tumours (Trojanowski *et al.*, 1982; Vaquero *et al.*, 1990b) is most likely a neoplastic exaggeration of this relationship between differentiating pineocytes and the vascular stroma.

Ultrastructural studies of pineocytomas have documented pineocytic differentiation with the presence of well-developed annulate lamellae, vesicle-crowned rodlets and associated structures, characteristic 9 + 0 neurosensory cilia and centriolar formations, in addition to processes containing dense-core and clear vesicles, sheaths of microtubules and bundles of intermediate filaments (Kline *et al.*, 1979; Markesbery *et al.*, 1981; Hassoun *et al.*, 1983, 1984; Hassoun and Gambarelli, 1989). None of these features need invoke the concept of general "neuronal" differentiation. The spectrum of synaptic junctions accompanying neoplastic neuronal differentiation in other CNS tumours has not been convincingly documented in the pineocytomas.

Pineoblastoma

Pineoblastomas occupy a position at the most primitive end of the spectrum and comprise approximately 3%–17% of pineal region tumours in children (D'Andrea *et al.*, 1987; Edwards *et al.*, 1988). Although the reported age at initial diagnosis ranges from 6 weeks to 52 years (Borit *et al.*, 1980; Sreekantaiah *et al.*, 1989), 50% of the tumours in Rubinstein's series occurred in the first decade with a male to female ratio 2:1 (Russell and Rubinstein, 1989). Two familial cases have been reported, the significance of which remains obscure (Lesnick *et al.*, 1985); however, a deletion at chromosome 11q13, which has also been described in neuroblastomas (Bigner *et al.*, 1988b), has been detected in a congenital pineoblastoma (Sreekantaiah *et al.*, 1989).

Macroscopically, the tumours are usually poorly defined with a gelatinous and often haemorrhagic appearance on cut-section. Necrosis is common but calcification is variable. Neuroimaging, in distinction to the pineocytomas, typically shows poor delineation between tumour and brain. Local infiltrative growth may also be associated with a perifocal oedema. T1 weighted MR images would show intense enhancement with contrast. Local invasion and leptomeningeal/CSF spread may be expected, although systemic metastases are not common (Banerjee and Kak, 1974). Most patients have a survival of less than two years (Herrick and Rubinstein, 1979; Neuwelt *et al.*, 1979; Borit *et al.*, 1980), but a 6-year survival has been reported (Uematsu *et al.*, 1988).

Pineoblastomas are highly cellular and are composed of small cells with round to oval nuclei containing coarse, hyperchromatic chromatin. The cells are typically arranged in patternless sheets (Fig. 3.23) but can form Homer Wright rosettes that recall medulloblastomas. Special silver carbonate impregnations for pineocytes can usually demonstrate a small, unipolar blunt process which commonly has a terminal expansion (Fig. 3.23a, inset). Rarely, a distinctive mosaic pattern, similar to the fetal pineal (Russell and Rubinstein, 1989), may be present. As in the pineocytomas, multinucleated cells may be found.

The exceptional retinoblastomatous differentiation, manifested by Flexner rosettes and photosensory fleurettes (Stefanko and Manschot,

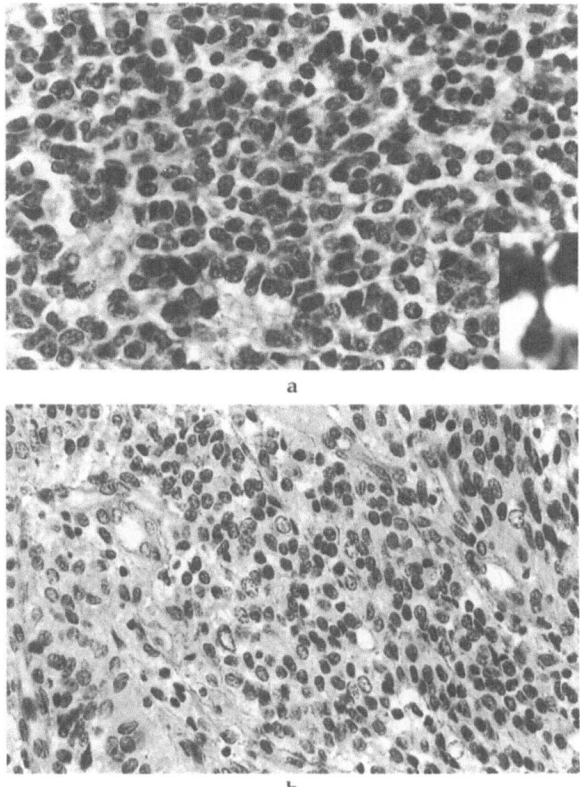

Fig. 3.23 Pineoblastoma. **a** This tumour is characterized by small, poorly differentiated tumour cells in an inconspicuous lacy matrix similar to the cerebellar medulloblastoma. Frequent micronecrosis and mitotic figures attest to the aggressive nature of these neoplasms (Haematoxylin and eosin, original magnification ×400). Inset: A silver carbonate stain demonstrates short, knob-like tumour cell processes (Silver carbonate original magnification ×800). **b** The cytoarchitectural features of both pineocytoma and pineoblastoma are combined in the mixed pineal parenchymal tumours with areas that recall both histological patterns. Less well-defined architectural features and increased nuclear pleomorphism are more frequent than in pineocytomas. Note the mitotic figure in the centre of the field (Haematoxylin and eosin, original magnification ×400)

1979; Sobel *et al.*, 1981; Russell and Rubinstein, 1989) reflect the early photoreceptor ontogeny of the human pineal gland (Sivak, 1974; Zimmerman and Tso, 1975; Reiter, 1981). During normal development, photoreceptor phenotypic expression in the form of cellular morphology is early and transient, although expression of specific retinal matrix is continued in adult tissue. To extend the analogy, retinoblastomatous features in pineal parenchymal tumours should be considered as evidence of primitive pineal differentiation, indicative of pineoblastoma, on small-tissue specimens. The phylogenetic relationship between pineoblastomas and retinoblastomas is also emphasized by development of trilateral retinoblastomas (Bader *et al.*, 1982; Johnson *et al.*, 1985).

In one series, the poorly differentiated cells were immunoreactive for S-retinal antigen in two of four pineoblastomas and only in two of seven pineocytomas. In the normal gland, the highest expression, detectable by immunohistochemistry, occurred in the perinatal period with less reactivity in the adult gland (Perentes *et al.*, 1986). Rare evidence of neuronal differentiation with ganglionic cells has been reported (Sobel *et al.*, 1981).

Ultrastructurally, pineoblastomas demonstrate features of a poorly differentiated neuroepithelial neoplasm with early and variable photoreceptor phenotypic expression with polarization of cytoplasmic organelles, 9 + 0 cilia, annulate lamellae and well-developed SER with whorl formation (Kline *et al.*, 1979; Markesbery *et al.*, 1981).

Mixed Pineocytoma/Pineoblastoma

The frequency of these tumours is hard to evaluate, since many tumours previously reported as poorly differentiated or malignant pineocytomas would fall into this category. The tumours have a more unpredictable clinical behaviour but are usually aggressive in local infiltration and distant CSF dissemination. Histological features are transitional between the well-differentiated pineocytomas and the extremely cellular, relatively monomorphic pineoblastomas (Fig. 3.23B). Areas identical with pineoblastomas may be admixed with rather sharp delineation from more differentiated areas, including the lobulated architecture associated with immature pineal tissue. Necrosis and cellular pleomorphism are common and mitotic activity is usually present. Pineocytomatous rosettes are rare.

Embryonal Tumours

The embryonal tumours, as primitive and clinically aggressive neoplasms that usually occur during the first decade of life, are distinct from tumours that commonly arise in adults (see Table 3.3). Although there may be common histopathological features between these paediatric and anaplastic adult tumours, there is abundant evidence that embryonal neoplasms arise from different populations of transformed cells and may not have the same molecular basis for malignant transformation or progression (Bigner *et al.*, 1990; Bigner and Vogelstein, 1990; Wasson *et al.*, 1990; Thomas and Raffel, 1991). Therefore, glioblastoma multiforme, as a poorly differentiated and anaplastic glioma occurring most commonly in adults, is appropriately separated from the primitive tumours that most frequently arise in the paediatric brain. The rare polar spongioblastoma and gliomatosis cerebri have been excluded from the embryonal tumours and are more appropriately considered glial tumours of uncertain origin.

The term *embryonal* does not connote that these tumours are biologically equivalent to the embryonal or immature cells that normally populate the developing nervous system. This designation, however, does indicate that these tumours arise from the transformation of undifferentiated and immature neuroepithelial cells, and thus necessitates special consideration with respect to their histopathological classification and assessment of their biological potential (VandenBerg *et al.*, 1987b; Rorke, 1989; Rubinstein, 1989; Rubinstein, 1991). Foremost, there are more diverse cell populations which are vulnerable targets for neoplastic transformation in the immature nervous system and these populations have different capacities for differentiation (Kleihues *et al.*, 1990).

Neoplastic differentiation differs from normal neurocytogenesis due to disruption of the complex interactions between the extrinsic milieu and the altered intrinsic molecular processes regulating growth and differentiation. Recent data that implicate regional extrinsic effects in the normal CNS on the expression of either glial and neuronal phenotypes from undifferentiated progenitor cells emphasize the importance of these interactions (Renfranz *et al.*, 1991). Thus, a greater potential exists for either heteroplastic phenotypes, such as myofibres in medulloblastomas (see Rao *et al.*, 1990b, for review), or incomplete phenotypic differentiation. Despite these caveats, a number of embryonal tumours may be characterized on the basis of distinctive histopathological features. However, all tumours, regardless of histogenesis, share the common features of high cellularity, numerous mitoses and at least focal necrosis. These features manifest the clinically aggressive behaviour that corresponds to a "Grade IV" tumour. There is a common propensity for leptomeningeal invasion and subsequent metastasis in the cerebrospinal fluid pathways.

The current classification scheme recognizes a number of embryonal tumours with restricted and relatively defined histogenetic potentials along neuronal, ependymal and retinal cell lineages, the cerebral neuroblastoma, ependymoblastoma and retinoblastoma, respectively. A very rare embryonal neoplasm which also merits a separate identification is the medulloepithelioma. In contrast to the other aforementioned tumours, this neoplasm has the greatest potential for expressing divergent neural cell types.

The new designation of primitive neuroectodermal tumour (PNET) embraces the concept that such tumours arise from primitive neuroepithelial progenitor cells which are equivalent throughout the neuroaxis. Accordingly, these tumours would have similar histopathological features and biological behaviours. Implicit in this concept is that a potentially diverse phenotypic differentiation can occur in a PNET after the transformation of an undifferentiated neuroepithelial cell (Gould *et al.*, 1990). This premise contrasts with the view of phenotypic differentiation as a manifestation of a stable, more restricted genotypic character which was established prior to neoplastic transformation. However, the concept of PNET, while valuable as an intriguing theory for investigation, does not readily enhance the basis for classifying the embryonal tumours with a defined histogenesis. Therefore, the designation of PNET is at present limited to those tumours which share the morphological features of the medulloblastoma. The inclusion of the medulloblastoma in this category reflects the uncertainty about the histogenetic potential of this heterogeneous, usually poorly differentiated tumour.

Neuroimaging Correlations. The embryonal tumours share many CT neuroimaging features (Pigott *et al.*, 1990) including a significant mass effect with variable demarcation, iso- to hyperdensity compared to adjacent brain, variable calcification, infrequent intratumoral haemorrhage, cystic regions and significant, but heterogeneous contrast enhancement. Neuroblastomas usually have been described as hypo- to isodense with a distinct cystic component, frequent calcification, variable but detectable necrosis, and, uncommonly, haemorrhage. Contrast enhancement is pronounced and heterogeneous (Berger *et al.*, 1983; Bennett and Rubinstein, 1984). Signal hyperdensity has also been described (Pigott *et al.*, 1990). The typical spread of ependymoblastomas is characterized by poor CT demarcation (Langford, 1986; Pigott *et al.*, 1990) and isodensity compared to adjacent brain. Cystic regions, necrosis, and calcification are variable. Haemorrhage has not been specifically reported.

Medulloepithelioma

This very rare tumour usually occurs early in the first decade of life. Although it may apparently arise throughout the neuroaxis, the cerebral hemispheres were the most frequent of the sites in the 20–25 cases which have been described (Russell and Rubinstein, 1989; Caccamo *et al.*, 1989). The cerebral tumours, in particular, may be associated with a lateral ventricle and are characteristically large and macroscopically well delineated.

The hallmark histopathological feature of these neoplasms is the mitotically active, pseudostratified columnar epithelium often arranged in ribbons of tubules or papillary rosettes with variable interposition of delicate stromal elements. These structures recall the primitive epithelium of the embryonal neural tube. Approximately 50% of the cases exhibit neuroblastic/neuronal, astroglial and/or ependymal cell populations which are either intimately admixed with the tubules or are present in more well-demarcated fields (Rubinstein, 1991). Silver stains confirm neuroblastic and ganglionic differentiation and PTAH staining highlights the astroglial/ependymal differentiation in these tumours.

Immunohistochemical studies of four cerebral tumours (Caccamo *et al.*, 1989) demonstrated that the primitive neuroepithelium contained abundant vimentin (Plate 3.31) with only rare, sporadic expression of neuroblastic and glial cytoskeletal proteins (class III β-tubulin isotype and GFAP, respectively) in morphologically identical neuroepithelial cells. In contrast, immunoreactivity for insulin-like growth factor I and basic fibroblastic growth factor is abundant in this primitive epithelium (Shiurba *et al.*, 1991). Immunoreactivity for neuroblastic and glial cell-associated cytoskeletal proteins was readily demonstrated in the more differentiated cell populations, contrasting to the relative scarcity in the neuroepithelium. Ultrastructural studies of the primitive neuroepithelium in two tumours revealed a paucity of cytoplasmic organelles, moderately developed juxta-luminal zonulae adherentes, and a conspicuous lack of apical differentiation (Pollak and Friede 1977; Troost *et al.*, 1990).

The predominance of primitive neuroepithelium accompanied by the common occurrence of divergent differentiation suggests that the medulloepithelioma has a histogenetic potential corresponding to the primitive ventricular zone cells of the embryonal neural tube. As to the significance of primitive neoplastic neural cells forming epithelial structures, previous data from our laboratory using an animal model for embryonal CNS tumours provide an interesting observation. In this experimental system, the capacity of the most primitive neural cells to form polarized medullo-epithelioma-like tubules was very frequently accompanied by ependymo-blastoma and ependymal differentiation. Loss of the primitive cells which could form the tubules also resulted in the loss of ependymal differentiation and a drastic reduction in the amount of the more mature, divergent neural phenotypes (VandenBerg *et al.*, 1981).

Neuroblastoma

This rare embryonal neoplasm is most common in children less than ten years old, with the greatest frequency of cases occurring during the first half of the decade. There appears to be no sex predilection. The tumours may arise throughout the central nervous system, including the pons, spinal cord and cauda equina, but the majority of the reported cases are supratentorial (Horten and Rubinstein, 1976; Berger *et al.*, 1983; Bennet and Rubinstein, 1984). The frontal and frontoparietal regions are commonly involved with these usually large, cystic tumours with macroscopically discrete borders. A variably desmoplastic stroma may confer a relatively firm texture and partially lobulated appearance, notable when intraoperative smears are prepared. Most tumours also have a cystic component, focal necrosis and infrequent haemorrhage.

Central neuroblastomas are highly cellular with frequent mitoses. The cytoarchitecture is highly variable, ranging from small cells with poorly defined cytoplasm and round to ovoid hyperchromatic nuclei to larger cells with more well-defined, polar processes and vesicular nuclei. A fibrillated matrix of both delicate neurites and more indistinct processes is common and more readily observed than discrete neuroblastic (Homer Wright) rosettes (Fig. 3.24A). In addition, parallel arrangements of compact cellular groups with rhythmic nuclear palisading that recalls a similar pattern in medulloblastomas may also be present (Ojeda *et al.*, 1987; Dehner *et al.*, 1988). These cells demonstrate ultrastructural features of neuroblastic differentiation. Therefore, diagnosis on small biopsy samples without special stains, immunohistochemistry or electron microscopy may be uncertain.

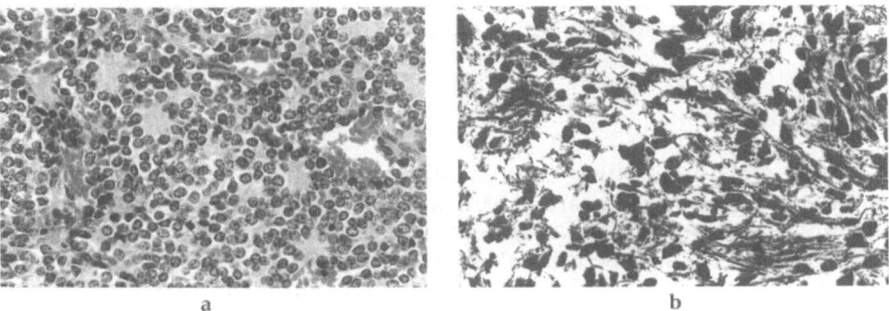

a   b

Fig. 3.24 Cerebral neuroblastoma. **a** Homer Wright rosettes are a typical histopathological feature and are observed here in metastatic tumour implants in the leptomeninges (Haematoxylin and eosin, original magnification ×400). **b** Bielschowsky silver impregnation highlights the neuritic processes of the tumour cells (Bielschowsky stain, original magnification ×400)

78   S.R. VandenBerg *et al.*

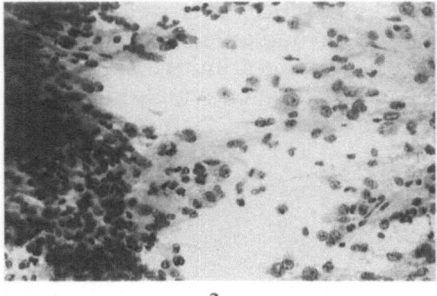 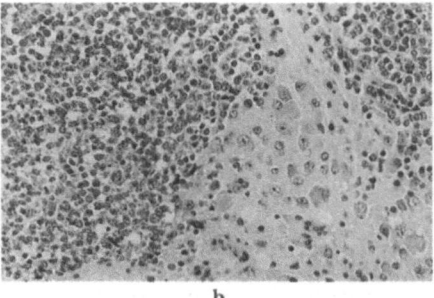

a   b

Fig. 3.25 Ganglioneuroblastoma. **a** Both neoplastic neurons and a more primitive cellular component can be identified in smear preparations (Haematoxylin and eosin, original magnification ×200). **b** The populations of maturing ganglion cells and primitive neuroblasts in this cerebral ganglioneuroblastoma recall similar features in differentiating neuroblastomas of the sympathetic nervous system (Haematoxylin and eosin, original magnification ×200)

.Both silver impregnation stains (Fig. 3.24B) and immunohistochemistry for neuron-associated cytoskeletal proteins, such as class III β-tubulin epitopes (Plate 3.32), are necessary to demonstrate these neuritic processes more specifically. These special studies are optimal with frozen sections of formalin-fixed tissue or with fixed tissue embedded in low melting temperature paraffin. In contrast to the neuroblastic/neuronal differentiation, both PTAH staining and GFAP immunohistochemistry confirm the absence of a neoplastic glial component in these primitive tumours.

The detection of various neuron-associated growth factor receptors, such as nerve growth factor, may also be important adjuncts to current histopathological techniques to recognize neuroblastomas. Specific identification of tumours with receptors to hormonal/growth factors may also have therapeutic implications (Baker *et al.*, 1990).

Neoplastic ganglionic differentiation occurred with a frequency of approximately 50% in a series of 70 cases, but was never an abundant component of individual tumours (Bennett and Rubinstein, 1984). Nevertheless, neoplastic ganglionic cells can be recognized in intraoperative smear preparations (Fig. 3.25). Ultrastructural studies confirm the neuronal maturation with increased organelle density and differentiation and synaptic junctions (Russell and Rubinstein, 1989).

Ependymoblastoma

This rare, ependymal tumour is usually massive by clinical presentation at a median age of two years (Mørk and Rubinstein, 1985). All reported examples are intracranial with a supratentorial predominance. Relationship to the ventricles may be difficult to ascertain due to the tumour size, and the masses tend to have defined macroscopic borders but commonly have demonstrable infiltration of adjacent brain and leptomeninges. The microscopic features, as for all the embryonal CNS tumours, include high cellularity and mitotic activity with poorly differentiated small cells forming amorphous arrangements or the characteristic rosettes and tubules. These are commonly pseudostratified with multiple layers and frequent, juxtalu-

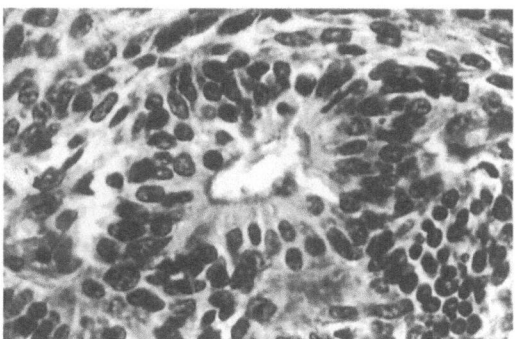

Fig. 3.26 Ependymoblastoma. These highly cellular and primitive tumours demonstrate ependymal differentiation, including rosettes with mitoses, in the absence of the anaplastic features used to distinguish anaplastic ependymomas (Haematoxylin and eosin, original magnification ×400)

minal mitoses. Although the cells that compose these rosettes are polarized with cytoplasmic processes extended to the apical lumen, the luminal space may be poorly defined (Fig. 3.26). Well-developed perivascular rosettes which are common in ependymomas are relatively rare. Immuno-histochemical localization of GFAP may be useful to define further the primitive cellular processes (Cruz-Sanchez *et al.*, 1988; Pigott *et al.*, 1990).

Ultrastructural features attest to differentiation that is restricted to ependymal features (Langford, 1986). In rosettes, the cells have luminal cytoplasmic specialization including microvilli and 9+2 cilia. Juxtaluminal zonulae adherentes are present. In cellular arrangements apart from rosettes, cytoplasmic polarization may persist with the formation of basal lamina-lined labyrinths.

These tumours should be distinguished from anaplastic ependymomas on the basis of several microscopic features, including minimal or absent endothelial proliferation, which is common and may be conspicuous in anaplastic ependymomas. In addition, ependymoblastomas seldom have the degree of cytoplasmic or nuclear pleomorphism which is characteristic of anaplastic ependymomas and do not have well-developed pseudo-rosettes. Necrosis may be found in both types of tumours, although geo-graphic and pseudopalisading patterns of necrosis are only found in anaplastic ependymomas. Microscopic invasion of adjacent parenchyma may occur with both types of tumour. Clinically, the ependymoblastomas commonly present in children below the age of five years in supratentorial regions, whereas the anaplastic ependymomas are more common in adults.

Retinoblastoma

Retinoblastomas, as the most common intraocular tumour of childhood, are the only embryonal tumours derived from the central nervous system for which the genetic basis of neoplastic transformation is known. The transforming event occurs in the immature retina when both alleles of the rb tumour suppressor gene are inactivated within a single cell (Gennett

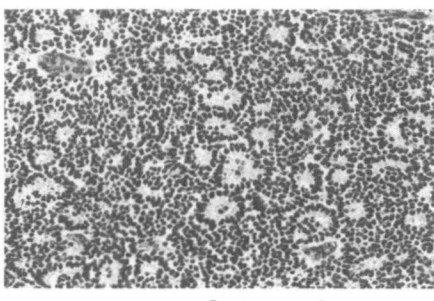 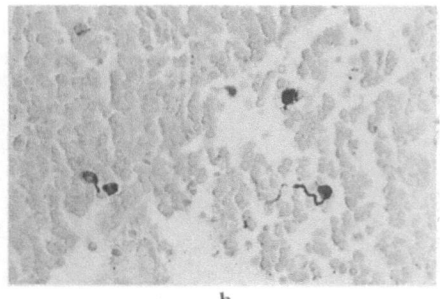

a   b

Fig. 3.27   Retinoblastoma. **a** Compared to the patternless sheets of primitive cells often found in retinoblastomas, Flexner–Wintersteiner rosettes imply an early stage of cellular differentiation. These structures, however, do not appear to correlate specifically with the expression of photosensory or neuronal intermediate filament phenotypes (Haematoxylin and eosin, original magnification ×200). **b** Immunohistochemical staining for retinal S-antigen can often delineate cells expressing this retinal protein which by conventional staining appear to be undifferentiated (Retinal S-antigen–peroxidase–antiperoxidase immunostaining without counterstain, original magnification ×400)

and Cavenee, 1990; Gallie *et al.*, 1990). Although the tumour commonly exhibits fields of poorly differentiated small cells with scant, ill-defined cytoplasm, the majority of retinoblastomas display typical rosettes of the Flexner–Wintersteiner and Homer Wright types. These may be either focal or diffusely distributed (Fig. 3.27A). A small fraction of cases also display fleurettes as a more well-developed manifestation of photosensory phenotypic differentiation. Although retinal S-antigen (Fig. 3.27B), neurofilament, neuron-associated class III β-tubulin and photoreceptor opsin immunoreactivity is not uncommon in rosettes, there is no constant correlation between the presence and number of rosettes or fleurettes and the expression of these neuronal/photoreceptor epitopes in the tumour cells (Lopes *et al.*, 1991). Although there is no predictive value for the clinical behaviour of these tumours which can be based on the differentiation of rosettes or fleurettes, rare tumours with distinctly lower cell density that are entirely composed of rosettes appear to have a reduced malignant potential. These neoplasms have been designated as "retinocytomas" (Margo *et al.*, 1983) to distinguish the different clinical behaviour.

Unequivocal documentation of neoplastic gliogenesis in retinoblastomas with GFAP expression has been technically problematic. This may be due, in part, to the restricted gliogenesis which can occur in the normal retina. Müller glial cells, in contrast to extrinsic astrocytes migrating from the optic nerve, are the only glia originating from retinal progenitor cells (Ling and Stone, 1984; Stone and Dreher, 1987; Schnitzer, 1988b; Watanabe and Raff, 1988). GFAP consistently labels reactive "stromal" astrocytes which are invariably associated with blood vessels, as in normal retina (Schnitzer, 1988a) and are often present in the interface zone between retina and the infiltrating tumour. The astroglial cell population in retinoblastomas labelled by GFAP histochemistry are most probably reactive and infiltrate from the vascular stroma of the residual retina. In contrast, cellular retinaldehyde binding protein (CRAlBP) is not normally expressed by retinal astrocytes (Lewis *et al.*, 1988) and is uniquely synthesized by the retinal pigmented epithelium and Müller cells (Saari *et al.*, 1982; Bunt-Milam and

Saari, 1983). Retinoblastoma cells expressing CRAlBP were identified in 50% of the cases and in 69% of tumours demonstrating photoreceptor cell gene expression in a recent study from our laboratory (Gonzalez-Fernandez et al., 1992).

The histogenetic potential of the retinoblastoma has been variously described as equivalent to: (a) a primitive neuroepithelial cell that is multi-potential along astroglial and retinal pigmented and neuroepithelial lines (Taylor et al., 1979; Jiang et al., 1984; Kyritsis et al., 1984; Terenghi et al., 1984; Tsokos et al., 1986; Rodrigues et al., 1987; Schroder, 1987; Campbell and Chader, 1988; Detrick et al., 1988; Tarlton and Easty, 1990); (b) a neuroretinal progenitor cell limited to neuronal differentiation (Perentes et al., 1987; Katsetos et al., 1991), or (c) a photosensory cell-specific progenitor restricted to selective photosensory cell types (Tso et al., 1969, 1970; Popoff and Ellsworth, 1971). Recent data from 22 cases of retinoblastoma in situ which were studied in our laboratory suggest that the retinoblastomas have a histogenetic potential analogous to the immature retinal neural epithelium with the presence of photosensory (IRBP and cone/rod opsins) and Müller cell (CRAlBP) phenotypes within the same tumour. As the inner layer of the developing optic cup, this mitotically active epithelium normally maintains the potential for divergent differentiation into photoreceptors, neurons or retinal glia (Müller cells), following the final mitotic cycle (Turner and Cepko, 1987; Holt et al., 1988; Wetts and Fraser, 1988). Müller glial cells, in contrast to extrinsic astrocytes migrating from the optic nerve, are the only glia originating from retinal progenitor cells (Ling and Stone, 1984; Schnitzer, 1988b; Watanabe and Raff, 1988). In contrast, retinoblastomas do not appear to be pluripotential in the sense of demonstrating unrestricted retinal epithelial or neuroepithelial differentiation to produce either pigmented cell or astroglial phenotypes (Lopes et al., 1991; Gonzalez-Fernandez et al., 1992).

Medulloblastoma

This neoplasm is the most common of the embryonal tumours and comprises approximately one-quarter of all intracranial tumours in children, second in frequency after the astrocytomas arising in the posterior fossa. The peak incidence occurs near the end of the first decade, with males slightly more affected. The majority of tumours arise in the cerebellar midline (Fig. 3.28d). Although tumours occurring in the paediatric group may also appear laterally in the cerebellar hemispheres, this location is more frequent in the adult cases (Russell and Rubinstein, 1989). Macroscopically, the soft, often friable, tumours are nominally discrete from the adjacent brain. Necrosis is invariably present and varies from diffusely distributed punctate foci to more expansive zones, sometimes creating central cavities. Cyst formation, in comparison to the pilocytic astrocytomas of this region, is rare. In contrast, the lobular or desmoplastic variants, which are usually more prevalent laterally in the cerebellar hemispheres of older patients, are well demarcated with a markedly firmer consistency. The latter feature is easily appreciated during the preparation of intraoperative smears.

The invariant histopathological feature of the medulloblastoma is an

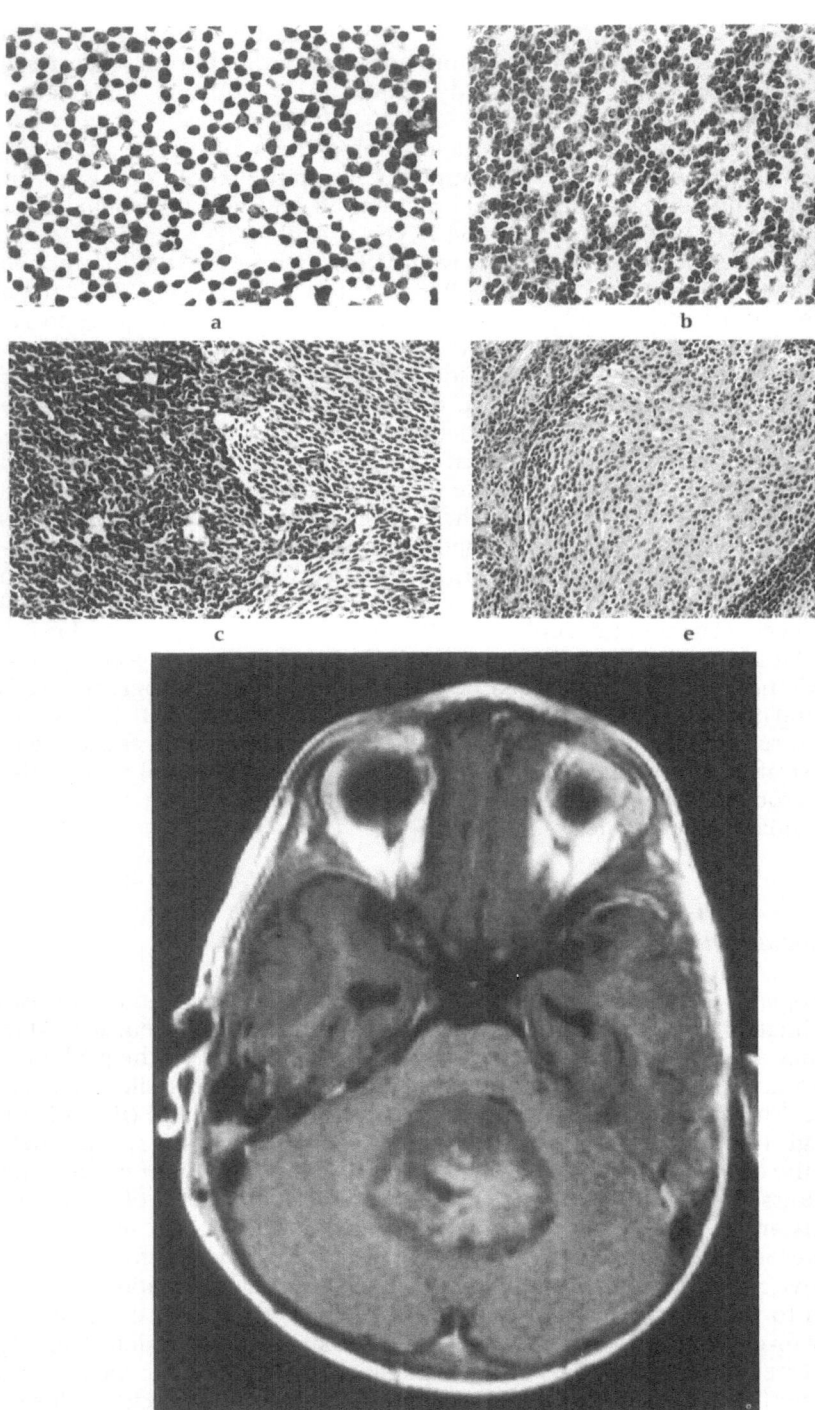

intense cellularity with relatively small cells exhibiting scant cytoplasm and ill-defined cell borders. Mitoses are usually found without difficulty. The nuclei are hyperchromatic and characteristically irregular with an angular to ovoid geometry, features which readily permit discrimination from the normal granular cells in intraoperative smear preparations (Fig. 3.28a). Focal or geographic necrosis is common, but endothelial/pericytic proliferation is rare. Medulloblastomas are commonly regarded as poorly differentiated by routine microscopy. However, in a recent series of 246 cases (Kleihues et al., 1989), approximately 42% of the tumours showed varying numbers of fields with neuroblastic rosettes; 7% of the tumours with neuroblastic differentiation also had a neoplastic ganglionic component. The areas of neuroblastic differentiation, in addition to rosettes, commonly are marked by a cell population with more delicate fibrillated processes and paler, more vesicular nuclei (Fig. 3.28b, c). These areas often demonstrate cell bodies and processes that are immunoreactive for an epitope of neurofilaments (Plate 3.34) and the early neuron-associated tubulin of the class III β-tubulin isotype (Plate 3.35).

The desmoplastic variant of medulloblastoma comprises approximately 10%–12% of cases (Burger et al., 1987b; Kleihues et al., 1989) and is distinctive for a biphasic architecture with a follicular arrangement of tumour cells (Fig. 3.28e). Highly cellular sheets and trabeculae of typical tumour cells encompass islands characterized by lower cellularity and cells with finely fibrillated processes. This architecture is particularly well highlighted by reticulin deposition only in the peripheral cellular areas. The reticulin-free islands prominently demonstrate neuronal tubulin and neurofilament immunoreactivity (Katsetos et al., 1989). It is interesting to note that the characteristic architectural pattern of the desmoplastic variant is not always preserved in the recurrent tumour.

Other rare variants of medulloblastoma include tumours with striated muscle cell differentiation (medullomyoblastoma), with populations of more primitive cells that recall a rhabdomyosarcoma (see Rao et al., 1990b, for review). In addition, pigmented papillary forms not dissimilar to the histopathological appearance of the melanotic neuroectodermal tumour of infancy have also been identified as medulloblastomas (Russell and Rubinstein, 1989, for review). The relationship of these variants in relation to the more common types of medulloblastoma is not yet well understood.

◀ **Fig. 3.28**   Medulloblastoma. **a** Smear preparations typically demonstrate a relatively homogeneous population of small, undifferentiated cells with ill-defined cell borders (Morris stain, original magnification ×400). **b** In contrast to smears, tissue sections often demonstrate a wider variety of cellular patterns including palisading in a more fibrillated matrix and Homer Wright rosettes (Arrowhead) (Haematoxylin and eosin, original magnification ×400). **c** Two histological patterns are discernible in this medulloblastoma with the more primitive cells adjacent to a population with delicate processes forming a fibrillated matrix. Immunohistochemical studies shown in Plates 3.34 and 3.35 indicate that these cells are neuroblastic with neuronal-type intermediate filaments (Haematoxylin and eosin, original magnification ×200). **d** A midline medulloblastoma demonstrates non-homogeneous signal intensity and enhancement on a T_1 weighted magnetic resonance image following administration of intravenous gadolinium–DTPA [Axial scan: SE 0.6 s/20 ms (0.1 T)]. **e** The desmoplastic or lobular variant of the medulloblastoma is invariably characterized by two histologically distinct cell populations forming a lobulated pattern by the formation of "pale islands" (Haematoxylin and eosin, original magnification ×200)

The histogenetic potential of the medulloblastoma is complex and may reflect a heterogeneous category of tumours. Phenotypic differentiation with neuroblastic/neuronal features in over 50% of medulloblastomas (Kleihues *et al.*, 1989) suggests that this tumour may be a form of neuroblastoma. However, in contrast to cerebral neuroblastomas, photoreceptor differentiation as demonstrated by opsin and S-Ag immunoreactivity can be demonstrated in approximately one-third to one-half of the cases (Korf *et al.*, 1987; Bonnin and Perentes, 1988).

The interpretation of neoplastic astrogliogenesis is problematical. Even when strict criteria are used to eliminate the usual component of "stromal" astrocytes (Plate 3.33) (Schindler and Gullotta, 1983; Coffin *et al.*, 1983; Kleihues *et al.*, 1989), approximately 10% of cases appear to have neoplastic cells with GFAP immunoreactivity. A similar divergence of neuronal and glial phenotypic differentiation was noted using the DAOY, D-283 MED and D-341 MED cell lines established from medulloblastomas (Bigner, 1989). The reliance on immunohistochemistry or immunofluorescence to analyse phenotypic differentiation of otherwise poorly differentiated cells introduces a number of problems for interpretation, including transient or heteroplastic expression of cell type-associated proteins, sensitivity of specific epitopes to tissue processing, and variations in the antibody reagents themselves. Greater understanding of the histogenesis and classification of the embryonal tumours arising in the posterior fossa awaits further studies.

Tumours of the Meninges

Tumours of Meningothelial Cells

Approximately 13%–19% of intracranial tumours are "meningothelial" neoplasms that presumably arise from the arachnoidal cap cells on the outer surface of the arachnoidal villi and membranes, and in the stroma of the perivascular zones and choroid plexus (Kepes, 1982; Russell and Rubinstein, 1989). The neoplasms usually become clinically apparent in middle age with a clear female predominance (3 : 1), and multiple tumours occur in approximately 8% of cases (Nakasu *et al.*, 1987). In contrast to adults, neoplastic transformation of meningothelial cells during childhood has a low incidence, no sex predilection and is associated with neurofibromatosis in approximately 25% of the cases. These paediatric tumours are also more likely to exhibit malignant behaviour and generally have a higher proportion of the papillary morphological variant (Deen *et al.*, 1982). Familial cases occur primarily in association with von Recklinghausen disease (Russell and Rubinstein, 1989); however, other genetic abnormalities have been described in affected families (Smidt *et al.*, 1990). Cytogenetic studies of sporadic meningiomas have demonstrated a loss of heterozygosity for loci on chromosome 22 in 40%–80% of tumours (Zankle and Zang, 1972; Zang, 1982; Dumanski *et al.*, 1990; Poulsgard *et al.*, 1990) and the genetic defect has been localized to 22q12.3 → qter (Lekanne Deprez *et al.*, 1991).

Although meningiomas can arise at all locations of arachnoidal cells, there are preferential sites. Within the cranial cavity, the anterior half is more often involved, with neoplasms frequently in a parasagittal location, laterally over the Sylvian fissure, at the parietal eminences or near the lateral sinuses. The sphenoidal ridge, olfactory grooves, tuberculum sellae and parasellar region are frequent sites. In the posterior fossa, cerebello–pontine angle tumours are often situated on the posterior aspect of the petrous bone. In the spinal canal, meningiomas occur in the thoracic, cervical and lumbar regions in order of decreasing frequency. An interesting finding is that epidural tumours are more often found in children with a male predominance, while intradural examples demonstrate a 90% female predominance (Kepes, 1982). Meningiomas can also occur in the ventricles, arising presumably from arachnoidal cells within the tela choroidea or stroma of the choroid plexus. The left lateral ventricle is the most frequent intraventricular site while 15% arise in the third ventricle (Criscuolo and Symon, 1986). Meningiomas comprise 3% of expanding lesions of the orbit (Reese, 1976), and can also be found in the temporal bone (intrapetrous) (Nager 1964; Nager and Masica 1970; Nager et al., 1983). Various other extracalvarial and ectopic sites for these tumours have been described (Russell and Rubinstein, 1989; Burger et al., 1991).

Macroscopically, meningiomas most often present as spherical or globular growth with a nodular surface and a thin investing capsule that is readily separated from the brain or dura. A less common form is the "meningioma en plaque" – a flat, carpet-like growth which typically occurs at the skull base over the sphenoid ridge. The cut surfaces of meningiomas present a varied appearance which is partially affected by the predominant histological variant, such as a gritty texture with abundant psammoma bodies or a more fatty colour and composition with the more lipidic forms. The tumours frequently may expand to adjacent structures without infiltrative behaviour. Invasion of the adjacent dura often leads to occlusion of vascular sinuses, and erosive bone invasion may produce hyperostosis. In the cavernous sinus, involvement of the vasa vasorum or adventitia of the carotid may render it more susceptible to rupture with surgical manipulation. This expansive growth of meningiomas, however, is not associated with any particular histological variant, or with a more aggressive or malignant phenotype. Unfortunately, however, complete surgical removal is often not possible in these cases, resulting in a higher incidence of recurrence.

Neuroimaging studies are particularly important complements to the histopathological diagnosis of meningiomas, especially in those cases in which the morphological picture is unusual. Angiography often demonstrates a characteristic tumour blush corresponding to the high vascularity of these tumours and, likewise, appropriate CT and MR studies show diffuse contrast enhancement. Magnetic resonance imaging is well established in the diagnosis of meningiomas and may also be useful in partially discriminating the predominant histological pattern of the tumour, especially on T_2 weighted images where the syncytial and angiomatous components are hyperdense compared to fibroblastic variants (Elster et al., 1989). The capacity for MR imaging also to distinguish the various types of tumour–brain relationships, including well-developed collagenous capsules, CSF spaces or brain adhesion and invasion may prove useful in preoperative surgical planning.

Three histopathological categories (see Table 3.4) are defined by an increase of cytological features of malignancy and aggressive growth, such as brain invasion: (a) meningiomas, (b) atypical meningiomas, and (c) anaplastic (malignant) meningiomas. Numerous histological variations have been described (Scheithauer, 1990), reflecting the complex mesenchymal and epithelial histogenetic potential attributed to arachnoidal cap cells, and may variably compose tumours in any of these categories. The importance of these separate designations is solely in the explicit recognition of diverse histogenesis in meningothelial neoplasms. All morphological variations have a similar biological behaviour, although some may be associated with distinctive clinical features, such as polyclonal gammopathies with lymphoplasmocyte-rich tumours (Horten *et al.*, 1979) or Castleman's disease with the chordoid variant (Kepes *et al.*, 1988).

Meningioma

The hallmark "meningothelial" features common to most meningiomas are defined ultrastructurally as conspicuous membrane interdigitations, abundant intermediate filaments, the production of basement membrane and the presence of desmosomes and hemidesmosomes (Kepes, 1982). Immunoelectron microscopic studies have shown that the intermediate filaments are composed of vimentin, which can be seen attached to desmosomes (Schwechheimer *et al.*, 1984). Indeed, the coexpression of vimentin and desmosomal proteins is a characteristic unique to meningothelial cells (Kartenbeck *et al.*, 1984; Parrish *et al.*, 1987). The tendency for meningothelial cells to form whorls, frequently around vascular and other stromal elements, is a cardinal histological pattern which is most abundant in the meningothelial variant of the tumour. In these tumours, thin collagen septa, often associated with an abundant microvasculature, separate the whorled and "syncytial" arrangements without a reticulin stroma into incomplete lobules. Principal cytological features include a polygonal geometry, poorly defined cell borders and nuclei with delicate, evenly dispersed chromatin and inconspicuous nucleoli (Fig. 3.29A). Cytoplasmic invaginations are readily observed as "nuclear vacuoles".

The mesenchymal and epithelial phenotypes which compose the myriad of histological variants appear to result from an "operative" mimicry combined with a relatively broad histogenetic potential (Table 3.4). The "mesenchymal" forms include the fibrous (fibroblastic) (Fig. 3.29B), angiomatous (Fig. 3.30) and the metaplastic variants which elaborate specific extracellular matrices and a reticulin-positive stroma. The microcystic, secretory (Plate 3.38), clear cell (glycogen-rich), chordoid (Plates 3.36 and 3.37) and papillary (see Fig. 3.33 and Plate 3.40) meningiomas are all examples of "epithelial" phenotypes. The ultrastructural features of the secretory meningiomas have been extensively documented (Radley *et al.*, 1989) and include intracytoplasmic lumina with microvilli.

Immunohistochemistry confirms the admixture of epithelial and mesenchymal phenotypes. Vimentin, as the principal intermediate filament, is

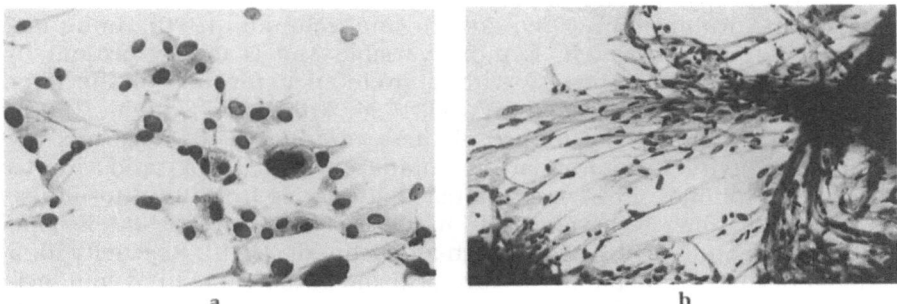

Fig. 3.29 Meningioma. **a** A smear preparation of a meningothelial meningioma shows cells with abundant cytoplasm forming broad processes with indistinct borders. The nuclei have the characteristic powdery chromatin staining and conspicuous chromatin nodes. Note the characteristic whorl formation (Morris stain, original magnification ×400). **b** Fibroblastic meningiomas contain spindle shaped cells with delicate, bipolar processes. The nuclei, however, have the chromatin pattern that is typical for meningiomas (Morris stain, original magnification ×200)

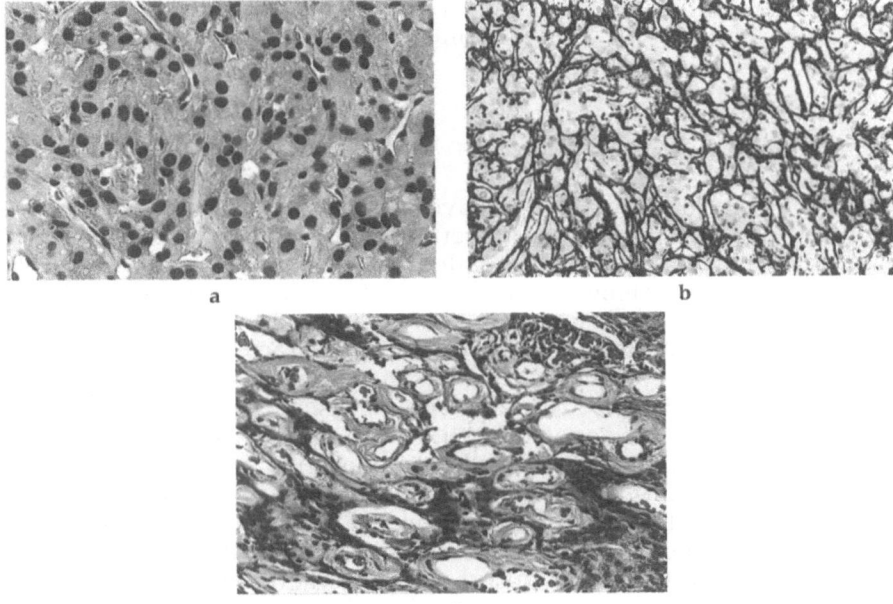

Fig. 3.30 Angiomatous meningioma. **a** The vascular elements of these highly vascular tumours have a wide spectrum of cytoarchitectural features. In this field, clusters of meningothelial cells are enmeshed in an abundant network of delicate capillaries (Haematoxylin and eosin, original magnification ×400). **b** The capillary network is well delineated by a reticulin-positive basal lamina (Reticulin stain, original magnification ×200). **c** Another angiomatous meningioma is composed primarily of thickened, hyalinized vessels which surround small groups of typical meningothelial cells (Haematoxylin and eosin, original magnification ×200)

readily demonstrated and appears to occur in close to 100% of cases (Russell and Rubinstein, 1989; Artlich and Schmidt, 1990). Immunoreactivity for S-100 protein is more variable and is focally present in approximately 50% of cases, with a preference for fibroblastic over meningothelial areas (Schnitt and Vogel, 1986; Artlich and Schmidt, 1990). In contrast, the epithelial nature of these tumours is demonstrated by immunoreactivity for epithelial membrane antigen (EMA) and various cytokeratin epitopes. EMA immunoreactivity varies from focal to diffuse with a reported incidence ranging from 50% to 100% of cases (Meis *et al.*, 1986; Schnitt and Vogel, 1986; Artlich and Schmidt, 1990). Reactivity for a variety of cytokeratin epitopes is also focal and varies from an overall incidence of 20% to a peak of nearly 40% in some epithelial variants (Meis *et al.*, 1986; Theaker *et al.*, 1986; Artlich and Schmidt, 1990). Cytokeratin immunoreactivity is usually most common in either meningothelial and secretory variants, often associated with cells producing hyaline bodies (Winek *et al.*, 1989) and absent in cells expressing vimentin. GFAP is invariably undetectable by immunohistochemistry.

A number of extracellular matrix proteins, including laminin, fibronectin and collagen types IV and V have been detected, with laminin expression more prevalent in fibroblastic types (McComb and Bigner, 1985; Bellon *et al.*, 1985; Rutka *et al.*, 1986). Meningiomas are invariably negative for GFAP.

Atypical Meningioma

Several large retrospective studies have validated the concept of a subpopulation of meningiomas, without overt evidence of malignant transformation, which have an increased tendency for recurrence and aggressive clinical behaviour (Jellinger and Slowick, 1975; de la Monte *et al.*, 1986; Jääskeläinen *et al.*, 1986; Jellinger, 1989; Chen and Liu, 1990). The most important histopathological features which characterize these atypical tumours include hypercellularity with increased mitotic activity, more diffuse or sheet-like growth, increased nuclear pleomorphism with nucleolar prominence, and micronecrosis. These features may be identified in any histological variant without specific predilections (Fig. 3.31) and are usually not uniform throughout the tumour. These considerations emphasize the necessity for careful examination of multiple tissue sections for appropriate evaluation of atypical tumours. Additional features, such as focal papillary patterns, may also be significant (Chen and Liu, 1990). Despite the limitation of interpreting cellularity and precise histoarchitectural patterns of tumours with intraoperative smears, there is a good correlation between the intraoperative and final pathological diagnosis of atypical meningioma.

One of the most important parameters for the designation of atypical meningiomas as operationally defined in terms of recurrence is a quantitative assessment of mitotic activity beyond routine histological evaluation. Studies of tumour cell proliferation, based on the labelling index with the immunohistochemical uptake of BUdR and flow cytometric evaluation, are

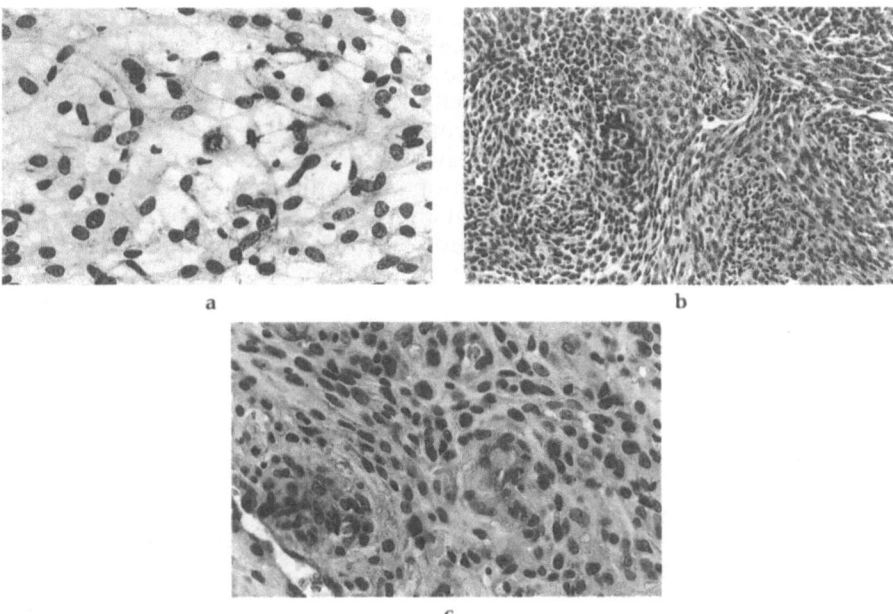

Fig. 3.31 Atypical meningioma. **a** Abundant, atypical mitotic figures on smear preparations are strong evidence for an atypical meningioma (Morris stain, original magnification ×400). **b** The appearance of the smear preparation corresponds to the increased cellularity, prominent nucleoli and necrosis in paraffin-embedded tissue (Haematoxylin and eosin, original magnification ×200). **c** At higher magnification, numerous mitotic figures and significant nuclear pleomorphism are evident (Haematoxylin and eosin, original magnification ×400)

important for identifying recurrent tumours (Cho *et al.*, 1986; May *et al.*, 1989; Lee *et al.*, 1990) and a labelling index of >1% appears to distinguish atypical meningiomas (Jellinger, 1989; Lee *et al.*, 1990). Additionally, the AgNOR technique provides a direct correlation to the LI (Orita *et al.*, 1990) and also appears to distinguish tumours with an intermediate potential for recurrence, analogous to the designation of atypical meningiomas (Chin and Hinton, 1991).

Anaplastic (Malignant) Meningioma

There are no specific histopathological features which can qualitatively discriminate between atypical meningiomas and tumours designated as anaplastic or malignant. Rather, the intensity of some histopathological features, including a high mitotic index, marked cellular and nuclear pleomorphism and geographic necrosis are regarded as indicative of malignancy. However, the assessment of malignant potential cannot always be based solely on these atypical features. Significant brain invasion, as indicated by transgression of the pial membrane with unequivocal infiltration and entrapment of brain parenchyma, may be regarded as

sufficient evidence for malignant behaviour in the absence of severe cytological anaplasia (Fig. 3.32; Plate 3.39). Recurrent and invasive meningiomas may also undergo karyotypic evolution in addition to monosomy or
deletions of chromosome 22 (Casartelli *et al.*, 1989; Poulsgard *et al.*, 1990).
These observations suggest that karyotypic analysis, in addition to quantitation of the proliferative activity, may be useful to identify anaplastic progression and malignant transformation in meningiomas. Although c-sis
expression may be elevated, PDGF-like growth factor may be secreted and
β-PDGF receptors have been immunohistochemically documented in
meningiomas (Kazumoto *et al.*, 1990; Smidt *et al.*, 1990; Mauro *et al.*,
1991), the identification of aberrant growth factor and/or receptor
expression as an indication of malignant clinical behaviour has not been
demonstrated.

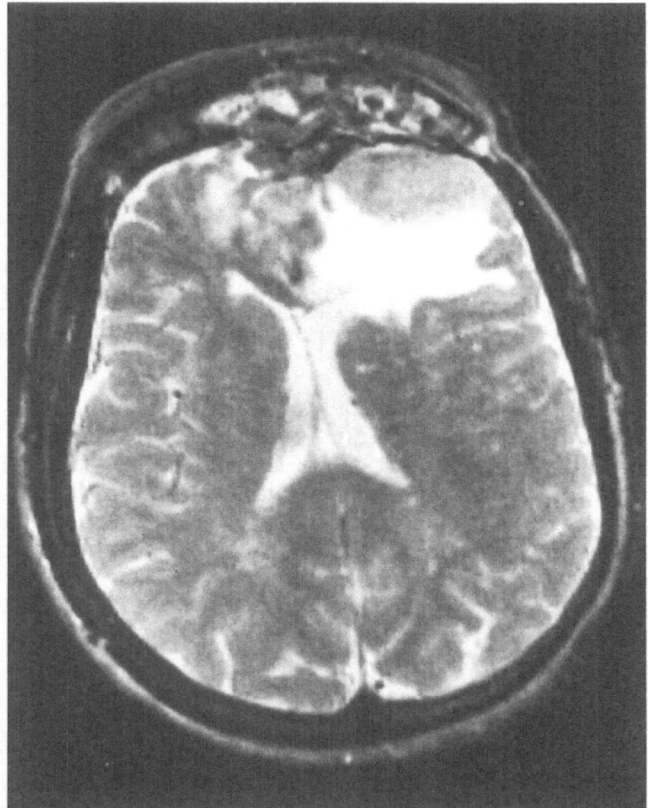

Fig. 3.32 Malignant meningioma. A T_2 weighted magnetic resonance image demonstrates
extensive circumferential oedema in a tumour with histologically documented brain invasion. [Axial scan: SE 2.5 s/90 ms (1.0 T)] (See Plate 3.39)

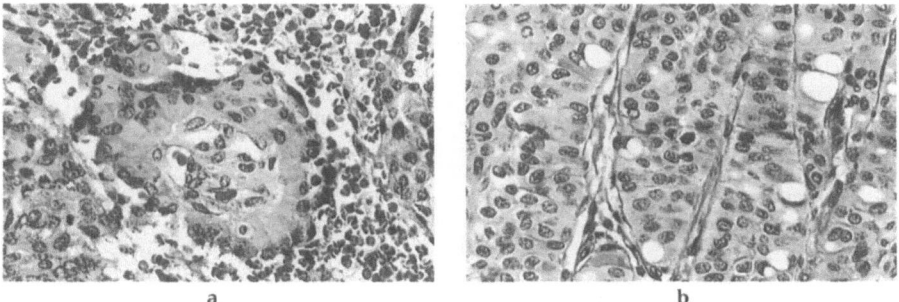

a   b

Fig. 3.33 Papillary meningioma. **a** The tumour cells commonly form a papillary structure around vascular elements (Haematoxylin and eosin, original magnification ×400). **b** Columnar cells line up in parallel arrays between the intervening vascular channels in another papillary meningioma. Note the mitotic figure, a frequent feature in these invariably aggressive tumours (Haematoxylin and eosin, original magnification ×400)

Papillary Meningioma

This malignant variant of meningioma is more commonly found in children and young adults with a tendency to metastasize (Ludwin *et al.*, 1975). These tumours are rare, with only 46 examples reported in the literature to date (Pasquier *et al.*, 1986). Papillary meningiomas demonstrate an epithelial architecture in which the cells range from cuboidal to columnar geometry with usually well-defined cell borders. These columnar cells often contain processes and fibrils with a remarkable affinity to PTAH (Plate 3.40). This corresponds to the ultrastructural demonstration of bundles of 10 nm filaments, which are most likely composed of vimentin (Piatt *et al.*, 1986). The tumour cells commonly form papillary structures around blood vessels (Fig. 3.33) and can also form epithelial ribbons which merge into more solid sheets, sometimes simulating an adenocarcinomatous growth (Kepes *et al.*, 1983). The papillary patterns of the perivascular arrangements are accentuated by vascular hyalinization and separation of tumour cells to produce peripapillary spaces. High cellularity and mitotic activity with evidence of brain invasion are often encountered.

Mesenchymal, Non-meningothelial Tumours

A number of mesenchymal neoplasms may occasionally arise in the meninges (Table 3.5). Generally, these tumours have similar features to their counterparts outside the CNS. The more common benign tumours in this category include chondroma, osteochondroma, osteoma, lipoma and fibrous histiocytoma. The most commonly encountered sarcomas include haemangiopericytoma, chondrosarcoma, malignant fibrous histiocytoma and rhabdomyosarcoma. From the perspective of changes in the WHO classification, it is particularly relevant to note that the haemangio-

Table 3.5   Non-meningothelial tumours of the meninges

Mesenchymal tumours	Tumours of uncertain histogenesis
1. Benign neoplasms	Haemangioblastoma (capillary
Osteocartilagenous tumours	haemangioblastoma)
Lipoma	
Fibrous histiocytoma	
2. Malignant neoplasms	
Haemangiopericytoma	
Chondrosarcoma	
Mesenchymal chondrosarcoma	
Malignant fibrous histiocytoma	
Rhabdomyosarcoma	
Meningeal sarcomatosis	

pericytoma has been removed from tumours of meningothelial origin (in the context of angioblastic meningiomas) and is now classified as a non-meningothelial, mesenchymal tumour. Therefore only this tumour will be described in detail.

Haemangiopericytoma

Haemangiopericytomas of the meninges have presented a persistent noso-logical problem among neuropathologists, but are currently considered to be mesenchymal tumours arising in the leptomeninges. These tumours are considered by some investigators to be identical to haemangiopericytomas of the soft tissue found elsewhere in the body (Scheithauer, 1990; Burger *et al.*, 1991; Jellinger and Paulus, 1991), and by others as "angioblastic meningiomas", which share a common origin with other meningiomas from multipotential mesenchymal cells originating in the meninges (Horten *et al.*, 1977; Russell and Rubinstein, 1989). Meningiomas in which papillary and haemangiopericytic patterns occur within the same tumour provide evidence for this point of view (Horten *et al.*, 1977; Russell and Rubinstein, 1989). The extensive evidence regarding this issue is beyond the scope of this review (Holden *et al.*, 1987; Nakamura *et al.*, 1987; Winek *et al.*, 1989; D'Amore *et al.*, 1990; Theunissen *et al.*, 1990); however, there is general agreement on the histological features and aggressive clinical behaviour of these tumours.

Haemangiopericytomas represent between 1% and 7% of all meningeal tumours and have an aggressive behaviour, as evidenced by a rate of recur-rence between 20% and 80% with 10%–55% developing metastases. Overall survival is between 58 and 84 months (Jellinger *et al.*, 1991; Mena *et al.*, 1991). These tumours usually occur in adults and the majority of cases are males (Mena *et al.*, 1991). Histologically, these tumours are iden-tical to those found in the soft tissues and are characterized by high cellu-larity, indistinct cytoplasmic borders, and a haphazard arrangement of pleomorphic nuclei with a high mitotic activity (Fig. 3.34b). Many areas show an abundance of slit-like vascular channels in a "staghorn" pattern, although the occurrence of this phenomenon is variable. Likewise, reticu-

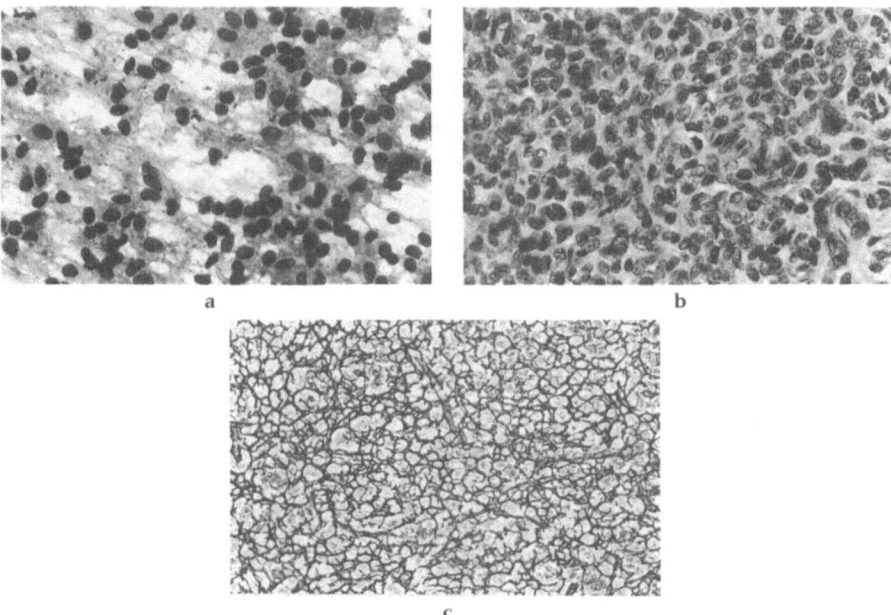

Fig. 3.34 Haemangiopericytoma. **a** The smear preparations of these tumours highlight the more discrete cell borders in comparison to meningiomas. The nuclei are round to oval with a coarsely stippled and more dense chromatin pattern than is seen in meningiomas (Haematoxylin and eosin, original magnification ×400). **b** The histological section of the same tumour as in A shows identical histological features as soft-tissue hemangiopericytoma (Haematoxylin and eosin, original magnification ×400). **c** Reticulin may be variably demonstrated around individual cells (Reticulin stain, original magnification ×200)

lin stains demonstrate a variable pattern, although the majority of these tumours will contain reticulin-positive areas, with reticulin fibres often surrounding individual cells (Fig. 3.34c). Intraoperative smears of meningeal haemangiopericytomas readily highlight the distinct cell borders and the absence of syncytial whorling patterns, in comparison to meningiomas (Fig. 3.34a). Nuclei are hyperchromatic with a round to oval shape and indistinct nucleoli.

Tumours of Uncertain Histogenesis

Haemangioblastoma

Haemangioblastomas usually arise below the tentorium, with the majority over the cerebellum and less common cases at the brainstem and spinal cord (Myers *et al.*, 1961; Nakamura *et al.*, 1985; Sanford and Smith, 1986). Supratentorial examples are rare, and invariably associated with von Hippel–Lindau disease. They occur most often in the fourth or fifth decades, and are most common in males (Palmer, 1972). The tumours commonly present with increased intracranial pressure (Palmer, 1972;

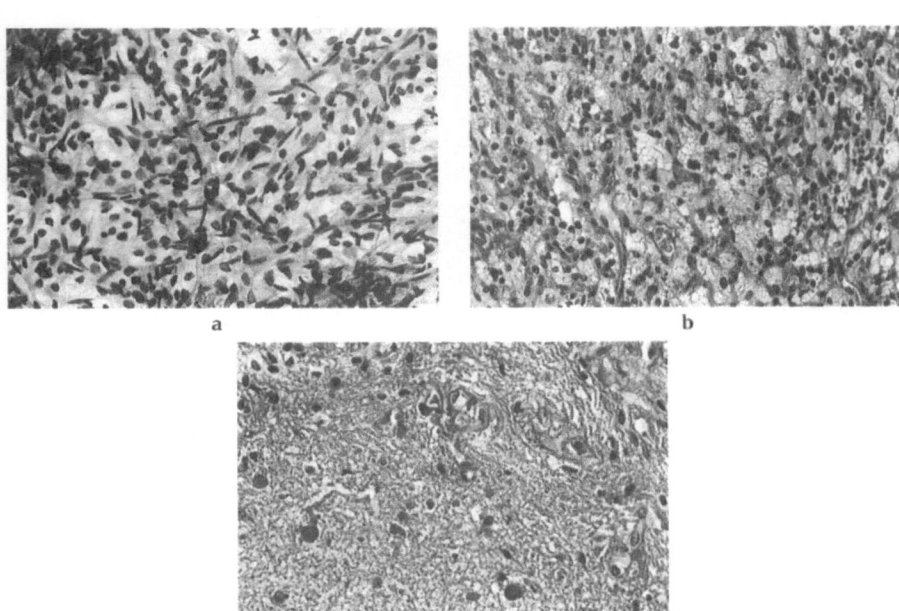

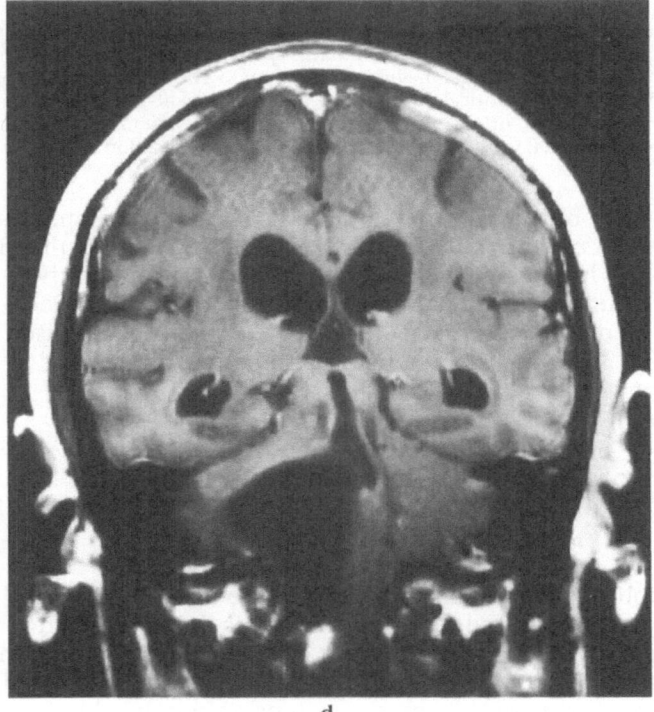

Neumann *et al.*, 1989) and the radiographic features of a cystic lesion with an enhancing mural nodule (Fig. 3.35d). Haemangioblastomas are predominantly discrete masses, although some infiltration associated with abnormal vessels may occur in the adjacent parenchyma. Abundant reactive astrocytosis with Rosenthal fibre formation is frequent (Fig. 3.35c) and may mislead the diagnosis, if this marginal tissue is submitted for intraoperative smear. In addition, the stromal cells may demonstrate variable immunoreactivity for GFAP, a feature which appears only with the intraparenchymal lesions. In contrast, rare extradural tumours which may occur in the spinal region demonstrate no GFAP immunoreactivity, suggesting that stromal cell GFAP may be non-specifically taken up from degenerating processes of adjacent, reactive astrocytes.

The tumours are relatively cellular and composed of heterogeneous cell populations – endothelial cells and pericytes admixed with commonly lipid-laden, NSE-immunoreactive stromal cells (Fig. 3.35b). The general histopathological architecture may superficially resemble renal cell carcinomas, but the lack of stromal cell immunoreactivity for EMA serves to distinguish haemangioblastomas (Hufnagel *et al.*, 1989; Mills *et al.*, 1990). Frozen sections often impart a considerable amount of distortion to the appearance of these tumours (Burger *et al.*, 1991) and intraoperative smears thus offer an important alternative. The smear appearance is characteristic and demonstrates an abundance of thin-walled capillaries with small to plump endothelial cells admixed with stromal elements (Fig. 3.35a). The latter have an abundant, pale eosinophilic cytoplasm and small hyperchromatic nuclei. The histological pattern combined with the distinctive cytological features and radiographic data assure a correct intraoperative diagnosis.

◀ **Fig. 3.35** Haemangioblastoma. **a** A smear preparation clearly demonstrates a dual population of tumour cells. The majority are endothelial cells with plump oval to fusiform nuclei and finely dispersed chromatin. The stromal cells have small, hyperchromatic nuclei and abundant, pale cytoplasm (Morris stain, original magnification ×300). **b** The smear appearance corresponds to the typical pattern of stromal cells with foamy cytoplasm admixed with multiple vascular channels in tissue sections (Haematoxylin and eosin, original magnification ×300). **c** The adjacent cerebellar cortex has a marked astrocytosis with Rosenthal fibres (Haematoxylin and eosin, original magnification ×400). **d** A large cyst with an enhancing mural nodule are typical features on a T_1 weighted magnetic resonance image following administration of intravenous gadolinium–DTPA [Coronal scan: SE 0.6 s/20 ms (1.5 T)]

References

Abe K, Hasegawa H, Kobayashi Y *et al.* (1990) A gemistocytic astrocytoma demonstrated high intensity on MR images. Neuroradiology 32: 166–167

Albrecht S, Rouah E, Becker LE, Bruner J (1991) Immunoreactivity for transthyretin in choroid plexus neoplasms and brain metastases. J Neuropathol Exp Neurol 50: 366

Allegranza A, Pileri S, Frank G, Ferracini R (1990) Cerebral ganglioglioma with anaplastic oligodendroglial component. Histopathology 17: 439–441

Ariza A, Fernandez LA, Inagami T, Kim JH, Manuelidis EE (1988) Renin in glioblastoma multiforme and its role in neovascularization. Am J Clin Pathol 90: 437–441

Artlich A, Schmidt D (1990) Immunohistochemical profile of meningiomas and their historical subtypes. Hum Pathol 21: 843–849

Ashley DG, Zee C-S, Chandrasoma PT, Segall HD (1990) Lhermitte–Duclos disease: CT and MR findings. J Comp Assist Tomogr 14: 984–987

Azzarelli B, Rekate HL, Roessman U (1977) Subependymoma: a case report with ultrastructural study. Acta Neuropathol 40: 279–282

Bader JL, Meadows AT, Zimmerman LE *et al.* (1982) Bilateral retinoblastoma with ectopic intracranial retinoblastoma: trilateral retinoblastoma. Cancer Genet Cytogenet 5: 203–213

Bailey P, Beiser H (1947) Concerning gangliogliomas of the brain. J Neuropathol Exp Neurol 6: 24–37

Baker DL, Reddy UR, Pleasure S *et al.* (1990) Human central nervous system primitive neuroectodermal tumor expressing nerve growth factor receptors: CHP707m. Ann Neurol 28: 136–145

Bancel B, Belin MF, Meiniel A *et al.* (1990) Contribution à l'étude de l'histogenèse des gliomes sous-épendymaires de la sclérose tubéreuse de bourneville. Ann Pathol 10: 109–116

Banerjee AK, Kak VK (1974) Pineoblastoma with spontaneous intra and extracranial metastasis. J Pathol 114: 9–12

Banerjee AK, Gleathill CA (1979) Lhermitte–Duclos disease (diffuse cerebellar hypertrophy): prolonged post-operative survival. Ir J Med Sci 148: 97–99

Barbosa MD, Balsitis M, Jaspan T, Lowe J (1990) Intraventricular neurocytoma: a clinical and pathological study of three cases and review of the literature. Neurosurgery 26: 1045–1054

Barnard RO, Bradford R, Scott T, Thomas DGT (1986) Gliomyosarcoma. Report of a case of rhabdomyosarcoma arising in a malignant glioma. Acta Neuropathol 69: 23–27

Beahr T, Movotny E, Barker J, McMorris SA, Dubois-Dalacq M. (1987) O-2A progenitor cells sorted from postnatal rat brain closely resemble their optic nerve counterpart. J Cell Biol 105: 318A

Bellon G, Caulet T, Cam Y *et al.* (1985) Immunohistochemical localization of macromolecules of the basement membrane and extracellular matrix of human gliomas and meningiomas. Acta Neuropathol 66: 245–252

Bennett JP Jr, Rubinstein LJ (1984) The biologic behavior of primary cerebral neuroblastoma: a reappraisal of the clinical course in a series of 70 cases. Ann Neurol 16: 21–27

Berger MS, Edwards MSB, Wilson CB (1983) Primary cerebral neuroblastoma. Follow-up analysis of eleven cases. In: Concepts in Pediatric Surgery, Vol 3. American Society for Pediatric Neurosurgery, S, Karger, Basel, pp 35–43

Betsholtz C, Nister M, Rorsman F, Heldin CH, Westermark B (1989) Structural and functional aspects of platelet-derived growth factor and its role in the pathogenesis of glioblastoma. Mol Chem Neuropathol 10: 27–36

Beyer C, Epp B, Fassberg J, Reisert I, Pilgrim C (1990) Region-and sex-related differences in maturation of astrocytes in dissociated cell cultures of embryonic rat brain. Glia 3: 55–64

Bigner DD, Bigner SH, Pontén J, Westermark B *et al.* (1981) Heterogeneity of genotypic and phenotype characteristics of fifteen permanent cell lines derived from human gliomas. J Neuropathol Exp Neurol 40: 201–229

Bigner SH, Wong AJ, Mark J *et al.* (1987) Relationship between gene amplification and chromosomal deviations in malignant human gliomas. Cancer Genet Cytogenet 29: 165–170

Bigner SH, Vogelstein B, Bigner DD (1988a) Chromosomal abnormalities and gene amplification in malignant gliomas. ISI Atlas of Sci: Biochem 1: 333–336

Bigner SH, Mark J, Friedman HS, Biegel JA, Bigner DD (1988b) Structural chromosomal abnormalities in human medulloblastoma. Cancer Genet Cytogenet 30: 91–101

Bigner DD (1989) Phenotypic analysis of medulloblastoma with monoclonal antibodies. In: Fields WS (ed) Primary Brain Tumours. A Review of Histologic Classification, Vol. 5. Springer Verlag, New York, pp 70–78

Bigner SH, Mark J, Bigner DD (1990) Cytogenetics of human brain tumours. Cancer Genet Cytogenet 47: 141–154

Bigner SH, Vogelstein B (1990) Cytogenetics and molecular genetics of malignant gliomas and medulloblastoma. Brain Pathol 1: 12–18

Bishop M, de la Monte SM (1989) Dual lineage of astrocytomas. Am J Pathol 135: 517–527

Bonnin JM, Colon LE, Morawetz RB (1987) Focal glial differentiation and oncocytic transformation in choroid plexus papilloma. Acta Neuropathol 72: 277–280

Bonnin JM, Perentes E (1988) Retinal S-antigen immunoreactivity in medulloblastomas. Acta Neuropathol 76: 204–207

Bonnin JM, Rubinstein LJ (1989) Astroblastomas: a pathological study of 23 tumors, with a postoperative follow-up in 13 patients. Neurosurg 25: 6–13

Borit A, Blackwood W, Mair WGP (1980) The separation of pineocytoma from pineoblastoma. Cancer 45: 1408–1418

Bunt-Milan AH, Saari JC (1983) Immunocytochemical localization of two retinoid-binding proteins in vertebrate retina. J Cell Biol 97: 703–712

Burgart LJ, Robinson RA, Haddad SF, Moore SA (1991) Oncogene abnormalities in astrocytomas: EGF-R gene alone appears to be more frequently amplified and rearranged compared with other protooncogenes. Mod Pathol 4: 183–186

Burger PC (1985) Use of cytological preparations in the frozen section diagnosis of central nervous system neoplasia. Am J Surg Pathol 5: 344–354

Burger PC, Vogel FS, Green SB, Strike TA (1985) Glioblastoma multiforme and anaplastic astrocytoma. Pathologic criteria and prognostic implications. Cancer 56: 1106–1111

Burger PC, Shibata T, Kleihues P (1986) The use of the monoclonal antibody Ki-67 in the identification of proliferating cells: application of surgical neuropathology. Am J Surg Pathol 10: 611–617

Burger PC, Rawlings CE, Cox EB et al. (1987a) Clinicopathologic correlations in the oligodendroglioma. Cancer 59: 1345–1352

Burger PC, Grahmann FC, Bliestle A, Kleihues P (1987b) Differentiation in the medulloblastoma. A histological and immunohistochemical study. Acta Neuropathol 73: 115–123

Burger PC (1989) The grading of astrocytomas and oligodendrogliomas. In: Fields WS (ed) Primary Brain Tumors. A Review of Histologic Classification, Vol.13. Springer Verlag, New York, pp 171–180

Burger PC, Scheithauer BW, Vogel FS (1991) Surgical Pathology of the Nervous System and Its Coverings, 3rd ed. Churchill Livingstone, New York

Caccamo DV, Herman MM, Rubinstein LJ (1989) An immunohistochemical study of the primitive and maturing elements of human cerebral medulloepitheliomas. Acta Neuropathol 79: 248–254

Campbell M, Chader GJ (1988) Retinoblastoma cells in tissue culture. Ophthal Paed Genetics 9: 171–199

Cardozo J, Cepeda F, Quintero M, Mora E (1985) Choroid plexus papilloma containing bone. Acta Neuropathol 68: 83–85

Carter LP, Beggs J, Waggener JD (1972) Ultrastructure of three choroid plexus papillomas. Cancer 30: 1130–1136

Casartelli C, Rogatto SR, Barbieri Neto J (1989) Karyotypic evolution of human meningioma — progression through malignancy. Cancer Genet Cytogenet 40: 33–45

Chamak BA, Fellows J, Glowinski J, Prochiantz A (1987) MAP2 expression and neurite outgrowth and branching are co-regulated through region-specific neuroastroglial interactions. J Neurosci 7: 3163–3170

Chandrasome PT and Apuzzo MLJ (1989) Stereotactic Brain Biopsy, Igaku-Shoin, New York

Changaris DG, Powers JM, Perot PL, Hungerford GD, Neal GB (1981) Subependymoma presenting as subarachnoid hemorrhage: case report. J Neurosurg 55: 643–645

Chen WY, Liu HC (1990) Atypical (anaplastic) meningioma: relationship between histologic features and recurrence – a clinicopathologic study. Clin Neuropathol 9: 74–81

Chin LS, Hinton DR (1991) The standardized assessment of argyrophilic nucleolar organizer regions in meningeal tumors. J Neurosurg 74: 590–596

Cho KG, Hoshino T, Nagashima T, Murovic JA, Wilson CB (1986) Prediction of tumor-doubling time in recurrent meningiomas: cell kinetic studies with bromodeoxyuridine labeling. J Neurosurg 65: 790–794

Choi BH, Kim RC (1985) Expression of glial fibrillary acidic protein by immature oligodendroglia and it implications. J Neuroimmunol 8: 215–235

Cockram CS (1990) Growth factors, astrocytes and astrocytomas. Sem Dev Biol 1: 421–435

Codd MB, Kurland LT (1985) Descriptive epidemiology of primary intracranial neoplasms. Prog Exp Tumor Res 29: 1–11

Coffin CM, Mukai K, Dehner LP (1983) Glial differentiation in medulloblastomas. Histogenetic insight, glial reaction, or invasion of brain? Am J Surg Pathol 7: 555–565

Collins VP (1987) Pineocytoma with neuronal differentiation demonstrated immunocytochemically. Acta Pathol, Microbiol Immunol Scand (Sect. A) 95: 113–117

Cosgrove M, Fitzgibbons PL, Sherrod A et al. (1989) Intermediate filament expression in astrocytic neoplasms. Am J Surg Pathol 13: 141–145

Courville CB (1930) Gangliogliomas; tumor of the central nervous system. Review of the literature and report of 2 cases. Arch Neurol Psychiat 24: 439–491

Courville CB (1931) Gangliogliomas. Further report with special reference to those occurring in the temporal lobe. Arch Neurol Psychiat 25: 309–326

Courville CB, Anderson FF (1941) Neuro-gliogenic tumors of the central nervous system. Bull Los Angeles Neurol Soc 6: 154–176

Criscuolo GR, Symon L (1986) Intraventricular meningioma. A review of 10 cases of the National Hospital, Queen Square (1974–1985) with reference to the literature. Acta Neuropathol 83: 83–91

Cruz-Sanchez FF, Haustein J, Rossie ML, Cervos-Navarro J, Hughs JT (1988) Ependymoblastoma: a histological, immunohistochemical and ultrastructural study of five cases. Histopathology 12: 17–27

Daita G, Yonemasu Y, Muraoka S, Nakai H, Maeda T (1991) A case of anaplastic astrocytoma transformed from pleomorphic xanthoastrocytoma. Brain Tumor Pathol 8: 63–66

D'Amore ESG, Manivel JC, Sung JH (1990) Soft-tissue and meningeal hemangiopericytomas: an immunohistochemical and ultrastructural study. Hum Pathol 21: 414–423

D'Andrea AD, Packer RJ, Rorke LB et al. (1987) Pineocytomas of childhood: a reappraisal of natural history and response to therapy. Cancer 59: 1353–1357

Daumas-Duport C, Scheithauer B, O'Fallon J, Kelly P (1988a) Grading of astrocytomas. A simple and reproducible method. Cancer 62: 2152–2165

Daumas-Duport C, Scheithauer B, Chodkiewicz J-P, Laws ER, Vedrenne C (1988b) Dysembryoplastic neuroepithelial tumor: a surgically curable tumor of young patients with intractable partial seizures: report of thirty-nine cases. Neurosurgery 23: 545–556

Davis RL (1989) Grading of gliomas. In: Fields WS (ed) Primary Brain Tumors. A Review of Histologic Classification, Vol.11. Springer Verlag, New York, pp 150–158

De Chadarévian JP, Hollenberg RD (1979) Subependymal giant cell tumor of tuberose sclerosis. A light and ultrastructural study. J Neuropathol Exp Neurol 38: 419–433

De Chadarévian J-P, Pattisapu JV, Faerber EN (1990) Desmoplastic cerebral astrocytoma of infancy. Light microscopy, immunocytochemistry, and ultrastructure. Cancer 66: 173–179

de la Monte SM, Flickinger J, Linggood RM (1986) Histopathologic features predicting recurrence of meningiomas following subtotal resection. Am J Surg Pathol 10: 836–843

de la Monte SM (1989) Uniform lineage of oligodendrogliomas. Am J Pathol 153: 529–540

Dean BL, Drayer BP, Bird CR et al. (1990) Gliomas: classification with MR imaging. Radiol 174: 411–415

DeClerck YA, Lee C (1985) Collagen synthesis by human neuroblastoma cells. In: Evans AE, D'Angio GJ, Seeger RC (eds) Advances in Neuroblastoma Research. Alan R Liss, New York, pp 239–247

Deen HG Jr, Scheithauer BW, Ebersold MJ (1982) Clinical and pathological study of meningiomas of the first two decades of life. J Neurosurg 56: 317–322

Dehner LP, Abenoza P, Sibley RK (1988) Primary cerebral neuroectodermal tumors: neuroblastoma, differentiated neuroblastoma, and composite neuroectodermal tumor. Ultrastr Pathol 12: 479–494

Denis-Donini S, Glowinski J, Prochiantz A (1984) Glial heterogeneity may define the three-dimensional shape of mouse mesencephalic dopaminergic neurons. Nature 307: 641–643

Detrick B, Chader GJ, Rodrigues M et al. (1988) Coexpression of neuronal, glial, and major histocompatibility complex Class II antigens on retinoblastoma cells. Cancer Res 48: 1633–1641

Disclafani A, Hudgins RJ, Edwards MS et al. (1989) Pineocytomas. Cancer 63: 302–304

Dropcho EJ, Wisoff JH, Walker RW, Allen JC (1987) Supratentorial malignant gliomas in childhood: a review of fifty cases. Ann Neurol 22: 355–364

Dumanski JP, Rouleau GA, Nordenskjold M, Collins VP (1990) Molecular genetic analysis of chromosome 22 in 81 cases of meningioma. Cancer Res 50: 5863–5867

Edwards MSB, Hudgins RJ, Wilson CB, Levin VA, Wara WM (1988) Pineal region tumors in children. J Neurosurg 68: 689–697

Ekstrand AJ, James CD, Cavenee WK *et al.* (1991) Genes for epidermal growth factor receptor, transforming growth factor alpha, and epidermal growth factor and their expression in human gliomas in vivo. Cancer Res 51: 2164–2172

Elster AD, Challa, VR, Gilbert TH, Richardson DN, Contento JC (1989) Meningiomas: MR and histopathologic features. Radiology 170: 857–862

Faillot T, Sichez J-P, Brault, J-L *et al.* (1990) Lhermitte-Duclos disease (dysplastic gangliocytoma of the cerebellum). Report of a case and review of the literature. Acta Neurochir 105: 44–49

ffrench-Constant C, Raff MC (1986) Proliferating bipotential glial progenitor cells in adult rat optic nerve. Nature 319: 499–502

Fried RL (1978) Gliofibroma. A peculiar neoplasia of collagen forming glia-like cells. J Neuropathol Exp Neurol 37: 300–313

Friedberg E, Katsetos CD, Reidy J *et al.* (1991) Immunolocalization of protease inhibitors α-1-antitrypsin and α-1-antichymotrypsin in eosinophilic granular bodies of cerebral juvenile pilocytic astrocytomas. J Neuropathol Exp Neurol 50: 293

Fu Y-S, Chen ATL, Kay S, Young HF (1974) Is subependymoma (subependymal glomerate astrocytoma) an astrocytoma or ependymoma? A comparative ultrastructural and tissue culture study. Cancer 34: 1992–2008

Fults D, Pedone CA, Thomas GA, White R (1990) Allelotype of human malignant astrocytoma. Cancer Res 50: 5784–5789

Gallie BL, Squire JA, Goddard A *et al.* (1990) Biology of disease. Mechanism of oncogenesis in retinoblastoma. Lab Invest 62: 394–408

Gambarelli D, Hassoun J, Choux M, Toga M (1982) Complex cerebral tumor with evidence of neuronal, glial and Schwann cell differentiation: a histologic immunocytochemical and ultrastructural study. Cancer 49: 1420–1428

Gennett IN, Cavenee WK (1990) Molecular genetics in the pathology and diagnosis of retinoblastoma. Brain Pathol 1: 25–32

Gherardi R, Baudrimont M, Nguyen JP *et al.* (1986) Monstrocellular heavily lipidized malignant glioma. Acta Neuropathol 69: 28–32

Giangaspero F, Doglioni C, Rivano MT *et al.* (1987) Growth factor in human brain tumors defined by the monoclonal antibody Ki-67. Acta Neuropathol 74: 179–182

Goldstein GW (1988) Endothelial cell-astrocyte interactions. A cellular model of the blood-brain barrier. Ann NY Acad Sci 529: 31–39

Gonzalez-Fernandez F, Lopes MBS, Garcia-Fernandez JM *et al.* (1992) Expression of developmentally-defined retinal phenotypes in the histogenesis of retinoblastoma. Am J Pathol (in press)

Gould VE, Rorke LB, Jansson DS *et al.* (1990) Primitive neuroectodermal tumors of the central nervous system express neuroendocrine markers and may express all classes of intermediate filaments. Hum Pathol 21: 245–252

Grant JW, Steart PV, Aguzzi A, Jones DB, Gallagher PJ (1989) Gliosarcoma: an immunohistochemical study. Acta Neuropathol 79: 305–309

Haddad SF, Moore SA, Schelper RL, Goeken J (1991) Smooth muscle cells can comprise the sarcomatous component of gliosarcomas. J Neuropathol Exp Neurol 50: 291

Halmagyi GM, Bignold LP, Allsop JL (1979) Recurrent subependymal giant-cell astrocytoma in the absence of tuberous sclerosis. J Neurosurg 50: 106–109

Hansson E (1988) Astroglia from defined brain regions as studied with primary cultures. Prog Neurobiol 30: 369–397

Hassoun J, Gambarelli D, Grisoli F *et al.* (1982) Central neurocytoma: an electron microscopic study of two cases. Acta Neuropathol 56: 151–156

Hassoun J, Gambarelli D, Peragut JC, Toga M (1983) Specific ultrastructural markers of human pinealomas: a study of four cases. Acta Neuropathol 62: 31–40

Hassoun J, Devictor B, Gambarelli D *et al.* (1984) Paired twisted filaments: a new ultrastructural marker of human pinealomas? Acta Neuropathol 65: 163–165

Hassoun J, Gambarelli D (1989) Pinealomas: need for an ultrastructural diagnosis. In: Fields WS (ed) Primary Brain Tumors. A Review of Histologic Classification. Springer Verlag, New York, pp 82–85

Herpers MJHM, Budka H (1984) Glial fibrillary acidic protein (GFAP) in oligodendroglial tumors: gliofibrillary oligodendroglioma and transitional oligoastrocytoma as subtypes of oligodendroglioma. Acta Neuropathol 64: 265–272

Herrick MK, Rubinstein LJ (1979) The cytological differentiating potential of pineal parenchymal neoplasms (true pinealomas): a clinicopathologic study of 28 tumours. Brain 102: 289–320

Hessler RB, Lopes MBS, Frankfurter A, Katsetos CD, VandenBerg SR (1991) Cytoskeletal immunohistochemistry of central neurocytomas. J Neuropathol Exp Neurol 50: 363

Ho K-L (1990) Histogenesis of sarcomatous component of the gliosarcoma: an ultrastructural study. Acta Neuropathol 81: 178–188

Holden J, Dolman CL, Churg A (1987) Immunohistochemistry of meningiomas including the angioblastic type. J Neuropathol Exp Neurol 46: 50–56

Holt CD, Bertsch TW, Ellis HM (1988) Cellular determination in the Xenopus retina is independent of lineage and birth date. Neuron 1:15–25

Horten BC, Rubinstein LJ (1976) Primary cerebral neuroblastoma. a clinicopathologic study of 35 cases. Brain 99: 735–756

Horten BC, Urich H, Rubinstein LJ, Montague SR (1977) The angioblastic meningioma: a reappraisal of a nosological problem. J Neurol Sci 31: 387–410

Horten BC, Urich H, Stefoski D (1979) Meningiomas with conspicuous plasma cell-lymphocytic components. Cancer 43: 258–264

Hoshino T, Prados M, Wilson CB et al. (1989) Prognostic implications of the bromodeoxyuridine labeling index of human gliomas. J Neurosurg 71: 335–341

Hosokawa Y, Tsuchihashi Y, Okabe H et al. (1991) Pleomorphic xanthastrocytoma. Ultrastructural, immunohistochemical, and DNA cytofluorometric study of a case. Cancer 68: 853–859

Hufnagel TJ, Kim JH, True ID, Manuelidis EE (1989) Immunohistochemistry of capillary hemangioblastoma. Immunoperoxidase-labeled antibody staining resolves the differential diagnosis with metastalic renal cell carcinoma, but does not explain the histogenesis of capillary hemangioblastoma. Am J Surg Pathol 13: 207–216

Humphrey PA, Wong AJ, Vogelstein B et al. (1990) Anti-synthetic peptide antibody reacting at the fusion junction of deletion-mutant epidermal growth factor receptors in human glioblastoma Proc Natl Acad Sci USA 87: 4207–4211

Iglesias JR, Richardson EP Jr, Collia F et al. (1984) Prenatal intramedullary gliofibroma. A light and electron microscopic study. Acta Neuropathol 62: 230–234

Jääkeläinen J, Haltia, M, Servo A (1986) Atypical and anaplastic meningiomas: radiology, surgery, radiotherapy, and outcome. Surg Neurol 25: 233–242

James CD, Carlbom E, Dumanski JP et al. (1988) Clonal genomic alterations in glioma malignancy stages. Cancer Res 48: 5546–5551

James CD, Carlbom E, Nordenskjöla M, Collins VP, Cavenee WK (1989) Mitotic recombination of chromosome 17 in astrocytomas. Proc Natl Acad Sci USA 86: 2858–2862

Jellinger K, Seitelberger F (1970) Zur Neuropathologie der Hirngeschwulste im Kindesalter. Wien Med Wschr 120: 855

Jellinger K, Slowik F (1975) Histological subtypes and prognostic problems in meningiomas. J Neurol 208: 279–298

Jellinger KL (1989) Biologic behavior of meningiomas. In: Field WS (ed) Primary Brain Tumours. A Review of Histologic Classification, Vol. 17. Springer Verlag, New York, pp 231–238

Jellinger K, Paulus W (1991) Mesenchymal, non-meningiothelial tumors of the central nervous system. Brain Pathol 1: 79–87

Jellinger K, Paulus W, Slowik F (1991) The enigma of meningeal hemangiopericytoma. Brain Tumor Pathol 8: 33–43

Jiang Q, Lim R, Blodi FC (1984) Dual properties of cultured retinoblastoma cells: immunohistochemical characterization of neuronal and glial markers. Exp Eye Res 39: 207–215

Johnson DL, Chandra R, Fisher WS et al. (1985) Trilateral retinoblastoma: ocular and pineal retinoblastomas. J Neurosurg 63: 367–370

Jones NR, Rossi ML, Gregoriou M, Hughes JT (1990) Investigation of the expression of epidermal growth factor receptor and blood group A antigen in 110 human gliomas. Neuropathol Appl Neurobiol 16: 185–192

Kajiwara K, Nishizaki T, Orita T et al. (1990) Silver colloid staining technique for analysis of glioma malignancy. J Neurosurg 73: 113–117

Kalyan-Raman UP, Olivero WC (1987) Ganglioglioma: a correlative clinicopathological and radiological study of ten surgically treated cases with follow-up. Neurosurgery 20: 428–433

Kartenbeck J, Schwechheimer K, Moll R, Franke WW (1984) Attachment of vimentin filaments to desmosomal plaques in human meningiomal cells and arachnoidal tissue. J Cell Biol 98: 1072–1081

Katsetos CD, Herman MM, Frankfurter A et al. (1989) Cerebellar desmosplastic medulloblastomas. A further immunohistochemical characterization of the reticulin free pale islands. Arch Pathol Lab Med 113: 1019–1029

Katsetos CD, Herman MM, Frankfurter A et al. (1991) Neuron-associated class III beta tubulin isotype, microtubule-associated protein 2, and synaptophysin in human retinoblastomas in situ. Lab Invest 64: 45–54

Kawai K, Takahashi H, Ikuta F et al. (1987) The occurrence of catecholamine neurons in a parietal lobe ganglioglioma. Cancer 60: 1532–1536

Kawano N (1991) Pleomorphic xanthoastrocytoma (PXA) in Japan: its clinico-pathologic features and diagnostic clues. Brain Tumor Pathol 8: 5–10

Kazumoto K, Tamura M, Hoshino H, Yuasa Y (1990) Enhanced expression of the sis and c-myc oncogenes in human meningiomas. J Neurosurg 72: 786–791

Keohane ME, Hall SW, VandenBerg SR, Gonias SL (1990) Secretion of α2-macroglobulin, α2-antitrypsin, and plasminogen activator inhibitor-1 by glioblastoma multiforme in primary organ culture. J Neurosurg. 73: 234–241

Kepes JJ (1970) "Xanthomatous" changes in a papilloma of the choroid plexus. Acta Neuropathol 16: 367–369

Kepes JJ, Rubinstein LJ, Eng LF (1979) Pleomorphic xanthoastrocytoma: a distinctive meningocerebral glioma of young subjects with relatively favorable prognosis; a study of 12 cases. Cancer 44: 1839–1852

Kepes JJ, Rubinstein LJ (1981) Malignant gliomas with heavily lipidized (foamy) tumor cells: a report of three cases with immunoperoxidase study. Cancer 47: 2451–2459

Kepes JJ (1982) Meningiomas. Biology, Pathology, and Differential Diagnosis. Masson, New York

Kepes JJ (1983) Oncocytic transformation of choroid plexus epithelium. Acta Neuropathol 62: 145–148

Kepes JJ, Goldware S, Leoni R (1983) Meningioma with pseudoglandular pattern. J Neuropathol Exp Neurol 42: 61–68

Kepes JJ, Rubinstein LJ, Chiang H (1984) The role of astrocytes in the formation of cartilage in gliomas. An immunohistochemical study of four cases. Am J Pathol 117: 471–483

Kepes JJ, Chen WY-K, Connors MH, Vogel FS (1988) "Chordoid" meningeal tumors in young individuals with peritumoral lymphoplasmacellular infiltrates causing systemic manifestation of Castleman syndrome. A report of seven cases. Cancer 62: 391–406

Kepes JJ, Rubinstein LJ, Ansbacher L, Schreiber DJ (1989) Histopathological features of recurrent pleomorphic xanthoastrocytomas: further corroboration of the glial nature of this neoplasm. Acta Neuropathol 78: 585–593

Kim TS, Halliday AL, Hedley-Whyte ET, Convery K (1991) Correlates of survival and the Daumas–Duport grading system for astrocytomas. J Neurosurg 74: 27–37

Kleihues P, Aguzzi A, Shibata T, Wiestler OD (1989) Immunohistochemical assessment of differentiation and DNA replication in human brain tumors. In: Fields WS (ed) Primary Brain Tumors. A Review of Histologic Classification, Vol. 8. Springer Verlag, New York, pp 123–132

Kleihues P, Aguzzi A, Wiestler OD (1990) Cellular and molecular aspects of neurocarcinogenesis. Toxicol Pathol 18: 193–203

Kline KT, Damjanov I, Katz SM, Schmidek H (1979) Pineoblastoma: an electron microscopic study. Cancer 44: 1692–1699

Korf HW, Klein DC, Zigler JS, Gery I, Schachenmayr W (1986) S-antigen-like immunoreactivity in a human pineocytoma. Acta Neuropathol 69: 165–167

Korf HW, Czerwionka M, Reiner et al. (1987) Immunocytochemical evidence of molecular photoreceptor markers in cerebellar medulloblastomas. Cancer 60: 1763–1766

Kros JM, Van Eden CG, Stefanko SZ et al. (1990) Prognostic implications of glial fibrillary acidic protein containing cell types in oligodendrogliomas. Cancer 66: 1204–1212

Kros JM, de Jong AAW, van der Kwast Th.H (1992) Ultrastructural characterization of transitional cells in oligodendrogliomas. J Neuropathol Exp Neurol 51: 186–193

Krouwer HGJ, Davis RL, Silver R, Prados M (1991) Gemistocytic astrocytomas: a reappraisal. J Neurosurg 74: 399–406

Kubota T, Hayashi M, Kawano H et al. (1991) Central neurocytoma: immunohistochemical and ultrastructural study. Acta Neuropathol 81: 418–427

Kurihara M, Ochi A, Kawaguchi T *et al.* (1990) Localization and characterization of endothelin receptors in human gliomas. Neurosurgery 27: 275–281

Kyritsis AP, Tsokos M, Triche TJ, Chader GJ (1984) Retinoblastoma – origin from a primitive neuroectodermal cell? Nature 307: 471–473

Kyritsis AP, Tsokos M, Triche TJ, Chader GJ (1986a) Retinoblastoma: a primitive tumor with multipotential characteristics. Invest Ophthal Vis Sci 27: 1760–1764

Kyritsis AP, Tsokos M, Chader GJ (1986b) Control of retinoblastoma cell growth by differentiating agents: current work and future directions. Anticancer Res 6: 465–474

Langford LA (1986) The ultrastructure of the ependymoblastoma. Acta Neuropathol 71: 136–141

Laurence KM (1979) The biology of choroid plexus papilloma in infancy and childhood. Acta Neurochir 50: 79–90

Laws ER Jr, Taylor WF, Clifton MB, Okazaki H (1984) Neurosurgical management of low-grade astrocytoma of the cerebral hemispheres. J Neurosurg 61: 665–673

Lee KS, Hoshino T, Rodriguez LA *et al.* (1990) Bromodeoxyuridine labeling study of intracranial meningiomas: proliferative potential and recurrence. Acta Neuropathol 80: 311–317

Leifer D, Moore T, Ukena T *et al.* (1989) Multifocal glioblastoma with liver metastases in the absence of surgery. J Neurosurg 71: 772–776

Lekanne Deprez RH, Groen NA, Van Biezen NA *et al.* (1991) A t(4;22) in a meningioma points to the localization of a putative tumor-suppressor gene. Am J Genet 48: 783–790

Lesnick JE, Chayt KJ, Bruce DA *et al.* (1985) Familial pineoblastoma. Report of two cases. J Neurosurg 62: 930–932

Lewis GP, Erickson PA, Kaska DD, Fisher SK (1988) An immunocytochemical comparison of Muller cells and astrocytes in the cat retina. Exp Eye Res 47: 839–853

Lewis P (1967) Carcinoma of the choroid plexus. Brain 90: 177–186

Libermann TA, Friesel R, Jaye M *et al.* (1987) An angiogenic growth factor is expressed in human glioma cells. EMBO J 6: 1627–1632

Lillien LE, Raff MC (1990) Differentiation signals in the CNS: type-2 astrocyte development in vitro as a model system. Neuron 5: 111–119

Ling T, Stone J (1984) The development of astrocytes in the cat retina: evidence of migration from the optic nerve. Dev Brain Res 4: 73–85

Liwnicz BH, Rubinstein LJ (1979) The pathways of extraneural spread in metastasizing gliomas. A report of two cases and review of the literature. Human Path 10: 453–467

Liwnicz BH, Wu SZ, Tew JM Jr (1987) The relationship between the capillary structure and hemorrhage in gliomas. J Neurosurg 66: 536–541

Lombardi D, Scheithauer BW, Meyer FB *et al.* (1991) Symptomatic subependymoma: a clinicopathological and flow cytometric study. J Neurosurg 75: 583–588

Lopes MBS, Gonzalez-Fernandez F, Rosemberg *et al.* (1991) Expression of temporally specific and divergent cell lineage in retinoblastoma. J Neuropathol Exp Neurol 50: 292

Louis DN, Edgerton S, Thor AD, Hedley-Whyte ET (1991) Proliferating cell nuclear antigen and Ki-67 immunohistochemistry in brain tumors: a comparative study. Acta Neuropathol 81: 675–679

Ludwin SK, Rubinstein LJ, Russell DS (1975) Papillary meningioma: a malignant variant of meningioma. Cancer 36: 1363–1373

Mapstone T, McMichael M, Goldthwait D (1991) Expression of platelet-derived growth factors, transforming growth factors, and the ros gene in a variety of primary human brain tumors. Neurosurgery 28: 216–222

Marano SR, Johnson PC, Spetzler RF (1988) Recurrent Lhermitte-Duclos disease in a child. J Neurosurg 69: 599–603

Margetts JC, Kalyan-Raman UP (1989) Giant-celled glioblastoma of brain: a clinico-pathological and radiological study of ten cases (including immunohistochemistry and ultrastructure). Cancer 63: 524–531

Margo C, Hydayat A, Kopelman J, Zimmerman LE (1983) Retinocytoma: a benign variant of retinoblastoma. Arch Ophthalmol 101: 1519–1531

Markesbery WR, Haugh RM, Young AB (1981) Ultrastructural of pineal parenchymal neoplasms. Acta Neuropathol 55: 143–149

Matson DD (1953) Hydrocephalus in a premature infant caused by papilloma of the choroid plexus: with report of surgical treatment. J Neurosurg 11: 416–420

Matsushima T, Inoue T, Takeshita I *et al.* (1988) Choroid plexus papillomas: an immunohistochemical study with particular reference to the coexpression of prealbumin. Neurosurg 23: 384–389

Mauro A, Bulfone A (1990) Oncogenes and growth factors in gliomas. J Neurosurg Sci 34:

171–173

Mauro A, Bulfone A, Di Sapio, A, Schiffer D (1991) Meningioma cells in culture secrete a PDGF-like growth factor and express functional PDGF β-receptors. J Neuropathol Exp Neurol 50: 373

May PL, Broome JC, Lawry J, Buxton RA, Battersby RDE (1989) The prediction of recurrence in meningiomas. A flow cytometric study of paraffin-embedded archival material. J Neurosurg 71: 347–351

McComb RD, Bigner DD (1985) Immunolocalization of laminin in neoplasms of the central and peripheral nervous systems. J Neuropathol Exp Neurol 44: 242

McComb RD, Moul JM, Bigner DD (1987) Distribution of type VI collagen in human gliomas: comparison with fibronectin and glioma-mesenchymal matrix glycoprotein. J Neuropathol Exp Neurol 46: 623–633

McGirr SJ, Ebersold MJ, Scheithauer B, Quast LM, Shaw EG (1988) Choroid plexus papillomas: long term follow-up of a surgically treated series. J Neurosurg 69: 843–849

Meis JM, Ordonez NG, Bruner JM (1986) Meningiomas: an immunohistochemical study of 50 cases. Arch Pathol Lab Med 110: 934–937

Meis JM, Martz KL, Nelson JS (1991) Mixed glioblastoma multiforme and sarcoma. A clinicopathologic study of 26 radiation therapy oncology group cases. Cancer 67: 2342–2349

Mena H, Ribas JL, Pezeshkpour GH, Cowan DN, Parisi JE (1991) Hemangiopericytoma of the central nervous system: a review of 94 cases. Hum Pathol 22: 84–91

Milbouw G, Born JD, Martin D et al. (1988) Clinical and radiological aspects of dyplastic gangliocytoma (Lhermitte–Duclos disease): a report of two cases with review of the literature. Neurosurg 22: 124–128

Miller LL, Ostrow PT, Chau R (1991) Characterization of gliosarcomas by image analysis of superimposed serial sections. J Neuropathol Exp Neurol 50: 365

Miller RH, French-Constant C, Raff MC (1989) The macroglial cells of the rat optic nerve. Ann Rev Neurosci 12: 517–534

Mills SE, Ross G.W, Perentes E, Nakagawa Y, Scheithauer BW (1990) Cerebellar hemangioblastoma: immunohistochemical distinction from metastatic renal cell carcinoma. Surg Pathol 3: 121–132

Morantz RA, Feigin I, Ransohoff J III (1976) Clinical and pathological study of 24 cases of gliosarcoma. J Neurosurg 45: 398–408

Morantz RA, Kepes JJ, Batnitzky S, Masterson BJ (1979) Extraspinal ependymomas: report of three cases. J Neurosurg 51: 383–391

Mørk SJ, Lindegaad K-F, Halvorsen TB et al. (1985) Oligodendroglioma: incidence and biological behavior in a defined population. J Neurosurg 63: 881–889

Mørk SJ, Rubinstein LJ (1985) Ependymoblastoma. A reappraisal of a rare embryonal tumor. Cancer 55: 1536–1542

Mørk SJ, Halvorsen TB, Lindegaard K-F, Eide GE (1986) Oligodendroglioma. Histologic evaluation and prognosis. J Neuropathol Exp Neurol 45: 65–78

Mørk SJ, Rubinstein LJ, Kepes JJ, Perentes E, Uphoff DF (1988) Patterns of epithelial metaplasia in malignant gliomas II. Squamous differentiation of epithelial-like formations in gliosarcomas and glioblastomas. J Neuropathol Exp Neurol 47: 101–118

Muller W, Afra D, Schroder R (1977) Supratentorial recurrences of gliomas: morphological studies in relation to time intervals with 544 astrocytomas. Acta Neurochir 37: 75–91

Myers J, Scott M, Silverstein A (1961) Cystic hemangioblastoma of pons. J Neurosurg 18: 694–697

Nager GT (1964) Meningiomas Involving the Temporal Bone: Clinical and Pathological Aspects. Charles C Thomas, Springfield, Ill.

Nager GT, Masica DN (1970) Meningiomas of the cerebello-pontine angle and their relation to the temporal bone. Laryngoscope 80: 863

Nager GT, Heroy J, Hoeplinger M (1983) Meningiomas invading the temporal bone with extension to the neck. Am J Otolaryngol 4: 297–324

Nakagawa Y, Perentes E, Rubinstein LJ (1986) Immunohistochemical characterization of oligodendrogliomas: an analysis of multiple markers. Acta Neuropathol 72: 15–22

Nakamura Y, Becker LE (1983) Subependymal giant-cell tumor: astrocytic or neuronal? Acta Neuropathol 60: 271–277

Nakamura N, Sekino H, Taguchi Y, Fuse T (1985) Successful total extirpation of hemangioblastoma originating in the medulla oblongata. Surg Neurol 24: 87–94

Nakamura M, Inoue HK, Ono K, Kunimine H, Tamada J (1987) Analysis of hemangiopericytic meningiomas by immunohistochemistry, electron microscopy and cell culture. J Neuropathol Exp Neurol 46: 57–71

Nakasu S, Hirano A, Shimura T, Llena, JF (1987) Incidental meningiomas in autopsy study. Surg Neurol 27: 319–322

Nelson JS, Tsukada Y, Schoenfeld D *et al.* (1983) Necrosis as a prognostic criterion in malignant supratentorial, astrocytic gliomas. Cancer 52: 550–554

Neumann HPH, Eggert HR, Weigel K *et al.* (1989) Hemangioblastomas of the central nervous system. J Neurosurg 70: 24–30

Neuwelt EA, Glasberg M, Frenkel E, Clark WK (1979) Malignant pineal region tumors: a clinico-pathologic study. J Neurosurg 51: 597–607

Ng THK, Poon WS (1990) Gliosarcoma of the posterior fossa with features of a malignant fibrous histiocytoma. Cancer 65: 1161–1166

Ng THK, Furg CF, Ma LT (1990) The pathological spectrum of desmoplastic infantile gangliogliomas. Histopathol 16: 235–241

Nishio S, Tashima T, Takeshita I, Fukui M (1988) Intraventricular neurocytoma–clinicopathological features of six cases. J Neurosurg 68: 665–670

Norton WT, Farooq M (1989) Astrocytes cultured from mature brain derive from glial precursor cells. J Neurosci 9: 769–775

Ojeda VJ, Spagnolo DV, Vaughan RJ (1987) Palisades in primary cerebral neuroblastoma simulating so-called polar spongioblastoma. A light and electron microscopy study of an adult case. Am J Surg Pathol 11: 316–322

Orita T, Kajiwara K, Nishizaki T *et al.* (1990) Nucleolar organizer regions in meningioma. Neurosurgery 26: 43–46

Palmer J.J (1972) Haemangioblastomas: a review of 81 cases. Acta Neurochir 27: 125–148

Parrish EP, Steart PV, Garrod DR, Weller RO (1987) Antidesmosomal monoclonal antibody in the diagnosis of intracranial tumours. J Pathol 153: 265–273

Pasquier B, Kojder I, Labat F *et al.* (1985) Le xanthoastrocytome du sujet jeune: revue de la litterature à propos de deux observations d'évolution discordante. Ann Pathol 5: 29–43

Pasquier B, Gasnier F, Pasquier D *et al.* (1986) Papillary meningioma. Clinicopathologic study of seven cases and review of the literature. Cancer 58: 299–305

Patil AA, McComb RD, Gelber B, McConnell J, Sasse S (1990) Intraventricular neurocytoma: a report of two cases. Neurosurg 26: 140–144

Paulus W, Peiffer J (1988) Does the pleomorphic xanthoastrocytoma exist? Problems in the application of immunological techniques to the classification of brain tumors. Acta Neuropathol 76: 245–252

Paulus W, Jänisch W (1990) Clinicopathologic correlations in epithelial choroid plexus neoplasms: a study of 52 cases. Acta Neuropathol 80: 635–641

Pearl GS, Takei Y, Stefanis GS, Hoffman JC (1981) Intraventricular neuroblastoma in a patient with von Hippel–Lindau's disease: an electron microscopic study. Acta Neuropathol 53: 253–256

Pearl GS, Takei Y, Bakay RAE, Davis P (1985) Intraventricular primary cerebral neuroblastoma in adults: report of three cases. Neurosurg 16: 847–849

Percy AK, Elveback LR, Okazaki H, Kurland LT (1972) Neoplasms of the central nervous system. Epidemiologic considerations. Neurology 22: 40–48

Perentes E, Rubinstein L (1986) Immunohistochemical recognition of human neuroepithelial tumors by anti-Leu 7 (HNK-1) monoclonal antibody. Acta Neuropathol 69: 227–233

Perentes E, Rubinstein LJ, Herman MM, Donoso LA (1986) S-antigen immunoreactivity in human pineal glands and pineal parenchymal tumors. A monoclonal antibody study. Acta Neuropathol 71: 224–227

Perentes E, Rubinstein LJ (1987) Recent applications of immunoperoxidase histochemistry in human neuro-oncology. Arch Pathol Lab Med 111: 796–812

Perentes E,CP, Herbort, LJ Rubinstein MM *et al.* (1987) Immunohistochemical characterization of human retinoblastomas in situ with multiple markers. Am J Ophthal 103: 647–658

Piatt JH Jr, Campbell GA, Oakes WJ (1986) Papillary meningioma involving the oculomotor nerve in an infant: case report. J Neurosurg 64: 808–812

Pigott TJD, Punt JAG, Lowe JS, Henderson MJ, Beck A, Gray T (1990) The clinical, radiological and histopathological features of cerebral primitive neuroectodermal tumours. Br J Neurosurg 4: 287–298

Pilkington GJ, Lantos PL (1982) The role of glutamine synthetase in the diagnosis of cerebral tumours. Neuropathol Appl Neurobiol 8: 227–236

Pixley SKR, DeVellis J (1984) Transition between immature radial glia and mature astrocytes studied with a monoclonal antibody to vimentin. Develop Brain Res 15: 201–209

Plate KH, Ruschoff J, Mennel HD (1991) Cell proliferation in intracranial tumours: selective silver staining of nucleolar organizer regions (AgNORs). Application to surgical and experimental neuro-oncology. Neuropathol Appl Neurobiol 17: 121–132

Poirier J, Gray F, Gherardi R, Favolini M (1985) Histopathologie des Tumeurs du Systeme Nerveux. Masson, Paris, p 9

Pollak A, Friede RL (1977) Fine structure of medulloepithelioma. J Neuropathol Exp Neurol 36: 712–725

Popoff NA, Ellsworth RM (1971) The fine structure of retinoblastoma. Lab Invest 25: 238–402

Poulsgard L, Schrøder HD, Rønne M (1990) Cytogenetic studies of 11 meningiomas and their clinical significance. II. Anticancer Res 10: 535–538

Probst A, Ulrich J, Zdrojewski B, Hirt HR (1979) Cerebellar ganglioglioma in a child. J Neuropathol Exp Neurol 38: 57–71

Pulitzer DR, Martin PC, Collins PC, Ralph DR (1988) Subcutaneous sacrococcygeal ("myxopapillary") ependymal rests. Am J Surg Pathol 12: 672–677

Radley, MG, Di Sant'Agnese PA, Eskin TA, Wilbur DC (1989) Epithelial differentiation in meningiomas. An immunohistochemical, histochemical and ultrastructural study — with review of literature. Am J Clin Pathol 92: 266–272

Raff MC, Miller RH, Noble M (1983) A glial progenitor cell that develops in vitro into an astrocyte or an oligodendrocyte depending on culture medium. Nature 303: 390–396

Rao JS, Suzuki R, Festoff BW (1990a) Serpins and brain tumors: roles in pathogenesis. In: Festoff BW (ed) Serine Proteinases and their Serpin Inhibitors in the Nervous System. Plenum Press, New York, pp 301–311

Rao C, Friedlander ME, Klein E, Anzil AP, Sher JH (1990b) Medullomyoblastoma in an adult. Cancer 65: 157–163

Rawlinson DG, Herman MM, Rubinstein LJ (1973) The fine structure of a myxopapillary ependymoma of the filum terminale. Acta Neuropathol 25: 1–13

Reeder RF, Saunders RL, Roberts DW, Fratkin JD, Cromwell LD (1988) Magnetic resonance imaging in the diagnosis and treatment of Lhermitte–Duclos disease (dysplastic gangliocytoma of the cerebellum). Neurosurg 23: 240–245

Reese AB (1976) Tumors of the Eye, 3rd edn. Harper and Row, Hagerstown MD, p 148

Reimund EJ, Sitton JE, Harkin JC (1990) Pigmented choroid plexus papilloma. Arch Pathol Lab Med 114: 902–905

Reiter RJ (1981) The Pineal Gland, Vol. I. Anatomy and Biochemistry. CRC Press, Boca Raton, pp 121–154

Renfranz PJ, Cunningham MG, McKay RDG. (1991) Regio-specific differentiation of the hippocampal stem cell line HiB5 upon implantation into the developing mammalian brain. Cell 66: 713–729

Reznik M, Schoenen J (1983) Lhermitte–Duclos disease. Acta Neuropathol 59: 88–94

Rodrigues MM, Bardenstein DS, Donoso LA, Rajagopalan S, Brownstein S (1987) An immunohistopathologic study of trilateral retinoblastoma. Am J Ophthal 103: 776–781

Roessmann U, Wongmongkolrit T (1984) Dysplastic gangliocytoma of cerebellum in a newborn. J Neurosurg 60: 845–847

Rorke LB (1989) Primitive neuroectodermal tumor — a concept requiring an apologia? In: Fields WS (ed) Primary Brain Tumors. A Review of Histologic Classification, Vol. 1. Springer Verlag, New York, pp 5–15

Ross GW, Rubinstein LJ (1989) Lack of histopathological correlation of malignant ependymomas with postoperative survival. J Neurosurg 70: 31–36

Rubinstein LJ (1972) Tumors of the Central Nervous System. Atlas of Tumor Pathology (Fascicle 6). Armed Forces Institute of Pathology, Washington, DC

Rubinstein LJ, Herman MM (1989) The astroblastoma and its possible cytogenetic relationship to the tanycyte. Acta Neuropathol 78: 472–483

Rubinstein LJ (1989) Justification for a cytogenetic scheme of embryonal central neuroepithelial tumors. In: Fields WS (ed) Primary Brain Tumors. A Review of Histologic Classification, Vol. 2. Springer Verlag, New York, pp 16–27

Rubinstein LJ (1991) Glioma cytology and differentiation viewed through the window of neoplastic vulnerability. In: Salcman M (ed) Neurobiology of Brain Tumors. Williams and Wilkins, Baltimore, pp 35–51

Rushing EJ, Mena H, Ribas JL (1991) Primary pineal parenchymal lesions: a review of 53 cases. J Neuropath Exp Neurol 50: 364

Russell DS, Rubinstein LJ (1989) Pathology of Tumours of the Nervous System, 5th edn.

Edward Arnold, London, pp 124–125

Rutka JT, Giblin J, Dougherty DV et al. (1986) An ultrastructural and immunocytochemical analysis of leptomeningeal and meningioma cultures. J Neuropathol Exp Neurol 45: 285–303

Saari JC, Bredberg L, Garwin GG (1982) Identification of the endogenous retinoids associated with three cellular retinoid-binding proteins from bovine retina and retina pigment epithelium. J Biol Chem 257: 13329–13333

Sage H, Iruela-Arispe ML (1990) Type VIII collagen in murine development. Association with capillary formation in vitro. Ann NY Acad Sci 580: 17–31

Salazar J, Vaquero J, Aranda IF et al. (1986) Choroid plexus papilloma with chondroma: case report. Neurosurgery 18: 781–783

Salgaller M, Agius L, Yates A et al. (1990) Application of automated image analysis to demonstrate the correlation between ras p21 expression and severity of gliomas. Biochem Biophys Res Comm 169: 482–491

Sanford RA, Smith RA (1986) Hemangioblastoma of the cervicomedullary junction: report of three cases. J Neurosurg 64: 317–321

Sawaya R (1990) Presence and significance of α_1-antitrypsin in human brain tumours. In Festoff BW (ed): Serine Proteinases and their Serpin Inhibitors in the Nervous System. Plenum Press, New York, pp 293–299

Scheithauer BW (1978) Symptomatic subependymoma: report of 21 cases with review of the literature. J Neurosurg 49: 689–696

Scheithauer BW (1985) Neuropathology of pineal region tumors. Clin Neurosurg 32: 351–383

Scheithauer BW (1990) Tumours of the meninges: proposed modifications of the World Health Organization classification. Acta Neuropathol 80: 343–354

Scherer HJ (1940) Cerebral astrocytomas and their derivatives. Am J Cancer 40: 159–198

Schiffer D, Giordana MT, Mauro A et al. (1986) Immunohistochemical demonstration of vimentin in human cerebral tumors. Acta Neuropathol 70: 209–219

Schindler E, Gullotta F (1983) Glial fibrillary acidic protein in medulloblastomas and other embryonic CNS tumours of children. Virchows Arch [A] 398: 263–275

Schnitt SJ, Vogel H (1986) Meningiomas: diagnostic value of immunoperoxidase staining for epithelial membrane antigen. Am J Surg Pathol 10: 640–649

Schnitzer J (1988a) Astrocytes in the guinea pig, horse and monkey retina: their occurrence coincides with the presence of blood vessels. Glia 1: 74–89

Schnitzer J (1988b) Immunocytochemical studies on the development of astrocytes, Müller (glial) cells and oligodendrocytes in the rabbit retina. Dev Brain Res 44:59–72

Schroder HD (1987) Immunohistochemical demonstration of glial markers in retinoblastomas. Virchows Arch [A] 411: 67–72

Schwechheimer K, Kartenbeck J, Moll R, Franke WW (1984) Vimentin filament-desmosome cytoskeleton of diverse types of human meningiomas: a distinctive diagnostic feature. Lab Invest 51: 584–591

Shapiro JR, Yung, W-KA, Shapiro WR (1981) Isolation, karyotype and clonal growth of heterogeneous subpopulations of human malignant gliomas. Cancer Res 41: 2349–2359

Shapiro JR, Shapiro WR (1984) Clonal tumor cell heterogeneity. Prog Exp Tumor Res 27: 49–66

Shepherd CW, Scheithauer BW, Gomez MR, Altermatt HJ, Katzmann JA (1991) Subependymal giant cell astrocytoma: a clinical, pathological, and flow cytometric study. Neurosurgery 28: 864–868

Shiurba RA, Gessaga EC, Eng LF et al. (1988) Lhermitte–Duclos disease: an immunohistochemical study of the cerebellar cortex. Acta Neuropathol 75: 474–480

Shiurba RA, Buffinger NS, Spencer EM, Urich H (1991) Basic fibroblastic growth factor and somatomedin C in human medulloepithelioma. Cancer 68: 798–808

Sima AAF, Robertson DM (1979) Subependymal giant-cell astrocytoma. Case report with ultrastructural study. J Neurosurg 50: 240–245

Sivak, JG (1974) Historical note: the vertebrate median eye. Vis Res 14: 137–140

Slooff, JL, Kernohan JW, MacCarty CS (1964) Primary Intramedullary Tumors of the Spinal Cord and Filum Terminale. Saunders, Philadelphia

Smidt M, Kirsch I, Ratner L (1990) Deletion of Alu sequences in the fifth c-sis intron in individuals with meningiomas. J Clin Invest 86: 1151–1157

Smith MT, Ludwig CL, Godfrey AD, Armbrustmacher VW (1983) Grading of oligodendrogliomas. Cancer 52: 2107–2114

Sobel, RA, Trice JE, Nielsen SL, Ellis, WG (1981) Pineoblastoma with ganglionic and glial differentiation: report of two cases. Acta Neuropathol 55: 243–246

Specht CS, Smith TW, DeGirolami U, Price JM (1986) Myxo-papillary ependymoma of the filum terminale: a light and electron microscopic study. Cancer 58: 310–317

Sreekantaiah C, Jockin H, Brecher ML, Sandberg AA (1989) Interstitial deletion of chromosome 11q in a pineoblastoma. Cancer Genet Cytogenet 39: 125–131

Stefanko SW, Manschot WA (1979) Pinealoblastoma with retinomatous differentiation. Brain 102: 321–332

Stone J, Dreher Z (1987) Relationship between astrocytes, ganglion cells and vasculature of the retina. J Comp Neurol 255: 35–49

Strom EH, Skullerud, K (1983) Pleomorphic xanthoastrocytoma: report of 5 cases. Clin Neuropathol 2: 188–191

Sugawa N, Ekstrand AJ, James CD, Collins VP (1990) Identical splicing of aberrant epidermal growth factor receptor transcripts from amplified rearranged genes in human glioblastomas. Proc Natl Acad Sci USA 87: 8602–8606

Sutton LN, Packer RJ, Rorke LB, Bruce DA, Schut L (1983) Cerebral gangliogliomas during childhood. Neurosurgery 13: 124–128

Takahashi H, Wakabayasyi K, Kawai K et al. (1989) Neuroendocrine markers in central nervous system neuronal tumors (gangliocytoma and ganglioglioma). Acta Neuropathol 77: 237–243

Takahashi JA, Mori HI, Fukumoto M et al. (1990) Gene expression of fibroblast growth factors in human gliomas and meningiomas: demonstration of cellular source of basic fibroblastic growth factor mRNA and peptide in tumor tissues. Proc Natl Acad Sci USA 87: 5710–5714

Tang TT, Harb JM, Mørk SJ, Sty JR (1985) Composite cerebral neuroblastoma and astrocytoma. A mixed central neuroepithelial tumor. Cancer 56: 1404–1412

Taratuto AL, Monges J, Lylyk P, Leiguarda R (1984) Superficial cerebral astrocytoma attached to the dura. Report of six cases in infants. Cancer 54: 2505–2512

Tarlton JF, Easty DL (1990) Immunohistochemical characterization of retinoblastoma and related ocular tissue. Br J Ophthal 74: 144–149

Taylor HR, Carroll N, Jack I, Crock GW (1979) A scanning electron microscopic examination of retinoblastoma in tissue culture. Br J Ophthal 63: 551–559

Terenghi G, Polak JM, Ballesta J et al. (1984) Immunocytochemistry of neuronal and glial markers in retinoblastoma. Virchows Arch [A] 404: 61–73

Theaker JM, Gatter KC, Esiri MM, Fleming KA (1986) Epithelial membrane antigen and cytokeratin expression by meningiomas: an immunohistological study. J Clin Pathol 39: 435–439

Theunissen PHMH, Baerts MD-T, Blaauw G (1990) Histogenesis of intracranial haemangiopericytoma and haemangioblastoma. An immunohistochemical study. Acta Neuropathol 80: 68–71

Thomas GA, Raffel C (1991) Loss of heterozygosity on 6q, 16q and 17p in human central nervous system primitive neuroectodermal tumors. Cancer Res 51: 639–643

Tomita T, Naidich TP (1987) Successful resection of choroid plexus papillomas diagnosed at birth: report of two cases. Neurosurgery 20: 774–779

Townsend JJ, Seaman JP (1986) Central neurocytoma — a rare benign intraventricular tumor. Acta Neuropathol 71: 167–170

Trojanowski JQ, Tascos NA, Rorke LB (1982) Malignant pineocytoma with prominent papillary features. Cancer 50: 1789–1793

Trojanowski JA, Lee VM -Y, Schlaepfer, WW (1984) An immunohistochemical study of human central nervous system tumors, using monoclonal antibodies against neurofilaments and glial filaments. Hum Pathol 15: 248–257

Trombley IK, Mirra SS (1981) Ultrastructure of tuberous sclerosis: cortical tuber and subependymal tumor. Ann Neurol 9: 174–181

Troost D, Jansen GH, Dingemans KP (1990) Cerebral medulloepithelioma — electron microscopy and immunohistochemistry. Acta Neuropathol 80: 103–107

Tso MO, Fine BS, Zimmerman LE (1969) The Flexner–Wintersteiner rosettes in retinoblastoma. Arch Pathol 88: 664–671

Tso MO, Zimmerman LE, Fine BS (1970) The nature of retinoblastoma. I. Photoreceptor differentiation. A clinical and histopathologic study. Am J Ophthal 69: 339–349

Tsokos M, Ross RA, Triche TJ (1985) Neuronal, Schwannian and melanocytic differentiation of human neuroblastoma cells in vitro. In: Evans AE, D'Angio GJ, Seeger RC (eds) Advances in Neuroblastoma Research. Alan R. Liss, New York, pp 55–68

Tsokos M, Kyritsis AP, Chader GJ, Triche TJ (1986) Differentiation of human retinoblastoma in vitro into cell types with characteristics observed in embryonal or mature retina. Am J Pathol 123: 542–552

Tsokos M, Scarpa S, Ross RA, Triche TJ (1987) Differentiation of human neuroblastoma recapitulates neural crest development. Study of morphology, neurotransmitter enzymes, and extracellular matrix proteins. Am J Pathol 128: 484–496

Turner DL, Cepko CL (1987) A common progenitor for neurons and glia persists in rat retina late in development. Nature 328: 131–136

Uematsu Y, Itakura T, Hayashi S, Komai N (1988) Pineoblastoma with an unusually long survival. Case report. J Neurosurg 69: 287–291

VandenBerg SR, Chatel M, Griffiths OM *et al.* (1981) Neural differentiation in the OTT-6050 mouse teratoma. Virchows Arch [A] 392: 281–294

VandenBerg SR, May EE, Rubinstein LJ *et al.* (1987a) Desmoplastic supratentorial neuroepithelial tumors of infancy with divergent differentiation potential ("desmoplastic infantile gangliogliomas"). Report on 11 cases of a distinctive embryonal tumor with favourable prognosis. J Neurosurg 66: 58–71

Vandenberg SR, Herman MM, Rubinstein LJ (1987b) Embryonal central neuroepithelilal tumors: current concepts and future challenges. Cancer Metast Rev 5: 343–364

Vandenberg SR (1991) Desmoplastic infantile ganglioglioma: a clinicopathologic review of sixteen cases. Brain Tumor Pathol 8: 25–31

Vaquero J, Ramiro J, Martínez R, Coca S, Bravo G (1990a) Clinicopathological experience with pineocytomas: report of five surgically treated cases. Neurosurgery 27: 612–619

Vaquero J, Coca S, Martínez R, Escandón J (1990b) Papillary pineocytoma: case report. J Neurosurg 73: 135–137

von Deimling A, Janzer R, Kleihues P, Wiestler OD (1990) Patterns of differentiation in central neurocytoma: an immunohistochemical study of eleven biopsies. Acta Neuropathol 79: 473–479

Wasson JC, Saylors RL, Zeltzer P *et al.* (1990) Oncogene amplification in pediatric tumors. Cancer Res 50: 2987–2990

Watanabe T, Raff MC (1988) Retinal astocytes are immigrants from the optic nerve. Nature 332: 834–837

Watanabe K, Nagai M, Wakai S, Arai T, Kawashima K (1990) Loss of constitutional heterozygosity in chromosome 10 in human glioblastoma. Acta Neuropathol 80: 251–254

Weldon-Linne GM, Victor TA, Groothuis DR, Vick NA (1983) Pleomorphic xanthoastrocytoma: ultrastructural and immunohistochemical study of a case with a rapidly fatal outcome following surgery. Cancer 52: 2055–2063

Wetts R, Fraser SE (1988) Multipotent precursors can give rise to all major cell types of the frog retina. Science 239: 1142–1145

Whittle IR, Gordon A, Misra BK, Shaw JF, Steers AJ (1989) Pleomorphic xanthoastrocytoma: report of four cases. J Neurosurg 70: 463–468

Wilkinson M, Hume R, Strange R, Bell JE (1990) Glial and neuronal differentiation in the human fetal brain 9–23 weeks of gestation. Neuropathol Appl Neurobiol 16: 193–204

Wilson AJ, Leaffer DH, Kohout ND (1985) Differentiated cerebral neuroblastoma: a tumor in need of discovery. Hum Pathol 16: 647–649

Winek RR, Scheithauer BW, Wick MR (1989) Meningioma, meningeal hemangiopericytoma (angioblastic meningioma), peripheral hemangiopericytoma and acoustic Schwannoma. A comparative immunohistochemical study. Am J Surg Pathol 13: 251–261

Wolf A, Morton BF (1937) Ganglion cell tumors of the central nervous system. Bull Neurol Inst NY 6: 453

Wolfson WL, Brown WJ (1977) Disseminated choroid plexus papilloma: an ultrastructural study. Arch Pathol Lab Med 101: 366–368

Yachnis AT, Trojanowski JQ, Memmo M, Schlaepfer WW (1988) Expression of neurofilament proteins in the hypertrophic granule cells of Lhermitte–Duclos disease: an explanation for the mass effect and the myelination of parallel fibers in the disease state. J Neuropathol Exp Neurol 47: 206–216

Zang KD (1982) Cytological and cytogenetical studies on human meningioma. Cancer Genet Cytogenet 6: 249–274

Zankle H, Zang KD (1972) Cytological and cytogenical studies on brain tumors. IV. Identification of the missing G chromosome in human meningiomas as no. 22 by fluorescence technique. Humangenetik 14: 167–169

Zimmerman HM (1969) Brain tumors: their incidence and classification in man and their experimental production. Ann NY Acad Sci 159, Art 2: 337–359

Zimmerman BL, Tso MOM (1975) Morphologic evidence of photoreceptor differentiation of pinealocytes in the neonatal rat. J Cell Biol 66: 60–75

Zülch KJ (1986) Brain Tumors. Their Biology and Pathology, 3rd edn. Springer-Verlag, Berlin

4 General Introduction to the Clinical Features of Malignant Brain Tumours

M.F. Pell and D.G.T. Thomas

Epidemiology

The incidence of primary brain tumours is difficult to assess accurately as the frequency of different tumour types depends on the source of the series which is analysed, but ranges from 4.2 per 100 000 to 12.8 per 100 000 (Brewis et al., 1966; Liebowitz and Atler, 1969). Of these, approximately 30% are astrocytomas (Percy et al. 1972).

A histogenic approach to classification of brain tumours has developed over the years. Currently in use is the classification based primarily on the microscopic characteristics of tumours determined by an international panel of the World Health Organization (see Chap. 3)

Cerebral gliomas are not uniformly distributed throughout the cerebral hemispheres. The frontal lobes are the most commonly involved followed by the temporal and parietal, while the occipital lobes are rarely the primary site of origin or an intracerebral glioma (McKeran and Thomas, 1980). Astrocytoma and medulloblastoma occur commonly in the infratentorial compartment, especially in children.

Different tumour types present at different age groups (Bailey et al. 1948; Zulch and Borck, 1965). Medulloblastoma, cerebellar astrocytoma, ependymoma, pineal region tumours, teratoma and craniopharyngioma all showed a peak incidence in childhood and young adults (less than 20 years of age), while the middle decades of life (third and fourth) were characterized by the frequent occurrence of gliomas of the cerebral hemispheres (astrocytoma and oligodendroglioma), meningioma, pituitary adenoma, neurinoma of the cerebellopontine angle and astrocytoma of the cerebellum. By the fifth and sixth decades of life, malignant glioblastoma and metastatic tumours were much more common while oligodendroglioma, astrocytoma and meningioma were still encountered.

In children under 18 years of age, brain tumours represent 40%–50% of all solid tumours and 60%–70% of these are gliomas (Battistella et al., 1990). The most common location is infratentorial (60% of cases) with cerebellar astrocytomas, medulloblastomas and brain stem gliomas occurring in similar proportion. Astrocytomas are the predominant form in the supratentorial compartment.

There is a slight preponderance of males over females with regard to cerebral gliomas, unlike meningiomas and neurinomas. In a series reported from the authors' institution (McKeran and Thomas, 1980) of cerebral gliomas, 55.3% were male and 44.7% female ($n = 653$).

Clinical Presentation of Primary Malignant Brain Tumours

Patients with cerebral gliomas present with one or more of the following features:

1. Symptoms and signs of raised intracranial pressure.
2. Non-localizing symptoms such as headache, epilepsy or mental symptoms.
3. Focal symptoms and signs.

Specific clinical patterns emerge, depending on the histological type and grading of glioma and its site of origin. The relative frequency of such manifestations in adults with cerebral gliomas is shown in Table 4.1 (McKeran and Thomas, 1980).

Table 4.1   Frequency of symptoms at presentation compared to that at assessment and diagnosis in 653 patients with cerebral gliomas investigated at the National Hospital, London (1955–1975) (From McKeran and Thomas, 1980, with permission)

Symptom	Relative frequency as initial symptom (%)	Relative frequency at assessment and diagnosis
Epilepsy	38.3	53.9
Grand mal	15.9	20.4
Focal	14.7	22.8
Temporal lobe	7.2	8.6
Minor absence	0.5	2.1
No epilepsy	61.7	46.1
Headache	35.2	71.4
Mental change	16.5	52.2
Hemiparesis	10.3	43.3
Vomiting	7.5	31.5
Dysphasia	6.9	27.0
Impaired consciousness	4.6	24.8
Visual failure	4.3	17.9
Hemianaesthesia	3.4	13.6
Hemianopia	1.8	8.1
Cranial nerve palsy	1.8	10.9
Miscellaneous	2.3	7.0

Symptoms and Signs of Raised Intracranial Pressure

The headache associated with intracranial tumours is due to the effect of raised intracranial pressure, producing vascular distension and distortion, as well as of the local effect of pain-sensitive intracranial structures from pressure or infiltration by tumour (vessels, dura, certain cranial nerves). A headache associated with raised intracranial pressure has a throbbing quality and is initially intermittent, occurring in the early morning. As time progresses it becomes more severe and lasts longer. There may be severe exacerbations and any manoeuvre that raises intracranial pressure produces a paroxysm of pain. There is some correlation as to the side of the headache compared to the laterality of the tumour, but the anteroposterior correlation is less precise, with the majority of headaches localized in the frontal and occipital regions (McKeran and Thomas, 1980).

Headache is the most common initial symptom, and is found most frequently in association with frontal, temporal and parietal gliomas, reflecting the greater number of tumours in these sites. By contrast, the proportion of cases of frontal, temporal and parietal gliomas associated with initial headache was less than half those in the occipital lobe or cerebellum. Cerebellar tumours, by obstruction of the fourth ventricle or aqueduct of Sylvius, may cause hydrocephalus. The pain due to hydrocephalic attacks may be excruciating.

Vomiting associated with raised intracranial pressure often occurs in the morning and may be preceded by retching and nausea, particularly in children. It may be precipitated or exacerbated by exertion and stooping, both of which further raise the intracranial pressure. Vomiting is generally more frequent and severe when the ventricular system is dilated, particularly with infratentorial tumours. At presentation, vomiting is found in approximately 30% of cerebral gliomas, rising to 50% in cerebellar gliomas.

The triad of headaches, vomiting and ocular disturbance are the hallmarks of raised intracranial pressure. Papilloedema, often with exudates, haemorrhages and optic atrophy, may be noted as a sign of symptomatic visual failure. Visual field defects, most commonly bitemporal hemianopia due to chiasmal compression or homonymous hemianopia, may be seen. Visual obscurations and sixth nerve palsies are frequently encountered in raised intracranial pressure.

Epilepsy

Epilepsy, particularly of late onset, is a common feature of brain tumours and almost 40% of the National Hospital series of 653 patients with cerebral gliomas presented with epilepsy (McKeran and Thomas, 1980). It is an early symptom and frequently predates the onset of focal neurological deficit and papilloedema (Lund, 1981). The growth of an intracranial tumour may be associated with a changing pattern in the clinical features of an epileptic attack (Strobos et al., 1958). The features of the clinical history which may indicate an underlying tumour include a change in the character of the epilepsy, resistance to drug therapy, the appearance of status epilepticus, post-ictal paresis, headaches and focal neurological signs or onset after 25 years of age (Ketz and Xanthakos, 1969; Hess, 1970). Olfactory hallucinations in psychomotor attacks are thought to be particularly suggestive of an underlying neoplasm, as has status epilepticus (Janz, 1969).

The character of the epilepsy depends on the location of the glioma. The deeper a tumour penetrates into the temporal lobe, the more complex the epileptic attack tends to become (Ketz, 1968). Frontal lobe tumours tend to cause grand mal seizures.

The likelihood of developing epilepsy from an intracranial tumour is related to the site of origin within the brain and the nature of the neoplasm. The closer that a brain tumour is situated to the cortex, the more likely is epilepsy. More slowly growing tumours near or on the cortex in the centroparietal regions are the most likely to give rise to epilepsy (Zulch, 1951).

Seizures are the most common presenting symptom in patients with low-grade tumours and there is a direct correlation between the chronicity of the growth of the neoplasm and the incidence of seizures (Cascino, 1990). Penfield's series of 230 patients with glioma showed seizures occurring in 70% of patients with astrocytoma, in 92% patients with oligodendroglioma and in 37% of patients with glioblastoma (Penfield et al. 1940). The shorter duration of disease in patients with glioblastoma is associated with a lower incidence of epilepsy (Ketz, 1974).

Epilepsy of late onset (after the age of 25 years) should be investigated by neuroradiological means as 60% of these will reveal abnormalities, with brain tumours accounting for 13%–18% of these (Henry et al., 1990; Zhu, 1990).

Focal Signs

Focal neurological symptoms and signs, such as visual field loss, diplopia, deafness, ataxia, hemisensory loss or hemiparesis and dysphasia, may provide strong evidence of the site of a cerebral tumour. The progressive nature of the symptoms suggests an expanding and locally compressive and infiltrative lesion. Prior to the advent of CT scan and MRI, over 50% of patients had one or more of papilloedema, mental deterioration, hemiparesis and cranial nerve involvement (usually an upper motor seventh nerve weakness (McKeran and Thomas, 1980).

Specific Clinical Syndromes

Against this general background of symptoms and signs, several characteristic clinical patterns typical of the development and natural history of the different histological types can be distinguished. To illustrate this, the clinical features typical of glioblastoma will be described here.

Glioblastoma

Glioblastoma multiforme (astrocytoma grade 4) accounts for 15%–20% of all intracranial tumours and approximately 50% of all gliomas. The peak incidence occurs between the ages of 48 and 52 years and males are twice as often affected as females (Zulch and Borck, 1965).

There is a predilection for the frontal and temporal regions to be affected, while the occipital lobe is very rarely the site of origin of the tumour. Intrinsic tumours of the frontal lobe commonly cause mental and personality changes, with emotional lability as an early symptom. As the posterior frontal region becomes involved by tumour, a spastic hemiparesis develops. In the dominant hemisphere this is accompanied by dysphasia. Epilepsy may be a manifestation of the tumour, particularly in the frontal lobe.

One characteristic pattern of spread in these tumours is a butterfly distribution spreading into the white matter of both hemispheres through the anterior corpus callosum. Dementia, bilateral pyramidal signs and frequency of micturition with occasional incontinence commonly result (Maurice-Williams, 1974).

With parietal lobe tumours, involvement of the sensory cortex leads to neglect of the contralateral side of the body, often accompanied by visual inattention. Tumours of the non-dominant (usually right) parietal lobe will affect the ability to orientate the body image as well as disordered left–right discrimination. In the dominant hemisphere, speech is affected by a receptive dysphasia.

Temporal lobe tumours may be accompanied by hemiparesis, homonymous hemianopia and, in the dominant hemisphere, dysphasia. Temporal lobe epilepsy may result in feelings of fear or pleasure, olfactory or gustatory hallucinations or repetitive psychomotor movements. The deeper a tumour penetrates into the temporal lobe, the more complex the epileptic attacks become, and dream-like states are particularly associated with temporal lobe tumours (Bingley, 1958).

The occipital lobe is the least common primary site of origin of glioblastoma, and tumours in this region lead to homonymous hemianopia and later to features of raised intracranial pressure.

The highly malignant nature of this tumour determines the progression of symptoms and signs, with rapidly developing raised intracranial pressure and progressive focal neurological deficit. Extensive oedema may occur at a stage when the tumour is relatively small, producing an early extensive focal neurological disturbance and symptoms of raised intracranial pressure with loss of alertness and early depression of consciousness. An ictal onset has been described in up to 4% of cases (Frankel and German, 1958).

In the following chapters the specific clinical syndromes found in a wide variety of malignant brain tumours will be described in detail. Thus, the clinicopathological correlations are considered in Chapter 5, while low-grade gliomas, paediatric brain tumours, tumours related to the acquired immunodeficiency syndrome (AIDS), and metastatic tumours are covered respectively in Chapters 6–9.

References

Bailey P, Buchanan DN, Bucy PC (1948) Intracranial Tumours of Infancy and Childhood, 2nd edn. University of Chicago Press, Chicago

Battistella PA, Ruffille R, Viero F, Benelagli B, Condini A (1990) Brain tumours: classification and clinical aspects. Pediatr Med Chir 12: 33–39

Brewis M, Poskanzer DC, Rolland C, Miller H (1966) Neurological disease in an English city. Acta Neurol Scand 42: 1

Bingley T (1958) Mental symptoms in TLE and in temporal lobe gliomas. Acta Psychiat Scand, Suppl 120

Cascino GD (1990) Epilepsy and brain tumours: implications for treatment. Epilepsia 31, Suppl 3: 537–544

Frankel SA, German WJ (1958) Glioblastoma multiforme – a review of 219 cases with regard to natural history, pathology, diagnostic methods and treatment. J Neurosurg 15: 489

Henry C, Despland PA, Regli F (1990) Initial epileptic crisis after the age of 60: aetiology, clinical aspect and EEG. Schweiz Med Wochenschr 120: 787–792

Hess R (1970) Die epileptogenen Hirntumoren. Mod Probl Pharmacopsych 4: 200

Janz D (1969) Die Epilepsien. Georg Thiem, Stuggart

Ketz E (1968) Zum Klinischen Aspekt der Psychomotorischen Epilepsie Dr Alfred Huthig, Heidelberg

Ketz E, Xanthakos D (1969) Die Bedeutung epileptischer Anfalle dei schiafeniappengeschwulsten. Med Weh 20: 638

Ketz E (1974) Brain tumours and epilepsy. In: Vinken PJ, Brwyn GW (eds) Handbook of Clinical Neurology, Vol. 16. North-Holland Amsterdam, p 254

Leibowitz W, Atler M (1969) Tumours of the nervous system: incidence and population selectivity. Neurology (Minneap) 19: 292

Lund M (1981) Epilepsy in association with intracranial tumour. Acta Psychiat Scand Suppl 81

Maurice-Williams R (1974) Micturition symptoms in frontal lobe tumours. J Neurol Neurosurg Psych 37: 43

McKeran RO, Thomas DGT (1980) The clinical study of gliomas. In: Thomas DGT, Graham DI (eds) Brain Tumours: Scientific Basis, Clinical Investigation and Current Therapy. London, Butterworths, pp 194–230

Penfield W, Erikson TC, Tarlov I (1940) Relation of intracranial tumours and symptomatic epilepsy. Arch Neurol Psychiat 44: 300–315

Percy AK, Elvaback LR, Okasaki H, Kurland LT (1972) Neoplasms of the central nervous system: epidemiologic considerations. Neurology (Minneap) 22: 40

Strobos RRJ, Alexander E, Maslund RL (1958) Brain tumour presenting as convulsive disorder. Dis Nerv Syst 19: 518

Zhu PG (1990) Delayed epileptic seizures in adults. Chung Hua Shen Ching Ching Shen Ko Tsa Chih 23: 286–288

Zulch KJ (1951) Röntgen diagnostik beim cerebralen Anfall. Verh Ditsch Ges Med. 56: 24

Zulch KJ, Borck WF (1965) Tafeln über die relative Haufigkeit der Hirngeschwulste in Verschiedenen Altersklassen. Zentrabl Neurochir 12: 93

5 Complications of Primary Malignant Brain Tumours

N.V. Todd and D.I. Graham

There are a large number of potential complications of malignant brain tumours (Table 5.1). These complications may arise from characteristics of the tumour itself, such as its progressive growth, its tendency to produce brain oedema, or haemorrhage within its substance. Alternatively, a complication may follow treatment of a malignant brain tumour, for example an intracerebral haematoma following tumour biopsy or leucocytopenia resulting from chemotherapy.

The two commonest intracranial complications caused by malignant brain tumours are *raised intracranial pressure* (ICP), which is predominantly due to the mass effect of both the tumour and its associated brain oedema, and *focal or site-specific complications* which may be irritative, causing epilepsy, or destructive or compressive phenomena producing a focal neurological deficit.

Complications of the Tumour Itself

Raised Intracranial Pressure

Clinical Effects

The clinical effects of raised ICP include headache, visual obscuration, nausea or vomiting, papilloedema, drowsiness, nuchal rigidity or 6th nerve palsy. At the time of diagnosis, symptoms of raised ICP are found in over three-quarters of patients with a malignant glioma (Table 5.2). The freqeuncy of symptoms of raised ICP are similar in patients with brain metastases, although the nature differs. *Headache* is typically worse on awakening, probably because of mild carbon dioxide retention during sleep. The headache tends to become nearly constant, although there are remissions, and exacerbations which are associated with activities that raise ICP such as stooping or coughing. *Papilloedema*, which used to be thought to be due to compression of the central retinal vein as it crosses the subarachnoid space of the optic nerve sheath (Hayreh, 1964), is now thought

Table 5.1   Complications of malignant brain tumours

Complications of the tumour itself
Raised intracranial pressure
   Clinical effects
   Brain compliance
   Tumour volume
   Brain oedema
   Hydrocephalus
   Pathology of intracranial expanding lesion
Focal or site-specific
   Epilepsy
   Focal neurological deficit
Spontaneous haemorrhage
Metastasis
   CSF
   Extraneural
Complications of treatment
Surgery
   Risks and benefits
   Freehand biopsy
   Stereotaxic biopsy
   Tumour decompression
Radiotherapy
   Acute reactions
   Early delayed reactions
   Late delayed reactions
Chemotherapy
   Systemic complications
Steroids
   CNS complications

Table 5.2   Symptoms of raised intracranial pressure at the time of diagnosis of malignant brain tumour

	Glioblastoma multiforme	Metastases
Headache	77%	55%
Vomiting	31%	
Papilloedema	58%	10%
Impaired consciousness	37%	
Visual failure	37%	

Data from McKeran and Thomas (1980); Roth and Elvidge (1960); Frankel and German (1958); Jelsma and Bucy (1969).

to be due to the accumulation of axoplasm in the optic papilla due to blockage of its flow from the ganglion cells of the retina along the optic nerve (Tso and Hayreh, 1977). Disc swelling is accompanied by swelling of unmyelinated nerve fibres around the papilla and, later, atrophy and gliosis of the peripheral retina occur.

Brain Compliance

An increase in intracranial volume eventually causes an increase in ICP because, except in infants, the skull cannot expand to accommodate the increased intracranial volume. The relationship between intracranial volume and pressure is shown in Fig. 5.1. Growth of tumour can initially be *compensated* for by displacement from the skull of cerebrospinal fluid (CSF) and intracranial blood volume. This, however, leads to reduced brain compliance, and further growth of the tumour produces a progressive rise in intracranial pressure. Eventually a point is reached where very small further increases in intracranial volume, such as may occur as a result of vasodilatation due to mild hypercapnia, will trigger a marked rise in intracranial pressure. This may cause ICP waves and may precipitate abrupt deterioration in the patient's clinical condition. *Progressive decompensation* then occurs and there is increasing distortion of the brain and herniation through the opening in the tentorium cerebelli (tentorial hernia) or the foramen magnum (tonsillar hernia). Initially, distortion of the brain can occur without any elevation in the ICP, but as the expanding lesion becomes larger there is eventually impaction of the brain at either the tentorial opening or the foramen magnum.

The *mechanisms of raised ICP* are multiple. The tumour has an intrinsic mass effect. Peritumoral brain oedema increases this effect – indeed the effect of oedema is often greater than that of the tumour itself. Intracranial pressure is also increased where there is obstruction of cerebral blood flow (CBF) through the aqueduct of Sylvius with a consequent increase in CSF

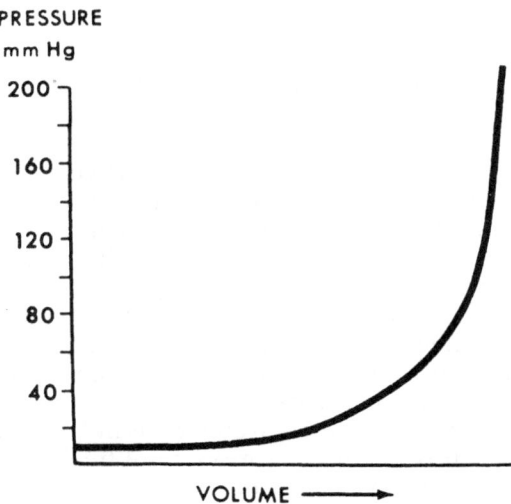

Fig. 5.1   Raised intracranial pressure. Relationship between intracranial pressure and volume of intracranial contents. The time scale may be from minutes to months, but once compensating mechanisms have been exhausted, a small increase in volume will produce a large increase in pressure (From Graham, 1990, with permission from Churchill Livingstone)

volume. Occlusion of the subarachnoid space at the tentorial hiatus (tentorial hernia) or at the level of the foramen magnum (tonsillar hernia) increases ICP and may set up pressure gradients between the supra- and infratentorial compartments or between the brain and the spinal compartment. These pressure gradients may cause abrupt deterioration in the patient's clinical condition. Eventually the ICP approaches the systemic arterial pressure and there is a reduction in CBF. Complete cessation of brain function, "brain death", is brought about by the arrest of the cerebral circulation and this may be confirmed by the failure of contrast medium to enter the skull during angiography – pseudo-occlusion (Langfitt and Kassell, 1966).

An increase in ICP is greatest with tumours which grow rapidly; the brain can often accommodate a large tumour, such as a meningioma, that grows slowly. The age of the patient modifies these responses; raised ICP occurs later in infants where the cranial vault can expand and it also occurs later in older patients where there is pre-existing brain atrophy.

Tumour Volume

Tumour volume often constitutes a considerable proportion of, for example, a lobe of the brain at post-mortem examination. The volume of malignant gliomas at the time of diagnosis has been calculated (Kelly et al., 1978a, b). In 14 patients with grades III or IV malignant gliomas, the mean volume of tumour calculated from computed tomography (CT) scan data for the contrast enhancing rim and necrotic centre was 2201 mm^3.

Brain Oedema

Brain oedema is a common accompaniment of malignant brain tumours and the volume of oedema is often greater than the volume of the tumour. The oedema of malignant brain tumours is protein-rich, "vasogenic oedema" (Klatzo, 1967) and its source is the defective blood – brain barrier which allows protein and fluid leakage from tumour capillaries into brain. The degree of damage to the barrier is heterogeneous both between different tumours of a similar histological grade and within different regions of the same tumour. The blood – brain barrier in brain adjacent to tumour is normal (for reviews see Neuwelt et al., 1986). Oedema probably does not have any adverse effect upon brain function but causes problems by increasing the effective volume of a focal mass lesion, producing a global rise in ICP and local brain or vascular shifts.

When brain oedema is present, water accumulates preferentially in white matter, through which it can spread more freely than through grey matter (Clasen et al., 1962; Meinig et al., 1973; Hochwald et al., 1976). The rate of formation of oedema can be accelerated by agents that cause cerebral arterial or arteriolar vasodilatation. Once formed, protein, water and other constituents of the oedema fluid move at different rates through the extracellular space. Ultimately, the oedema fluid either drains into the ventricular CSF or is absorbed into the blood stream through cerebral blood vessels.

Whether localized or generalized, space occupation by tumour and oedema can be recognized post mortem by an increase in the weight of the brain, by narrowing of sulci, flattening of gyri and a reduction in the size of the ventricles. Upon sectioning the brain, an increase in the volume of the affected brain tissue is seen. After fixation, the oedematous region often has a yellowish-green tinge.

Histologically, the oedematous tissue appears paler than normal. In the cortex the extracellular space remains small, much of the oedema fluid being taken up by astrocytes and their processes. A different appearance, however, is seen in the white matter, in which oedema fluid spreads in the extracellular space between the myelinated fibres. As white matter can accommodate the most oedema fluid, it is usually more swollen than grey matter. Reactive astrocytes appear and serum proteins may be demonstrated within them by immunohistochemistry (Seitz and Weschler, 1987). The oligodendrocytes are also swollen and their enlarged nuclei are surrounded by halo-like clear areas. In experimental preparations, perfusion fixation, the combined use of tracer materials and electron microscopy have confirmed that, during the first 48 h, oedema fluid follows the path of least resistance and spreads relatively freely along the extracellular pathways in white matter, whereas in grey matter the extracellular spread of the tracer is limited by the various adhesions between cells.

These structural abnormalities can be demonstrated in life with X-ray, CT or magnetic resonance imaging (MRI). On CT images, brain surrounding a malignant tumour is usually oedematous as shown by reduced attenuation compared to normal brain. The oedema is confined to white matter and tends to have sharply demarcated margins. Abnormalities in the blood–brain barrier can be demonstrated with intravenous contrast enhancement which is found in 97% of glioblastomas; in 84% of these it is a thick, irregular, peripheral rim of enhancement (Moseley, 1986). Metastatic tumours show enhancement in 90% of cases, which tends to be homogeneous when the tumour is less than 2 cm in diameter and heterogeneous in larger metastases.

There is considerable variability in the amount of oedema produced by otherwise apparently similar tumours. Perhaps this, in part, reflects differences in the degree of damage to the barrier between different tumours. Rapid spread of oedema may follow minor trauma (Russell and Rubinstein, 1989) and most neurosurgeons will have seen massive hemispheric oedema precipitated by an otherwise uncomplicated burr-hole tumour biopsy.

Hydrocephalus

Hydrocephalus is commonly the consequence of obstruction of CSF circulation and may be the major mechanism of raised ICP (see Miller and Adams, 1984). Posterior fossa tumours may cause obstruction by invading the fourth ventricle or by distorting the cerebral aqueduct. Patients with the common posterior fossa tumours such as medulloblastomas in children and metastatic tumours in adults frequently present with hydrocephalus. Third ventricular malignant tumours may also cause CSF obstruction that may be intermittent. Both primary and secondary intracranial tumours may cause hydrocephalus by diffuse invasion of the leptomeninges; this will be discussed in detail later.

Pathology of Intracranial Expanding Lesions

Initially there is deformation or destruction of brain tissue around or within the tumour, with a reduction in the volume of CSF within the skull as the mass enlarges. Shift and distortion of the brain occur and, in an intact skull, internal herniae eventually form (Fig. 5.2). A tumour within a cerebral hemisphere may result in herniation of the cingulate gyrus under the free edge of the falx, or herniation of the parahippocampal gyrus of the medial part of the temporal lobe through the opening of the tentorium cerebelli. Tumours in the cerebellum may result in herniation of the cerebellar tonsils through the foramen magnum. The internal herniae may produce specific clinical features such as bilateral leg weakness with herniation of the cingulate gyrus, or neck pain and nystagmus with tonsillar herniation. As the internal herniae form and the subarachnoid space and cisterns are obliterated, pressure gradients between the various intracranial compartments develop. In addition, vascular lesions such as haemorrhage and ischaemic necrosis of the brain are important secondary complications of internal herniae.

Supratentorial Expanding Lesions

With a unilateral tumour, the affected hemisphere expands and the surface of the brain is thus pressed against the unyielding dura. The dura becomes tight, cerebral sulci are narrowed, gyri are flattened and CSF is displaced from the surface of the brain. The ipsilateral ventricle and third ventricle become smaller and there is displacement of the midline structures away

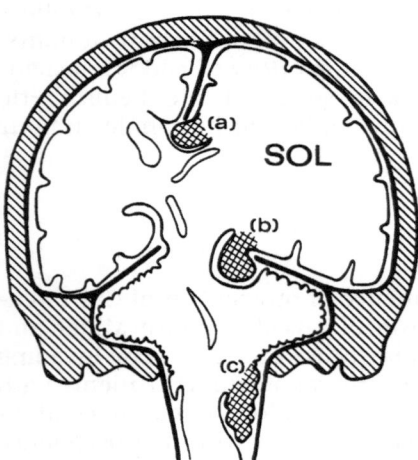

Fig. 5.2 Raised intracranial pressure. Diagrammatic representation of distortion and herniation of the brain caused by a space-occupying lesion (SOL), in one cerebral hemisphere. There is displacement of the midline structures and ventricles, and supracallosal (a), tentorial (b) and tonsillar (c) herniae have developed (From Graham, 1990, with permission from Churchill Livingstone)

from the lesion. In some cases the contralateral ventricle becomes enlarged due to obstruction of the interventricular foramen of Monro. Other external features may include a groove on the undersurface of the frontal lobe as an impression of the lesser wing of the sphenoid bone, and downward displacement of the mammillary bodies into a narrowed interpeduncular fossa. These features are followed by the development of internal herniae as parts of the brain are displaced from one intracranial compartment to another.

Supracallosal Hernia. Known also as subfalcine or cingulate hernia, the supracallosal hernia is particularly associated with frontoparietal tumours, which account for 35% of all gliomas and 62% of malignant gliomas (McKeran and Thomas, 1980). Supracallosal herniation occurs when the cingulate gyrus on the same side as a mass lesion herniates under the free edge of the falx cerebri. If the displacement is sufficiently severe, circulation through the pericallosal artery may be reduced and infarction of the medial surfaces of the frontal and parietal lobes, together with the corpus callosum, may ensure.

Tentorial Hernia. Tentorial herniation frequently occurs with malignant tumours of the medial temporal lobe. Temporal lobe tumours represent 34% of all malignant gliomas (McKeran and Thomas, 1980). One clinical effect of tentorial herniation is compression of the midbrain which causes reduction in the level of consciousness; this is present at the time of diagnosis in 37% of patients with supratentorial glioblastoma (see Table 5.2). Compression of the ipsilateral descending pyramidal tract produces hemiparesis contralateral to the tentorial hernia. Seventy per cent of patients with malignant gliomas have a contralateral hemiparesis at the time of diagnosis (McKeran and Thomas, 1980), although some will be secondary to direct involvement of the motor cortex rather than tentorial herniation. As the hernia develops, the ipsilateral oculomotor nerve becomes compressed and the midline is narrowed in its transverse axis with resulting compression of the aqueduct. Ipsilateral signs may cause "false localization" as the opposite crus is thrust against the tentorial edge.

Known also as an uncal or lateral transtentorial hernia, tentorial herniation occurs when the uncus in the medial part of the parahippocampal gyrus is displaced downwards and medially through the tentorial incisura (Fig. 5.3a). Nevertheless, it may be difficult to establish post mortem whether the ICP has been high during life because some degree of shift and herniation of the brain can occur during the period of spatial compensation before a rise in ICP has occurred. The best indication of high ICP due to a supratentorial expanding lesion, is the presence of a wedge of pressure necrosis in one or both of the parahippocampal gyri (Adams and Graham, 1976; Graham *et al.*, 1987). Such lesions are present only in patients in whom the recorded ICP is greater than 5.3 kPa (40 mmHg) during life. Significant tentorial herniation can therefore be recognized by the presence of a wedge of pressure necrosis in the parahippocampal gyrus (Fig. 5.3b). Not all such lesions are identifiable macroscopically and some are not haemorrhagic; a past episode of herniation can be recognized by the presence of a small organized gliotic wedge of pressure necrosis. A tentorial hernia will eventually cause obliteration of the subarachnoid space and of

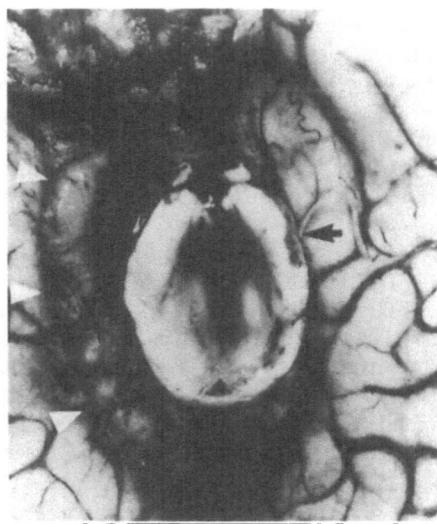

Fig. 5.3a Tentorial hernia. There is a tentorial hernia along the full length of the parahip-pocampal gyrus (white triangles). There is also angulation of the oculomotor nerves, extensive secondary haemorrhage in the midline of the upper brain stem and haemorrhagic necrosis in the contralateral cerebral peduncle (black arrow) (From Graham, 1990, with permission from Churchill Livingstone)

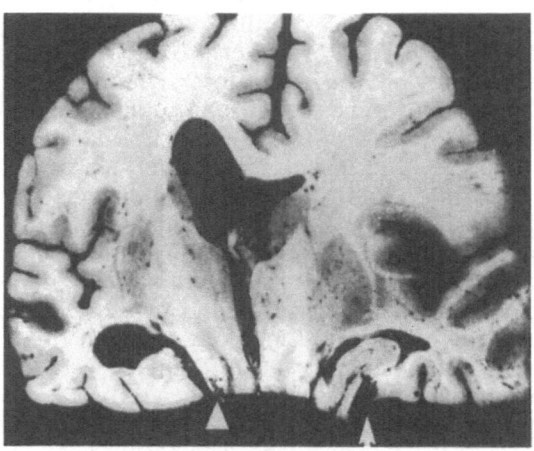

Fig. 5.3b Tentorial hernia. Lateral shift by mass lesion on the right. There is displacement of the midline structures to the left, a supracallosal hernia, enlargement of the contralateral lateral ventricle, a wedge of haemorrhagic necrosis in relation to tentorial hernia on the right (white arrow), and haemorrhagic necrosis in the contralateral cerebral peduncle (white triangle) (From Graham, 1990, with permission from Churchill Livingstone)

the tentorial incisura, leading to a CSF pressure gradient between the supratentorial compartment and the posterior fossa.

Tentorial herniae may obstruct the circulation of the anterior choroidal artery leading to infarction, particularly in the caudate and amygdaloid nuclei, the medial segment of the globus pallidus, the subthalamic region and the anterior half of the hippocampus. Compression of the posterior cerebral artery and its branches results in necrosis in the thalamus, the posterior part of the hippocampus, the cortex of the undersurface of the temporal lobe and the inferior and medial surfaces of the occipital lobe including the visual cortex. The superior cerebellar artery may be compressed and lead to necrosis of the superomedial portions of the cerebellum. The posterior cerebral artery and its branches are the vessels most commonly affected by tentorial herniation and the visual cortex frequently undergoes infarction.

Central Transtentorial Hernia. Central transtentorial hernia develops particularly when ICP increases rapidly either due to frontal or parietal lobe tumours, or to a bilateral expanding lesion. Herniation of both parahippocampal gyri occurs through the tentorial incisura to form a circular or ring hernia.

Haemorrhage and Infarction in the Midbrain and Pons. These features are common and are often terminal events in patients with supratentorial tumours, high ICP and tentorial herniation. They are found mainly in the midline of the midbrain and upper pons and are thought to be due to displacement and elongation of the upper brain stem (Hassler, 1967). The pathogenesis of the haemorrhagic infarction is uncertain, but important contributory factors include caudal displacement and anteroposterior elongation of the rostral brain stem and relative immobility of the basilar artery. The presence of these factors results in obstruction of venous drainage and stretching of arteries.

Infratentorial Expanding Lesions

Tonsillar Hernia. The early tonsillar herniation may be asymptomatic. Later there is neck stiffness and, less commonly, ataxia and nystagmus with spasticity of legs and arms. Terminally, there is medullary compression which causes respiratory irregularities, apnoea and death.

Reversed tentorial herniation was thought to be a rare, usually fatal, complication of posterior fossa tumours. It was diagnosed at post mortem where the vermis of the cerebellum lay above the tentorial hiatus causing compression of the brain stem at that level. Reversed herniation can be diagnosed during life with sagittal MRI scanning and may be more common than was previously thought. We have recently identified reverse herniation in 17 of 29 children (59%) with posterior fossa tumours imaged with MRI (N.V. Todd, unpublished observation).

Because of the considerable variation in the normal configuration of the cerebellar tonsils, a significant hernia may be difficult to recognize post mortem. Incontrovertible evidence of a hernia is seen, however, when the tonsils become impacted and their tips undergo necrosis, which is sometimes haemorrhagic. A transverse groove may also be present on the ventral aspect of the medulla where it has become compressed against the anterior edge of the foramen magnum. Although most commonly found in association with a mass lesion in the posterior fossa, a tonsillar hernia may also develop as a result of an expanding mass above the tentorium.

Expanding lesions in the posterior fossa characteristically produce obstructive hydrocephalus. Tonsillar herniation may be severe and occasionally one or both posterior inferior cerebellar arteries may be compressed, resulting in infarction and necrosis of the undersurfaces of the cerebellar hemispheres.

In a patient with an intracranial expanding lesion, lumbar puncture can precipitate tonsillar or tentorial herniation with serious consequences and even death of the patient. Even if only a small amount of CSF is withdrawn, more may leak into the spinal extradural space via the puncture wound in the meninges. Lumbar puncture is, therefore, contraindicated in any patient with suspected increased ICP until the presence of an intracranial expanding lesion has been excluded by appropriate imaging. An exception to the rule is a suspected case of bacterial meningitis without focal signs, when lumbar puncture is an essential step in establishing the diagnosis.

Other Features of the Pathology of Intracranial Expanding Lesions

Pressure Effects Upon the Skull. Any chronic increase in ICP causes local effects upon the bone of the skull. Brain tumours usually cause bone erosion but occasionally hyperostosis may occur. The generalized rise in ICP produces erosion of the dorsum sellae and of the posterior clinoid processes. Other changes include thinning of the lesser wings of sphenoid or of the skull vault. Changes in the skull vault may be extreme in children, producing the classical "copper-beating" appearance on skull X-ray. The incidence of these findings is low in clinical series.

External Herniation. If the ICP remains high after a neurosurgical operation, an external hernia may develop through a defect in the skull. This may amount simply to a protrusion of small pieces of cortex through burr holes, but if an internal tumour decompression has been undertaken, a large part of the cerebral hemisphere may herniate as an external hernia cerebri through a craniotomy or craniectomy defect.

Pituitary. Necrosis of the anterior lobe of the pituitary gland may occur as a consequence of raised ICP and distortion of the brain (Wolman, 1956). The mechanism of anterior lobe necrosis is presumably that of interference with the flow of blood through the long hypothalamo-hypophyseal portal vessels in the pituitary stalk. Previous studies have shown that every patient with a large or medium-sized infarct had had at some stage raised intracranial pressure (Harper *et al.*, 1986). It is suggested that mechanical factors such as lateral displacement of the hypothalamus could have a role in the pathogenesis of anterior lobe infarction.

Focal or Site-Specific Complications

Epilepsy

Epilepsy occurs in one-third of patients with malignant gliomas and in 15%–20% of patients with metastases at the time of diagnosis (McKeran and Thomas, 1980). The likelihood of a patient with a malignant glioma developing epilepsy is predominantly dependent upon the location of the tumour. For all tumours, epilepsy is most frequent where the tumour involves the motor cortex and least frequent when the occipital cortex is involved (Penfield and Jasper, 1954). Cortical tumours are more likely to produce seizures than tumours located within white matter (Jackson, 1958). Malignant gliomas are more likely to produce brain destruction than irritation, and the incidence of seizures at the time of diagnosis in patients with benign brain tumours is double that of patients with malignant tumours. Temporal lobe tumours often cause complex seizures; frontal lobe tumours are more likely to cause grand mal seizures. The patient's age appears to be important; the risk of developing epilepsy in patients with malignant gliomas is greater in patients aged 30–40 than in the commoner age group for presentation of 50–60 years (Ketz, 1974). Similarly the type of seizure found in patients with malignant brain tumours varies with the age of presentation (Ketz, 1974).

Epilepsy is the second commonest symptom at the time of diagnosis – it is frequently the initial symptom and often predates a neurological deficit; 5% of patients with epilepsy and intracranial tumours have had seizures for over 10 years prior to tumour diagnosis (Penman and Smith, 1954). Nevertheless the incidence of brain tumours in patients with epilepsy is low. Probably less than 1% of patients presenting with epilepsy to general medical or neurological clinics will have an intracranial tumour. The incidences of intracranial tumours in patients attending neurosurgical clinics may be higher. Late onset status epilepticus is rather more commonly associated with an intracranial tumour.

Focal Neurological Deficit

A focal neurological deficit due to *destruction and compression* of brain is a feature of many brain tumours and we have earlier alluded to the fact that malignant brain tumours are more likely to produce brain destruction than irritative phenomena such as epilepsy. Symptoms and signs of brain destruction or compression include hemiparesis, dysphasia, hemianaesthesia or hemianopia (Table 5.3). These features obviously relate to the site of the tumour. Large tumours in "silent" brain regions may be relatively asymptomatic. Mental deterioration is present in over 50% of patients with malignant gliomas at the time of diagnosis and has been present for an average of 8 months prior to diagnosis (Roth and Elvidge, 1960). Although mental deterioration may be caused by globally raised intracranial pressure, it more commonly reflects the tendency of malignant gliomas to occur in the frontal lobe (20%) and, in particular, to invade the corpus callosum (Matsukado et al., 1961). Brain stem tumours frequently cause lower cranial nerve palsies by invading the nuclei and fibre tracts of cranial nerves.

Macroscopically malignant gliomas are often roughly spherical, although they may be irregular as in the "butterfly" glioma. The central part of the tumour is characteristically variegated with mottled opaque and creamy-yellow areas caused by necrosis. The peripheral rim tends to be pinkish-grey and this region contains viable and dividing tumour cells. The necrotic

Table 5.3   Clinical features of brain destruction or compression in malignant glioma at the time of diagnosis

Symptom	Incidence	Sign	Incidence
Hemiparesis	43%	Hemiparesis	67%
Dysphasia	29%	Hemianaesthesia	38%
Hemianaesthesia	14%	Hemianopia	32%
Hemianopia	1%	Dysphasia	29%

Data from McKeran and Thomas (1980); Roth and Elvidge (1960); Jelsma and Bucy (1969).

zone of malignant brain tumours usually produces a complete neurological deficit which is not improved by steroids and there is no prospect of functional recovery. Brain that is adjacent to the tumour periphery is more likely to have functional impairment because of compression rather than invasion. The neurological deficit in these cases is usually incomplete and improves with steroid administration; this is particularly true for oedematous brain regions. These features can be determined from the CT or MR image. Low-density central regions represent necrosis; the contrast-enhancing peripheral rim seen on CT scans contains tumour alone in 61% of cases and tumour cells heavily infiltrating "normal" brain in a further 21%; the peritumoral low-density areas (oedema) surrounding malignant gliomas are regions of normal brain in which malignant cells can always be identified (Burger *et al.*, 1983; Kelly *et al.*, 1987b). Attempting to distinguish brain destruction or compression as the mechanism causing neurological deficits has important implications for the likelihood of alleviating neurological deficit in patients with malignant brain tumours.

Spontaneous Haemorrhage

Malignant brain tumours are frequently characterized by abnormal blood vessels, and haemorrhage is not unusual. Major intracerebral haemorrhage, which may be seen post mortem in metastatic carcinoma of bronchus and choriocarcinoma, was thought to be an uncommon complication of malignant brain tumours (Fig. 5.4), but with the advent of widespread CT scanning a higher incidence may well become established. Small foci of haemorrhage are the commonest secondary change identified in pathological specimens of capillary haemangioblastomas and are also common in lymphomas, leukaemias and glioblastomas.

Prior to CT scanning, the incidence of haemorrhage into primary cerebral neoplasms was thought to be around 3%. Oldberg (1933) reported a pathological series of 832 patients with gliomas in which 31 intracerebral haematomas were found (3.7%). Other series of brain tumours have found massive intratumoral haemorrhage in 9 of 94 (10%), 7 of 162 (5%) and 8 of 590 (1.3%) of patients with cerebral neoplasms (Globus and Saperstein, 1942; Glass and Abbott, 1955; Scott, 1975). In all of these series the commonest primary malignant tumour was the glioblastoma multiforme. Data from the CT era in fact give a similar incidence of intratumoral haemor-

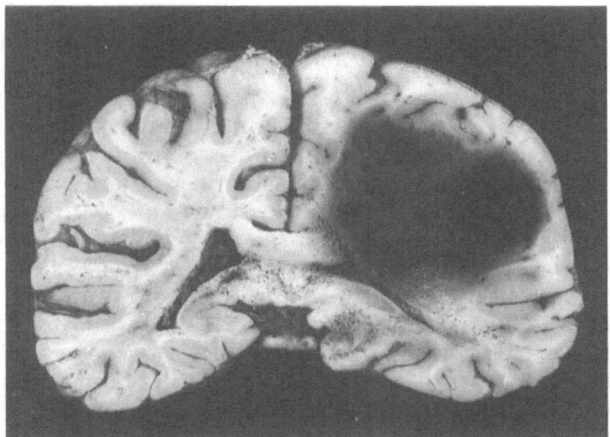

Fig. 5.4 Tumour "stroke". There is extensive spontaneous haemorrhage into a glioblastoma centred on the partial lobe of the right cerebral hemisphere. Although haemorrhage has obscured the main mass of the tumour, glioblastoma can be seen in the splenium of the corpus callosum and the posterior part of the parahippocampal gyrus

rhage in primary brain tumours. In a series of 973 intracranial tumours consecutively diagnosed by CT scanning, 35 cases (3.4%) of intratumoral bleeding were identified (Zimmerman and Bilanink, 1980). Little *et al.* (1979) reported 13 cases of CT-proven intratumoral haemorrhage in 172 consecutive brain tumours diagnosed by CT scanning (7.6%); 8 of these tumours were gliomas, 5 were metastases.

Wakai *et al.* (1982) reported 94 cases of haemorrhage in 1861 patients with brain tumours (5.1%). Haemorrhage was diagnosed clinically, at operation or at post mortem, and the 1861 cases were of all tumour types including pituitary adenoma. Excluding the pituitary tumours, haemorrhage was found in 45 of 1550 patients (2.9%). Haemorrhage was found in 10 of 129 cases (7.8%) of glioblastoma, 3 of 43 cases (7%) of oligodendroglioma and 3 of 104 cases (2.9%) of metastatic tumours. In 26 of these 45 patients (58%) there was a history typical of intracranial haemorrhage; in the others there was no such history. In 11 cases (24%) the haemorrhage was the presenting symptom of the intracranial tumour. Haemorrhage occurred significantly more commonly in young patients (Wakai *et al.*, 1982). Liwnicz *et al.* (1987) have recently reported over 100 cases of histologically-proven glioma in which clinically obvious haemorrhage was found in 5 (5%). Three of these were glioblastomas, 2 were oligodendrogliomas. The incidence of pre-mortem haemorrhage diagnosed pathologically, including microscopic haemorrhage, was very much higher, being identified in 53% of glioblastomas, 57% of oligodendrogliomas and 10% of astrocytomas (Liwnicz *et al.*, 1987).

The incidence of haemorrhage into metastases appears to be greater. Mandybur (1977) reported 13 cases of intratumoral haemorrhage in 93 cases of metastatic carcinoma (14%) of which carcinoma of bronchus accounted for over half of the cases. Secondary tumours most prone to develop spontaneous haemorrhage appear to be metastatic choriocarcinoma, melanoma, carcinoma of bronchus or hypernephroma (Mandybur, 1977).

The association of haemorrhage and brain tumours has more commonly been assessed in patients presenting with intracerebral haemorrhage (ICH). Between 2% and 10% of patients with spontaneous ICH haemorrhage will have an underlying neoplasm as the cause of the haemorrhage (Kase et al., 1985). Russell (1954) reported 461 post-mortem cases of "spontaneous" ICH and found a brain tumour in 9 (2%). Mutlu et al. (1963) reported a similar series of 225 post-mortem cases of spontaneous ICH, and found a brain tumour in 2 (<1%). The data of Scott (1975) quoted earlier give an incidence of brain tumour of 10% (8 tumours) in his clinical series of 80 patients with spontaneous ICH. Of ICH diagnosed by CT scanning, 2% will be associated with gliomas and 4% with metastases (Little et al., 1979). It is therefore reasonable to biopsy the wall of atypical intracerebral haematomas undergoing evacuation to exclude tumour.

Intracranial tumours may also cause subarachnoid haemorrhage. Twelve of the 2092 (0.6%) patients in the Co-operative Study had brain tumours (Locksley et al., 1966). Yasargil (1969) has quoted an incidence of brain tumour in 1%–2% of subarachnoid patients coming to angiography.

Haemorrhage may be the first sign of an intracranial neoplasm and, despite treatment, has a high mortality. In various series, the 30-day mortality for intratumoral haemorrhage was 10 of 50 (20%), 12 of 13 (92%), 2 of 8 (25%) and 7 of 15 (47%) (Scott, 1975; Mandybur, 1977; Little et al., 1979; Albert, 1986), which yield an average mortality of 36%.

Metastasis of Primary Brain Tumours

Malignant primary brain tumours characteristically spread by local infiltration. Typically the brain tissue is invaded but local spread to leptomeninges, ependyma or venous sinuses may occur where the tumour is adjacent to these structures. Metastasis within the CNS is rare, particularly for certain malignant brain tumours. Extraneural metastasis occurs but is extremely rare and in certain instances may be monitored by hormone assay.

CSF metastasis in medulloblastoma is common. The incidence of this phenomenon depends upon the criteria used for diagnosis. An elevated cell count in the CSF will be found in up to 70% of patients with medulloblastoma and this increases to 91% if repeated sampling is undertaken (Balhuizen et al., 1978). In only about one-third of these cases can tumour cells be clearly identified. The floor of the third ventricle, the orbital surface of the frontal lobe and the spinal cord are sites of predilection for intraneural metastasis in medulloblastoma. Spinal metastases can be demonstrated myelographically in about 40% of cases (Deutsch and Reigel, 1980), although there is a wide range quoted for clinical incidence from 7% to 51%. If spinal imaging is performed after spinal irradiation, the incidence of spinal metastasis is less than 10% (Tomita and McLure, 1986). Generalized CSF spread of medulloblastoma may occur as a late phenomenon and the incidence of spinal metastasis in post-mortem series ranges from 50% to 90% (Jacobi and Kornhuber, 1987). Supratentorial metastasis occurs in 6%–10% of patients, although an incidence of 14.6% has been reported (Park *et al.*, 1983). A similar incidence of metastasis occurs in the posterior fossa. Where arachnoid biopsies of the cisterna magna are performed at the time of decompressive surgery, two-thirds will show meningeal invasion by medulloblastoma (Tomita and McLure, 1986).

The high incidence of widespread CSF metastasis in medulloblastoma necessitates the need for whole brain radiotherapy, spinal radiotherapy and/or chemotherapy which may all be required to control the disorder (Fig. 5.5). The main cause of death, however, in children with medulloblastoma is local recurrence, despite treatment, which is present in all patients who die of this disease.

Extraneural metastasis in medulloblastoma is rare and is even more rare in most other CNS tumours. Kleinman *et al.* (1981) reviewed 101 reported cases and found an overall incidence of around 5%, although isolated series have found incidence rates of up to 20% (Park *et al.*, 1983). Most cases have had previous surgery to the primary tumour or ventriculoperitoneal shunting. Extraneural metastasis has been the initial clinical presentation in 3 cases (McComb *et al.*, 1980; Pollak *et al.*, 1981). In the absence of a CSF shunt, the desmoplastic variant, particularly in adults, seems to be particularly associated with the development of systemic metastases (Spencer *et al.*, 1984). Ventriculoperitoneal shunting produces a favourable pathway for the extraneural spread of medulloblastoma (Hoffman *et al.* 1976). In order to reduce the risk of this complication,

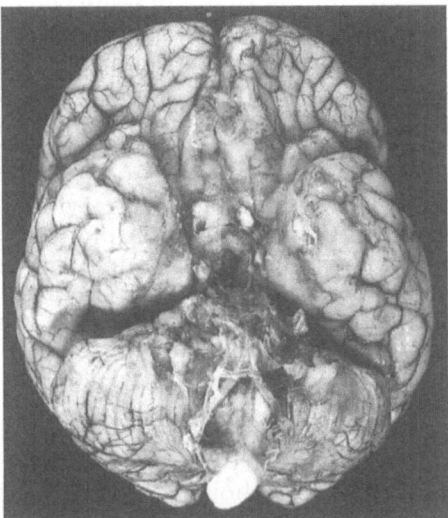

Fig. 5.5 Meningeal gliomatosis. Much of the anatomical detail of the base of the brain is obscured by diffuse infiltration of the meninges by glioma. These changes are specially marked in relation to the hypothalamus and over the ventral aspect of the brain stem

some surgeons routinely place a filter in the line of the shunt. This almost certainly increases the risk of shunt malfunction, but in one series has been shown to reduce the risk of extraneural metastasis from 20% to 4.2% (Park et al., 1983).

Extraneural metastases of medulloblastoma are located mainly in lymph nodes and bone marrrow where there is no CSF shunt, and in peritoneum, bone marrow and perhaps pleura where there is a CSF diversion in place. Clinical detection of extraneural metastasis occurs at an average of 2 years (range 4 months to 7 years) following diagnosis of the primary tumour; survival is poor, with 50%–70% dead within one year and 70%–90% dead within two years (Kleinman et al. 1981).

CSF metastasis by glioma is not uncommon. Penetration of the ependyma allows tumour cells to be shed into the CSF; penetration of the leptomeninges or basal cisterns is a mechanism of meningeal metastasis. CSF cytology will demonstrate malignant cells in 15%–40% of patients with malignant gliomas (Balhuizen et al., 1978; Bigner and Johnston, 1984), although few of these cells will be viable and capable of forming metastases. Although tumour cells tend to seed downward, and indeed cytology upon ventricular CSF is much less rewarding than that performed on lumbar CSF, nevertheless seeding of malignant glioma in a rostral direction has been reported in a number of cases (e.g. Eade and Urich, 1971). The incidence of spinal metastasis from malignant glioma is high in postmortem series. Cairns and Russell (1931) reported spinal metastasis in 8 of 22 (36%) consecutive necropsies in patients with glioma. Others have reported between 25% and 45% of patients with spinal metastasis dying from malignant gliomas (Erlich and Davis, 1978). Choucair et al. (1986) found 15 cases of spinal metastasis in 1047 patients with malignant glioma undergoing radiotherapy and/or chemotherapy, giving an incidence of 0.14% of spinal metastases developing during treatment.

It is a matter of debate as to whether malignant gliomas occurring at *multifocal sites* within brain parenchyma are separate glomas or are connected via white matter tracts; or whether truly separate lesions, if they occur, represent separate foci of malignancy in an abnormal field of brain or whether they represent metastasis from an original monoclonal tumour focus. We note, however, that multiple lesions within brain parenchyma are not uncommon and their incidence probably reflects how thoroughly they are looked for. Pathological series have demonstrated an incidence of multifocal malignant glioma in 2.3% 7.5% of patients (Batzdorf and Malamud, 1963; Barnard and Geddes, 1987). A similar incidence of 4%–6% has been found by CT scanning during life with post-mortem confirmation of tumour (Hochberg and Pruitt, 1980; Barnard and Geddes, 1987). Choucair et al. (1986) noted that 8.6% of patients with malignant glioma developed a separate lesion during treatment, one-third in the hemisphere ipsilateral to the primary tumour and two-thirds contralateral. Taking a number of clinical and pathological series, which do not appear to have gross selection bias, the incidence of all metastases within the CNS is 4%–16%.

Extraneural metastasis of malignant gliomas is much less common. Two excellent reviews summarize a total of 95 cases (Smith et al., 1969;

Pasquier *et al.*, 1980). This phenomenon is being recorded with increasing frequency and Cerame *et al.* (1985) have estimated that 20–25 new cases are reported per 5-year period. In almost all cases the malignant glioma has been given access to extraneural tissues by surgical intervention. This most commonly follows craniotomy, perhaps repeated craniotomies, with the sequence of exophytic extension of the tumour into the scalp, local lymphatic penetration, and subsequently the appearance of nodular secondary deposits in the scalp and cervical lymph nodes. This mechanism accounted for extraneural metastasis in 44% of 59 cases (glioma excluding medulloblastoma) reported by Russell and Rubinstein (1989). Tumour may cause bone erosion with direct extension of tumour into extraneural tissues. This route accounted for 14 of the 59 cases (24%) reported by Russell and Rubinstein (1989) and erosion of the temporal bone by temporal lobe tumours accounted for two-thirds of these cases.

Ventriculoperitoneal shunting has been a route for extraneural metastasis in 5 cases of malignant glioma (e.g. Brust *et al.*, 1968). A necessary prerequisite is the presence of viable tumour cells within CSF. Both CSF metastasis and the need for a CSF shunt are uncommon in patients with malignant gliomas so the relative rarity of shunt-induced extraneural metastasis is not surprising. At least 8 cases of extraneural metastasis in the absence of any surgical intervention have been reported (Pasquier *et al.*, 1980). In these cases, tumour penetration of either a great vein or of the tumour's intrinsic vessels appears to have been the mechanism of spread.

The type of glioma most likely to lead to extraneural metastasis (85 cases) is: glioblastoma 56%, malignant astrocytoma 14%, ependymoma 22%, ologodendroglioma 7%. Sites of extraneural metastasis (85 gliomas excluding medulloblastoma) are: lung 43%, pleura 20%, liver 15%, lymph nodes 43%, bone 25%.

Complications of Treatment

Surgery

Risks and Benefits

The risks and benefits of intracranial surgery have to be considered. Any complication of surgery is unacceptable if the procedure offers no benefit to the patient. In many countries not only is a histological diagnosis considered mandatory, but tumour decompression is attempted in the majority of cases. In Britain a more conservative view has often been taken. Wroe *et al.* (1986) found that only 43% of patients presenting to neurologists with a CT diagnosis of malignant glioma were referred for biopsy. Todd *et al.* (1987) found a similar biopsy rate of 47% in 142 patients with a CT diagnosis of solitary brain tumour (either presumed glioma or metastasis). Diagnosis of malignant brain tumour by CT is correct in 90%–95% of cases (Kendall *et al.*, 1979; Wroe *et al.*, 1986; Todd *et al.*, 1987). Histological confirmation of the diagnosis may alter management for the 5%–10% in whom the CT diagnosis is incorrect and it will alter management where subsequent treatment such as radiotherapy, surgery or chemotherapy are determined by the tumour histology.

The surgical management of malignant brain tumour includes *tumour biopsy* (Adams *et al.*, 1981) to achieve a histological diagnosis and *decompressive operations* designed to relieve symptoms and to improve the quality and length of survival. *Tumour biopsy* can be achieved by *freehand burr-hole biopsy, stereotaxic biopsy* or at an *open operation*.

Freehand Biopsy

Freehand biopsy is a simple, rapid and inexpensive technique and is particularly suited to the biopsy of large superficial tumours of the cerebral hemispheres. The particular risks of freehand biopsy are failure to locate tumour or to obtain a histological diagnosis from the biopsy taken, or haemorrhage which may be impossible to control without undertaking a craniotomy.

The risks of burr-hole biopsy at the present time are probably overestimated. In the pre-steroid era, biopsy was often undertaken in very ill patients with unfortunate/disastrous results. With the routine use of pre-operative steroids, an operative mortality of around 2% can be expected. Marshall *et al.* (1974) reported 60 patients undergoing burr-hole biopsy for suspected malignant glioma with 2 deaths (3.3%) – both in moribund patients – and 1 transient hemiparesis. Seizures are also a risk of brain biopsy and an abnormal EEG can be found in the majority of patients following biopsy (Elian, 1975). Whether these EEG abnormalities are specific to the biopsy procedure or the underlying brain abnormality is not clear. Tumour localization may be aided by peroperative ultrasound guidance. This technique may improve the diagnostic yield and reduce the risks of freehand biopsy.

Stereotaxic Biopsy

Stereotaxic biopsy permits the target point of a needle biopsy to be calculated from a CT scan taken with a stereotaxic frame attached to the patient's head. Stereotaxic biopsy is time-consuming and requires further CT scanning. It is, however, highly accurate and biopsies can be performed with minimal mortality and morbidity. One of the largest current series is that of Apuzzo *et al.* (1987). They performed 741 stereotaxic biopsies for 500 brain masses. In most patients a twist drill hole was made under local anaesthetic. The surgical objectives which included biopsy or cyst aspiration were achieved in 96% of procedures. There was one death (0.2%) from haemorrhage. The morbidity rate was 1.5%: 2 extracerebral haematomas, 1 wound infection, 1 increased neurological deficit and 1 peroperative seizure. Kelly *et al.* (1987a, b) performed 195 stereotaxic biopsies in 40 patients (an average of 5–6 biopsies in one trajectory per patient) without complication. Even in regions such as the brain stem, where biopsy has a high risk of producing complications, stereotaxic biopsy is relatively safe. Hood *et al.* (1986) reported 14 biopsies in 12 patients with brain stem mass lesions, 11 of which were shown to be tumour. A permanent deficit was produced in one patient and a transient deficit in another patient.

Internal Decompression of Tumour

Potential complications of tumour resection include all of the standard risks of craniotomy. These include:

1. *The position of the patient* – the sitting position, for example, may be associated with hypotension, air embolus or aerocele.
2. *The incision* – problems can arise simply because the incision is misplaced; attention should be paid to the blood supply of the flap or it may become necrotic, and closure may be difficult, such as in transverse posterior fossa incisions which are probably best avoided.
3. *Blood loss* may be considerable in raising a scalp and bone flap. This is a particular hazard of surgery for meningiomas, but may occur in gliomas with a large dural attachment.
4. *Infection* – significant wound infection will occur in 2–3% of clean craniotomies. There is considerable debate as to whether routine peroperative antibiotics can reduce this risk, but the use of antibiotics is becoming standard. The risk of infection is increased by re-exploration, for example to remove a postoperative haematoma, by prolonged operative exposure and by the use of drains.
5. *Postoperative CSF leak* may occur, particularly if the ventricle has been entered, and adequate dural closure is recommended to reduce this risk.
6. *Intraoperative seizures* are uncommon.

Identification of tumour may be difficult even if the bone flap and dural opening are in the correct position. Malignant brain tumours commonly do not present on the surface of the brain and the site of tumour may be difficult to identify. Expansion of a gyrus, discoloration or the presence of many fine cortical blood vessels will help but all are frequently absent. It is usually necessary to probe with a fine, blunt cannula. The consistency of gliomas is variable and distinguishing tumour from oedematous or normal brain may not be possible. In a few cases, despite open operation, the tumour cannot be justified. Immediate diagnostic cytology/histology are usually performed before undertaking major decompressive operations.

The *approach to the tumour* may be by lobectomy if the tumour is confined to a frontal, temporal or occipital pole. Malignant brain tumours are less commonly found in these regions and the approach to the tumour is usually via an incision into an appropriate gyrus. Trans-sulcal incisions have their proponents but carry increased risks of haemorrhage from the many fine vessels lying in the sulcus. The approach must consider "eloquent" brain regions. A postoperative neurological deficit may be avoided if the surgeon takes a less direct approach that avoids important gyri.

The *extent of resection* is determined by a number of factors. In general terms one would wish to remove as much tumour as possible. Malignant tumours infiltrate surrounding brain and edges between the tumour itself, and tumour in otherwise functioning brain cannot be defined at operation for malignant gliomas; hence the occasional use of multiple smears. Very extensive tumour resections carry the risk of increased postoperative deficit. Electrical stimulation mapping techniques have been used to define safe limits of resection in cortical regions serving language and motor functions. Partial tumour resections have their own hazards. Bleeding may be very difficult to control within an incompletely resected tumour and in many cases, particularly tumours with a basal dural attachment, bleeding can only be controlled by complete tumour removal. Tumour swelling is a further hazard of incomplete resection and surgical mortality is lowest for patients who have had either lobectomy or a radical excision (Frankel and German, 1958).

The surgical (30-day) mortality for glioma surgery has been radically altered by the use of preoperative steroids. Prior to the introduction of steroids, operative mortalities of 25% were common (Jelsma and Bucy, 1969). Other factors that influence surgical mortality include the location of the tumour within the cerebral hemisphere, the extent and depth of the tumour, the presence of multicentric lesions and the neurological state of the patient at the time of operation. The overall surgical mortality of decompressive surgery for gliomas is currently around 2% (Laws *et al.*, 1984). The commonest cause of surgical death is postoperative haemorrhage (Roth and Elvidge, 1960) which accounts for around one-third of the surgical deaths.

Radiotherapy

Radiotherapy is the mainstay of the treatment of malignant brain tumours. Ionizing radiation is usually given as a gamma ray beam from radioactive sources such as cobalt-60 or caesium-137, or it is given as megavoltage X-ray beams. The dose is usually fractionated, that is the total dose is divided into equal fractions, and typical conventional fractionation might be 1.8–2.0 Gy/day, 5 times per week to a total dose of 60 Gy. The most important CNS complications result from the effects of ionizing radiation and the concept that the human brain is resistant to the effect of therapeutic irradiation is no longer held. The type, frequency and extent of post-radiotherapy complications will depend on many factors which include the total dose of irradiation, the number of fractions, the dose per fraction, the total treatment time, the volume irradiated, the elimination of "hot spots" by the use of multiple fields, and the use of other treatments such as steroids, radiosensitizers or antineoplastic chemotherapy.

Other methods of irradiating tumours include the use of high-energy electron or neutron beams, protons, focused X-ray beams and the implantation of radioisotopes into tumours ("interstitial radiotherapy"). A higher incidence of local complications may be expected from the high doses given to tumours by these forms of treatment if not appropriately focused.

The main *effect of irradiation on the tumour* itself is to produce tumour necrosis. This is particularly true for radiosensitive tumours such as pineal germinoma or lymphoma where large fluid-filled cysts lined with glial scar tissue can be found following treatment. With less radiosensitive tumours including gliomas, extensive central necrosis may be produced, although peripheral tumour may remain (Burger *et al.*, 1979). Tumour irradiation may also lead to a change in tumour cytology. Typically multinucleated giant cells with irregular hyperchromatic nuclei are found; the number of mitotic figures is reduced. The pathological effects of irradiation on cerebral blood vessels include thickening, hyalinization and occlusion of vessels, with proliferation of collagen in perivascular regions and in distant parenchyma.

Irradiation can cause *injury to scalp and bone*. Although megavoltage radiation is relatively skin sparing, the exit dose to the skin must be considered. Skin erythema and hair loss are common. Hair loss may be patchy or extensive and may be permanent. The combination of irradiation with certain chemotherapeutic regimens may exacerbate skin reactions and can increase the risk of permanent hair loss. Bone flap necrosis is a potential risk but is not often a clinical problem, but impaired wound healing does occur.

The *effects of irradiation upon normal brain* have been considered according to the time of their manifestation: *acute reactions* occur during the course of irradiation, *early delayed reactions* appear a few weeks to 2–3 months, and *late delayed reactions* appear from a few months to many years after irradiation (Sheline, 1986; Leibel and Sheline, 1987).

Acute Reactions

Acute reations are usually minor with conventional doses and fractionation, particularly when steroids are used. They cause symptoms and signs of raised intracranial pressure including headache and reduced conscious level, with an increase in the patient's pre-existing neurological deficit. Acute reactions include tiredness and occasional nausea. Vomiting is rare and other causes must be sought if this occurs. This reaction is dose-related, is probably due to increased oedema and usually responds to increasing the dose of steroids, although osmotic diuretics may be required. Total doses of irradiation as high as 80 Gy have been given to limited brain volumes under steroid cover with fractionation (Salazar *et al.*, 1976). Single fractions of 3, 4 or 6 Gy (total dose 30, 20 and 12 Gy, respectively) are well tolerated in patients with brain metastases (Kramer *et al.*, 1977). However, increasing the dose to 2 fractions of 7.5 Gy has been shown to produce a dramatic increase in the number of acute reactions (Young *et al.*, 1974). Forty-nine per cent of these patients with brain metastases suffered headache, vomiting and in some cases acute brain herniation. Hindo *et al.* (1970) used a single dose of 10 Gy in 54 patients, with 5 deaths occurring within 1 week of treatment. Acute reactions, once controlled, do not preclude completion of the intended treatment.

Early Delayed Reactions

Early delayed reactions are usually transient and disappear without treatment. Symptoms include lethargy, somnolence and increase in pre-existing symptoms and signs. Somnolence occurring 6–8 weeks after treatment was initially reported in 3% of children irradiated for ringworm of the scalp (Druckmann, 1929). Transient somnolence and lethargy were subsequently reported in 80% of children given 25 Gy whole-brain radiotherapy as prophylactic CNS treatment for acute lymphoblastic leukaemia. Hoffman *et al.* (1979) treated a group of adult patients with 1.8 Gy daily fractions to a total dose of 60 Gy. Patients also received carmustine (BCNU). Forty-nine per cent deteriorated within 18 weeks of treatment, but two-thirds of these patients subsequently improved without any specific therapy. Deterioration in neurological function at this time was therefore twice as likely to be due to an acute delayed reaction as it was to tumour progression.

Although usually transient, an acute delayed reaction may lead to the patient's death and from a few of these cases neuropathological data are available (Lampert *et al.*, 1959; Lampert and Davis, 1964; Monro and Mair, 1968). Multiple small foci of demyelination have been found in some cases with perivascular infiltration by lymphocytes and plasma cells, while in others vascular damage was also present (Monro and Mair, 1968; Palmer, 1972). The pathological features are not dissimilar to those found in multiple sclerosis patients and, given the absence of vascular damage, the acute delayed reaction probably results from injury to oligodendrocytes either primarily or secondary to autoimmune damage. The latent interval of clinical recovery is consistent with the time required for myelin replacement which supports the concept of demyelination as the mechanism for the acute delayed reaction.

Late Delayed Reactions

Late delayed reactions are caused either by diffuse white matter loss with ventricular enlargement, a *leucoencephalopathy* or as a space-occupying glial reaction known as *radionecrosis*. Radiation necrosis presents clinically as a space-occupying mass lesion. There are symptoms and signs of raised intracranial pressure with increased focal neurological deficit. Seizures are a prominent feature. These features occur in the cerebral hemispheres from about 1 year to many years after treatment. In the cerebral midline, the latency may be shorter. In the cerebral hemispheres, radionecrosis rarely occurs before 12 months after conventional dosage/fractionation and cases have been reported over 30 years after treatment (Mussini *et al.*, 1980). There are many examples of radionecrosis following irradiation of intracranial tumours, or extracranial tumours such as nasopharyngeal carcinoma, parotid carcinoma or basal cell carcinoma of the scalp (see Russell and Rubinstein, 1989). Late delayed reactions are a particular complication of high linear energy transfer (LET) radiation which has virtually no place in the treatment of brain tumours for this reasons.

Neuropathology of Radionecrosis. The naked eye appearances of radionecrosis in the cerebrum are similar to those of a malignant astrocytoma. The white matter is expanded and replaced by focally cystic, waxy, pale yellow tissue in which there may be petechial haemorrhages (Fig. 5.6a,b). In long-standing cases the affected tissue becomes granular and is apt to crumble. Anatomical definition is blurred, but, in the main, grey matter is spared. Histologically, the process is characterized by appearances that range from coagulative necrosis (Fig. 5.6c) around which there is no or minimal reactive change, to foci of demyelination, loss of axons and infiltration by lipid-containing macrophages, lymphocytes and plasma cells. The most important change, however, is fibrinoid necrosis and hyalinization of the walls of blood vessels (Fig. 5.6d), and proliferation of endothelium, which may be sufficient to cause an obliterative endarteritis and thrombotic occlusion of small vessels. Additional features include the formation of telangiectatic vessels, the proliferation of perivascular fibroblasts with the formation in some cases of large amounts of relatively acellular collagen and an associated astrocytosis, often with bizarre multinucleated cells (Van der Kogel, 1986; Myers *et al.*, 1986; Russell and Rubinstein, 1989).

The brain stem may occasionally be the principal site of radionecrosis (Manz *et al.*, 1979) and radiation effects differing from typical radionecrosis in the cerebellum have also been described (Russell and Rubinstein, 1989). The earliest changes in the cerebellar cortex consist of the formation of small cysts in the Purkinje cell layer with some loss of both Purkinje and granule cells, with subsequent atrophy of folia, demyelination and gliosis. Eventually the Purkinje cells are replaced by a series of confluent cystic spaces, associated with which there is insudation of fibrin that spreads readily along the molecular layer, hyaline thickening of vessels and focal calcification (Fig. 5.6d).

Similar changes may follow radiation directed to the spinal cord either for tumours related to the spinal cord or, less commonly, radiation to tumours arising in the mediastinum or the neck. Referred to as radiation myelopathy, this complication is said to occur in between 1% and 2% of radiated patients (Henson and Urich, 1982). As in the cerebral hemispheres, the lesion is space occupying, with all the features of a malignant astrocytoma. Histologically, there appears to be selective involvement of white matter and marked changes in the blood vessels. In the latest stages the segments of cord above and below the lesion are characterized by Wallerian degeneration in ascending and descending fibre tracts, respectively.

The pathogenesis of radionecrosis remains uncertain but vascular change, direct effect of radiation on the glia and immunological mechanisms have all been proposed (Van der Kogel, 1986; Myers *et al.*, 1986). It is possible, however, that multiple mechanisms operate and that their relative importance varies with the radiation dose and interval between exposure and occurrence of damage.

The *risks of developing radionecrosis* are not clearly defined. Kramer (1968) identified 57 cases and concluded that daily fractions of 1.8–2.0 Gy to a

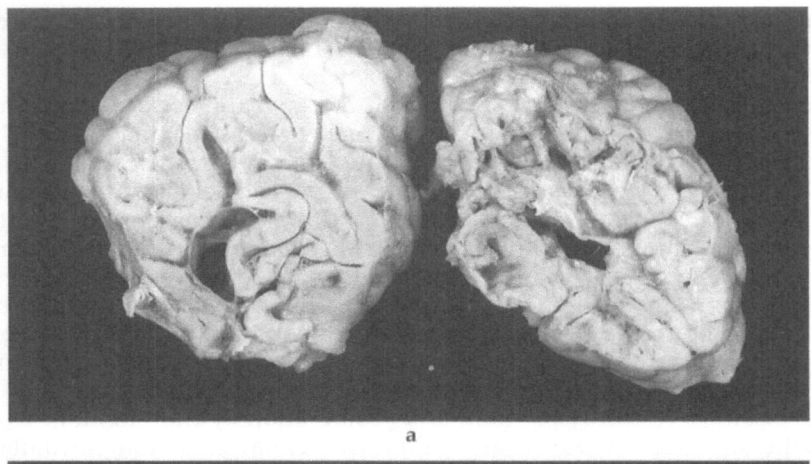

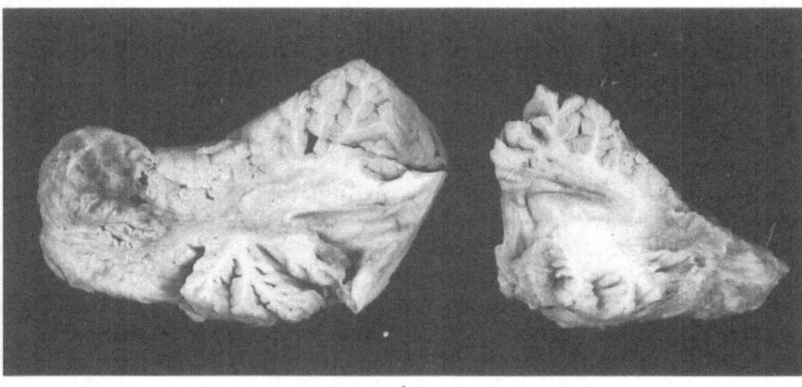

Fig. 5.6 Delayed radiation necrosis. **a** There is extensive involvement of the white matter of each occipital lobe. **b** There is also marked involvement of the hemisphere of the cerebellum.

(*Parts c and d opposite*)

total dose of 65–70 Gy will normally be well tolerated, although occasional cases of radiation necrosis will occur. Sheline *et al.* (1980) reviewed 80 cases and concluded that the thresholds for brain necrosis were 35 Gy for 10 fractions, 60 Gy for 35 fractions, or 70 Gy for 60 fractions. The incidence of radiation necrosis is hard to establish. Many patients who deteriorate 12–18 months after treatment of a malignant astrocytoma and who are found to have a mass lesion on follow-up CT scan are assumed to have local recurrence and are not investigated further. Marks *et al.* (1981) reported a consecutive series of 139 patients receiving 1.8–2.0 Gy per day, 5 days per week, 1 field per day, to at least 45 Gy. Nine cases of proven radiation necrosis were found (6%). The high incidence of necrosis in part reflects the use of 1 field per day which leads to high incident doses. In this

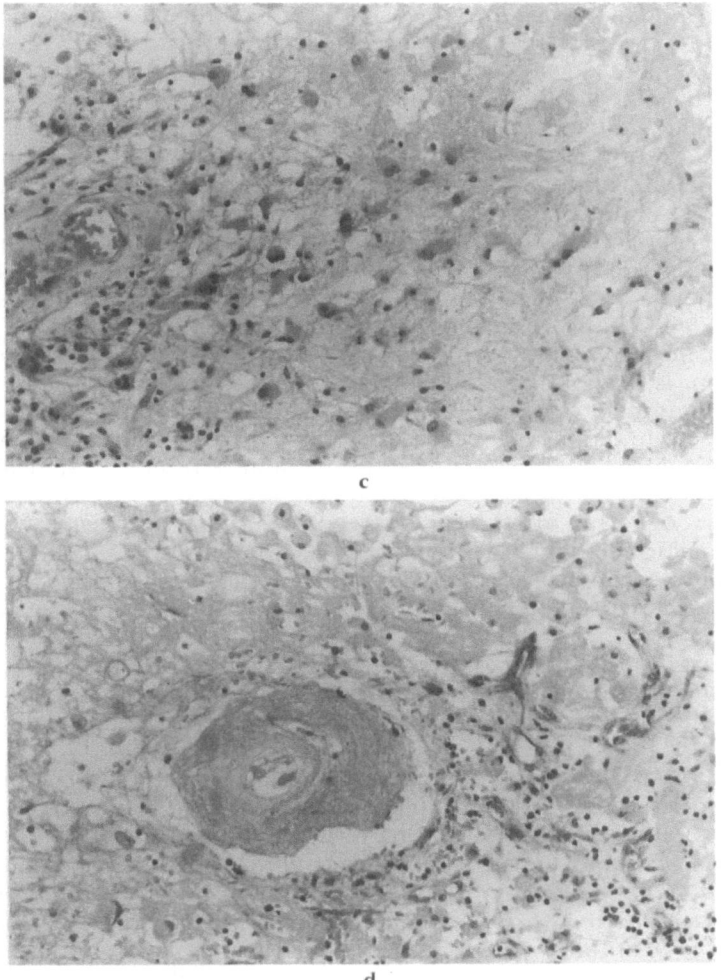

Fig. 5.6 **c** There is coagulative necrosis with little in the way of a glial or mesodermal reaction (Haematoxylin and eosin, original magnification ×130). **d** Note striking changes in the small blood vessels in the form of fibrinoid necrosis and endothelial proliferation (Haematoxylin and eosin, original magnification ×130).

study the incidence of necrosis was progressively greater with higher radiation doses: below 57 Gy no cases were seen, between 57 and 65 Gy 2 cases were seen in 60 patients (3%), and above 65 Gy 5 cases were seen in 28 patients (18%). The risk of radiation necrosis is probably increased by adjuvant chemotherapy (see e.g. Pratt *et al.*, 1977; Burger *et al.*, 1979).

Treatment of radionecrosis includes the use of steroids. Where the lesion is accessible, resection of the region of focal necrosis relieves symptoms and may be life-saving (Edwards and Wilson, 1980).

Diffuse leucoencephalopathy, which may be necrotizing, is associated with radiotherapy. Most cases have been simultaneously treated with various chemotherapeutic agents and it is likely that this is primarily a complication of drug treatments and will be discussed further below.

Intellectual deficits may follow irradiation in both children and adults. Low doses of around 25 Gy produce little or no intellectual disturbance, although mild psychomotor dysfunction and general "slowing down" may be seen in children (Harten *et al.*, 1984). However, moderate or severe deficits will be found in about half of children receiving 40–60 Gy and followed up 5–47 years after treatment (Li *et al.*, 1984). These complications of radiotherapy are much greater in the developing brain and radiotherapy to the CNS is usually delayed as long as possible in young children.

Hypothalamic–pituitary dysfunction, especially of growth hormone regulation, is becoming increasingly recognized. Mechanick *et al.* (1986) reported a number of cases following irradiation for primary glial tumours. Pituitary dysfunction can be demonstrated in the majority of patients receiving 40–85 Gy for extracranial head and neck tumours (Samaan *et al.*, 1975).

Blindness of abrupt onset has been reported in patients treated with low-dose cranial irradiation and lomustine (CCNU) (Wilson *et al.*, 1987). This is presumably secondary to demyelination of the visual pathways and is similar to changes seen in the spinal cord or peripheral nerve. As already noted, damage to small blood vessels is a characteristic feature of radionecrosis and is likely to be a major pathogenetic factor.

Radiation-induced damage to both *large extra- and intracranial arteries* has also been described (Henson and Urich, 1982; Benson and Sung, 1989).

Induction of brain tumours is an uncommon complication of cerebral irradiation. The commonest examples are meningeal fibrosarcomas which have generally been complications of irradiation of pituitary adenomas (Russell and Rubinstein, 1989). Intracerebral sarcoma has been found 6.5 years after low-dose (22 Gy) irradiation for acute lymphocytic leukaemia (Tiberin *et al.*, 1984). A cerebellar chondrosarcoma has been reported 16 years after 46 Gy to the posterior fossa following subtotal excision of a malignant astrocytoma (Bernstein *et al.*, 1984). Meningiomas can be induced either by high-dose irradiation for the treatment of intracranial tumours (Henson and Urich, 1982), or following low-dose treatment of benign scalp disorders such as tinea capitis (e.g. Modan *et al.*, 1974; Soffer *et al.*, 1983). Radiation-induced meningiomas have a high frequency of multiple lesions and a high incidence of recurrence following apparently complete excision.

An increasing number of children are surviving after radiotherapy for intracranial tumour. There have been reports of further primary neuroectodermal tumours developing in these children (see e.g. Zuccarello *et al.*, 1986; Ushio *et al.*, 1987). Approximately 50 cases have been reported. Doses of radiation have been very wide, from 1.5 to 60 Gy, and the latency period has been 5–25 years. Most of these gliomas were anaplastic astrocytomas or glioblastomas and have appeared within the first three decades of life.

Alternative techniques are now available which can be used to maximize the local radiation dose and minimize the dose to the surrounding brain. These include *interstitial radiotherapy* and focused or stereotaxic radiotherapy. Local complications are likely to be higher with all of these techniques. Interstitial radiotherapy, or brachytherapy, involves the placement of radioactive sources within the tumour. For the treatment of gliomas, catheters are temporarily placed into the tumour and typically a total of 50 Gy might be given over 50 h. This radiation dose is at the tumour periphery and the dose within the centre may be very much higher. Not surprisingly, acute brain necrosis may occur and this may necessitate urgent resection of necrotic swollen tumour. *Stereotaxic radiotherapy* uses multiple sources to give a single dose of radiation to a point source in the brain. This has mainly been used for the treatment of arteriovenous malformations and some benign tumours. Two cases of radiation necrosis following stereotaxic radiotherapy have recently been reported (Statham *et al.*, 1990).

Chemotherapy

Treatment with antineoplastic agents may be given as a primary treatment or, more commonly, as adjunctive therapy following surgery and/or radiotherapy. Several factors limit the effectiveness of chemotherapy on malignant brain tumours. *Drug access* into the brain is limited by the blood–brain barrier for agents that are not highly lipid soluble. Increasing drug dosage increases brain concentrations of antineoplastic drugs but also increases the systemic complications of treatment. Intra-arterial chemotherapy and treatment after osmotic blood–brain barrier disruption have been used to increase local drug concentrations without increasing systemic toxicity. *Tumour sensitivity* to single agent chemotherapy is often low. Multiple agent treatments are commonly used which increases the number of potential drug-induced complications, but since a lower dose of each individual agent can be used, this will reduce the risk of any specific complication arising. The *sensitivity of normal brain* to chemotherapeutic agents is the cause of most CNS complications. These include an acute or chronic encephalopathy or myelopathy, and an acute cerebellar syndrome. Treatment schedules such as intra-arterial chemotherapy that increase the concentrations of drugs in the brain may improve the antitumour effect but also increase the risks of damage to normal brain (Henson and Urich, 1982; Shapiro and Young, 1984). Most chemotherapeutic schedules include a nitrosourea, BCNU, or CCNU, which are highly lipid soluble. Other agents commonly used include procarbazine, bleomycin, vincristine, cyclophosphamide, VM-26 and cis-platinum, although many other drugs have been used.

Systemic complications will vary with the drug or drugs used and the dosage and timing of treatment. The highest possible dose of an antineoplastic agent is often the most effective, so that these drugs are commonly given in doses at or around levels that cause systemic toxicity. Rapidly dividing tissues, particularly bone marrow, bear the brunt of this effect and leucopenia and thrombocytopenia are common.

CNS complications are usually due to a direct effect of the drug, although CNS infection may occur secondary to the leucopenic complications of treatment. The toxic effects can be broadly grouped as a number of clinical and/or pathological syndromes. *Seizures* may complicate treatment with any agent in sufficient dosage, but is most commonly seen with methotrexate, vincristine, AraC, nitrosoureas and cis-platinum.

An *acute cerebellar syndrome* has been described following treatment with 5-fluorouracil, procarbazine, BCNU and vincristine (Riehl and Brown, 1964). Nystagmus, dysarthria, dizziness, truncal and gait ataxia develop a few days after starting treatment. This complication rarely occurs when the weekly dose of 5-fluorouracil is in the range 7.5–15 mg kg^{-1}. At weekly doses of 15–20 mg kg^{-1}, 3–7% of patients develop this syndrome which seems to be secondary to high plasma levels of drug rather than a cumulative effect (Weiss *et al.*, 1974). The syndrome is reversible on stopping the drug. Pathologically there is loss of neurons in the olive, dentate nucleus and in the granular layer.

Acute encephalopathy may complicate intravenous administration of many agents and has been found in 40% of patients treated with high-dose L-asparginase and 5-fluorouracil (Weiss *et al.*, 1974). Subtle personality change may be an early sign, with the later development of a state similar to Korsakoff's psychosis. More commonly, lethargy, drowsiness, confusion and hallucinations develop. Focal deficits are uncommon although 5-fluorouracil may cause hemiparesis and brain stem dysfunction. This acute encephalopathy is often reversible on stopping treatment, but progressive deterioration to coma and death have been described (Weiss *et al.*, 1974).

A *chronic encephalopathy* was originally described following methotrexate administration and has since been recognized as a disseminated necrotizing leucoencephalopathy. The clinical onset is often insidious and may occur months or years after stopping treatment. Confusion, drowsiness, irritability, ataxia, tremor, seizures and dementia may all occur. Most patients stabilize with a permanent neurological deficit. A significant number progress to coma and death.

Neuropathology of Disseminated Necrotizing Leucoencephalopathy. The morbid anatomical and histological features have been well described in the past 15 years (Price and Jamieson, 1975; Smith, 1975; Garcia *et al.*, 1977; Liu *et al.*, 1978; Nakazato *et al.*, 1980).

Macroscopically, there may be abnormalities in the white matter of the centrum semiovale, cerebellum and brain stem, and rarely of the spinal cord; grey matter is spared. The changes vary from grey-pink, randomly distributed foci of necrosis to larger areas of creamy-white necrosis in which there may be petechial haemorrhage or cavitation, depending upon the survival of the patients.

The histological features are distinctive. Early cases are characterized by foci of disseminated coagulative necrosis that appear to be unrelated to blood vessels. There is loss of myelin, extensive damage to axons with bulb formation and an astrocytosis, but all in the apparent absence of either an inflammatory cell or macrophage response. These changes may occur in the absence of any vascular damage: on the other hand, the features of irradiation with fibrinoid necrosis may be seen in the vessels in the larger lesions. Viable tumour may or may not be seen. In some instances the lesions may become chronic and are then characterized by periventricular cavitation of white matter, calcification of the centrum semiovale and compensatory ventricular enlargement (Liu et al., 1978; Muller et al., 1981).

Various treatment schedules that increase delivery of chemotherapeutic agents into the CNS also increase the risk of necrotizing leucoencephalopathy. These schedules include intra-arterial and intraventricular chemotherapy, and chemotherapy plus osmotic blood–brain barrier opening. *Intrathecal or intraventricular* methotrexate can cause a severe leucoencephalopathy characterized by demyelinating and necrotizing lesions in the periventricular white matter bilaterally (Shapiro et al., 1973; Norrell et al., 1974; Smith, 1975). Intrathecal methotrexate has been combined with intravenous methotrexate and cranial irradiation for the treatment of childhood leukaemia and 45% of these patients developed a necrotizing leucoencephalopathy (Bleyer and Griffin, 1980). *Intracarotid infusion* of either BCNU or ACNU was used by Papavero et al. (1987) to treat 30 patients with recurrent malignant glioma following surgery and radiotherapy. Early complications included eye pain (100%), permanent blindness or hemiplegia (7%), temporary hemiplegia (10%) and marrow toxicity (17%). A permanent neurological deficit will occur in about a quarter of patients treated with intra-arterial chemotherapy (Kapp and Sanford, 1986). Intracarotid chemotherapy may also cause an acute diffuse encephalopathy within the hemisphere ipsilateral to the infusion (Burger et al., 1981; Kleinschmidt-DeMasters, 1986; Mahaley et al., 1986). Superselective carotid catheterization can place the intra-arterial catheter above the level of the ophthalmic artery. Very high doses of chemotherapeutic agents can be infused as the toxic effects on the retina are avoided. However, the risks of leucoencephalopathy may be very high. Foo et al. (1986) treated 5 patients with glioblastoma with 200 mg m^{-2} BCNU and found clinical evidence of leucoencephalopathy in 3 of the 5 and CT evidence of leucoencephalopathy in all 5 patients.

Osmotic blood–brain barrier opening with mannitol can also increase drug delivery into the CNS. Neuwelt and co-workers have treated a number of patients with malignant brain tumours in this way (Neuwelt et al., 1986; Neuwelt and Dahlborg, 1987). In 38 patients with glioblastoma, barrier opening was successful in about two-thirds of patients. Subsequent treatment with methotrexate, cyclophosphamide and procarbazine was associated with seizures in 55%, transient neurological deficit in 58% and a permanent neurological deficit in 8% (Neuwelt et al., 1986).

In summary, many of the antineoplastic agents have a neurotoxic effect and potentially cause a variety of clinical problems, the most serious of which is disseminated necrotizing leucoencephalopathy. Certain treatment schedules that increase drug delivery into the CNS seem to have a high incidence of neurotoxic complications. The addition of radiotherapy to certain chemotherapeutic schedules, particularly those including adriamycin, increase the risk of neurotoxicity.

Other complications of antineoplastic agents include white matter gliosis (Hendin *et al.*, 1974), diffuse cortical atrophy (Crosley *et al.*, 1978), cerebellar sclerosis (Wizniter *et al.*, 1987), neuroaxonal dystrophy (Liu, 1978), peripheral neuropathy (Henson and Urich, 1982) and spinal myelopathy (Mena *et al.*, 1981).

Steroids

Corticosteroids are the single most important drug used in the management of patients with malignant brain tumours. They reduce cerebral oedema and possibly have a direct oncolytic effect if used in very high dosage. Relief of symptoms of raised intracranial pressure and of neurological signs is often rapid and dramatic. Neurosurgeons have conventionally used dexamethasone, although there are theoretical advantages to methylprednisolone; the clinical response is independent of the type of steroid used (Jamous, 1989). Dexamethasone 16–24 mg daily would be considered to be maximal conventional treatment. A recent trial has examined the use of much higher doses of steroids (Capildeo, 1989; Jamous, 1989). Following surgical decompression, a maintenance dose of 0.5 mg per day may be satisfactory. Patients who deteriorate on full conventional doses of steroids may gain further palliation if very high doses ("megadoses") of steroids are used: their use, however, is associated with complications (Table 5.4).

The complications of prolonged treatment with corticosteroids include iatrogenic Cushing's syndrome with hypertension, diabetes mellitus and "moon facies", mood disorders including frank psychosis, steroid myopa-

Table 5.4 Complications of high-dose steroid therapy*

		Early (first month)	Late (ninth month)
Number of patients		49	22
Moon facies:	mild	14%	59%
	moderate/severe	2%	18%
Oedema:	mild	10%	5%
	moderate/severe	4%	0%
Nausea:	mild	50%	9%
	moderate/severe	2%	0%
Indigestion:	mild	29%	23%
	moderate/severe	6%	9%
Depression:	mild	29%	18%
	moderate/severe	6%	5%
Mean blood pressure		no change	no change
Weight		no change	no change
Blood glucose		no change	no change
Triglycerides		increased	increased

*Methylprednisolone 500 mg on alternate days.

thy, gastrointestinal bleeding, osteoporosis, skin changes including acne, and fungal and yeast infections.

The incidence of steroid-induced side effects will depend on the dose of steroid used. Prolonged steroid therapy is not a "benign" treatment and is associated with significant morbidity. Most of the side effects are reversible, provided that steroids can be withdrawn which may or may not be possible without deterioration in the patient's neurological state.

References

Adams JH, Graham DI (1976) The relationship between ventricular fluid pressure and the neuropathology of raised intracranial pressure. Neuropathol Appl Neurobiol 2: 323–332

Adams JH, Graham DI, Doyle D (1981) Brain Biopsy. The Smear Technique for Neurosurgical Biopsies. Chapman and Hall, London

Albert FK (1986) Tumour hemorrhages in intracranial tumours. Neurochir (Stuttg) 29: 67–74

Apuzzo MLJ, Chandrasema PT, Cohen D, Zee C-S, Zelman V (1987) Computed imaging stereotaxy: experience and perspective related to 500 procedures applied to brain masses. Neurosurgery 20: 930–937

Balhuizen JC, Bots GTh, Schaberg A, Bosman FT (1978) Value of cerebrospinal fluid cytology for the diagnosis of malignancies in the central nervous system. J Neurosurg 48: 747–753

Barnard RO, Geddes JF (1987) The incidence of multifocal cerebral gliomas. A histologic study of large hemisphere sections. Cancer 60: 1519–1531

Batzdorf U, Malamud N (1963) The problem of multicentric gliomas. J Neurosurg 20: 122–136

Benson PS, Sung JH (1989) Cerebral aneurysms following radiotherapy for medulloblastoma. J Neurosurg 70: 545–550

Bernstein M, Perrin RG, Platts ME, Simpson WJ (1984) Radiation-induced cerebellar chondrosarcoma. J Neurosurg 61: 174–177

Bigner SH, Johnston WW (1984) The diagnostic challenge of tumors manifested initially by the shedding of cells into cerebrospinal fluid. Acta Cytologica 28: 29–36

Bleyer WA, Giffin TW (1980) White matter necrosis, mineralising microangiopathy and intellectual abilities in survivors of childhood leukaemia: associations with central nervous system irradiation and methotrexate therapy. In: Gilbert HA, Kagan AR (eds) Radiation Damage to the Nervous System. Raven Press, New York, pp 155–174

Brust JCM, Moiel RH, Rosenberg RN (1968) Glial tumor metastases through a ventriculo-pleural shunt. Resultant massive pleural effusion. Arch Neurol 18: 649–653

Burger PC, Mahaley MS, Dudka L, Vogel FS (1979) The morphologic effects of radiation administered therapeutically for intracranial gliomas. A post mortem study of 25 cases. Cancer 44: 1256–1272

Burger PC, Kamenar E, Schold SC, Fay JW, Phillips GL, Herzig GP (1981) Encephalomyelopathy following high-dose BCNU therapy. Cancer 48: 1318–1322

Burger PC, Dubois PJ, Schold C et al. (1983) Computerized tomographic and pathologic studies of the untreated, quiescent, and recurrent glioblastoma multiforme. J Neurosurg 58: 159–169

Cairns H, Russell DS (1931) Intracranial and spinal metastases in gliomas of the brain. Brain 54: 377–420

Capildeo R (1989) High-dose methylprednisolone for the treatment of malignant brain tumours. In: Capildeo R (ed) Steroids in Diseases of the Central Nervous System. Wiley, Chichester, pp 103–112

Cerame MA, Guthikonda M, Kohli CM (1985) Extraneural metastases in gliosarcoma: a case report and review of the literature. Neurosurgery 17: 413–418

Choucair AK, Levin VA, Gutin PH et al. (1986) Development of multiple lesions during radiation therapy and chemotherapy in patients with gliomas. J Neurosurg 65: 654–658

Clasen RA, Cooke PM, Pandolfi S, Boyd D, Raimondi AJ (1962) Experimental cerebral edema produced by focal freezing. J Neuropathol Exp Neurol 21: 579–596

Crosley CJ, Rorke LB, Evans A, Nigro M (1978) Central nervous system lesions in childhood leukemia. Neurology 28: 678–685

Deutsch M, Reigel DM (1980) The value of myelography in the management of childhood medulloblastoma. Cancer 45: 2194–2197

Druckmann A (1929) Schlafsucht als Folge der Rontgenbestrahlung. Beitrag zur Strahle empfindlichkeit des Gehirns. Strahlentherapie 33: 382–384

Eade OE, Urich H (1971) Metastasising gliomas in young subjects. J Pathol 103: 245–256

Edwards MS, Wilson CB (1980) Treatment of radiation necrosis. In: Gilbert HA, Kagan AR (eds) Radiation Damage to the Nervous system. Raven Press, New York, pp 129–144

Elian M (1975) Late effects of brain biopsy. J Neurol 211: 95–104

Erlich SS, Davis RL (1978) Spinal subarachnoid metastases from primary intracranial glioblastoma multiforme. Cancer 42: 2854–2864

Foo SH, Choi IS, Berenstein A, Wise A et al. (1986) Supraopthalmic intracarotid infusion of BCNU for malignant glioma. Neurology 36: 1437–1444

Frankel SA, German WJ (1958) Glioblastoma multiforme – a review of 219 cases with regard to natural history, pathology, diagnostic methods and treatment. J Neurosurg 15: 489–503

Garcia JH, Sandbank V, Gutin P (1977) Multifocal leukoencephalopathy in adult leukemia: histologic features and etiologic considerations. Acta Neuropathol 40: 273–279

Glass B, Abbott KH (1955) Subarachnoid hemorrhage consequent to intracranial tumors. Review of the literature and report of seven cases. Arch Neurol Psychiat 73: 369–379

Globus JH, Saperstein M (1942) Massive hemorrhage into brain tumor. J Am Med Ass 120: 348–352

Graham DI (1990) The pathophysiology of raised intracranial pressure. In: Weller RO (ed) Systemic Pathology, 3rd edn, Vol. 4. Churchill Livingstone, Edinburgh, pp 64–77

Graham DI, Lawrence AE, Adams JH, Doyle D, McLellan DR (1987) Brain damage in non-missile head injury secondary to high intracranial pressure. Neuropathol Appl Neurobiol 13: 209–217

Harper CG, Doyle D, Adams JH, Graham DI (1986) Analysis of abnormalities in pituitary gland in non-missile head injury: study of 100 consecutive cases. J Clin Path 39: 769–773

Harten G, Stephani U, Henze G et al. (1984) Slight impairment of psychomotor skills in children after treatment of acute lymphoblastic leukaemia. Europ J Paediatr 142: 189–197

Hassler O (1967) Arterial pattern of human brain stem. Normal appearance and deterioration in expanding supratentorial conditions. Neurology 17: 368–375

Hayreh SS (1964) Pathogenesis of oedema of the optic disc (papilloedema). Br J Ophthalmol 48: 522–542

Hendin B, DeVivo DC, Torack R et al. (1974) Parenchymatous degeneration of the central nervous system in childhood leukemia. Cancer 33: 468–472

Henson RA, Urich H (1982) Cancer and the Nervous System. The Neurological Manifestations of Systemic Malignant Disease. Blackwell Scientific, Oxford

Hindo WA, DeTrana FA, Lee MS, Hendrickson FR (1970) Large dose increment irradiation in treatment of cerebral metastases. Cancer 26: 138–141

Hochberg FH, Pruitt A (1980) Assumptions in the radiotherapy of glioblastoma. Neurology 30: 907–911

Hochwald GM, Marlin AE, Wald A, Malhan C (1976) Movement of water between blood, brain and CSF in cerebral oedema. In: Pappius HM, Feindel W (eds) Dynamics of Brain Edema. Springer-Verlag, Berlin, pp 129–137

Hoffman HJ, Hendrick EB, Humphreys RP (1976) Metastasis via ventriculoperitoneal shunt in patients with medulloblastoma. J Neurosurg 44: 562–566

Hoffman WF, Levin VA, Wilson CB (1979) Evaluation of malignant glioma patients during the postirradiation period. J Neurosurg 50: 624–628

Hood TW, Gebarski SS, McKeever PE, Venes JL (1986) Stereotaxic biopsy of intrinsic lesions of the brain stem. J Neurosurg 65: 172–176

Jackson H (1958) On convulsive seizures. In: Taylor J (ed) Selected Writings of John Hughlins Jackson, Vol. 1. Basic Books, New York

Jacobi G, Kornhuber B (1987) Malignant brain tumors in children. In: Jellinger K. (ed) Therapy of Malignant Brain Tumours. Springer-Verlag, Berlin, pp 396–493

Jamous MA (1989) High-dose methylprednisolone for malignant gliomas: safety and side effects. In: Capildeo R (ed) Steroids in Diseases of the Central Nervous System. Wiley, Chichester, pp 113–124

Jelsma R, Bucy PC (1969) Glioblastoma multiforme. Its treatment and some factors affecting survival. Arch Neurol 20: 161–171

Kapp JP, Sanford RA (1986) Neurological deficit after carotid infusion of cisplatin and 1,3-bis(2-chloroethyl)-1-nitrososurea (BCNU) for malignant glioma: an analysis of risk factors. Neurosurgery 19: 779–783

Kase CS, Robinson RK, Stein RW *et al.* (1985) Anticoagulant-related intracerebral hemorrhage. Neurology 35: 943–948

Kelly PJ, Daumas-Duport C, Scheithauer BW, Kall BA, Kispert DB (1987a) Stereotactic histologic correlations of computed tomography- and magnetic resonance imaging-defined abnormalities in patients with glial neoplasms. Mayo Clin Proc 62: 450–459

Kelly PJ, Daumas-Duport C, Kispert DB *et al.* (1987b) Imaging-based stereotaxic serial biopsies in untreated intracranial glial neoplasms. J Neurosurg 66: 865–874

Kendall BE, Jakubowski J, Pullicino P, Symon L (1979) Difficulties in the diagnosis of glioma by cat scan. J Neurol Neurosurg Psychiat 42: 485–492

Ketz E (1974) Brain tumors and epilepsy. In: Vinken PJ, Bruyn GW (eds) Handbook of Clinical Neurology, Vol. 16. North-Holland, Amsterdam, pp 254–269

Klatzo I (1967) Neuropathological aspects of brain edema. J Neuropathol Exp Neurol 26: 1–14

Kleinman GM, Hochberg FH, Richardson EP Jr (1981) Systemic metastases from medulloblastoma: report of two cases and review of the literature. Cancer 48: 2296–2299

Kleinschmidt-DeMasters BK (1986) Intracarotid BCNU leukoencephalopathy. Cancer 57: 1276–1280

Kramer S, (1968) The hazards of therapeutic irradiation of the central nervous system. Clin Neurosurg 15: 301–318

Kramer S, Hendrickson F, Zelen M, Schotz M (1977) Therapeutic trials in the management of metastatic brain tumors by different time/dose fractionation schemes of radiation therapy. Natl Cancer Inst Monogr 46: 213–221

Lampert PW, Tom MI, Rider WD (1959) Disseminated demyelination of the brain following Co-60 (gamma) radiation. Arch Pathol 68: 322–330

Lampert PW, Davis RL (1964) Delayed effects of radiation on the human central nervous system. Neurology 14: 912–917

Langfitt TW, Kassell NF (1966) Non-filling of cerebral vessels during angiography: correlation with intracranial pressure. Acta Neurochir 14: 96–104

Laws ER, Taylor WF, Clifton MB, Okazaki H (1984) Neurosurgical management of low-grade astrocytoma of the cerebral hemispheres. J Neurosurg 61: 665–673

Leibel SA, Sheline GE (1987) Radiation therapy for neoplasms of the brain. J Neurosurg 66: 1–22

Li FP, Winston KR, Gimbrere K (1984) Follow-up of children with brain tumours. Cancer 54: 135–138

Little JR, Dial B, Belanger G, Carpenter S (1979) Brain hemorrhage from intracranial tumor. Stroke 3: 283–287

Liu HM (1978) Reactive neuroaxonal dystrophy in children. Clinical pathological correlation. Acta Neuropathol 42: 237–241

Liu HM, Maurer HS, Vongsvivut S, Conway JJ (1978) Methotrexate encephalopathy. A neuropathologic study. Human Pathol 9: 635–641

Liwnicz BH, Wu SZ, Tew JM (1987) The relationship between the capillary structure and hemorrhage in gliomas. J Neurosurg 66: 536–541

Locksley HB, Sahs, AL, Sandler R (1966) Report on the Cooperative Study of Intracranial Aneurysms and Subarachnoid Hemorrhage. Section 3. Subarachnoid hemorrhage unrelated to intracranial aneurysm and A-V malformation. A study of associated diseases and prognosis. J Neurosurg 24: 1034–1056

Mahaley MS, Whatley RA, Blue M, Bertsch L (1986) Central neurotoxicity following intracarotid BCNU chemotherapy for malignant gliomas. J Neuro-oncol 3: 297–302

Mandybur TI (1977) Intracranial hemorrhage caused by metastatic tumors. Neurology 27: 650–655

Manz HJ, Woolley PV, Ornitz RD (1979) Delayed radiation necrosis of brainstem related to fast neutron beam irradiation. Case report and literature review. Cancer 44: 473–478

Marks JE, Baglan RJ, Prassad SC, Bland WF (1981) Cerebral radionecrosis: incidence and risk in relation to dose, time, fractionation and volume. Int J Radiat Oncol Biol Phys 7: 243–252

Marshall LF, Jennett B, Langfitt TW (1974) Needle biopsy for the diagnosis of malignant glioma. J Am Med Ass 288: 1417–1418

Matsukado Y, MacCarthy CS, Kernohan JW (1961) The growth of glioblastoma multiforme (astrocytomas grades III and IV) in neurosurgical practice. J Neurosurg 18: 636–644

McComb JG, Davis RL, Isaacs H Jr (1981) Extraneural metastatic medulloblastoma during childhood. Neurosurgery 9: 548–551

McKeran RO, Thomas DGT (1980) The clinical study of gliomas. In: Thomas DGT, Graham DI (eds) Brain Tumours. Scientific Basis, Clinical Investigation and Current Therapy. Butterworth, London, pp 194–230

Mechanick JI, Hochberg FH, La Rocque A (1986) Hypothalamic dysfunction following whole-brain irradiation. J Neurosurg 65: 490 494

Meinig G, Revlen HJ, Magauly C (1973) Regional cerebral blood flow and cerebral perfusion pressure in global brain edema induced by water intoxication. Acta Neurochir 29: 1–13

Mena H, Garcia JH, Velandia F (1981) Central and peripheral myelinopathy associated with systemic neoplasia and chemotherapy. Cancer 48: 1724–1728

Miller JD, Adams JH (1984) The pathophysiology of raised intracranial pressure. In: Adams JH, Corsellis JAN, Duchen LW (eds) Greenfield's Neuropathology, 4th edn. Edward Arnold, London, pp 53–84

Modan B, Baidatz D, Mart H et al. (1974) Radiation-induced head and neck tumours. Lancet 1: 277–279

Monro P, Mair WGP (1968) Radiation effect on the human central nervous system 14 weeks after x-irradiation. Acta Neuropathologica 11: 267–274

Moseley I (1986) Imaging techniques in the investigation of cerebral tumours. In: Bleehen NM (ed) Tumours of the Brain. Springer-Verlag, Berlin, pp 35–50

Muller K-K, Menne R, Bachmann KD, Grobe H (1981) Calcified cerebral necrosis following ALL therapy. J Cancer Res Clin Oncology 102: 81–86

Mussini JM, Friol M, de Kersaint-Gilly A et al. (1980) Radionécrose cérébrale découverte 32 ans aprés irradiation conventionelle pour adénome hypophysaire. Revue Neurol 136: 43–58

Mutlu N, Berry RG, Alpers BJ (1963) Massive cerebral hemorrhage. Arch Neurol 8: 644–661

Myers R, Rogers MA, Hornsey S (1986) A reappraisal of the roles of glial and vascular elements in the development of white matter necrosis in irradiated rat spinal cord. Br J Cancer 53, Supp 7, 221–223

Nakazato Y, Ishida Y, Morimatsu M (1980) Disseminated necrotizing leukoencephalopathy. Acta Pathol Jap 30: 659–670

Neuwelt EA, Frenkel EP, Gumerlock MK et al. (1986) Developments in the diagnosis and treatment of primary CNS lymphoma: a prospective series. Cancer 58: 1609–1620

Neuwelt EA, Dahlborg SA (1987) Chemotherapy administered in conjunction with osmotic blood–brain barrier modification in patients with brain metastases. J Neuro-oncol 4: 195–207

Norrell H, Wilson CB, Slagel DE, Clark DB (1974) Leukoencephalopathy following the administration of methotrexate into the cerebrospinal fluid in the treatment of primary brain tumors. Cancer 33: 923–927

Oldberg E (1933) Hemorrhage into gliomas. Arch Neurol Psychiat 30: 1061–1073

Palmer AJ (1972) Radiation myelopathy. Brain 95: 109–122

Papavero L, Loew F, Jaksche H (1987) Intracarotid infusion of ACNU and BCNU in adjuvant therapy of malignant gliomas: clinical aspects and critical considerations. Acta Neurochir 85: 128–137

Park TS, Hoffman HJ, Hendrick EB, Humphreys RP, Becker LE (1983) Medulloblastoma: clinical presentation and management. J Neurosurg 58: 543–552

Pasquier B, Pasquier D, N' Golet A, Panh MH, Couderc P (1980) Extraneural metastases of astrocytma and glioblastoma. Clinicopathological study of two cases and review of literature. Cancer 45: 112–118

Penfield W, Jasper H (1954) Epilepsy and the Functional Anatomy of the Human Brain. Little Brown, Boston, Mass.

Penman J, Smith MC (1954) Intracranial Gliomata, PMRC Report No. 248, HMSO, London

Pollak ER, Miller HJ, Vye MV (1981) Medulloblastoma presenting as leukemia. Am J Clin Pathol 76: 98–103

Pratt RA, Di Chiro G, Weed JC (1977) Cerebral necrosis following irradiation and chemotherapy for metastatic choriocarcinoma. Surg Neurol 7: 117–120

Price RA, Jamieson PA (1975) The central nervous system in childhood leukaemia. II. Subacute leukoencephalopathy. Cancer 35: 306–318

Riehl JL, Brown WJ (1964) Acute cerebellar syndrome secondary to 5FU therapy. Neurology 14: 961–967

Roth JG, Elvidge AR (1960) Glioblastoma multiforme: a clinical survey. J Neurosurg 17: 736–750

Russell DS (1954) The pathology of spontaneous intracranial haemorrhage. Proc Roy Soc Med 47: 689–693

Russell DS, Rubinstein LJ (1989) Pathology of tumours of the nervous system, 5th edn. Edward Arnold, London, pp 438, 855, 873

Salazar OM, Rubin P, McDonald JV, Feldstein ML (1976) High dose radiation therapy in the treatment of glioblastoma multiforme. Int J Radiat Oncol Biol Phys 1: 717–727

Samaan NA, Dakdash MM, Caderao JB (1975) Hypopituitarism after external irradiation: evidence for both hypothalamic and pituitary origin. Ann Intern Med 83: 771–777

Scott M (1975) Spontaneous intracerebral hematoma caused by cerebral neoplasms. Report of eight verified cases. J Neurosurg 42: 338–342

Seitz RJ, Wechsler W (1987) Immunohistochemical demonstration of serum proteins in human cerebral gliomas. Acta Neuropath 73: 145–152

Shapiro WR, Chernik NL, Posner JB (1973) Necrotizing encephalopathy following intraventricular instillation of methotrexate. Arch Neurol 28: 96–102

Shapiro WR, Young DF (1984) Neurological complications of antineoplastic therapy. Acta Neurol Scand (Suppl) 100: 125–132

Sheline GE, Wara WM, Smith V (1980) Therapeutic irradiation and brain injury. Int J Radiat Oncol Biol Phys 6: 1215–1228

Sheline GE (1986) Normal tissue tolerance and radiation therapy of gliomas of the adult brain. In: Bleehen NM (ed) Tumours of the Brain. Springer-Verlag, Berlin, pp 147–160

Smith B (1975) Brain damage after intrathecal methotrexate. J Neurol Neurosurg Psychiat 38: 810–815

Smith DR, Hardman JM, Earle KM (1969) Metastasizing neuroectodermal tumors of the central nervous system. J Neurosurg 31: 50–58

Soffer D, Pittaluga S, Feiner M, Beller AJ (1983) Intracranial meningiomas following low-dose irradiation to the head. J Neurosurg 59: 1048–1053

Spencer CD, Weiss RB, Van Eys J, Cohen P, Edwards B (1984) Medulloblastoma metastatic to the marrow. Report of four cases and review of the literature. J Neuro-oncol 2: 223–226

Statham P, Macpherson P, Johnston R, Forster DM, Adams JH, Todd NV (1990) Cerebral radiation necrosis complicating stereotactic radiosurgery for arteriovenous malformation. J Neurol Neurosurg Psychiat 53: 476–478

Tiberin P, Maor E, Zaizov R et al. (1984) Brain sarcoma of meningeal origin after cranial irradiation in childhood acute lymphocytic leukemia. J Neurosurg 61: 772–776

Todd NV, McDonagh T, Miller JD (1987) What follows diagnosis by computed tomography of solitary brain tumour? Lancet i, 611–612

Tomita T, McLure DG (1986) Medulloblastoma in childhood: results of radical resection and low-dose neuraxis radiation therapy. J Neurosurg 64: 238–242

Tso MOM, Hayreh SS (1977) Optic disc edema in raised intracranial pressure. III. A pathologic study of experimental papilledema. Arch Ophthalmol 95: 1148–1157

Ushio Y, Arita N, Yoshimine T, Nagatani M, Mogami H (1987) Glioblastoma after radiotherapy for craniopharyngioma. Neurosurgery 21: 33–38

Van der Kogel AJ (1986) Radiation-induced damage in the central nervous system: an interpretation of target cell responses. Br J Cancer 53, Suppl 7, 207–209

Wakai S, Yamakawa K, Manaka S et al. (1982) Spontaneous intracranial hemorrhage caused by brain tumor: its incidence and clinical significance. Neurosurgery 10: 437–444

Weiss HD, Walker MD, Wiernik PH (1974) Neurotoxicity of commonly used antineoplastic agents. New Engl J Med 291: 75–81

Wilson WB, Perez GM, Kleinschmidt, De Masters BK (1987) Sudden onset of blindness in patients treated with oral CCNU and low dose cranial irradiation. Cancer 59: 901–907

Wizniter M, Parker RJ, Rorke LB, Meadows AT (1987) Cerebellar sclerosis in pediatric cancer patients. J Neuro-oncol 4: 353–360

Wolman L (1956) Pituitary necrosis in raised intracranial pressure. J Pathol Bacteriol 72: 575–586
Wroe SJ, Foy PM, Shaw MDM et al. (1986) Differences between neurological and neurosurgical approaches in the management of malignant brain tumours. Br Med J 293: 1015–1018
Yasargil MG (1969) Subarachnoid hemorrhage. Schweiz Med Wochenschr 99: 1629–1632
Young DF, Posner JB, Chu F, Nisce L (1974) Rapid-course radiation therapy of cerebral metastases: results and complications. Cancer 34: 1069–1076
Zimmerman RA, Bilaniuk LT (1980) Computed tomography of acute intratumoral hemorrhage. Radiology 135: 355–359
Zuccarello M, Sawaya R, deCourten-Myers G (1986) Glioblastoma occurring after radiation therapy for meningioma: case report and review of literature. Neurosurgery 19: 114–119

6 Low-grade Gliomas

J. Koivukangas

Introduction

Low-grade gliomas are histologically relatively well-differentiated tumours of neuroepithelial origin. The term "glioma" was created by Virchow. He first described the supporting elements of the nervous system, the neuroglia, and the tumours arising from this cell type. The classification of brain tumours was later modified by Bailey and Cushing (1926), who combined the histogenetic basis with a correlated study of prognosis.

According to the WHO classification of central nervous system tumours (Zulch, 1979), low-grade gliomas include histologically well-differentiated grades of the following tumours:

Astrocytic tumours
  astrocytoma (all variants)
  pilocytic astrocytoma
Oligodendroglioma
Mixed oligo-astrocytoma
Ependymoma
Choroid plexus papilloma
Pineocytoma

Since the low-grade cerebral astrocytoma of adults has been the subject of most study, this chapter will concentrate on the clinical management of this tumour as a model for management of cerebral oligodendroglioma and mixed oligo-astroctyoma as well. Pilocytic astrocytoma, ependymoma, choroid plexus papilloma and pineocytoma, while usually candidates for surgical treatment, require additional special management considerations described elsewhere.

Low-grade Cerebral Astrocytoma

The term "low-grade" assumes a grading system. At present, several grading systems are in use. The Kernohan grading system (Kernohan and

Sayre, 1952) divides astrocytomas into four groups, grades 1 to 4, on the basis of the degree of malignancy or, conversely, the extent of astrocytic differentiation. Since grades 1 and 2 are associated with clearly better survival prognoses, it has become customary to combine these groups together under the term "low-grade". The WHO classification (Zulch, 1979), on the other hand, assigns a grade (1 to 4) based on the histological subtypes of astrocytoma. Thus, in the WHO classification, pilocytic astrocytomas of the cerebellum, third ventricle and optic nerve, and the subependymal giant cell astrocytoma are grade 1 astrocytomas. The fibrillary, protoplasmic, gemistocytic and mixed variants of astrocytoma are grouped together into grade 2. Daumas-Duport et al. (1988) have advanced the newest grading method for what they call "ordinary astrocytomas", i.e. fibrillary, protoplasmic, gemistocytic and anaplastic astrocytomas and glioblastomas. They grade these tumours on the basis of the presence or absence of four morphological criteria: nuclear atypia, mitoses, endothelial proliferation and necrosis. If a tumour shows none of these criteria, it is assigned into grade 1, whereas astrocytomas displaying one criterion (usually nuclear atypia) are grade 2 tumours, and so on. Similarly, various grading systems have been presented for other gliomas as well. (For further discussion see Chapter 3.)

Thus, one of the chief difficulties in the study of low-grade tumours is the lack of uniformity and consensus in grading in the literature. Indeed, whether to use a three- or four-tier system, or a numerical system at all, has not yet been resolved. Some, including Burger (1990), prefer a non-numerical classification such as the Ringertz system (Ringertz, 1950), which simply divides the astrocytic tumours into astrocytoma, anaplastic astrocytoma ("intermediate type") and glioblastoma multiforme. All studies of treatment effectiveness, however, should clearly define the tumour under consideration in terms of the most appropriate grading system, so that results can be compared at different centres.

This chapter will deal specifically with astrocytomas falling into grades 1 and 2 of the Kernohan or grade 2 of the WHO classification. These tumours coincide roughly with those of grades 1 and 2 in the Daumas-Duport system. Histologically, the tumours show increased cellularity with or without nuclear atypia, depending on the grading system, but do not show other signs of anaplasia, such as frequent mitoses, endothelial proliferation and necrosis. Where the cited literature differs from this definition, the appropriate comments will be made.

Incidence

Between 1% and 2% of post mortems have been found to have primary brain tumours (Green et al., 1976). In the clinical setting, low-grade (Kernohan grades 1 and 2) astrocytomas accounted for 8% of primary brain tumours in the epidemiological study of Schoenberg et al. (1976). Whereas astrocytomas as defined by the WHO classification constitute 20%–30% of gliomas, oligodendrogliomas account for only about 5%–6% (Graham, 1980). The most common histological type of astrocytoma is the fibrillary variant. Low-grade astrocytic tumours are most commonly found in the cerebral hemispheres. In the study of Laws *et al.* (1984), out of 500 cases of supratentorial low-grade astrocytoma, 208 (42%) were in the frontal lobe, 207 (41%) in the temporal lobe, 77 (15%) in the parietal lobe and only 8 (2%) in the occipital lobe. In the same study, 70% of patients were between 20 and 49 years of age.

Patient Series

In the present catchment series of patients treated in Oulu University Hospital, Finland, out of 200 consecutive operations for primary adult brain tumours, there were 27 low-grade tumours (grades 1 and 2 of the Kernohan classification) or 14% of surgically treated primary brain tumours in adults. These patients were retrospectively studied before and 1 year after surgery. Seventeen out of 27 (63%) were between the ages of 25 and 50 years. There were 17 females and 10 males. Sixteen out of 27 (59%) patients had had symptoms for less than 1 year. Most of the patients (74%) had undergone partial resection of the tumour and, after surgery, 17 out 27 (63%) received local tumour radiation therapy. New technology (laser, ultrasound aspirator, ultrasound imaging and/or open stereotaxic technique) was used in 63% of cases. One-third of the patients had developed recurrence of tumour within the follow-up year and 4 were reoperated during this time. Twenty-three (85%) were alive at the end of the follow-up year.

Nineteen (70%) of the patients had had epilepsy before surgery, compared with 12 out of 23 (52%) after treatment for the group as a whole during the follow-up year. Before surgery, the average Karnofsky score was 79, whereas 1 year later it was 77 for the 23 living patients. However, before surgery, 22 patients (81%) had had Karnofsky scores of 80 to 100 compared with 12 out of 23 (52%) at the end of the first follow-up year.

Clinical Presentation

In the retrospective study of Laws *et al.* (1986), 42% of patients with supratentorial low-grade astrocytoma had had symptoms for less than 1 year before surgery, while 58% had had symptoms for 1 year or longer. Sixty-six per cent of patients presented with seizures, 44% with headaches, less than 10% with nausea or vomiting, 22% with papilloedema, 7% with altered consciousness, 16% with personality change, 14% with language deficit and 16% with visual loss. About 40% of patients had no preoperative neurological deficits, with 77% of the rest having mild deficits. Seventy per cent of the patients were 20–49 years of age.

Thus, the typical low-grade cerebral astrocytoma patient in the retrospective study of Laws and co-workers was a young adult (70% were 20–49 years of age) presenting with seizures and/or headache. The study included patients treated at the Mayo Clinic between 1915 and 1976, i.e. before the era of X-ray computed tomography (CT). During that time, tentative diagnosis of low-grade astrocytoma and the decision to operate was based on such examinations as isotope brain scan, cerebral angiography and pneumoencephalography. These methods produced rough approximations of the site and extent of tumour involvement. The main imaging examination used for screening was based on radionuclide scanning, which was associated with a 44% incidence of false-negative scans in cases of tumours of unremarkable vascularity (low-grade astrocytomas) even when compared with cerebral angiography, the mainstay of neurosurgical tumour localization before CT (Fulghum J.S., et al., 1971). Furthermore, the threshold for the invasive examinations necessary for diagnosis was considerably higher than it is currently for CT. It is no surprise that CT and, later, magnetic resonance imaging (MRI) have made diagnosis easier, and today patients are in better condition when presenting to the neurosurgeon. The watershed effect of CT on therapeutic results must be stressed when comparing current results with those of the literature. Piepmeier (1987) and Vertosick et al. (1991) have also stressed the effect of CT and early diagnosis on longer survival rates in recent series.

In more recent practice, seizures are more commonly the presenting symptom with headache and papilloedema less common than in the study of Laws and colleagues. In the present study of 27 consecutive low-grade astrocytomas of adults, 70% presented with seizures, but only 15% had papilloedema. Headache was reported by 40% of patients, but was not necessarily associated with the tumour because of its irregular nature. In the recent study by Piepmeier (1987), 90% of patients had seizures, with only 5% reporting headache. In Piepmeier's paper, low-grade astrocytic tumours included also mixed astrocytoma/oligodendroglioma and pleomorphic xanthoastrocytoma. He defined low-grade astrocytic tumours as "without evidence of anaplasia, nuclear pleomorphism or coagulation necrosis". Furthermore, in the study of Vertosick et al. (1991), 92% of patients presented with seizures, 8% with papilloedema and 16% with headache.

In terms of functional ability, 78% of the patients in the present study were capable of working and 19%, while unable to work, were independent. By the end of the first year, 11 (48%) out of the 23 living patients were capable of working.

Compared with the study of Laws et al. (1986), the present study and that of Piepmeier reflect the smaller size of astrocytomas detected earlier in the course of the disease by less invasive CT and MRI. Indeed, in the present study 58% of the lesions (tumour and surrounding reaction) were 4 cm or less in maximal diameter.

The typical low-grade cerebral astrocytoma patient is a young adult in the 30s or 40s with subcortical or deep low-density non-contrast enhancing lesion on CT, with high-intensity signal on T_2-weighted MR images. At operation the tumour appears as a diffusely demarcated area of toughened white or slightly discoloured tissue. The patient often has had only a focal epileptic seizure and with adequate anti-epileptic medication is otherwise completely healthy, capable of working and neurologically and neuropsychologically intact. The tumour may be small and near important centres such as the motor strip.

Special Features

Several special characteristics of low-grade cerebral astrocytoma affect the surgical management of these tumours, including those described below.

Development of Malignancy

Laws *et al.* (1986) found that 49% of 79 supratentorial low-grade astrocytomas recurred and were documented at surgery or autopsy as having changed to astrocytoma grade 3 or 4. In the study of Muller *et al.* (1977), 62 out of 72 tumours that were well-differentiated astrocytomas (Kernohan grades 1 and 2) at first operation had become either anaplastic astrocytoma or glioblastoma multiforms at the second operation. This progression may be due to sampling error in the original biopsy, or it may be clinical evidence for the malignant transformation of the tumour. In the study of McCormack *et al.* (1992), 24 out of 41 patients (59%) who had had "gross or radical subtotal resection" of their tumour developed a recurrence of symptoms and CT mass with a median time to recurrence of 4.5 years. In almost all cases, the CT scan showed oedema and contrast enhancement suggestive of transformation to a higher grade of tumour malignancy. The present author also found, in a series of 26 consecutive low-grade astrocytoma patients distinct from the series reported above and followed for up to 6 years, that the Karnofsky score of 4 of the patients began to drop abruptly from a level of 80–100, with death ensuing within 2 years (unpublished results). Thus there seems to be clinical evidence for malignant transformation, i.e. the low-grade astrocytoma becomes a high-grade one with a corresponding change in clinical course. Vertosick *et al.* (1991) suggested that death from a low-grade astrocytoma in the absence of malignant transformation may be a rare event and that in this respect, early radical surgery may play a role in improved prognosis, since reduction of the neoplastic cell pool could decrease the statistical chance that dedifferentiation will occur.

The biological explanation for this behaviour is that while the low-grade astrocytoma initially grows slowly, being composed of isomorphic cells and having poor angiogenesis, it may later change its rate of growth. Cells pass from the non-proliferating pool to the proliferating one and the growth fraction increases. An increase in cell density and mitoses signals the beginning of the change, to be followed by necrosis and angiogenesis (Schiffer, 1991). Classically, this has been considered to be due to dedifferentiation of the low-grade astrocytoma cells, i.e. regression to a more primitive cell stage as the tumour cells multiply (the theory of tumour cell dedifferentiation) (Rubenstein, 1972). More recently, it has been found that astrocytomas display heterogeneous tumour cell populations, suggesting that different clones of cells replace each other dynamically (selection by competition). Hoshino *et al.* (1988) showed that the low-grade astrocytomas display considerable variation in proliferative potential, as assessed by bromodeoxyuridine cell kinetic studies.

The clinical implication is that astrocytomas need to be sampled adequately, whether by serial biopsy or resection of tumour, and the worst degree of anaplasia should be noted. On the other hand, the timing of surgery measured from the time of radiographic diagnosis, whether early or after a waiting period for various reasons, was not found to have any appreciable effect on the rate of malignant transformation (Recht *et al.*, 1992).

Infiltration of Astrocytomas

Astrocytomas typically grow by infiltration, invading the cortex or extending along white matter fibre tracts such as the capsula interna and corpus callosum. The tumour may start in one gyrus and begin to infiltrate along associative fibres to the neighbouring gyrus or to deeper regions. As the tumour grows, the extent of surrounding oedema seems also to increase. Kelly *et al.* (1987) found that tumour cell infiltration generally extended at least as far as the prolongation of the T_2 signal on the T_2-weighted MR images, in some cases even further.

The clinical implication here is that while the tumour proper, i.e. especially the bulk of relatively small astrocytomas, can often be removed radically, the paths of microscopic infiltration are impossible to assess at operation.

Neurosurgical Management

Debulking of large tumour masses for palliation and reduction of intracranial pressure is no longer the only indication for surgery. Since many tumours are small, definitive diagnosis and control of epilepsy and other neurological symptoms are also relevant indications for surgical management which includes maximal removal of the tumour. The surgical management of small tumours, not detected before the era of CT and MRI, is controversial, but an argument for gross total removal with the aid of new surgical technology is made in this chapter (Table 6.1).

Since the patient with low-grade glioma presents today with less severe symptoms and neurological signs than in earlier years, and since the radiological findings may be very slight, the challenges confronting the operating neurosurgeon are also different. The management of the patient may

Table 6.1 Surgical alternatives in cerebral low-grade astrocytoma

Small tumour without appreciable surrounding oedema	Biopsy Gross total removal with or without partial lobectomy
Large tumour with extensive infiltration and reactive oedema	Biopsy Partial removal with or without partial lobectomy

require follow-up studies. The surgical procedure may involve eloquent regions in a neurologically intact patient. The tumour may be difficult to find on the basis of visual and tactile senses alone. More and more, the essential question in neurosurgical management has moved past operative mortality and morbidity, always crucial in themselves, to quality of life issues.

Expectant Management

Expectant management by definition means that neurosurgical treatment is withheld for the time being in anticipation of possible changes in the tumour or the clinical state of the patient. It is assumed that there is a certain probability that the tumour will either remain dormant for a time or grow slowly, giving the patient maximal neurological function with no surgical risk. The patient is not given the possible benefits of surgery, but the surgical risks are also avoided.

Expectant management without biopsy confirmation is not really a policy decision on low-grade astrocytoma, but rather on certain radiological findings, such as "a subcortical low-density lesion" on CT. Such findings may indicate low-grade astrocytoma, but the differential diagnoses have not been ruled out. For example, later imaging studies serve to exclude a recent vascular insult, which may resolve within weeks, leaving a smaller, usually better defined lesion. On the other hand, a lesion which remains constant or shows signs of growth may indicate neoplasia, but again without biopsy there is no histological differentiation.

Expectant management following biopsy is often reserved for low-grade astrocytoma of eloquent subcortical regions or, especially in neurosurgical centres with stereotaxic facilities, of the deep brain areas. Further neurosurgical decisions depend on the growth rate of the tumour, the clinical state of the patient, the effect of adjuvant therapy and local referral policies. Radiation therapy specialists usually require histological confirmation before treatment. It should be stressed that optimum surgical practice also requires definitive knowledge about the nature of the lesion being managed expectantly.

Once surgery has been decided upon, the options are burr-hole and craniotomy procedures. Conventionally, burr-hole (or twist-drill) procedures facilitate stereotaxic biopsy methods, whereas craniotomy can be used for freehand biopsy or tumour resection. During the past decade or so, a number of new surgical methods have evolved, including open stereotaxic techniques and ultrasound-guided surgery.

Biopsy

In a biopsy procedure, an aspiration needle or probe such as small forceps or spiral needle is passed into the lesion to obtain cytological and histological tissue samples. Previously, there was a sharper distinction between stereotaxic and open procedures. A classical stereotaxic procedure was based on the use of one of several available stereotaxic frames and was performed by a stereotaxic neurosurgeon. An open procedure, on the other hand, implied formal craniotomy, dissection of brain tissue and exposure of tumour. Astrocytomas in deep cerebral regions, for example, are still often biopsied by stereotaxic methods, while more superficial tumours of eloquent regions, such as the motor strip, are often biopsied in an open procedure.

Stereotaxic Biopsy

Stereotaxic biopsy is still usually performed using mechanical frames based on the principles developed in the 1940s. These frames are very precise mechanical instruments and often employ the simple geometry of the sphere: the sphere is described, for example, by the movement of the biopsy needle guide along an arc which, in turn, is rotated on a collar. Point-targets are calculated manually or using computer software, and either the head or the frame is moved to accommodate the target. Single or serial biopsies can be obtained. During the 1980s, several of the conventional stereotaxic frames, originally designed for use with anteroposterior and lateral X-ray films, have been adapted for digital data derived from tomographic imaging methods such as CT, MRI and positron emission tomography (PET). Sophisticated imaging software has also been developed to graphically display the stereotaxic system and the imaging data registered together.

Open Biopsy

Open biopsy implies the use of craniotomy to gain visual and tactile contact with the tumour. Deeper lesions require brain dissection, while superficial subcortical tumours may be apparent at the surface of the brain. The general goal of maximal tumour resection may be compromised in cases where the tumour lies in or close to eloquent regions of brain in neurologically intact patients. In such cases, the procedure may be limited to biopsy, generally required for consideration of radiotherapy.

Open or stereotaxic biopsy can be aided with real-time intraoperative ultrasound (Knake *et al.* 1984; for review see Koivukangas, 1984). The advantages of using ultrasound are many:

1. The preoperative CT and MRI images of tumour can be *supplemented* with ultrasound images to help choose the most representative region of the lesion to be biopsied.
2. The approach to the tumour can be *guided* by showing the location of major (pulsating) cerebral arteries and their branches within the cerebrum. In open nonstereotaxic approaches, the approach to the tumour can also be followed during the dissection. In many cases the approach can be made by opening a sulcus that can be seen on the ultrasound scan to reach close to the tumour (sulcal approach).
3. After biopsy, the site of the biopsy can be *controlled* for correct location and to *rule out* complications, especially haematoma.

Biopsy of tumour should be made in regions best accessible, but also representative of the active region of the tumour: necrotic and cystic regions are usually absent in cerebral low-grade lesions, as often are contrast-enhancing regions. Most representative is the uniformly low-density region in CT scans and the hyperechogenic region in ultrasound scans. Serial biopsies along approach trajectories give several specimens from different parts of the tumour. Disturbingly, such studies have shown interstitial tumour cells (ITC) within otherwise normal surrounding brain tissue (Kelly et al., 1987).

Removal of Tumour

According to the largest study of low-grade astrocytomas so far, "the data generally support a philosophy of radical surgical removal whenever possible" (Laws et al., 1984). The main evidence for this prevailing argument in astrocytoma surgery has to do with survival curves of patients with low-grade astrocytoma treated by biopsy, subtotal resection and "total" resection. The main weakness of the argument has been the complete lack of randomized controlled clinical trials. It can be argued that the tumours which have been biopsied have been "only" biopsied because the tumour cannot be more radically treated surgically: e.g. the patient's clinical condition may be poor, the tumour may be inaccessible, i.e. the tumour may simply grow in a location which brings relatively early death. The "totally" removed tumours on the other hand are probably in less critical areas to begin with, in the frontal or non-dominant temporal lobe, relatively easily accessible and often radically removable, i.e. the tumour may simply grow in an area which allows fairly long survival. This bias was recognized by Laws et al. (1984) in their retrospective study.

In addition to survival statistics, radical tumour removal has been supported by cell kinetic studies (Hoshino, 1991). According to these studies, well-differentiated, i.e. low-grade, astrocytomas may proliferate conservatively, i.e. not all daughter cells retain mitotic activity after cell division. Since this non-cycling pool is the one that contributes to tumour growth, the cycling pool remaining fairly constant, one would expect resection actually to slow down, although not completely stop, the growth rate of the tumour.

The third rationale for maximal resection of tumour is based on the theory of immunological host-tumour response: the smaller the residual tumour cell population, the better the chances for immunological "rejection" or limitation of tumour growth. This is called the reduction of cell burden. One of the problems with this theory in the case of low-grade astrocytomas is the relatively intact blood–brain barrier, which impedes host responses in this immunologically "privileged" tissue. An associated hypothesis is that the less the residual tumour, the more effective is the action of radiotherapy and the adjuvant treatment.

As Laws et al. (1984) showed, surgical mortality has decreased since the 1920s from around 30% to 1%–2%. This is due to improved neuroanaesthesia, imaging and neurosurgical technique. At the same time, morbidity associated with surgery has also decreased. Thus alleviation and postpone-ment of neurological deficits has been a further indication for surgical intervention. During the 1980s, attention has been focused on improving the quality of life of the patients, something more subtle and more sub-jective than findings in the neurological examination.

Thus, in addition to statistically improved survival time, the notion of the quality of that survival is becoming an important criterion for neuro-surgical management of low-grade astrocytoma. In the present retrospec-tive series of 27 consecutive low-grade astrocytoma patients operated at the author's centre, the Karnofsky score was chosen as an indicator of quality of life. It was found that over a period of the first year the Karnofsky score of surviving patients usually stayed between 80 (independ-ent; subnormal activity with some symptoms and signs of disease) and 100 (normal). Based on these crude approximations of quality of life, one can present the following hypothetical reasoning.

The basic clinical hypothesis in surgical removal of low-grade astrocy-toma is that neurological symptoms (epilepsy, headache, nausea, etc.,) reduce the quality of life. Thus, assuming a score of 1.0 for normal healthy individuals, a patient with neurological symptoms would have upon pre-sentation to the neurosurgeon a score of less than 1.0, say 0.8, on an accepted quality of life scale. A reduction of symptoms, either in amount or degree, is assumed to improve disease-related quality of life. Furthermore, assuming that the theories presented above are correct, one would assume that the patient, once his quality of life is improved by reduction of cell burden, i.e. following maximal tumour resection, would also have a sus-tained quality of life that would be better than that of the patient who did not receive surgical care. As the low-grade tumour grows slowly over time, perhaps even showing periods of dormancy due to biological reasons or to

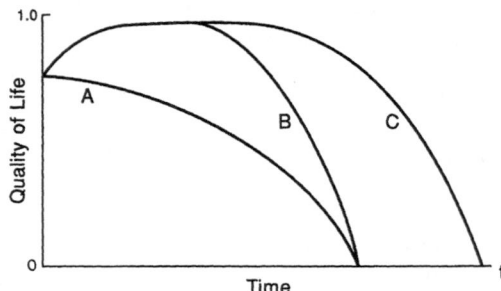

Fig. 6.1 Clinical hypothesis favouring radical resection of low-grade astrocytoma. Curve A assumes that surgical treatment has been withheld and the quality of life would eventually decrease over the survival time of the patient. Curve B assumes that radical tumour resection has been performed, that alleviation and postponement of neurological symptoms results in immediate improvement in quality of life, that the latter is sustained until growth of the tumour again results in lowered quality of life and death, and that the survival time does not change. Curve C assumes additionally that survival time actually is increased by the surgical intervention. The ordinate is a quality of life scale, with 1.0 meaning normal health, 0 meaning death. The differences in the areas under the curves give the net benefit of surgery expressed as quality-adjusted life-years (After Koivukangas and Koivukangas, 1986)

the effect of radiotherapy, a relatively long period of improved quality of life could be expected. However, except for rare cases of cure, a later worsening of symptoms would again worsen the quality of life and growth of the tumour would eventually lead to death. The clinical hypothesis behind the argument for radical tumour resection further implies that, without surgery, the quality of life of a low-grade astrocytoma patient would not increase, but actually worsen over time. Whether surgery would achieve an actual prolongation of survival time or not, the improvement in quality of life would make surgery worthwhile. Graphically, the clinical argument for radical tumour resection is presented in Fig. 6.1.

Stereotaxic Resection of Low-grade Astrocytoma

In the management of low-grade astrocytoma, conventional mechanical stereotaxic surgery has generally been used for biopsy of deep tumours. During the 1980s, following the adaptation of CT to many of these instruments, and the realization that imaging voxels can be used to construct three-dimensional targets, the concept of volumetric or open stereotaxic surgery was developed. Here, the objective is to model the astrocytoma, determine a safe route of approach and then remove the tumour more or less stereotaxically. Shelden et al. (1980) described the principal of tumour surgery simulation with pioneering computer graphics, but it was Kelly and co-workers (Kelly, 1988) who developed the most sophisticated and truly stereotaxic system for computer-assisted laser resection of brain tumours. Later, the author developed a method for ultrasound-based volumetric laser removal of small subcortical tumours, including low-grade astrocytomas, using a skull-mounted adapter to coordinate the imaged tumour to the surgical laser (Koivukangas and Louhisalmi, 1990).

Stereotaxic frames can also be used to guide the neurosurgeon to the tumour, after which it is resected with bipolar coagulation and irrigation–suction, for example, under the operating microscope. These procedures have also been referred to as "open stereotaxic surgery".

Non-stereotaxic Resection of Low-grade Astrocytoma

Non-stereotaxic brain tumour surgery is still the mainstay of low-grade astrocytoma management in many centres. For large tumours, the use of standard craniotomies and dissection and removal of the astrocytomas, with or without lobectomy, are described in detail in most textbooks of neurosurgery. Here, recent experience with the surgery of the small subcortical low-grade astrocytoma will be reviewed. "Small" is defined as localized within one gyrus, with no evidence of infiltration or contrast enhancement in CT and MRI scans. On CT scans, these tumours appear as superficial non-enhancing low-density lesions and MRI shows increased T_2 signal. Intraoperative ultrasound imaging shows the tumours as lesions of increased echogenicity. Often they are located near the motor strip, causing focal seizures or sensory or motor impairment (Fig. 6.2).

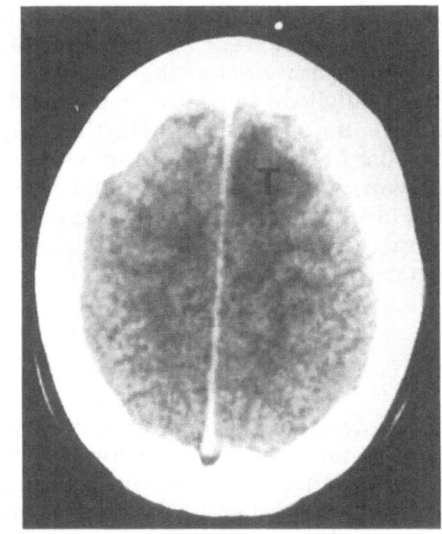

a

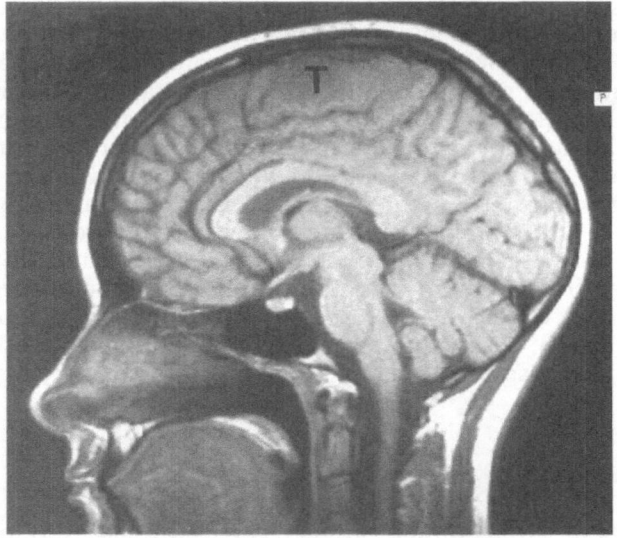

b

Fig. 6.2   a and b

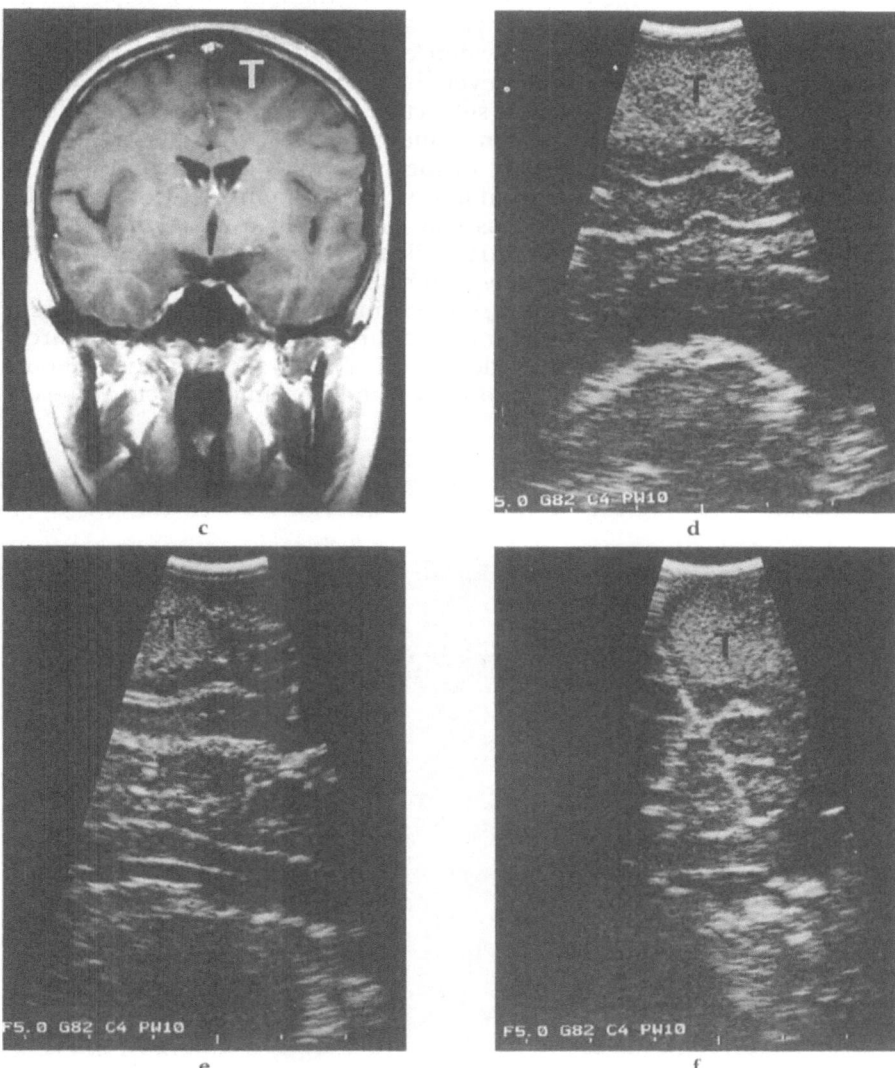

Fig. 6.2 Representative case of low-grade astrocytoma near precentral region of left hemisphere. Twenty-year-old female presenting with only epileptic seizure. CT (**a**) showed subcortical non-enhancing low-density tumour (T), MRI sagittal (**b**) and frontal (**c**) slices showed low signal in the tumour region on T_1-weighted images. The intraoperative ultrasound images (**d–f**) were taken in planes roughly identical to those of the MRI images. Note similar gyral architecture, lateral ventricles and uniformly hyperechogenic tumour. **d** displays the central part of the tumour and **e** the posterior border of the tumour in the sagittal plane, whereas **f** shows the relationship of the tumour to the falx in a frontal plane. The tumour was meticulously resected along the borders of the ultrasound image, with excellent postoperative outcome

So far, no randomized clinical trials on the surgical management of low-grade astrocytomas have been performed. On the basis of in-depth interview of patients before and 1 year after surgery (Koivukangas and Koivukangas, 1988), there was evidence supporting the policy of radical resection of these tumours whenever they can be reasonably delineated. The patients improved in terms of subjectively experienced quality of life.

Since small low-grade astrocytomas may resemble inflammatory or vascular lesions, and the growth rate of the tumours is low, a control CT or MRI scan can be taken following an interval of 1–2 months: by this time the other differential diagnostic lesions will have resolved. In addition to the principle of removing these tumours while they are small, existing tumour-associated paresis and epilepsy may be further indications for surgery.

Since the small astrocytomas are often in close proximity to eloquent brain areas, methods of precise intraoperative localization and demarcation of these tumours have been developed. The surgery of these tumours with neuronavigator guidance and ultrasound imaging is described below (Koivukangas *et al.*, 1993).

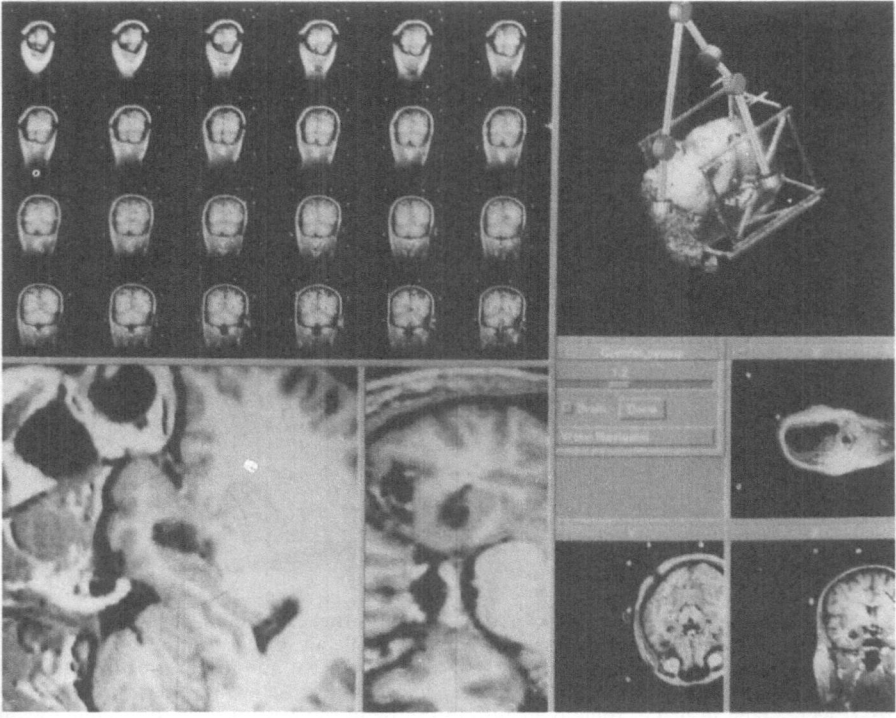

Fig. 6.3 Workstation monitor during neuronavigator surgery. Note representative MR images (upper left), neuronavigator arm pointing to patient's left temporal region, with CRW stereotaxic imaging frame used in this case for fiducials (upper right), orthogonal 3D images showing location of tip of neuronavigator (lower right), ultrasound image equivalent MRI reconstruction with operating axis of neuronavigator (lower centre) and MRI reconstruction perpendicular to the operating axis at a given depth (lower left). A small low-grade astrocytoma can be seen in the mesial temporal lobe within the circle of the left lower image

Rehearsal of Surgical Procedure. The images are examined to choose the approach to the tumour. For an intragyral tumour, the correct positioning of the craniotomy and the determination of the affected gyrus are important. For a deeper tumour, an additional consideration is whether the tumour may be exposed intergyrally using a sulcus that may reach close to the tumour. These aspects of surgical planning can be rehearsed by interacting with the imaging data on the workstation monitor (Fig. 6.3).

Placement of Craniotomy. Following calibration of the neuronavigator, the sterilized area of the scalp is investigated with the neuronavigator pointer, which shows where the tumour and possible sulci leading to it are situated. An S-shaped scalp incision and the craniotomy are then centred on the tumour (Fig. 6.4).

Ultrasound Imaging. Transdural ultrasound imaging confirms the site of the tumour and the dural flap is turned. Ultrasound imaging can be used repeatedly to follow the resection of tumour and to rule out complications such as haematoma. The ultrasound image can be compared to CT/MR images because they are shown in the same plane.

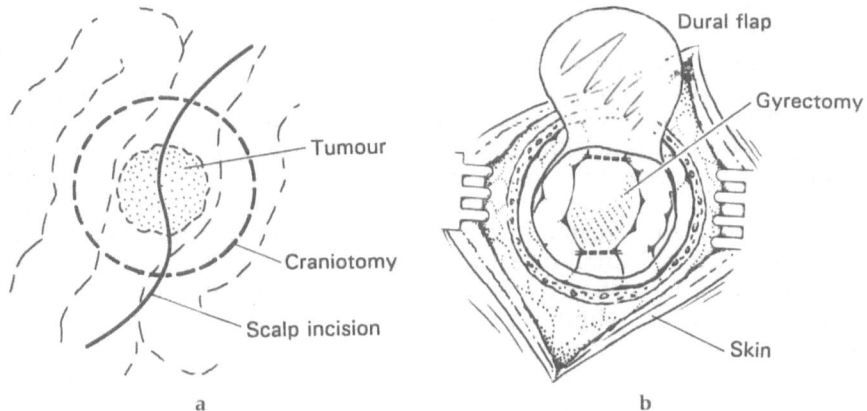

Fig. 6.4 a Superimposed-C-shaped scalp incision, craniotomy and intergyral tumour. **b** Dural flap has been turned, exposing tumour and site of gyrectomy

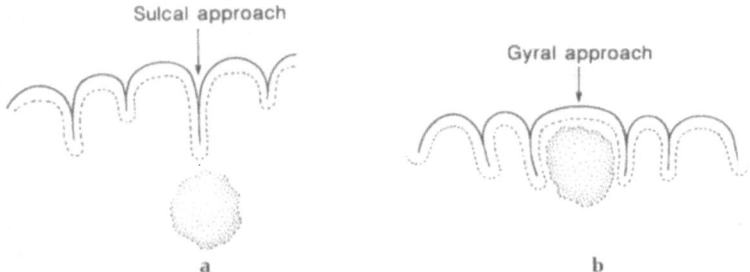

Fig. 6.5   a Principle of sulcal and **b** gyral approaches to small tumour

Microsurgical dissection. Using ultrasound aspirator, laser or conventional irrigation–suction–bipolar coagulation. For intragyral tumours, removal of the gyrus or intragyral dissection of the tumour; for deeper tumours, dissection of the sulcus leading to the tumour followed by its removal (Fig. 6.5).

Evaluation of Surgical Results

During the first half-century of surgical treatment of astrocytomas, reduction of *surgical mortality*, defined as death within 30 days after operation, was the prime consideration, as illustrated by the surgical mortality curve from the Mayo Clinic (Laws *et al.*, 1984). During early years, surgical mortality was 30%, but this has been reduced to well under 5% in most clinics. With improvement of these results, the next consideration was *surgical morbidity*, especially in terms of reduction of surgical complications such as paresis and other deficits. Improvement of *survival time* is a further consideration: surgical treatment must slow the natural progression of the disease, i.e. survival time must be longer in patients treated surgically than in those treated conservatively. Wroe *et al.* (1986) challenged this result in the case of malignant gliomas.

When comparing new survival studies with those in the literature, the effect of earlier diagnosis with CT and MRI during the past 15 years needs to be addressed. In assessing the effect of recent advances in surgical technique on survival time, one must account for the fact that astrocytomas, especially low-grade ones, are being treated earlier in the course of their disease. The imaging methods before CT simply could not detect the small tumours now being treated. Since astrocytomas grow over time, detection while they are still small directly implies that the patients are coming for treatment earlier in the course of their disease, which needs to be taken into consideration when comparing survival times of patients treated before and after the CT and MRI era.

While neurosurgeons have probably always attempted to enhance the subjective well-being of their patients, the most recent consideration in the surgical treatment of brain tumours has been an effort to evaluate scientifically results of surgical series in terms of *quality of life*. Lieberman *et al.* (1982) studied long-term survival of patients with malignant brain tumours. Koivukangas and Koivukangas (1986) found an improvement in quality of life in patients undergoing ultrasound-guided radical surgery for subcortical low-grade astrocytoma. The 'proxy' for quality of life in both studies was the Karnofsky score (Karnofsky *et al*, 1948), which assesses the patient's overall ability to perform certain physical activities.

Quality of life studies were initiated at the author's clinic in 1986. The reason for this interest was the development and application of new methods for radical resection of subcortical tumours (Koivukangas and Koivukangas, 1986). To assess the effectiveness of new treatment modalities, the primary concern is that the methods result in improved prognosis. Since survival time is fairly long in patients with low-grade astrocytomas, quality of life studies have been carried out to assure the safety and efficacy of new treatment. Two of the main questions under investigation at the author's centre are: (a) Should the quality of life instrument be a generic, i.e. generally applicable, questionnaire, or a disease-specific one, or a mixture of both? Specifically, how does one measure quality of life in low-grade astrocytoma patients? (b) Is early radical surgery for low-grade astrocytoma indicated on the basis of survival time and especially quality of life?

Internationally developed instruments for measuring quality of life have recently been standardized in Dulu, Finland. Two such questionnaires are generic in nature and are being used to provide data for a more comprehensive single measure of effectiveness, the MIMIC health index, described below. These two quality of life instruments are the Nottingham Health Profile (NHP) (Hunt *et al.* 1980) and the Health Measurement Questionnaire (HMQ) (Williams, 1988). These questionnaires give data on quality of life of tumour patients, for example, compared with the general population. An ongoing study of 191 brain tumour patients who had survived for 5 years after surgery included 27 patients with low-grade astrocytoma (Kernohan classification). Eighteen (66.7%) of these patients were alive; 14.8% had had biopsy, 77.8% partial removal and 7.4% gross total removal of the tumour. In the NHP, the patients had greatest difficulty in the dimension of energy, and less in pain and mobility dimensions. The HMQ index was 0.796 for the low-grade astrocytoma patients, compared with 0.517 for the grade 3 patients and 0.835 for pituitary adenoma patients. It is possible to use such instruments in studies of the relative effectiveness of alternative treatment protocols for low-grade astrocytomas.

To develop a suitable proxy for quality of life, an elaborate statistical tool based on the consideration of quality of life as a latent variable, i.e. a variable that cannot be directly measured, but that can be described in terms of other measurable variables, has been applied. This resulted in the development of a MIMIC (multiple indicators, multiple causes) model based on a retrospective study of 289 consecutive adult brain tumour patients surgically treated at the author's centre (Koivukangas, 1993). The quality of life of the patients was described in terms of a MIMIC health index, in which the lowest index value (0) represented the least healthy patient in the series and the highest index value (100) the patient with the best health. Each of the other patients in the series received an intermediate score before and 1 year after surgical treatment. In this analysis, the mean score for patients with histologically benign tumours, including low-grade astrocytomas, improved following surgery: whereas their average MIMIC health index was 53 before surgery, it was 74 one year after surgery. This supports the hypothesis of improvement of quality of life following surgery. Significant variables in the MIMIC model measuring health were Karnofsky score, functional ability as assessed by a physician, subjective self-assesssment by the patient, age, diagnosis, tumour location, tumour size, number of symptoms and mood. An ongoing prospective study of 100 successive surgically-treated brain tumour patients in the catchment area will trace in detail the course of quality of life during the first year following surgery using several quality of life instruments as well as the MIMIC model approach. The MIMIC index serves as a cardinal, not just ordinal, measure of quality of life, so it can be used in cost-effectiveness analysis. It can also be used to calculate QALYs (quality-adjusted life years). For example, in the above present series of 27 low-grade astrocytoma patients, 11 (41%) patients were alive after 6 years and of these, 4 still enjoyed a QALY score of 1.0, the highest score. It is suggested that in calculating QALYs for low-grade astrocytoma patients treated with different protocols, the relative effectiveness of treatment alternatives can be estimated.

The proper management of low-grade astrocytoma is more than an academic question. In many societies, economic considerations are also being addressed. Particularly in the UK, the societal implications of health care policy is being addressed in terms of QALYs. The underlying problem is allocation of resources on the basis of health outcomes. Williams (1985) has addressed his question in the economic analysis of coronary surgery. Extending his reasoning to the management of low-grade astrocytoma, the neurosurgical community must increasingly address the following questions: What is the best possible management for my patients as individuals? Is the treatment I am offering really adding to the patient's well-being over his survival time? And increasingly, is someone else suffering more than my patient is benefiting because of my management policy and the use of resources involved? These are some of the ethical challenges that all physicians, including those offering surgical services for low-grade gliomas, need to address rationally and scientifically in the near future. After all, it is the responsibility of the neurosurgeon to choose the method of treatment that best serves the individual patient in his care.

Future Considerations

Prospective multicentre-controlled clinical studies of the management of low-grade astrocytoma should follow consensus on criteria especially for the following issues:

Tumour scoring method.

Method for determination of extent of resection of tumour, based on visual or imaging data.

Evaluation of results in terms not only of mortality and morbidity but also of quality of life.

Acknowledgements

The author's studies summarized in this chapter form the neurosurgical part of multidisciplinary studies in outcomes research and ultrasound and neuronavigator technology. The input of Pirjo Koivukangas, PhD, Arto Ohinmaa, LicSc and Asko Niemela, MD (outcomes research), as well as Yrjo Louhisalmi, LicSc, Jarkko Oikarinen, MSc and Jyrki Alakuijala, MSc (neuronavigator technology), are greatly appreciated, and their contributions have been appropriately cited. Financial assistance from the Technological Development Center, the Academy of Finland, and the Inari and Reijo Holopainen Foundation is also acknowledged. The artwork in the figures was done by Mr Brian Beardsley, Department of Biomedical Graphics, University of Minnesota.

References

Bailey P, Cushing HA (1926) A Classification of the Tumors of the Glioma Group on a Histogenetic Basis with a Correlated Study of Prognosis. J.B. Lippincott, Philadelphia

Burger PC (1990) Classification and biology of brain tumors. In: Youmans JR (ed) Neurological Surgery, Vol. 5. W.B. Saunders, Philadelphia, pp 2967–2999

Daumas-Duport C, Scheithauer B, O'Fallon J, Kelly P (1988) Grading of astrocytomas: a simple and reproducible method. Cancer 62:2152–2165

Fulghum JS, Adcock DF, Guinto FC, Krigman MR, Radcliffe WB (1971) Radionuclide imaging and tumor vascularity in supratentorial gliomas. Invest Radiol 6: 388-391

Graham DI (1980) Primary malignant tumours of the cerebral hemispheres of adults. In: Thomas DGT, Graham DI (eds) Brain Tumours: Scientific Basis, Clinical Investigation and Current Therapy. Butterworths, London

Green JR Waggener JD, Kriegsfeld BA (1976) Classification and incidence of neoplasms of the central nervous system. Adv Neurol 15:51–55

Hoshino T, Rodriguez LA, Cho KG et al.(1988) Prognostic implications of the proliferative potential of low-grade astrocytomas. J Neurosurg 69: 839–842

Hoshino T (1991) Cell kinetics of brain tumours. In: Salcman M (ed) Neurobiology of Brain Tumours, Vol. A: Concepts in Neurosurgery. Williams and Wilkins, Baltimore

Hunt SM, McKenna SP, McEwen J et al. (1980) A quantitative approach to perceived health status: a validation study. J Epid Comm Hlth 34: 281–286

Karnofsky DA, Burchenal JH (1949) The clinical evaluation of chemotherapeutic agents. In: MaLeod CM (ed) Evaluation of Chemotherapeutic Agents, Columbia University Press, New York

Kelly PJ, Daumas-Duport C, Kispert DB et al. (1987) Imaging-based stereotaxic serial biopsies in untreated intracranial glial neoplasms. J Neurosurg 66: 865–874

Kelly PJ (1988) Volumetric sterotaxic and computer-assisted stereotactic resection of subcortical lesions. In: Lundsford LD (ed) Modern Stereotactic Neurosurgery. Martinus Nijhoff, Boston

Kernohan JW, Sayre GP (1952) Tumours of the Central Nervous System. Armed Forces Fascicle of Pathology, Washington

Knake JE, Chandler WF, Gabrielsen To et al. (1984) Intraoperative sonographic delineation of low-grade brain neoplasms defined poorly by computed tomography. Radiology 151: 735–739

Koivukangas J (1984) Ultrasound imaging in operative neurosurgery: an experimental and clinical study with special reference to ultrasound holographic B (UHB) imaging. Doctoral dissertation. Acta Universitatis Ouluensis D 115. University of Oulu Printing Center, Oulu

Koivukangas J, Koivukangas P (1986) Treatment of low-grade cerebral astrocytoma: new methods and evaluation of results. Ann Clin Res 18 (Suppl 47): 115–124

Koivukangas J, Louhisalmi Y (1990) Computer-guided laser for neurosurgery. Ann Chir Gynaecol. 79: 192–196

Koivukangas J, Louhisalmi Y, Alakuijala, J. Oikarinen J (1993) Ultrasound-controlled neuro-navigator-guided brain surgery. J Neurosurg (in press)

Koivukangas P, Koivukangas J (1988) Role of quality of life in therapeutic strategies in brain tumours. Hlth Policy 10: 241–257

Koivukangas P (1993) A two-wave MIMIC model for the construction of a health index based on the health production function: with special reference to the assessment of neurosurgical treatment of brain tumour patients. Doctoral dissertation. Acta Universitatis Ouluensis C 69. University of Oulu Printing Center, Oulu

Laws ER, Taylor WF, Clifton MB, Okazaki H (1984) Neurosurgical management of low-grade astrocytoma of the cerebral hemispheres. J Neurosurg 61: 665–673

Laws ER, Taylor WF, Bergstralh EJ. et al. (1986) The neurosurgical management of low-grade astrocytoma. Clin Neurosurg 33: 575–588

Lieberman AN, Foo SH, Ranshohoff J *et al.* Long term survival among patients with malignant brain tumors. Neurosurgery 10: 450–453

McCormack BM, Miller DC Budzilovich GN *et al.* (1992) Treatment and survival of low-grade astrocytoma in adults 1977–1988

Muller W, Afra D, Schroder R (1977) Supratentorial recurrences of gliomas. Morphological studies in relation to time intervals with astrocytomas. Acta Neurochir 37: 75–91

Piepmeier JM (1987) Observations on the current treatment of low-grade astrocytic tumours of the cerebral hemispheres. J Neurosurg 67: 177–181

Recht LD, Lew R, Smith TW (1992) Suspected low-grade glioma: is deferring treatment safe? Ann Neurol 31: 431–436

Ringertz N (1950) Grading of gliomas. Acta Pathol Microbiol Scand 27: 51–64

Rubinstein LJ (1972) Tumors of the Central Nervous System. Armed Forces Institute of Pathology, Washington

Schiffer D (1991) Patterns of tumor growth. In: Salcman M (ed) Neurobiology of Brain Tumors, Vol. 4: Concepts in Neurosurgery. Williams and Wilkins, Baltimore

Schoenberg BS, Christine BW, Whisnant JP (1976) The descriptive epidemiology of primary intracranial neoplasms: the Connecticut experience. Am J Epidemiol 104: 499–510

Shelden CH, McCann G Jacques S *et al.* (1980) Development of a computerized microstereo-taxic method for localization and removal of minute CNS lesions under direct 3-D vision. J Neurosurg 52: 21–27

Vertosick FT, Selker RG, Arena VC (1991) Survival of patients with well-differentiated astro-cytoma diagnosed in the era of computed tomography. Neurosurgery 28: 496–501

Williams AH (1985) Economics of coronary artery bypass grafts. Br Med J 291: 326–329

Williams A (1988) Applications in management. In: Teeling-Smith G (ed) Measuring Health: A Practical Approach. Wiley, London

Wroe SJ, Foy PM, Shaw MDM *et al.* (1986) Differences between neurological and neurosurgi-cal approaches in the management of malignant brain tumours. Br Med J 293: 1015–1018

Zulch KJ (1979) Histological Typing of Tumors of the Central Nervous System. World Health Organization, Geneva

7 Management of Brain Tumours in Childhood

Jonathan Punt

Introduction

Cancer is one of the four big killers in childhood, lying third after accidents and congenital malformations (OPCS, 1986) and afflicting one child in every 650 (Draper *et al.*, 1982). Brain tumours are second only to leukaemia in importance, accounting for 18.6% of all childhood cancers and 40%–50% of all solid tumours (Silverberg, 1980). The incidence is fairly consistent across populations at between two and five cases per 100 000 per year (Dohrmann and Farwell, 1976; Schoenberg *et al.*, 1976). There is, however, a considerable variation in tumour type between different racial groups, with cerebellar astrocytomas and medulloblastomas being more common in Europe and North America, craniopharyngiomas in Japan and Africa, ependymomas in India, and pineal germinomas in Japan (Dohrmann and Farwell, 1976).

There are highly significant differences between brain tumours in children and those occurring in adults and these must be fully acknowledged if management is to be successful. These differences are considered thus :

1. *The sites.* The predominance of midline and paraventricular tumours and the 55% incidence of posterior fossa tumours (Bruno and Schut, 1982; Amador, 1983) makes hydrocephalus a frequent concomitant feature and also dictates a variety of operative approaches. It may also place anatomical limits on the extent of surgical resection and put at risk intricate physiological functions.

2. *The cells.* Seventy per cent of childhood brain tumours arise from cells of the glial series (Bruno and Schut, 1982). Meningiomas, schwannomas and metastatic epithelial tumours are all very rare. The definition of the primitive neuroectodermal tumour (PNET) by Hart and Earle (1973) may yet prove fundamental to our understanding (Rorke, 1983)

3. *The patients.* Childhood brain tumours arise in an organ which is still developing in ultrastructural and functional terms, and which is central to the very development of the child's physical, intellectual, metabolic and social behaviour and progress. Not only does this supply clues, still

undisclosed and largely unexploited, as to the origin and behaviour of these tumours but it also carries hazards in terms of adverse long-term consequences of anti-cancer treatments, especially radiation therapy, on the development and future health of the host organ (Allen, 1978; Cumberlin *et al.* 1979; Danoff *et al.*, 1982). The immaturity of the patients imposes special responsibilities on their medical attendants to respect their special needs and their place in the family and in society.

Over the past 40 years there has been a marked improvement in survival in all forms of childhood cancer as reported from the UK (Birch *et al.*, 1988), from the USA (Miller and McKay, 1984) and from Italy (La Vecchia and Decarli, 1988). This is due to better treatment regimens, and in particular to the development of effective and survivable chemotherapy programmes. Relatively speaking, the improvement in the prospects of children with brain tumours has been less dramatic than for those with other cancers; the exception being cerebellar PNET (medulloblastoma) (Birch *et al.*, 1988). This probably reflects the relative ineffectiveness of chemotherapy in brain tumours to date. Furthermore, long-term survival prospects for most forms of childhood cancer have improved, with over 80% of those who survive 3 years being alive after 10 years with no or little excess of mortality over the expected for the population.

Long-term survival in children with brain tumours shows a much wider variability according to cell type: thus, 90% of those with juvenile astrocytoma of the cerebellum contrasts with just over 50% of those with cerebellar PNET surviving 10 years. In addition, for those surviving at least 10 years there is only one extra death per 100 survivors per year over the expected for the population (Hawkins, 1989).

This chapter is not intended to be comprehensive, but will consider the current management of a variety of childhood brain tumours, illustrating the problem and the controversies encountered. Thus those relatively benign tumours for which there is mostly a straightforward surgical solution, such as the juvenile cerebellar astrocytoma and the choroid plexus papilloma, are omitted; likewise, malignant cerebral glioma for which the poor results and relative unresponsiveness to radiation therapy and chemotherapy simply echoes that experienced in adults. The relative frequency of the different brain tumours encountered in childhood is shown in Table 7.1.

Table 7.1   The relative incidence of different brain tumours in childhood

Infratentorial 55%	
PNET/medulloblastoma	20%
Astrocytoma	15%
Brain stem glioma	15%
Ependymoma	5%
Supratentorial 45%	
Low-grade astrocytoma	15%
Anaplastic astrocytoma/glioblastoma	10%
Craniopharyngioma	10%
Pineal region tumours	5%
Ependymoma	5%

Primitive Neuroectodermal Tumour (PNET)

From the first description of medulloblastoma (Bailey and Cushing, 1925) the occurrence of tumours elsewhere in the brain, histologically identical to those found in the cerebellum, has been recognized. By advocating the term "primitive neuroectodermal tumour" for a collection of supratentorial tumours in young people, Hart and Earle (1973) paved the way for the modern concept of classification of neoplasms of the central nervous system in infants and children (Rorke, 1983) wherein there are subtypes of undifferentiated neuroectodermal round cell tumours grouped according to any features of glial, neuronal or ependymal differentiation that they may display. That the commonest site for such a tumour is the cerebellum remains an intriguing mystery. This section will consider principally the therapy of cerebellar PNET.

PNET is the single most common intracranial neoplasm of childhood, comprising 15%–20% of the total (Koos and Miller, 1971). It is highly malignant and advances by local invasion of the cerebellum and brain stem; by spread through the cerebrospinal fluid (CSF) pathways and occasionally by metastasis to extraneural sites such as bone, bone marrow and lymph nodes. Favourable prognostic factors include female gender (Raimondi and Tomita, 1979b; Park et al., 1983); Afro-Caribbean origin; and age over 6 years at diagnosis (Bloom, 1967; Mealey and Hall, 1977); age under 2 years is particularly unfavourable (Farwell et al., 1978; Raimondi and Tomita, 1979b). Extent of disease at diagnosis is important both in terms of local invasion and distant spread (Harisiadis and Chang, 1977; Allen and Epstein, 1982). The value of a detailed staging system, as in that of Chang (Chang et al., 1969), is disputed with some claiming value (Harisiadis and Chang, 1977; Kopelson et al., 1983) and others denying it (Berry et al., 1981; Silverman and Simpson, 1982; Tomita and McClone, 1986). In some studies, differentiation along neuronal, ependymal or glial lines has been a major adverse prognostic factor (Packer et al., 1984a). On this background of divided opinion and experience, the role of the three major treatment modalities will be considered.

Surgery

The place of surgery, as in most intracranial neoplasms, is to relieve raised intracranial pressure and neurological deficit and to provide tissue to establish a histological diagnosis. The majority of children have significant hydrocephalus at presentation and although there was a vogue for relieving this by ventricular shunting prior to definitive tumour excision (Abraham and Chandy, 1963), this is now rarely necessary and is indeed considered undesirable. Dexamethasone provides adequate preparation in the majority of cases. The occasional baby or infant with severe visual failure or nutritional depletion may, however, benefit from an initial ventriculoperitoneal shunt.

The danger of extraneuraxial metastasis via a shunt (Kessley et al., 1975; Hoffman et al., 1976) has previously been highlighted (Park et al., 1983) but has probably been overestimated (Berger *et al.*, 1991): the value of filtered shunts in preventing shunt-mediated metastasis remains debatable (Guthkelch and Taylor, 1983; Hoffman, 1983). There is also a significant risk of precipitating intratumoral haemorrhage and upward coning (Epstein and Murali, 1978). In the majority of cases a standard posterior fossa craniotomy is performed and macroscopic excision of the tumour attempted. Improved techniques of neuroanaesthesia, the control of raised intracranial pressure by dexamethasone and mannitol, the operating microscope and the ultrasonic aspirator have all contributed to enhanced and safer surgical resections. In a number of cases, invasion of the brain stem will dictate that macroscopic tumour must remain.

It is probably unjustifiable to create a serious neurological deficit in an attempt to gain more extensive resection. It does seem, however, that macroscopically total removal carries a more favourable prognosis (Raimondi and Tomita, 1979a, b; Tomita and McClone, 1986). In any event it must be acknowledged that in terms of cancer surgery no excision of any intrinsic brain tumour is strictly radical. It may be that the association between less than macroscopic resection and poorer prognosis really reflects the adverse factor of brain stem invasion, itself probably a feature of more aggressive multicentric disease at presentation. An unconventional, though not irrational, approach is CT- or MR-, guided stereotaxic biopsy of the most accessible lesion, with relief of raised intracranial pressure by steroids and/or ventricular shunt followed by palliative chemotherapy and radiation therapy. It is not surprising to discover that surgical resection alone is not adequate to control the disease, giving only 0.125 1-year and 0.042 5-year survival rates (Farwell *et al.*, 1984).

Radiation Therapy

Prior to the use of radiation therapy, cerebellar PNET was a uniformly fatal condition. The value of radiation therapy was established over 50 years ago (Cutler *et al.*, 1936) and subsequently the necessity for whole neuraxis irradiation became clear (Paterson and Farr, 1953; Jenkin, 1969; Smith *et al.*, 1973). This relates to the very strong tendency for spontaneous seeding of the tumour through the spinal and intracranial CSF pathways, observed in over 90% of cases when biopsy of the arachnoid of the cisterna magna is combined with CSF cytology (Tomita and McClone, 1983). The danger is further illustrated by the 36%–43% incidence of positive findings when myelography is performed in the postoperative period (Deutsch and Reigel, 1980; Allen and Epstein, 1982).

The need for a high dose of radiation to the posterior fossa is established and generally accepted (Bloom, 1982a), as low doses are associated with a very high incidence of local recurrence (Cumberlin *et al.*, 1979; Silverman and Simpson, 1982). The highest, and most durable, disease-free survival rates have been obtained with the highest doses of radiation to the posterior fossa (Berry *et al.*, 1981). The ideal dose to the rest of the neuraxis remains to be established, some workers obtaining satisfactory results with low-dose neuraxis radiation as long as a macroscopically complete resection of the primary cerebellar tumour has been achieved (Tomita and

McClone, 1986), whereas others insist that higher doses are essential (Jereb *et al.*, 1982). The technique of administration is crucial, with particular attention to avoidance of any overlap of the fields, and to the need to include the subfrontal meninges, especially in the midline, if subfrontal relapse is to be avoided (Vandyk *et al.*, 1977; Hardy *et al.*, 1978). Currently, resection and neuraxial radiation therapy can generally achieve an overall 50%–60% 5-year survival rate, falling to about 40% at 10 years (Krischer *et al.*, 1991). The upper limit attainable with these modalities alone, and incorporation of the highest radiation doses to the posterior fossa, is thought to be over 70% survival at 10 years (Berry *et al.*, 1981). Sadly these successes are often obtained at the price of serious consequences to the child.

The earliest studies signalled the hazards of radiation therapy to the immature brain in terms of impaired intellectual development (Paterson and Farr, 1953; Smith *et al.*, 1961). Later, more detailed assessments confirmed the presence of serious intellectual deficits and also of emotional disturbance (Hirch *et al.*, 1979; Raimondi and Tomita, 1979c; Park *et al.*, 1983; Silverman *et al.*, 1984). The risk of intellectual impairment is particularly high in children aged under 2 years (Jenkin, 1969; Bloom, 1971; Mealey and Hall, 1977; Chin and Maruyama, 1984). As there are very few survivors in this age group, radiation therapy cannot now be justified and alternative modalities of treatment, principally chemotherapy, should be employed. Growth retardation is increasingly recognized. Some of this may be due to the effects of radiation on the growing spine (Nevhauser *et al.*, 1952; Probert *et al.*, 1973). More serious, however, is the damage to the hypothalamus and pituitary which produces growth hormone deficiency and hypothyroidism in up to 65% of cases (Shalet *et al.*, 1975; Hirsh *et al.*, 1979; Park *et al.*, 1983). Long-term growth and endocrine follow-up is, therefore, obligatory. finally, there is a real risk of further tumours being induced by radiation therapy administered in childhood, especially meningeal tumours (Modan *et al.*, 1974).

Repeat radiation therapy is quite effective for intracranial recurrences at a distance from the primary site, providing good palliation, prolonged survival and even the occasional long-term remission. It is of little value in spinal subarachnoid metastases, which tend to appear earlier and are associated with more widespread neuraxial disease (Smith *et al.*, 1973).

Chemotheraphy

Evidence that cerebellar PNETS are sensitive to chemotherapy is found in reports of clinical responses to treatment of recurrent tumours, occasionally even with remissions up to 18 months (Park *et al.*, 1983). Thus the effectiveness has been established of vincristine (Lassmann *et al.*, 1965; Lampkin *et al.*, 1967), BCNU (Wilson *et al.*, 1970), CCNU (Fewer *et al.*, 1972), melphalan (Friedman *et al.*, 1989) and cisplatin (Bertolone *et al.*, 1989) as single agents, and of vincristine and cyclophosphamide (Friedman *et al.*, 1986) and vincristine, cyclophosphamide and adriamycin (Chamberlain *et al.*, 1988) used in combination. Large randomized multicentre phase III studies of adjuvant chemotherapy have however failed to demonstrate any statistically significant value of combination chemotherapy with CCNU and vincristine (Bloom, 1982a); CCNU, vincristine and prednisone (Finlay *et al.*, 1987); and CCNU, vincristine and prednisolone (Evans *et al.*, 1990), except for those children with more extensive disease invading the brain stem at diagnosis. A relatively small randomized trial of nitrogen mustard, vincristine, procarbazine and prednisone as adjuvant post-irradiation chemotherapy demonstrated a significant increase in overall 5-year survival but not in event-free survival (Krischer *et al.*, 1991); unfortunately a relatively large proportion of the patients had undergone subtotal excisions, of undefined extent, and it may be that the overall result merely demonstrated again the value of adjuvant chemotherapy in the patient with more extensive disease.

Therefore, although radiation therapy remains the mainstay of treatment there are sufficient grounds to pursue chemotherapy within the context of an organized clinical trial. There is real potential for chemotherapy in treating children aged under 3 years in whom radiation therapy is relatively ineffective and usually unacceptably damaging and, combined with minimal surgery, in the palliative approach to those who have advanced neuraxial disease at presentation.

PNETs at non-cerebellar sites should be managed on the same regimens as those for cerebellar lesions.

Brain Stem Glioma

Brain stem gliomas are now recognized as being as common as cerebellar PNETs and cerebellar astrocytomas (Bilaniuk *et al.*, 1980), accounting for 10%–20% of all childhood brain tumours (Farwell *et al.*, 1977; Bruno and Schut, 1982). Although the majority of children die within 2 years, however treated, a variable proportion of 20%–35% may survive longer and even enter permanent remission (Littman *et al.*, 1980; Albright *et al.*, 1983). It is possible to classify the tumours according to their CT scan characteristics (Stroink *et al.*, 1986): this has real value as there is correlation between the CT scan features and operability and also prognosis. For example, enhancing exophytic tumours may be controlled by surgical excision alone and are usually composed of low-grade astrocytoma (Hoffman *et al.*, 1980). Similarly intrinsic, well-defined enhancing masses may be amenable to resection using modern aids such as the operating microscope, the ultrasonic aspirator and the surgical laser (Walker and Storrs, 1985), and employing techniques similar to those in excising intramedullary spinal cord tumours. By contrast diffuse, hypodense non-enhancing tumours carry a poor prognosis and are often of anaplastic astrocytoma histology (Albright *et al.*, 1986). The adverse prognostic factors can therefore be summarized as cranial nerve involvement; hypodense appearance on CT, especially if the whole of the brain stem is involved; and anaplastic astrocytoma or glioblastoma histology (Albright *et al.*, 1986). The advent of image-guided stereotactic techniques has made biopsy feasible and relatively safe (Coffey and Lunsford, 1985). Despite previous reservations to the contrary (Epstein, 1985), it is held that additional diagnostic and prognostic information can be obtained and that biopsy is therefore justified: as the real extent of the tumour as portrayed by MRI may be considerably greater than on CT scan (Packer *et al.*, 1985), then ideally MR-guided stereotactic biopsy should be employed.

It is therefore suggested that these children with exophytic or intrinsic brightly enhancing lesions should undergo an attempt at excision and only be subjected to radiation therapy if the histology is unexpectedly of high-grade astrocytoma. All others should undergo image-guided stereotaxic biopsy and given radiation therapy if the histology is low-grade astrocytoma, and radiation therapy plus chemotherapy if the histology is anaplastic astrocytoma or glioblastoma. As no study has ever demonstrated any value from chemotherapy, regimens should only be employed which are incorporated in an evaluable study. In the terminal stages, the prolonged use of steroids is to be deprecated because of the unpleasant side effects.

Craniopharyngioma

This is one of the commonest supratentorial tumours in childhood and although it may arise anywhere between the basisphenoid and the third ventricle it typically grows, in children, from the region of the tuber cinereum. In childhood, the predominant symptom precipitating presentation is usually raised intracranial pressure, but on investigation most children are found to have a varying degree of pituitary insufficiency and visual disturbance. The tumour is typically of mixed solid and cystic consistency. The objectives of treatment are relief of raised intracranial pressure; decompression of the anterior visual pathways; correction of any endocrine dysfunction and long-term control of the disease.

Surgery

Prior to the availability of synthetic glucocorticoids in 1950, the results of surgical excision with or without radiation therapy were very poor and the outcome was almost universally fatal. With the advent of appropriate steroid replacement, surgical resection became feasible and aggressive primary excision at craniotomy was introduced (Matson and Crigler, 1969).

Review of Matson's cases after two decades (Katz, 1975) supported this approach: 22 out of 34 children in whom the surgeon felt that excision had been total were alive without clinical or radiological evidence of residual or recurrent tumour; 15 out of the 22 had excellent or good functional capacity and the majority had hormone deficiencies requiring replacement therapy. It was noted that even small tumours might recur and that, paradoxically, cystic tumours were more difficult to resect completely than solid ones. That not all the children had a satisfactory functional status was a warning that this was sensitive territory. Since then, advances in terms of better imaging by CT scan and MRI, and improved operative techniques afforded by the operating microscope, ultrasonic aspirators and the surgical laser, have encouraged surgeons to pursue aggressive primary excision wherever possible. The ability to achieve this objective has undoubtedly advanced. For example at the Hospital for Sick Children, Toronto, there were 17 total excisions out of 48 cases with 3 operative deaths reported in 1977 (Hoffman et al., 1977), compared to 45 total excisions out of 50 cases with 1 operative death reported in a series 15 years later (Hoffman et al., 1991). From the purely surgical aspect, 40 years of operative experience has produced an understanding that no one surgical approach will suffice for all craniopharyngiomas. Initially the standard approach was by sub-frontal craniotomy (Shillito, 1976), but the recognition of retrochiasmatic extensions led to alternatives such as extending a subfrontal approach to include the pterion (Hoffman et al., 1992); dividing the lamina terminalis, possibly opening the sphenoid sinus by drilling away the tuberculum sellae and accepting the risk of CSF fistula and meningitis (Patterson and Danylevich, 1989); or, as favoured by the present author, a temporal approach (Symon and Sprich, 1985). Preoperative MRI with careful examination of the midline sagittal cuts indicates the appropriate approach (Punt and Worthington, unpublished data).

Occasional cases will require a transcallosal route (Yasargil et al., 1990), or a trans-sphenoidal attack (Laws, 1980): for the relatively more intrasellar lesion a total excision is feasible by the trans-sphenoidal route and it is also of value in some recurrent tumours, being kind on the anterior visual pathways and the hypothalamus. The problem with aggressive excision is that the tumour may still recur even when postoperative CT or MR image confirms the surgeon's operative impression of total excision, as in 5 out of 13 cases of recurrence following microsurgical resection reported in one series (Hoffman et al., 1992). In addition, it is genuinely impossible for the surgeon to be sure of the extent of the resection. There is no doubt that enthusiastic surgery may cause serious damage with neurological, visual, endocrinological and neuropsychiatric sequelae. Such damage may relate to vascular insults to the internal carotid arteries or to the microvasculature of the optic chiasm and hypothalamus in an attempt to find a surgical plane through the gliotic margin around the tumour, the very extent and nature of which remains disputed (Hoffman, 1982).

Although gross neurological disability in terms of hemiplegia or dysphasia is relatively unusual unless there has been serious vascular damage, visual morbidity is not insignificant, deterioration in visual acuity and in visual fields being sometimes more common than improvement even in the presence of normal preoperative vision (Hoffman et al., 1992), although such morbidity is not invariable (Yasargil et al., 1990). Endocrinological deterioration is almost inevitable and the majority of children require multiple hormone replacement and very skilled supervision (Matson and Crigler, 1969; Grant and Lyen, 1982; Yasargil et al., 1990; Hoffman et al., 1992). Of particular importance is the precipitation of obesity in nearly half the patients, even those of normal preoperative habitus (Hoffman et al., 1992). This is an additional psychological burden, especially in adolescence, and requires sympathetic counselling.

Of great concern, however, are the neuropsychological sequelae with memory loss and behavioural disturbance which, when coupled with acquired obesity and the need to take multiple hormone replacements, can lead to considerable distress and unhappiness (Galatzer *et al.*, 1981; Cavazutti *et al.*, 1983).

Radiation Therapy

It has been established for some time that radiation therapy is effective against craniopharyngioma (Kramer *et al.*, 1968). This has been more recently confirmed in a large retrospective study (Manaka *et al.*, 1985) in which the median survival for 21 irradiated children was very significantly better at greater than 10 years compared to 3 years for 32 children not irradiated: the survival rates were 28.1% and 85.7% at 5 years and 21.7% and 63.4% at 10 years for non-irradiated and irradiated children, respectively. Unfortunately this useful study contains no comment on quality of life, visual deficits or endocrine deficiencies. The established efficacy of radiation therapy coupled with the recognized morbidity of surgical excision has led some centres to opt for minimal surgery combined with radiation therapy, and advocates of this policy claim good results; for example, Richard *et al.* (1980) experienced better survival and control of tumour growth with biopsy followed by radiation therapy than after subtotal or total excision whether or not combined with radiation therapy. Similarly, Fischer *et al.* (1985) reported a cohort of children treated between 1972 and 1981 for whom conservative surgery followed by radiation therapy was as effective as attempted gross excision in controlling tumour growth and was associated with less morbidity. A later review (Fischer *et al.*, 1990) of 37 children having conservative operations and radiation therapy and followed for more than 10 years demonstrated a recurrence rate and mortality equal to or better than more surgically orientated series of the same era. Radiation therapy was again shown to be effective in controlling in situ disease. Most telling was a lower morbidity in terms of successful jobs after education in those having conservative surgery.

Despite these studies there remains an unresolved polarization between the advocates of aggressive surgery and those for conservative surgery plus radiation therapy. It is difficult to see how this may be settled as it is unlikely that any one centre would have sufficient numbers to conduct a randomized prospective study even if there were a willingness to perform one. A multicentre investigation might encounter difficulties of standardization and verification of the extent and quality of surgery and even of radiation therapy. There is no doubt that craniopharyngioma is one of those conditions that challenges the surgical machismo in some surgeons and may invite over-enthusiastic onslaughts.

Radiation therapy has been delivered in a variety of modes besides conventional external beam and these have included stereotaxic introduction of intracystic yttrium-90 (Backlund *et al.*, 1973) and phosphorus-30 (Anderson *et al.*, 1989), and stereotaxic radiosurgery (Backlund, 1973). A recent long-term follow-up of the stereotaxic approach is extremely encouraging (Backlund *et al.*, 1989). It would, however, be incorrect to believe that radiation therapy is innocuous. Side effects include radionecrosis (Mikhael, 1978, 1979); hypothalamic damage with resultant endocrine and behavioural disturbance (Mechanick *et al.*, 1986), sometimes with an acute syndrome of fever, coma, collapse and even death (Grant and Lyen, 1982; Fischer *et al.*, 1990); hearing loss (Fischer *et al.*, 1985); and radiation-induced neoplasia in the form of malignant change in the craniopharyngioma (Nelson *et al.*, 1988), and by way of development of malignant gliomas in the cerebrum (Liwnwicz *et al.*, 1985) and brain stem (Fischer *et al.*, 1990).

Chemotherapy

Surprisingly, considering its histological features, craniopharyngioma has on occasion shown individual responses to chemotherapy administered systemically or instilled into a tumour cyst, usually when given to patients in recurrence resistant to surgery and radiation therapy.

Significant and persistent reduction in tumour size has been recorded in response to systemic therapy with BCNU, vincristine and procarbazine in combination (Bremer *et al.*, 1984) and to nitrogen mustard, vincristine, procarbazine and prednisolone in combination (Fischer *et al.*, 1990). Responses to intracavitary methotrexate (Constine *et al.*, 1989) and to bleomycin (Broggi *et al.*, 1989) have been observed and in one series seven children with cystic craniopharyngiomas showed durable responses to intracyst bleomycin given as the primary treatment (Takahashi *et al.*, 1985).

Conclusion

In conclusion, craniopharyngioma is an unpleasant, capricious and unpredictable tumour which can be very difficult to eradicate surgically and may recur despite apparent total excision. It may yield to radiation therapy and there are reports in small numbers of a response to chemotherapy. Radical excision carries the best chance of long-term remission, but may be associated with handicaps. A rational approach is to attempt total excision, but if this is not successful then radiation therapy should follow. For recurrent tumours there is the possibility of a further attempt at excision, but this may carry a high mortality: if radiation therapy has not been used already, then it can be exhibited at this stage. Simple cyst drainage by a CT- or MR-guided stereotaxic procedure is an effective palliative measure and can be repeated easily if a catheter and subgaleal reservoir are inserted: coupled with radiation therapy this can bring good long-term remission with little morbidity. The place of highly-focused radiation therapy using the stereotaxic technique needs further investigation, as does the place of interstitial irradiation and that of chemotherapy. An attempt should be made to resolve the relative merits of aggressive primary excision and conservative surgery plus radiation therapy, perhaps by way of a prospective comparative study between major centres committed to one particular policy but utilizing a panel of referees to ensure impartiality of assessment.

It seems unlikely that proponents of either approach would agree to a randomized prospective study. Whichever policy is used, these tumours are not for the occasional surgeon and can only effectively be managed by major paediatric centres where there is suitable surgical expertise supported by specialist paediatric endocrinology and radiation therapy facilities. There must also be an effective means of monitoring ophthalmological and neuropsychological sequelae and of communicating with educational institutions to ensure that appropriate schooling and training continues into adult life.

Ependymoma

Ependymomas account for about 10% of all paediatric CNS tumours (Dohrmann and Farwell, 1976) occurring in infratentorial, supratentorial and intraspinal locations with decreasing frequency: about 60% arise in the infratentorial compartment, usually in the midline (Kricheff et al., 1964a). Typically they present in children around the middle of the first decade of life (Dohrmann et at., 1976). Ependymomas excite considerable controversy even over histological classification (Zülch, 1979). A simple and workable division is between low-grade ependymoma and anaplastic ependymoma. The term "ependymoblastoma" is best reserved for a highly cellular embryonal tumour with a median postoperative survival between 12 and 20 months and universally fatal outcome by 3 years (Liu et al., 1976; Mørk and Rubinstein, 1985): this particularly lethal tumour is more logically grouped with PNET. There is even dispute about the correlation between histological grading and prognosis; some authors stress the unpredictable behaviour of ependymoma (Kricheff et al., 1964a, b; Mørk and Løken, 1977) and even within a group of frankly malignant features it appears that survival may differ very greatly (Ross and Rubinstein, 1989). Other reviewers regard the tumour grade as the most important prognostic factor (Leibel and Sheline, 1987). The risk of seeding throughout the neuraxis is controversial, ranging from zero (Barone and Elvidge, 1970) to 33% (Svien et al., 1953).

There is a correlation with both site and histology, the greatest risk of seeding, at 20%, being in children with high-grade infratentorial lesions, the least risk, at 4%, being for those with low-grade supratentorial tumours (Leibel and Sheline, 1987).

Surgery

Untreated, the natural history is for progression and death within 3 years (Mørk and Løken 1977). Over half a century the operative mortality has steadily declined from 40% to 17% (Dohrmann et al., 1976), although currently much lower operative mortality must appertain in major paediatric neurosurgical centres, especially since the introduction of modern surgical adjuncts such as the operating microscope, the ultrasonic aspirator and the surgical laser. In the cerebral hemisphere it may be possible to get a macroscopic total excision of the tumour from within the cerebral white matter which often completely separates the tumour from the ventricle (Ringertz and Reymond, 1949). Frequently, with infratentorial tumours, there may be intimate involvement, by attachment or invasion, with the floor of the fourth ventricle, brain stem and lower cranial nerves rendering complete excision impossible (Bouchard, 1980; Marks and Adler, 1982). Even with apparent complete excision, surgery alone is not sufficient to prevent recurrence (Shuman et al., 1975). With surgery alone, 5-year survival rates between 17% and 27% are quoted (Cushing 1932; Ringertz and Reymond, 1949; Fokes and Earle, 1969).

Radiation Therapy and Chemotherapy

Further therapy is therefore required regardless of the extent of surgical resection. Ependymomas are generally recognized as being sensitive to radiation therapy as judged by extended survival times (Salazar et al., 1977; Mørk and Løken, 1977). The breadth of success is, however, wide with 5-year survivals ranging from 37% to 87% for all intracranial ependymomas; from 33% to 100% from infratentorial sites (Phillips et al., 1964; Kim and Fayos, 1977; Mørk and Løken, 1977; Bouchard, 1980; Glansman et al., 1980; Chin et al., 1982; Marks and Adler, 1982; Garrett and Simpson, 1983).

Considerable thought has been devoted to the size of the radiation fields required and it appears that, regardless of histological grade, the commonest cause of treatment failure is local recurrence (Salazar et al., 1983; Leibel and Sheline, 1987). However, the local control rate for low-grade tumours is very much greater with whole-brain irradiation (78%) than with partial-brain irradiation (17%), as is the 10-year survival rate at 67% versus 12% (Salazar et al., 1983): even this experience is not universal, with some groups finding no difference in survival between partial- and whole-brain irradiation (Read, 1984). A strong case has been made for an aggressive approach geared to location, histology and extent of subarachnoid spread: a logical protocol in which low-grade supratentorial tumours received whole-brain irradiation and low-grade infratentorial tumours received whole-brain and cervical irradiation if neither had any subarachnoid spread, and all high-grade tumours and those low-grade tumours with subarachnoid spread received craniospinal axis irradiation, produced 10-year survival rates of 75% for low-grade tumours and 67% for high-grade ependymomas regardless of site (Salazar et al., 1983). Interestingly, patients treated according to these protocols achieved 5- and 10-year survival rates of 69% compared to only 6% for those treated, within the same institution over the same period of time, in whom the protocol was violated. Furthermore, long-term survival rates for those treated according to protocol showed little difference between low-grade (75%) and high-grade histologies (67%). In those under 3 years of age at diagnosis, there was a significant incidence of intellectual and hormonal sequelae even with reduction of radiation doses to 80%. Whereas craniospinal irradiation for high-grade tumours does seem effective in reducing the incidence of spinal metastasis (Marks and Adler, 1982; Garrett and Simpson, 1983), there is still a significant problem of failure of local control.

Perhaps this could be approached by higher local radiation doses with fields determined by more accurate modelling of tumour size derived from the newer generation of CT and MR scanners and their associated software packages. Chemotherapy has been employed sporadically and in wider clinical trials, often along with medulloblastoma, and has generally failed to produce any increase in survival rates (Van Eys et al., 1981; Bloom, 1982b). There are, however, anecdotal reports of stabilization of disease with BCNU (Wilson et al., 1976; Levin, 1981) and an encouraging observation of stabilization of recurrent disease in 75% of cases with dibromodulcitol with mean time to progression of 67 weeks and remission in 25% of cases up to 3 years (Levin et al., 1984).

It is of interest that although most series published by radiation therapists and medical oncologists show a better outcome for infratentorial locations than for supratentorial sites (Mørk and Løken, 1977), the reverse is the case for some neurosurgical series (Oi and Raimondi, 1982); this may be due to exclusion of early postoperative deaths from the non-surgical series.

Further progress in the management of this relatively rare and unpredictable tumour will only be made by large multicentre studies which should address the question of the optimal radiation schedules and explore chemotherapeutic regimens.

Pineal Region Tumours

Pineal region tumours account for 3%–8% of childhood brain tumours (Hoffman et al., 1983). The relatively high incidence in Japan, especially of germinomas, is well known (Dohrmann and Farwell, 1976; Sano, 1976).

The management continues to stimulate controversies, in the resolution of which it must not be forgotten that the range of histopathological entities encountered is considerable – up to 17 in one series (Edwards et al., 1988); and that between 36% and 50% of lesions may be histologically benign or radioresistant (Demakas et al., 1982; Hoffman et al., 1983). Improved diagnostic imaging, especially MRI with its ability to produce images of high quality in multiple planes, has led to increased identification of these lesions and to enhanced surgical planning; even MRI, however, is relatively non-specific with regard to histology, with the exception of delineating intrinsic tectal plate lesions for which no treatment is required beyond diversionary CSF shunting and MRI monitoring (May et al., 1991). Attempts to identify tumours, principally germinomas, by CSF cytology, although successful in Japan where this tumour type is so common, have frequently been disappointing elsewhere. Tumour markers in blood and CSF are useful in the minority of cases, but when there are grossly elevated levels of the β-subunit of human chorionic gonadotrophin (β-HCG), then the diagnosis of choriocarcinoma is undoubted (Edwards et al., 1988), although mildly elevated levels may indicate germinoma. Similarly, elevated levels of alpha-fetoprotein (AFP) indicate the presence of a non-germinoma germ-cell tumour, usually an endodermal sinus tumour (Edwards *et al.*, 1988; Hoffman *et al.*, 1991). With these few exceptions, all cases require histological diagnosis based upon tissue obtained at operation.

The argument in favour of treating by shunt placement and "blind" radiotherapy is now discredited: it was based upon the low morbidity compared with open operation – its alleged effectiveness and perceived ability to discriminate between germinoma and alternative pathologies according to the response to a "test" dose of radiation. Although some successes were claimed for such management – for example, 70% 5-year disease-free survival with mean survival times of 17 years (Abay *et al.*, 1981) – the underlying arguments are now recognized as fallacious. The "test" dose of radiation of 2000 rad is certainly not innocuous and carries significant risk of morbidity in terms of neurological sequelae when administered to the developing brain (Painter *et al.*, 1975; Hirsh *et al.*, 1979; Davis *et al.*, 1986; Asai *et al.*, 1989). Furthermore, only 33%–50% of pineal region tumours can be expected to be germinomas (Herrick, 1984; Edwards *et al.*, 1988). The now-recognized existence of mixed germinoma and non-germinoma germ-cell tumours and of other tumours that may show a partial response further vitiates the "test" dose approach which is no longer acceptable to paediatric neuro-oncologists (Packer *et al.*, 1984b). With the exception of the aforementioned situations, histological verification is therefore essential and the debate moves to the relative merits of open operation versus stereotactic biopsy.

Since the first surgical approach to the pineal gland by Sir Victor Horsley in 1910 (Horsley, 1910), a variety of open operations have been described and are well reviewed and their relative advantages and applications discussed elsewhere (Pendl, 1984). Although morbidity and mortality were for many years formidable, even in the hands of giants like Dandy (Dandy, 1921), it must now be accepted that the risks are acceptable when surgery is performed in units familiar with the microsurgical techniques involved (Yamamoto and Kageyama, 1980; Hoffman *et al.*, 1991). The surgical approach most frequently used is the transcallosal one; the occipital transtentorial and the infratentorial supracerebellar routes rarely being required.

Magnetic resonance imaging has replaced angiography in determining the approach. The advantages of open operation are that a more complete surgical excision can be achieved and a more extensive sampling of tissue can be obtained. Although the latter is undoubtedly correct and is an important argument as mixed tumour types are not uncommon, it should be noted that in one large series, from an undoubtedly experienced department, out of 29 cases total resection was only achieved in 5, compared with partial resection in 10 and biopsy in 14 cases (Hoffman *et al.*, 1991). Furthermore, the pathologies for which surgical resection will be the only treatment modality required, such as benign teratoma, dermoid and non-tumoral cysts, are relatively rare. Stereotaxic biopsy (Pecker *et al.*, 1979), especially when CT- or MR-guided (Apuzzo *et al.*, 1984), can be effective at establishing a diagnosis, with the proviso that the small specimens obtained may not be totally representative of all tissue types present in the tumour (Edwards *et al.*, 1988; Hoffman *et al.*, 1991). Hopefully, the accuracy of stereotaxic sampling will improve as MRI provides increasingly better images. In balance, the arguments still favour open operation, especially for lesions that are heterogeneous on CT or MRI, with image-guided stereotaxic biopsy being used for very small lesions and for cases where MRI has demonstrated extensive or multicentric neuraxial disease, for those with significantly elevated levels of AFP of β-HCG in blood or CSF, and for apparently non-tumoral cysts.

Radiation Therapy and Chemotherapy

With the exception of benign teratomas, and non-tumoral cysts, the vast majority of pineal region tumours will require adjuvant therapy in about 86% (Edwards *et al.*, 1988). Radiation therapy is used for germinoma, being confined to the locality of the tumour in the absence of any MRI, myelographic or cytological evidence of dissemination. For non-germi-noma germ-cell tumours, radiation therapy delivered to the craniospinal axis, together with appropriate adjuvant chemotherapy, is required. In these latter cases it is probable that chemotherapy should precede radiation therapy.

The outcome of germinoma is excellent, with 85% 5-year disease-free survival rates, teratoma being more in the order of 45% 5-year disease-free survival (Hoffman *et al.*, 1991). The occasional recurrent or extensive germinoma can usually be salvaged by craniospinal irradiation and chemotherapy. The need for accurate presurgical imaging with gadolin-ium-enhanced MRI of the whole neuraxis is emphasized. The future will certainly lie in the development of more effective chemotherapy which will probably oust radiation therapy. Currently, cisplatin and bleomycin seem to be the most useful agents (Japanese Intracranial Germ Cell Tumour Study Group, 1986; Kida *et al.*, 1986).

Conclusion

The management of a variety of childhood brain tumours has been described, each chosen to illustrate the problems, uncertainties and contro-versies that currently exist. But what of the way forward? Maximum use must be made of modern modes of imaging, principally MRI, to evaluate the extent of measurable disease accurately at all stages in active treatment. The range of prognostic factors must be extended, especially in the field of cytogenetic studies on tumour material. Basic scientists, especially experi-mental embryologists and neurodevelopmental workers, must pursue the developmental origins of paediatric CNS tumours. Clinical oncologists must be more aggressive and ambitious in the use of chemotherapy both in the treatment of frankly malignant neoplasms but also the more slowly-evolving lesions: there are, for example, some encouraging reports of responses in chiasmal and hypothalamic astrocytomas to nitrosoureas (Petronio *et al.*, 1991). Active investigations must be undertaken to reduce the morbidity using different therapeutic modalities in complementary ways so as to gain maximal control of disease with minimal morbidity: craniopharyngioma is a prime example here, as is PNET.

The management of children under 3 years requires special attention: present results are poor and successes often accompanied by unacceptable sequelae when radiation therapy is employed, a clear justification for pursuing chemotherapy as the major modality and withholding radiation therapy until the CNS is more mature and resilient (Van Eys *et al.*, 1985; Horowitz *et al.*, 1988; Strauss and Killmond, 1991). The optimum use must be made of the now highly sophisticated image-guided stereotaxic systems both in terms of biopsy and radiation therapies, external and interstitial. It must be abundantly clear that no single specialty holds all the answers, and progress will only be made by paediatric neurosurgeons working intimately at a local level with their colleagues in radiation therapy and medical oncology. There is no place for the occasional non-specialist involvement. At national and international level there must be active communication and collaboration between paediatric cancer study groups. At the centre, the needs of the individual child and his/her family must be finely balanced to ensure that optimal antineoplastic therapies are given alongside practical and humane support mechanisms. Again there is no place for the occasional dabbling. Only major centres providing the full range of necessary services – surgical, radiotherapeutic, oncological, endocrinological, rehabilitative and counselling – are fit to approach these challenges.

References

Abay EO II, Laws ER Jr, Grado GL *et al.* (1981) Pineal tumors in children and adolescents. Treatment by CSF shunting and radiotherapy. J Neurosurg 55: 889–895

Abraham J, Chandy J (1963) Ventriculo-atrial shunt in the management of posterior fossa tumours. J Neurosurg 20: 252–253

Albright AL, Price RA, Guthkelch AN (1983) Brainstem gliomas of children. A clinico-pathologic study. Cancer 52: 2313–2319

Albright AL, Guthkelch AN, Packer RJ *et al.* (1986) Prognostic factors in pediatric brainstem glioma. J Neurosurg 65: 751–755

Allen J (1978) The effects of cancer therapy on the nervous system. J Pediatr 93: 903–909

Allen JC, Epstein F (1982) Medulloblastoma and other primary malignant neuroectodermal tumors of the CNS. The effect of patient's age and extent of disease on prognosis. J Neurosurg 57: 446–451

Amador LV (1983) Brain neoplasms–infancy and childhood. In Brain Tumors in the Young. Charles C. Thomas, Springfield, Ill., pp 3–22

Anderson DR, Trobe SD, Taren SA *et al.* (1989) Visual outcome in cystic craniopharyngiomas treated with intracavitary phosphorus-32. Ophthalmology 96: 1786–1792

Apuzzo MLJ, Chandrasoma PT, Zelman V *et al.* (1984) Completed tomographic guidance stereotaxis in the management of lesions of the third ventricular region. Neurosurgery 15: 502–503

Asai A, Matsutani M, Takakura *et al.* (1989) Subacute brain atrophy after radiation therapy for malignant brain tumours. Cancer 63: 1962–1974

Backlund EO (1973) Studies on craniopharyngioma IV. Stereotaxic treatment with radio-surgery. Acta Chir Scand 139: 334–351

Backlund EO, Axelsson B, Bergstrand CG *et al.* (1973) Studies on craniopharyngioma III. Stereotaxic treatment with intracystic Yttrium-90. Acta Chir Scand 139: 237–247

Backlund EO, Axelsson B, Berstrand CG et al. (1989) Treatment of craniopharyngiomas – stereotactic approach in a 10 to 20 year perspective. Surgical, radiological and ophthalmological aspects. Acta Neurochir (Wien) 99: 11–19

Bailey P, Cushing H (1925) Medulloblastoma cerebelli, a common type of midcerebellar glioma of childhood. Arch Neurol Psychiat 14: 192–224

Barone BM, Elvidge AT (1970) Ependymomas. A clinical survey. J Neurosurg 33: 428–438

Berger MS, Baumeister B, Geyer Jr et al. (1991) The risks of metastases from shunting in children with primary central nervous system tumors. J Neurosurg 74: 872–877

Berry MP, Jenkin RDT, Keen CW et al. (1981) Radiation treatment for medulloblastoma. A 21-year review. J Neurosurg 55: 43–51

Bertolone SJ, Baum ES, Krivit W et al. (1989) A Phase II study of cisplatin therapy in recurrent childhood brain tumours. A report from the Children's Cancer Study Group. J Neuro-oncol 7: 5–11

Bilaniuk LT, Zimmerman RA, Litman P et al, (1980) Computed tomography of brainstem gliomas in children. Radiology 134: 89–95

Birch JM, Marsden HB, Morris-Jones PH et al. (1988) Improvements in survival from childhood cancer: results from a population based survey over 30 years. BMJ 296: 1372–1376

Bloom HJG (1967) Treating brain tumors in children. Radiation's role is becoming more significant. J Am Med Ass 200(12): 34–35

Bloom HJG (1971) Concepts in natural history and treatment of medulloblastoma in children. CRS Crit Rev Radiolog Sci: 2 89–143

Bloom HJG (1982a) Medulloblastoma in children: increasing survival rates and further prospects. Int J Radiat Oncol Biol Phys 8: 2023–2027

Bloom HJG (1982b) Intracranial tumors: response and resistance to therapeutic endeavours, 1970–1980. Int J Radiat Oncol Biol Phys 8: 1083–1113

Bouchard J (1980) Central nervous system. In: Fletcher G (ed) Textbook of Radiotherapy, 3rd edn. Lea and Febiger, Philadelphia, pp 444–498

Bremer AM, Nguyen TQ, Balsys R (1984) Therapeutic benefits of combination chemotherapy with vincristine, BCNU and procarbazine on recurrent craniopharyngioma. J Neuro-oncol 2: 47–51

Broggi G, Giorgi C, Franzini A et al. (1989) Preliminary results of intracavitary treatment of craniopharyngioma with Bleomycin. J Neurosurg Sci 33: 145–148

Bruno L, Schut L (1982) Survey of pediatric brain tumors. In: McLaurin RL (ed) Pediatric Neurosurgery. Surgery of the Developing Nervous System. Section of Pediatric Neurosurgery of the American Association of Neurological Surgeons. Grune and Stratton, New York, pp 361–365

Cavazutti V, Fischer EG, Welch K et al. (1983) Neurological and psychophysiological sequelae following different treatments of craniopharyngioma in children. J Neurosurg 59: 409–417

Chamberlain MC, Silver P, Edwards MJB et al. (1988) Treatment of extraneural metastatic medulloblastoma with a combination of cyclophosphamide, adriamycin, and vincristine. Neurosurgery 23: 476–479

Chang CH, Housepian EM, Herbert C Jr (1969) An operative staging system and a megavoltage radiotherapeutic technique for cerebellar medulloblastomas. Radiology 93(6): 1351–1359

Chin HW, Maruyama Y (1984) Age at treatment and long-term performance results in medulloblastoma. Cancer 53: 1952–1958

Chin HW, Maruyama Y, Markesbery W et al. (1982) Intracranial ependymoma. Results of radiotherapy at the University of Kentucky. Cancer 49: 2276–2280

Coffey RJ, Lunsford LD (1985) Stereotactic surgery for mass lesions of the midbrain and pons. Neurosurgery 17: 12–18

Constine LS, Randall SM, Rubin P et al. (1989) Craniopharyngioma: fluctuations in cyst size following surgery and radiation therapy. Neurosurgery 24: 53–59

Cumberlin RL, Luk KH, Wara WM et al. (1979) Medulloblastoma: treatment results and effect on normal tissues. Cancer 43: 1014–1020

Cushing H (1932) Intracranial Tumors. Notes Upon a Series of Two Thousand Verified Cases with Surgical Mortality Percentages Pertaining Thereto. Charles C. Thomas, Springfield, Ill., p 56

Cutler ED, Sosman MC, Vaughn WW (1936) The place of radiation in the treatment of cerebellar medulloblastomas. Report of twenty cases. Am J Roentgenol 35: 429–450

Dandy WE (1921) An operation for the removal of pineal tumors. Surg Gynecol Obstet 33: 113–119

Danoff BF, Cowchock FS, Marquette C *et al.* (1982) Assessment of the long-term effects of primary radiation therapy for brain tumors in children. Cancer 49: 1580–1586

Davis PC, Hoffman JC, Pearl GS *et al.* (1986) CT evaluation of effects of cranial radiation therapy in children. Am J Neuroradiol 7: 639–644

Demakas JJ, Sonntag VKH, Kaplan AM *et al.* (1982) Surgical management of pineal area tumors in early childhood. Surg Neurol 17: 435–400

Deutsch M, Reigel DH (1980) The value of myelography in the management of childhood medulloblastoma. Cancer 45: 2194–2197

Dohrmann GJ, Farwell JR (1976) Intracranial neoplasms in children: a comparison of North America, Europe, Africa and Asia. Dis Nerv Syst 37: 696–698

Dohrmann GJ, Farwell JR, Flannery JT (1976) Ependymomas and ependymoblastomas in children. J Neurosurg 45: 273–283

Draper GJ, Birch JM, Bitchell JR *et al.* (1982) Childhood Cancer in Britain: Incidence, Survival and Mortality (Studies on Medical and Population Subjects No. 37). HMSO, London

Edwards MSB, Hudgins RJ, Wilson CB *et al.* (1988) Pineal region tumors in children. J Neurosurg 68: 689–697

Epstein F, Murali R (1978) Pediatric posterior fossa tumors: hazards of the preoperative shunt. Neurosurgery 3: 348–350

Epstein F (1985) A staging system for brainstem gliomas. Cancer 56: 1804–1806

Evans AE, Jenkin RDT, Sposto R *et al.* (1990) The treatment of medulloblasoma. Results of a prospective randomised trial of radiation therapy with and without CCNU, vincristine and prednisone. J Neurosurg 72: 572–582

Farwell JR, Dohrmann GJ, Flannery JT (1977) Central nervous system tumours in children. Cancer 40: 3123–3132

Farwell JR, Dohrmann GJ, Flannery TJ (1978) Intracranial neoplasms in infants. Arch Neurol 35: 533–537

Farwell JR, Dohrmann GJ, Flannery JT (1984) Medulloblastoma in childhood: an epidemiological study. J Neurosurg 61: 593–664

Fewer D, Wilson CB, Boldrey EB *et al.* (1972) Phase II study of 1-(2-chloroethyl)-3-cyclohexyl-1-nitrosourea (CCNU; NSC-79037) in the treatment of brain tumours. Cancer Chemother Rep 56: 421–427

Finlay JL, Goins SC, Uteg R *et al.* (1987) Progress in the management of childhood brain tumors. Hematol Oncol Clin North America 1: 753–776

Fischer EG, Welch K, Belli JA (1985) Treatment of craniopharyngiomas in children: 1972–1981. J Neurosurg 62: 496–501

Fischer EG, Welch K, Shiluto J Jr (1990) Craniopharyngiomas in children – long-term effects of conservative surgical procedures combined with radiation therapy. J Neurosurg 73: 534–540

Fokes EL Jr, Earle KM (1969) Ependymomas: clinical and pathological aspects. J Neurosurg 30: 585–594

Friedman HJ, Mahaley MS Jr, Schold SC Jr *et al.* (1986) Efficacy of vincristine and cyclophosphamide in the therapy of recurrent medulloblastoma. Neurosurgery 18: 335–340

Friedman HJ, Schold SC Jr, Mahaley MS Jr *et al.* (1989) Phase II treatment of medulloblastoma and pineal blastoma with Melphalan: clinical therapy based on experimental models of human medulloblastomas. J Clin Oncol 7: 904–911

Galatzer A, Nofar E, Beit-Halachmi N *et al.* (1981) Intellectual and psychosocial functions before and after operation for craniopharyngioma. Child Care Health Devl 7: 307–316

Garrett PG, Simpson WJK (1983) Ependymomas: results of radiation treatment. Int J Radiat Oncol Biol Phys 10: 1709–1712

Glansman C, Horst W, Schiess K *et al.* (1980) Considerations in the radiation treatment of intracranial ependymoma. Prognosis in 24 own cases and results in published series after different techniques of radiation treatment. Strahlentherapie 156: 97–101

Grant DB, Lyen K (1982): Hypopituitarism after surgery for craniopharyngioma. Child's Brain 9: 201–204

Guthkelch AN, Taylor FH (1983) Filtered shunts in medulloblastoma? (letter). J Neurosurg 59: 364

Hardy DG, Hope-Stone HF, McKenzie CG *et al.* (1978): Recurrence of medulloblasoma after homogeneous field radiotherapy. J Neurosurg 49: 434–440

Harisiadis L, Chang CH (1977) Medulloblastoma in children: a correlation between staging and results of treatment. Int J Radiat Oncol Biol Phys 2: 833–841

Hart MN, Earle KM (1973) Primitive neuroectodermal tumors of the brain in children. Cancer 32: 890–897

Hawkins MM (1989) Long-term survival and care after childhood cancer. Arch Dis Child 64: 798–807

Herrick MK (1984): Pathology of pineal tumors In Neuwelt EA (ed) Diagnosis and Treatment of Pineal Region Tumors. Williams and Wilkins, Baltimore, pp 31–60

Hirsh JF, Renier D, Czernichow P et al. (1979) Medulloblastoma in childhood. Survival and functional results. Acta Neurochir 48: 1–15

Hoffman HJ, Hendrick EB, Humphreys RP (1976) Metastasis via ventriculoperitoneal shunt in patients with medulloblastoma. J Neurosurg 44: 562–566

Hoffman HJ, Hendrick EB, Humphreys RP et al. (1977) Management of craniopharyngioma in children. J Neurosurg 47: 218–227

Hoffman HJ, Becker L, Craven MA (1980) A clinically and pathologically distinct group of benign brainstem gliomas. Neurosurgery 7: 243–248

Hoffman HJ (1982) Craniopharyngioma: the continuing controversy on management. In: Humphreys RP (ed) Concepts in Pediatric Neurosurgery II. American Society for Pediatric Neurosurgery Karger, Basel, pp 15–28

Hoffman HJ (1983) Filtered shunts in treatment of medulloblastoma? (letter). J Neurosurg 59: 364

Hoffman HJ, Yoshidam, Becker LE et al. (1983) Pineal region tumors in childhood. Experience at the Hospital for Sick Children. In: Humphreys RP (ed) Concepts in Pediatric Neurosurgery 4. Karger, Basel, pp 360–386

Hoffman HJ, Otsubo H, Hendrick EB et al. (1991) Intracranial germ-cell tumors in children. J Neurosurg 74: 545–551

Hoffman HJ, De Silva M, Humphreys RP et al. (1992) Aggressive surgical management of craniopharyngiomas in children. J Neurosurg 76: 47–52

Horowitz ME, Mulhern RK, Kun LE et al. (1988) Brain tumors in the very young child. Postoperative chemotherapy in combined modality treatment. Cancer 61(3): 428–434

Horsley V (1910) Discussion. Proc Roy Soc Med 3:2

Japanese Intracranial Germ Cell Tumor Study Group (1986) Cisplatin, vinblastine, bleomycin (PVB) combination chemotherapy in the treatment of intracranial malignant germ cell tumours: a preliminary report of phase II study. Jpn J Cancer Clin 32: 1387 1393

Jenkin RDT (1969) Medulloblastoma in childhood: radiation therapy. Can Med Ass J 100: 51–54

Jereb B, Reid A, Ahuja RK (1982) Pattern of failure in patients with medulloblastoma. Cancer 50: 2941–2947

Katz EL (1975) Late results of radical excision of craniopharyngiomas in children. J Neurosurg 42: 86–90

Kessley LA, Dugan P, Concannon JP (1975) Systemic metastases of medulloblastoma promoted by shunting. Surg Neurol 3: 147–152

Kida Y, Kobayashi, Yoshida J et al. (1986) Chemotherapy with cisplatin for AFP-secreting germ-cell tumours of the central nervous system. J Neurosurg 65: 470–475

Kim YH, Fayos JY (1977) Intracranial ependymomas. Radiology 1245: 805–808

Koos WT, Miller MH (1971) Intracranial Tumors of Infants and Children. C.V. Mosby, St Loins

Kopelson G, Linggood RM, Kleinman GM (1983) Medulloblastoma. The identification of prognostic subgroups and implications for multimodality management. Cancer 5l: 312–319

Kramer S, Southard M, Mansfield CM (1968) Radiotherapy in the management of craniopharyngiomas. Further experience and late results. Am J Roentgenol Radium Ther Nucl Med 103(1): 44–52

Kricheff II, Becker M, Schneck GA et al. (1964a) Intracranial ependymomas: factors influencing prognosis. J Neurosurg 21: 7–14

Kricheff II, Becker M, Schneck GA et al. (1964b) Intracranial ependymomas: a study of survival in 65 cases treated by surgery and irradiation. Am J Roentgenol 91: 167–175

Krischer JP, Ragab AH, Kun L et al. (1991) Nitrogen mustard, vincristine, procarbazine, and prednisone as adjuvant chemotherapy in the treatment of medulloblastoma. A Pediatric Oncology Group study. J Neurosurg 74–905–909

Lampkin BC, Maver AM, McBride BH (1967) Response of medulloblastoma to vincristine sulfate: a case report. Pediatrics 39: 761–763

Lassman LP, Pearce GW, Ganz J (1965) Sensitivity of intracranial gliomas to vincristine sulphate. Lancet 1: 296–298

La Vecchia C, Decarli A (1988) Decline in childhood cancer mortality in Italy, 1955–1980. Oncology 45: 93–97

Laws ER Jr (1980) Trans-sphenoidal microsurgery in the management of craniopharyngioma. J Neurosurg 52: 661

Leibel SA, Sheline GE (1987) Radiation therapy for neoplasms of the brain. Review article. J Neurosurg 66: 1–22

Levin VA (1981) Chemotherapy of recurrent brain tumors. In: Prestayko AW, Crooke ST (eds) Nitrosoureas: Current Status and New Developments. Academic Press, New York, pp 159–167

Levin VA Edwards MSB, Gutin PH et al. (1984) Phase II evaluation of dibromodulcitol in the treatment of recurrent medulloblastoma, ependymoma, and malignant astrocytoma. J Neurosurg 61: 1063–1068

Littman P, Jarrett P, Bilaniuk LT et al. (1980) Pediatric brainstem gliomas. Cancer 45: 2787–2792

Liu HM, Boggs J, Kidd J (1976) Ependymomas of childhood I. Histological survey and clinico-pathological correlation. Child's Brain 2: 92–110

Liwnicz BH, Berger TS, Liwnicz RG et al. (1985) Radiation and associated gliomas: a report of four cases and analysis of post-radiation tumours of the central nervous system. Neurosurgery 17: 436–445

Manaka S, Teramoto A, Takakura K (1985) The efficacy of radiotherapy for craniopharyn-gioma. J Neurosurg 62: 648–656

Marks JE, Adler SJ (1982) A comparative study of ependymomas by site of origin. Int J Radiat Oncol Biol Phys 8: 37–43

Matson DD, Crigler JF (1969) Management of craniopharyngiomas in childhood. J Neurosurg 30: 377–390

May PL, Blaser SL, Hoffman HJ et al. (1991) Benign intrinsic tectal "tumors" in children. J Neurosurg 74: 867–871

Mealey J Jr, Hall PV (1977) Medulloblastoma in children. Survival and treatment. J Neurosurg 46: 56–64

Mechanick JL, Hochberg FH, Larocque A (1986) Hypothalamic dysfunction following whole brain irradiation. J Neurosurg 65: 490–494

Mikhael MA (1978) Radiation necrosis of the brain: correlation between computed tomogra-phy, pathology and dose distribution. J Comput Assist Tomogr 2: 71–80

Mikhael MA (1979) Radiation necrosis of the brain: correlation between patterns on com-puted tomography and dose of radiation. J Comput Assist Tomogr 3: 241–249

Miller RW, McKay FW (1984) Decline in US childhood cancer mortality: 1950 through 1980. J Am Med Ass 251: 1567–1570

Modan B, Baidatz D, Mart H et al. (1974) Radiation induced head and neck tumours. Lancet 1: 277–279

Mørk SJ, Løken AC (1977) Ependymoma. A follow-up study of 101 cases. Cancer 40: 907–915

Mørk SJ, Rubinstein LJ (1985) Ependymoblastoma. A reappraisal of a rare embryonal tumor. Cancer 55: 1536–1542

Nelson GA, Bastian FO, Schlitt M et al. (1988) Malignant transformation in craniopharyn-gioma. Neurosurgery 22: 427–429

Nevhauser EBD, Wittenborg MH, Berman CZ et al. (1952) Irradiation effects of Roentgen therapy on the growing spine. Radiology 59: 637–650

Oi S, Raimondi AJ (1982) Ependymoma. In: Bruno L, Schut L Pediatric Neurosurgery of the American Association of Neurological Surgeons. Grune and Stratton, New York, pp 419–427

OPCS (Office of Population Censuses and Surveys)(1986) OPCS Monitor. Deaths by Causes, 1985. DH2 86/2. Government Statistical Service, London, 16 September 1986

Packer RJ, Sutton LN, Rorke LB (1984a) Prognostic importance of cellular differentiation in medulloblastoma of childhood. J Neurosurg 61: 296–301

Packer RJ, Sutton LN, Rosenstock JG et al. (1984b) Pineal region tumor of childhood. Pediatrics 74: 97–102

Packer RJ, Zimmerman RA, Lverssen TG et al. (1985) Brain stem gliomas of childhood: mag-netic resonance imaging. Neurology 35: 397–401

Painter MJ, Chutorian AM, Hilal S (1975) Cerebrovasculopathy following irradiation in child-hood. Neurology 25: 189–194

Park TS, Hoffman HJ, Hendrick EB (1983) Medulloblastoma; clinical presentation and man-agement. Experience at the Hospital for Sick Children, Toronto, 1950–1980. J Neurosurg 58: 543–552

Patterson RH Jr, Danylevich A (1982) Surgical removal of craniopharyngiomas by a transcra-nial approach through the lamina terminalis and sphenoid sinus. Neurosurgery 7: 111–117

Patterson E, Farr RE (1953) Cerebellar medulloblastoma treatment by irradiation of the whole central nervous system. Acta Radiol 39: 323–336

Pecker J, Scarabin J, Vallee B et al. (1979) Treatment of tumors in the pineal region: value of stereotactic biopsy. Surg Neurol 12: 341–348

Pendl G (1984) The surgery of pineal lesions–historical perspective. In: Neuwelt EA (ed) Diagnosis and Treatment of Pineal Region Tumors. Williams and Wilkins, Baltimore, pp 139–154

Petronio J, Edwards MSB, Prados M et al. (1991) Management of chiasmal and hypothalamic gliomas of infancy and childhood with chemotherapy. J Neurosurg 74: 701–708

Phillips TL, Sheline GE, Boldrey E (1964) Therapeutic considerations in tumors affecting the central nervous system: ependymomas. Radiology 83: 98–105

Probert JC, Parker BR, Kaplan HS (1973) Growth retardation in children after megavoltage irradiation of the spine. Cancer 32: 634–639

Raimondi AJ, Tomita T (1979a) Medulloblastoma in childhood. Comparative results of partial and total resection. Child's Brain 5: 310–328

Raimondi AJ, Tomita T (1979b) Medulloblastoma in childhood. Acta Neurochir 50: 127–138

Raimondi AJ, Tomita T (1979c) The disadvantages of prophylactic whole CNS postoperative radiation therapy for medulloblastoma. In: Paoletti P, Walker MD, Butti G et al. (eds) Multidisciplinary Aspects of Brain Tumor Therapy. North-Holland, Amsterdam, pp 209–218

Read G (1984) The treatment of ependymoma of the brain or spinal canal by radiotherapy: a report of 79 cases. Clin Radiol 35: 163–166

Richard IL, Wara WM, Wilson CB (1980) Role of radiation therapy in the management of craniopharyngiomas in children. Neurosurgery 6: 513–517

Ringertz N, Reymond A (1949) Ependymomas and choroid plexus papillomas. J Neuropathol Exp Neurol 8: 355–380

Rorke LB (1983) The cerebellar medulloblastoma and its relationship to primitive neuroectodermal tumors. J Neuropathol Exp Neurol 42: 1–15

Ross GW, Rubinstein LJ (1989) Lack of histopathological correlation of malignant ependymomas with postoperative survival. J Neurosurg 70: 31–36

Salazar CM, Rubin P, Bassano D et al. (1977) Improved survival of patients with intracranial ependymomas treated by irradiation: dose selection and field extension. Cancer 35: 1563–1573

Salazar CM, Castro-Vita H, Vanhoutte P (1983) Improved survival in cases of intracranial ependymomas after radiation therapy. Later report and recommendations. J Neurosurg 59: 652–659

Sano K (1976) Pinealomas in children. Child's Brain 2: 67–72

Schoenberg BS, Schoenberg DG, Christine BE et al. (1976) The epidemiology of primary intracranial neoplasms of childhood. A population study. Mayo Clinic Proc 51: 51–56

Shalet SM, Beardwell CG, Morris-Jones PH et al. (1975) Pituitary function after treatment of intracranial tumours in children. Lancet 2: 104–107

Shillito J Jr (1976) Craniopharyngiomas: the subfrontal approach or none at all? Clin Neurosurg 23: 52–79

Shuman RM, Alvord EL Jr, Leech RW (1975) The biology of childhood ependymomas. Arch Neurol 32: 731–739

Silverberg E (1980) Cancer Statistics, 1980. Cancer J Clinicians 30: 23–38

Silverman CL, Simpson JR (1982) Cerebellar medulloblastoma: the importance of posterior fossa dose to survival and patterns of failure. Int J Radiat Oncol Biol Phys 8: 1869–1876

Silverman CL, Palkes H, Talent B et al. (1984) Late effects of radiotherapy on patients with cerebellar medulloblastoma. Cancer 54: 825–829

Smith RA, Lampe J, Kahn EA (1961) Prognosis of medulloblastoma in children. J Neurosurg 18: 91–97

Smith CE, Long DM, Jones TK Jr et al. (1973) Medulloblastoma: an analysis of time–dose relationships and recurrence problems. Cancer 32: 722–728

Strauss LC, Killmond TM (1991) Efficacy of postoperative chemotherapy using cisplatin plus otoposide in young children with brain tumors. Med Pediatr Oncol II: 16–21

Stroink ARM, Hoffman HJ, Hendrick EB et al. (1986) Diagnosis and management of pediatric brainstem gliomas. J Neurosurg 65: 745–750

Svien HJ, Mabon RF, Kernohan JW et al. (1953) Ependymoma of the brain: pathologic aspects. Neurology 3: 1–15

Symon L, Sprich W (1985) Radical excision of craniopharyngioma. Results in 20 patients. J Neurosurg 62: 174–181

Takahashi H, Nakazawa S, Shimura T (1985) Evaluation of postoperative infratentorial injection of bleomycin for craniopharyngioma in children. J Neurosurg 62: 120–127

Tomita T, McClone DG (1983) Spontaneous seeding of medulloblastoma: results of cerebrospinal fluid cytology and arachnoid biopsy from the cisterna magna. Neurosurgery 12: 265–267

Tomita T, McClone DG (1986) Medulloblastoma in childhood: results of radical resection and low-dose neuraxis radiation therapy. J Neurosurg 64: 238–242

Vandyk J, Jenkin RDT, Leung PMK et al. (1977) Medulloblastoma: treatment technique and radiation dosimetry. Int J Radiat Oncol Biol Phys 2: 993–1005

Van Eys J, Chen T, Moore T et al. (1981) Adjuvant chemotherapy for medulloblastoma and ependymoma using IV vincristine, intrathecal methotrexate, and intrathecal hydrocortisone: a Southwest Oncology Group study. Cancer Treat Rep 65: 681–684

Van Eys J, Cangir A, Coody D et al. (1985) MOPP regime as primary chemotherapy for brain tumours in infants. J Neuro-oncol 3: 237–243

Walker ML, Storrs B (1985) Surgical therapy for intrinsic brainstem gliomas. In: Humphreys RP (ed) Concepts in Pediatric Neurosurgery 5. Karger, Basel, pp 178–186

Wilson CB, Boldrey EB, Knot KJ (1970) 1,3-bis-(2-chloroethyl)-1-nitrosourea (NSC-409962) in the treatment of brain tumors. Cancer Chemother Rep 54: 273–281

Wilson CB, Gutin P, Boldrey EB et al. (1976) Single-agent chemotherapy of brain tumors. A five-year review. Arch Neurol 33: 739–744

Yamamoto I, Kageyama S (1980) Microsurgical anatomy of the pineal region. J Neurosurg 53: 205–221

Yasargil MG, Curck M, Kis M et al. (1990) Total removal of craniopharyngiomas. Approaches and long-term results in 114 patients. J Neurosurg 73: 3–11

Zülch KJ (1979) Histological Typing of Tumours of the Central Nervous System. World Health Organization, Geneva

8 Central Nervous System Neoplasms in Acquired Immunodeficiency Syndrome

Yuen T. So, Richard L. Davis and Jay H. Beckstead

Introduction

Infection by the human immunodeficiency virus (HIV) is responsible for the acquired immune deficiency syndrome (AIDS), a devastating disease associated with frequent life-threatening infectious and neoplastic complications. The epidemic has spread rapidly since its first recognition in the USA in the early 1980s. As of July 1989, over 170 000 cases of AIDS have been reported worldwide. Its full impact on the practice of medicine is only beginning to be appreciated.

Neurological symptoms present in at least one-third of patients with AIDS and sometimes precede the systemic manifestations of the disease (Snider et al., 1983; Levy et al., 1985). Moreover, pathology of the nervous system is encountered in an even higher proportion of patients who have died of AIDS-related disease (Gonzales and Davis, 1988). Viral, fungal or other opportunistic infections of the brain are among the most common neurological complications of HIV infection. Patients with AIDS are also at risk of developing two types of neoplasm – non-Hodgkin's lymphoma and Kaposi's sarcoma. Although Kaposi's sarcoma is the more common systemic malignancy in AIDS, central nervous system (CNS) complications are seen more frequently in lymphoma. Even patients with advanced Kaposi's sarcoma rarely develop metastases in the nervous system. In contrast, non-Hodgkin's lymphoma can present either as a primary brain tumour without systemic malignancy or as a systemic malignancy with secondary spread to the nervous system.

Primary CNS Lymphoma

Prior to the AIDS epidemic, primary CNS lymphoma was an uncommon neoplasm, accounting for only 1% of all intracranial tumours and less than 2% of all lymphomas. In contrast, primary CNS lymphoma accounts for

5%–10% of the CNS complications seen in AIDS (Levy *et al.*, 1988). In medical centres with large case loads of patients with AIDS, it is among the most commonly encountered primary intracranial tumours. Although most cases occur in adults (Jellinger, 1991), it has also been recognized in children with AIDS (Epstein *et al.*, 1988).

Primary CNS lymphoma has historically been designated by a variety of names, such as reticulum cell sarcoma, perithelial sarcoma, microgliomatosis and atypical granulomatous encephalitis. The confusing terminology reflects the rarity of the tumour in the past, and the lack of a consensus in regard to the cell of origin of this neoplasm. The present terminology, primary malignant lymphoma of the CNS, emerged from a study at the US Armed Forces Institute of Pathology which showed that the neoplasm was histologically identical to the extraneural non-Hodgkin's lymphomas (Henry *et al.*, 1974). Subsequent studies with more refined histological and immunological techniques have further shown that these tumours are composed of neoplastic B lymphocytes (Egerter and Beckstead, 1988; Adams and Howatson, 1990; Iglesias-Rozas *et al.*, 1991).

Clinical Manifestations

Primary CNS lymphoma is second only to cerebral toxoplasmosis as a cause of intracranial masses in patients with AIDS. The lymphoma generally presents as either solitary or multiple tumours (So *et al.*, 1986). The clinical deficits depend on the location of tumours and the degree of associated cerebral oedema. Hemiparesis, hemianopsia, ataxia, language disturbances, confusional state and seizures are among the most commonly encountered symptoms and signs. The clinical progression of neurological deficits can be very rapid, and some patients succumb before a pre-mortem diagnosis can be made. It appears that this neoplasm generally follows a more fulminant course in AIDS patients than in the general population.

Primary CNS lymphomas appear as discrete cerebral masses on computed tomographic (CT) or magnetic resonance imaging (MRI) studies of the brain. However, many opportunistic infections, notably cerebral toxoplasmosis, also present as mass lesions in patients with AIDS, and an unequivocal distinction between tumour and infection is difficult on clinical grounds alone. Another problem in AIDS is that opportunistic infections sometimes coexist with CNS lymphoma. Biopsy of the brain lesions is one of the few means to make a confident diagnosis. In some centres, the response of brain lesions to a short course of antibiotic therapy has been used tentatively to distinguish cerebral toxoplasmosis from lymphoma.

Pathology

As a rule, primary CNS lymphoma is restricted to the central nervous system and does not metastasize to extracranial sites. The tumour, however, is almost always multicentric within brain parenchyma (Fig. 8.1).

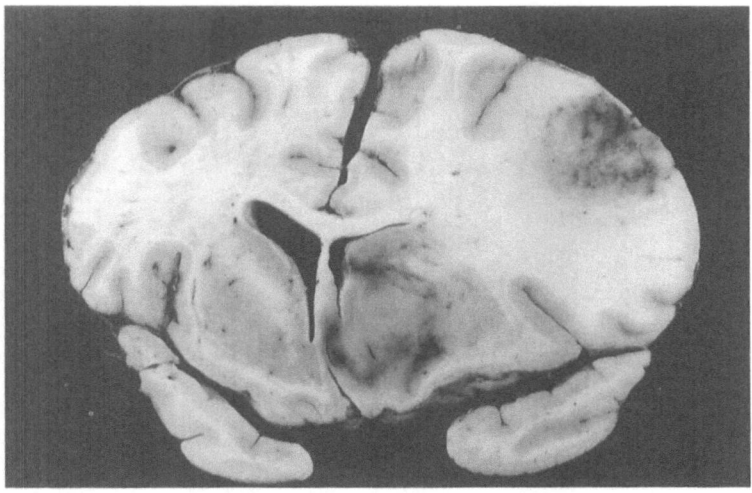

Fig. 8.1 Gross photograph showing primary lymphoma in AIDS patient. There is a large mass involving the left convexity cortex and white matter, and also the left basal ganglia

The cerebral hemispheres and the basal ganglia are the most common anatomical locations. Microscopically, the tumour appears as a highly pleomorphic mixture of neoplastic cells which are intermixed with reactive microglial cells and macrophages. There is a tendency for the abnormal cells to cluster along blood vessels (Fig. 8.2). Undifferentiated cell types, both large immunoblastic cell and small non-cleaved cell (Fig. 8.3), are seen in all the primary CNS lymphomas reported so far. Immunoperoxidase staining has established a B-cell origin for these transformed cells (Egerter and Beckstead, 1988; Adams and Howatson, 1990; Iglesias-Rozas *et al.*, 1991) (Fig. 8.4). Immunohistological typing with antisera against B- and T-cell markers should be routinely performed, and a significant presence of T-cells should raise doubt about the diagnosis of lymphoma.

Treatment and Prognosis

Primary CNS lymphoma in AIDS is a rapidly fatal disease without treatment. Almost all treatment protocols use some form of radiation therapy since the neoplasm is highly radiosensitive. Some centres combine radiotherapy with chemotherapy, and have reported improved survival (Hochberg and Miller, 1988). Unfortunately, relapse after an initial response is common. Especially in patients with AIDS-related lymphoma, survival beyond 6 months is rare. Most patients with AIDS who had an initial tumour response to treatment died from other fatal opportunistic infections.

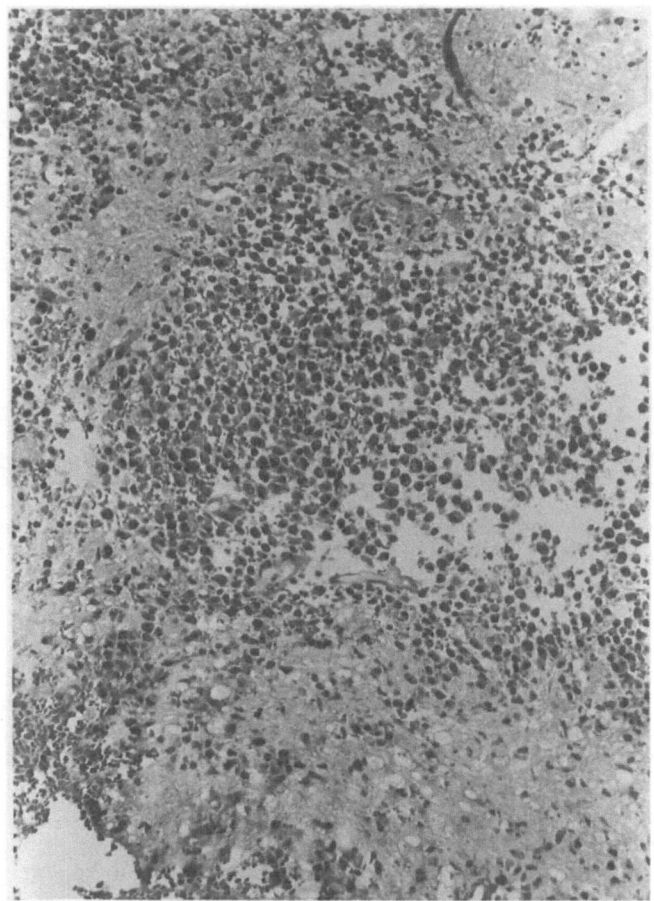

Fig. 8.2 Photomicrograph of immunoblastic lymphoma with typical angiocentric concentrations of tumour cells in plastic embedded tissue (Haematoxylin and eosin, original magnification ×40)

Systemic Non-Hodgkin's Lymphoma

Systemic lymphoma in patients with AIDS is a more fulminant disease than that seen in immunocompetent patients (Ziegler *et al.*, 1982, 1984). Widespread involvement of visceral organs and bone marrow as well as metastases to the CNS are common at the time of first presentation of the malignancy. Approximately 40% of patients with AIDS with systemic lymphoma have CNS involvement, as compared to 5%–29% reported in lymphoma in patients who do not have AIDS.

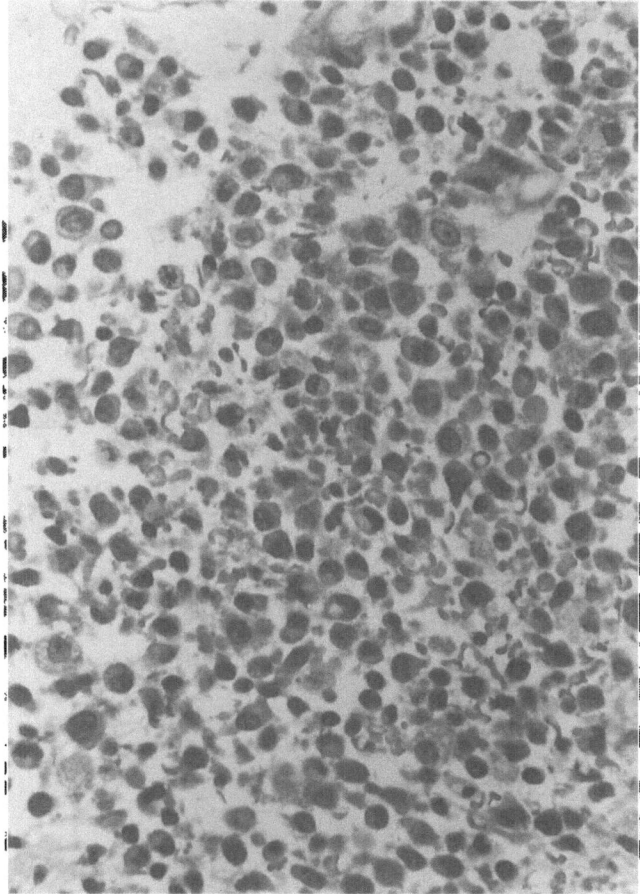

Fig. 8.3 High-power photomicrograph of immunoblastic lymphoma showing polymorphous population in same case as Fig. 8.2. Plastic embedded tissue (Haematoxylin and eosin, original magnification ×100)

Clinical Manifestations

Systemic lymphoma spreads to the CNS chiefly along the leptomeninges or the dura (So *et al.*, 1988). Neurological symptoms are usually due to lymphomatous involvement of cranial nerves or spinal roots producing focal motor or sensory deficits. Encephalopathy, seizures, headache and signs of meningeal irritation often accompany the neurological deficits. The neoplasm may also aggregate as epidural or subdural deposits, producing acute compression of the spinal cord or cauda equina. Back pain, paraplegia, sensory loss and sphincter dysfunction are common. Intracerebral masses and hydrocephalus can also be seen. Rarely, patients may develop brachial or lumbosacral plexopathy from local infiltration of extracranial neural structures.

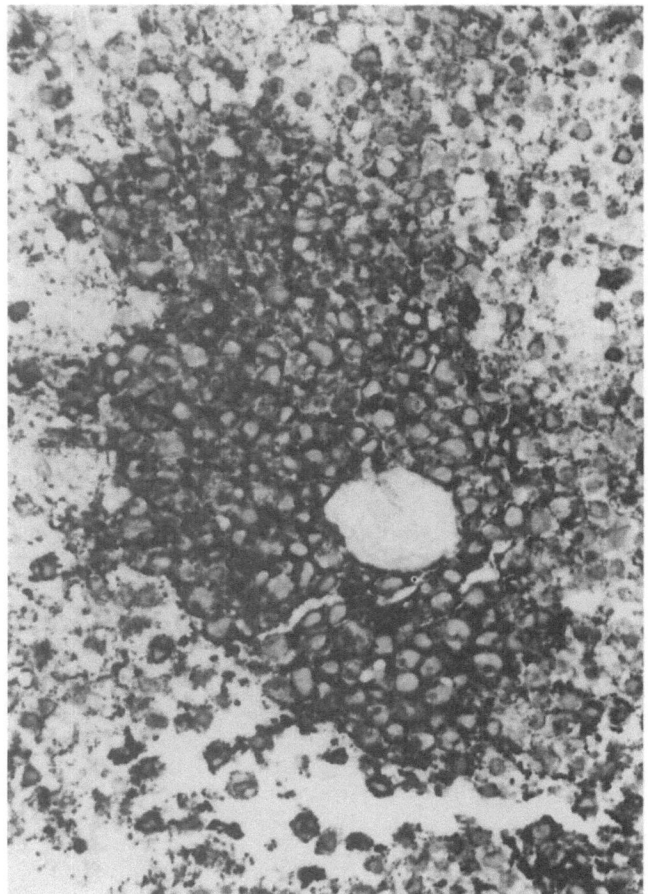

Fig. 8.4 High-power photomicrograph of same tumour as Fig. 8.2 stained with B-cell marker, LN-3, demonstrating the very dominant B-cell origin of the lymphoma (original magnification ×100)

Examination of the cerebrospinal fluid (CSF) is an important diagnostic test for leptomeningeal lymphoma. Increased mononuclear cell count and elevated CSF protein content are common findings. In addition, abnormal cells with large nuclei and prominent nucleoli can often be identified in cytological preparations. In patients with acute spinal cord compression, MRI or myelographic examination are necessary for anatomical localization and to direct treatment of the tumour.

Pathology

Two high-grade histological subtypes, small non-cleaved cell and large immunoblastic cell lymphomas, account for about 60% of the systemic lymphoma in patients with AIDS (Ziegler *et al.*, 1984). Diffuse large cell lymphoma, an intermediate-grade lymphoma, is next in frequency. Widely disseminated lymphoma is found at autopsy in most patients, although the sanctuary sites (brain and testes) may be the only areas involved in patients successfully treated with systemic chemotherapy.

Treatment and Prognosis

Intrathecal chemotherapy is the mainstay of treatment for leptomeningeal lymphoma. Patients with spinal cord compression are treated as emergencies with either radiation therapy or laminectomy for decompression. The prognosis is poor even with treatment, reflecting the advanced clinical stage of the systemic malignancy (Loareiro et al., 1988).

Kaposi's Sarcoma

Direct tumour invasion of the brain by Kaposi's sarcoma is rare. In large reviews of patients with AIDS-associated neurological symptoms, metastatic deposits of Kaposi's sarcoma in the nervous system were seen in less than 1% of the cases (Levy et al., 1985). Even in Africa, where the disease is prevalent, few cases of neurological involvement have been reported. Macroscopically, the tumours are well circumscribed and contain focal areas of recent and remote haemorrhages (Gorin et al., 1985). The lesions are highly vascular. The vascular spaces are irregular and slit-like, and are surrounded by pleomorphic spindle-shaped cells. Frequent mitotic figures can be seen.

Pathogenesis of Neoplasms in Immune Deficiency

The increased incidence of malignancies is well recognized in many immunodeficiency states other than AIDS. For instance, organ transplant recipients and patients with congenital immunodeficiency have risk of malignancy as high as 10–100 times that of the general population. Neoplasms of the haematopoietic and lymphoreticular systems are the most frequently implicated.

The immune system is probably partly responsible for the surveillance and control of neoplasms in the body. For example, tumour regression sometimes occurs in patients after an immunodeficiency state is corrected. In immunocompromised patients, therefore, malignantly transformed cells may multiply and spread more rapidly if unchecked by some immune surveillance mechanism. Whether this is sufficient to account for the vastly increased frequency of malignancy in these patients is unclear.

Unfortunately, very little is known about the pathogenesis of neoplasms. Viral infection as an initiating event in oncogenesis has probably received the most recent attention. Epstein–Barr viral genome has been identified in neoplastic cells of primary CNS lymphoma in patients with AIDS as well as those without AIDS (Hochberg and Miller, 1985; Rosenberg *et al.*, 1986; MacMahon *et al.*, 1991). Moreover, the vast majority of the general population has serological evidence of prior infection with Epstein–Barr virus. Persistent infection has also been demonstrated in a small number of lymphocytes. In contrast, there is little evidence to implicate other viruses in the pathogenesis of lymphoma. Human immunodeficiency virus belongs to the retroviral family lentivirus, and most lentiviruses have not been associated with neoplasms. Moreover, HIV genome has not been demonstrated in the neoplasms in AIDS.

The striking frequency of occurrence of primary CNS lymphoma in patients with immunodeficiency raises another important question about oncogenesis in the CNS. The brain has neither lymphoid tissue nor lymphatic circulation, and therefore does not have an endogenous source of neoplastic cells for lymphoid tumour. Two suggestions have been proposed to account for the origin of these neoplastic lymphocytes (Hochberg and Miller, 1985). First, lymphocytes may transform into neoplastic cells outside the brain, and later deposit themselves in the CNS. The brain provides a permissive sanctuary for cell proliferation and neoplastic transformation. This scheme, for example, potentially can explain the high incidence of multicentricity of this neoplasm. A second hypothesis proposes that the cellular proliferation begins as reactive lymphocytes to an infectious or inflammatory process in the CNS. A second, undefined event then leads to a local malignant transformation.

AIDS-related neoplasia remains a disease of unknown aetiology. Better understanding of the pathogenesis will be required to design a more rational approach to treatment. Meanwhile, despite the rapidly increasing incidence of these diseases, effective treatment has remained elusive.

References

Adams JH, Howatson AG (1990) Cerebral lymphoma: review of 70 cases. J Clin Path 43: 544–547

Egerter DA, Beckstead JH (1988) Malignant lymphomas in the acquired immunodeficiency syndrome. Arch Pathol Lab Med 112: 602–606

Epstein LG, DiCarlo FJ, Joshi VV *et al.* (1988) Primary lymphoma of the central nervous system in children with acquired immunodeficiency syndrome. Pediatrics 82: 355–363

Gonzales MF, Davis RL (1988) Neuropathology of acquired immunodeficiency syndrome. Neuropathol Appl Neurobiol 14: 345–363

Gorin FA, Bale JF, Halks-Miller M, Schwartz RA (1985) Kaposi's sarcoma metastatic to the CNS. Arch Neurol 42: 162–165

Henry JM, Reid RH, Dillard SH, Earle KM, Davis RL (1974) Primary malignant lymphomas of the central nervous system. Cancer 34: 1293–1302

Hochberg FH, Miller DC (1988) Primary central nervous system lymphoma. J Neurosurg 68: 835–853

Iglesias-Rozas JR, Bantz B, Adler T *et al.* (1991) Cerebral lymphoma in AIDS: clinical, radiological, neuropathological and immunopathological study. Clin Neuropath 10: 65–72

Jellinger KA (1991) Primary malignant lymphomas of the central nervous system. In: Ikuta F (ed) Neuropathology in Brain Research. Elsevier Science Publishers, Amsterdam, pp 187–200

Levy RM, Bredesen DE, Rosenblum ML (1985) Neurological manifestations of the acquired immunodeficiency syndrome (AIDS): experience at UCSF and review of the literature. J Neurosurg 62: 475–495

Loureiro C, Gill PS, Meyer PR et al. (1988) Autopsy findings in AIDS-related lymphoma. Cancer 62: 735–739

MacMahon EME, Glass JD, Hayward SD et al. (1991) Epstein–Barr virus in AIDS-related primary central nervous system lymphoma. Lancet 338: 969–973

Rosenberg NL, Hochberg FH, Miller G, Kleinschmidt-DeMasters BK (1986) Primary central nervous system lymphoma related to Epstein–Barr virus in a patient with acquired immune deficiency syndrome. Ann Neurol 20: 98–102

Snider WD, Nielsen S, Gold JWM, Metroka CE, Posner JB (1983) Neurological complications of acquired immune deficiency syndrome: analysis of 50 patients. Ann Neurol 14: 403–418

So YT, Beckstead JH, Davis RL (1986) Primary central nervous system lymphoma in acquired immune deficiency syndrome. A clinical and pathologic study. Ann Neurol 20: 566–572

So YT, Choucair A, Davis RL et al. (1988) Neoplasms of the Central Nervous System in Acquired Immunodeficiency Syndrome. Raven Press, New York.

Ziegler JL, Drew WL, Miner RC et al. (1982) Outbreak of Burkitt's-like lymphoma in homosexual men. Lancet 2: 631–633

Ziegler JL, Beckstead JH, Volberding PA et al. (1984) Non-Hodgkin's lymphoma in 90 homosexual men. Relation to generalized lymphadenopathy and the acquired immunodeficiency syndrome. N Engl J Med 311: 565–570

9 Metastatic Brain Tumours

Kintomo Takakura

Introduction

Metastatic tumours of the central nervous system (CNS) are manifestations of advanced cancers. The brain, physically and mentally controlling all human behaviours and functions, is unfortunately one of the predominant sites for metastasis of cancer cells. When the brain is so affected, the patient suffers grievous agonies such as headache, nausea, epileptic seizures or gradually developing paresis of nervous function. The number of patients with brain metastases has moreover increased recently concurrent with the increased number of general cancer patients. Since, in addition, the number of ageing people in the general population is also increasing, the metastatic brain tumour has become the most frequent tumour in all geriatric brain tumour patients.

Although many neurosurgeons have neglected the treatment of metastatic brain tumours, simply because they seemed to be too advanced and there appeared to be no clear benefit from the treatment, the longest survivor in the author's cases has been active for more than 17 years after craniotomy for the treatment of a brain metastasis, and the number of such long-term survivors is increasing. If we had abandoned treatment for those patients, they would not have been able to enjoy a prolonged useful and happy life. Those patients demonstrate to us the efficacy and importance of treatment for metastatic brain tumours.

Cancer metastasizes not only to brain but also to skull, spine and spinal cord, and causes many neurological signs and symptoms. It is evident that the successful treatment of those metastatic tumours is quite difficult to obtain, simply because there is no proper therapeutic method for cancer itself. It is, however, important for us to treat patients to improve their quality of life, with full knowledge of the pathophysiology of metastases in the brain.

General Aspects

Primary Tumours

All kinds of malignant neoplasms metastasize to brain. Lung cancer (Takakura *et al.*, 1982; The Committee of the Brain Tumor Registry in Japan,

1987) is, however, the most frequent primary cancer (50% in Japan). The frequency of primary cancer for metastatic brain tumours is shown in Table 9.1. The clinical data are based on the author's own cases (Department of Neurosurgery, University of Tokyo Hospital and National Cancer Center), and those reported in the Brain Tumor Registry in Japan (1987). The autopsied cases were accumulated in Montefiore Hospital, New York, by Professor Asao Hirano and the National Cancer Center in Tokyo (Takakura et al., 1982).

Among cancers of the gastrointestinal tract, the number of gastric cancers is gradually decreasing, while those of colon and rectal cancer have recently increased. Among cancers of genitourinary organs, renal cancer is the most frequent source of brain metastasis. The recent trend is the increasing number of hepatomas and malignant lymphomas. In head and neck cancers, those of the maxilla and nasal sinus invade the skull base and grow directly into the intracranial cavity.

A comparison of the primary source of metastatic brain tumours between Japan and the USA showed that the number of metastases due to breast cancer is higher in the USA (24%), simply because the incidence of breast cancer is much more frequent among Caucasians than Japanese. The same tendency is also noticeable in cases of melanoma.

The primary sources of brain secondary tumours have been reported by many investigators (Earle, 1954; Lesse and Netsky, 1954; Störtebecker, 1954; Perese, 1959; Simionescu, 1960; Chu and Hilaris, 1961; Richard and McKissock, 1963; Nisce et al., 1971; Takakura et al., 1982), but the ratio of

Table 9.1 Primary sources of intracranial metastatic tumours

	Clinical cases		Autopsy cases[†]	
Site of primary neoplasm	Author's series (1989)	All Japan Brain Tumor Registry (1987)*	Montefiore	National Cancer Center, Japan
Lung	315(44.6%)	1631(50.0%)	349(29.9%)	144(42.4%)
Breast	106(15.0)	375(11.5)	278(23.8)	44(12.9)
Gastrointestinal	52(7.4)	340(10.3)	98(8.4)	36(10.6)
Liver, biliary tract and pancreas	8(1.1)	40(1.2)	–	–
Genitourinary	55(7.8)	338(10.3)	110(9.4)	29(8.5)
Melanoma and skin	6(0.8)	37(1.1)	47(4.0)	5(1.5)
Head and neck	21(3.0)	152(4.7)	29(2.5)	10(2.9)
Thyroid	13(1.8)	49(1.5)	15(1.3)	4(1.2)
Leukaemia	26(3.7)	22(0.7)	92(7.9)	27(7.9)
Lymphoma	8(1.1)	30(1.0)	112(9.6)	17(5.0)
Sarcoma	23(3.3)	71(2.2)	20(1.7)	14(4.1)
Others	66(9.3)	177(5.4)	17(1.5)	8(2.3)
Unknown	7(1.0)	–	0(0.0)	3(0.9)
Totals	706	3262	1167	341

* From The Committee of Brain Tumor Registry in Japan (1987).
† From Takakura et al. (1982).

primary tumours is quite different depending on the background of each clinic.

Sites of Metastatic Involvement

Malignant tumours metastasize in every part of the brain. Almost 80% of intracranial metastatic tumours are located in the supratentorial space and about 12% in the infratentorial space. Other metastatic tumours are situated both in the supra- and infratentorial spaces. The number of metastatic tumours in the intracranial cavity depends on the stage of the cancer and also the biological characteristics of the primary neoplasm.

It has been noted that many metastatic tumours are located in the subcortical or cortical areas. The area of the middle cerebral artery is the predominant site for metastasis, since there the blood flow is maximal because of its anatomical situation.

The statistical data on locations of intracranial metastatic tumours are shown in Table 9.2. Parietal and frontal lobes are the most frequent sites. About one-tenth of metastatic tumours appear in the cerebellum. Several cancers also frequently invade the skull base.

Table 9.3 shows the location of metastases in the CNS. Among 3359 autopsied cases who died of cancer, 28.7% had metastases either in the intracranial space or vertebral canal. In cases of lung cancer, 40.8% of the patients died with intracranial metastasis and 6.6% with metastasis in the vertebral canal.

Lung cancers commonly metastasize to brain parenchyma (34.4%). In cases of breast cancer, 21.1% had metastases in brain parenchyma. Breast cancers also metastasize frequently to dura mater (31.2%) and pituitary gland (20.0%). Conversely, in cases of gastrointestinal cancers, only 5.5% of the patients had metastases to the brain parenchyma and a mere 2.2% to the vertebral canal.

Table 9.2 Location of intracranial metastatic tumour (From The Committee of Brain Tumor Registry in Japan, 1987)

Location	No. of patients
Frontal lobe	616 (19.3%)
Temporal lobe	231 (7.2%)
Parietal lobe	638 (20.0%)
Occipital lobe	246 (7.7%)
Basal ganglia and rostral brain stem	49 (1.5%)
Corpus callosum	23 (0.7%)
Chiasmal region	51 (1.1%)
Ventricular system	44 (1.5%)
Pineal region	11 (0.3%)
Cerebellum	284 (9.0%)
Pons and medulla	11 (0.3%)
Calvarium	65 (2.0%)
Skull base	199 (16.3%)
Others	719 (22.9%)
Total	3187 (100.0)

Table 9.3 Location of metastases in central nervous system (From Takakura *et al.*, 1982)

Location of metastasis	Lung	Breast	Gastroin-testinal	Urinary tract	All sites
Total number of autopsy	774	526	750	199	3359
Intracranial	40.8%	50.8%	8.3%	21.1%	25.6%
Parenchyma	34.4*	21.1	5.5	17.1	16.5
Leptomeninges	3.2	5.7	0.5	0,5	2.7
Cranial nerves	0.5	1.9	0.0	0.5	0.7
Dura mater	7.0	31.2	2.7	4.5	9.6
Pituitary gland	5.9	20.0	1.6	2.5	6.1
In vertebral canal	6.6	6.7	2.2	6.0	5.2
Spinal cord	0.8	0.4	0.0	0.0	0.3
Spinal nerves	1.3	0.8	0.5	1.5	0.8
Epidural tissue	5.3	4.9	2.1	6.0	4.2
Total CNS and its covering	45.1	52.9	9.8	26.1	28.7

* The frequency of metastases is expressed as a percentage value. For example, 34.4% of patients who died of lung cancer had metastases in the parenchyma of the brain.

Leptomeningeal involvement by carcinoma cells was first described by Eberth (1870) under the name of "endothelioma of the pia", but its metastatic nature was not clarified. The term "meningeal carcinomatosis" was first adopted by Beerman (1912) for diffuse leptomeningeal involvement by carcinoma cells. The ratio of metastatic leptomeningeal tumours is about 7.1% of all metastatic intracranial tumours collected by neurosurgical clinics (Takakura, *et al.*, 1982). Two-thirds of these were due to leukaemia, and others were due to stomach cancer, retinoblastoma, breast cancer, malignant lymphoma and other miscellaneous malignancies.

Histological Diagnosis

The histological diagnosis of metastatic brain tumours depends on the primary source. In Table 9.4, statistical data of the histological diagnosis are shown. More than 50% of the tumours were due to adenocarcinoma. These data were, however, accumulated from neurosurgical clinics and may be slightly different from the autopsied cases.

Age and Sex

The highest incidence of metastatic brain tumours was found between the ages of 55 and 65 years in males and between 45 and 60 years in females, the latter mainly due to the effects of breast cancer (Fig. 9.1).

Table 9.4 Histological diagnosis of the metastatic brain tumours (From The Committee of Brain Tumor Registry in Japan, 1987)

Histology	No. of cases
Adenocarcinoma	1689 (54.5%)
Squamous cell carcinoma	467 (15.1%)
Undifferentiated carcinoma	205 (6.6%)
Oat cell carcinoma	61 (2.0%)
Transitional carcinoma	29 (0.9%)
Acute myelocytic leukaemia	8 (0.3%)
Acute lymphocytic leukaemia	11 (0.4%)
Chronic myelocytic leukaemia	3 (0.1%)
Chronic lymphocytic leukaemia	2 (0.1%)
Other leukaemias	3 (0.1%)
Sarcoma	82 (2.6%)
Retinoblastoma	13 (0.4%)
Neuroblastoma	16 (0.5%)
Melanoma	42 (1.4%)
Malignant lymphoma	51 (1.6%)
Others	417 (13.5%)
Total	3099 (100.0)

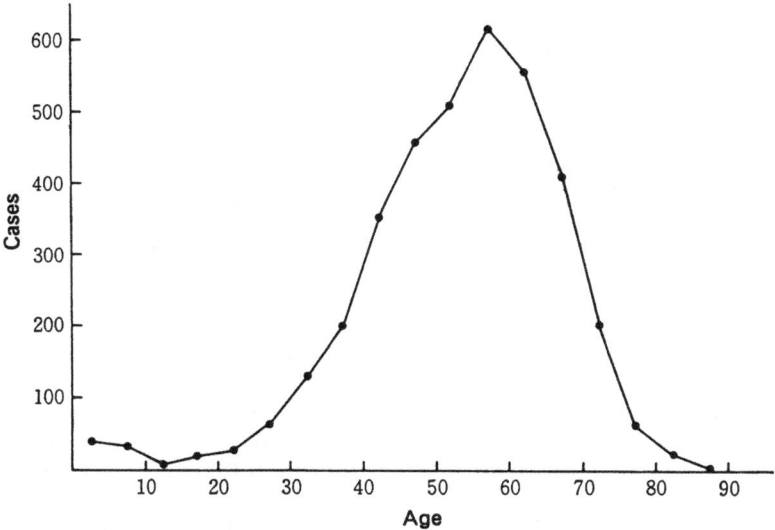

Fig. 9.1 Age distribution of intracranial metastases of malignant neoplasms (3709 cases) (From The Committee of Brain Tumor Registry in Japan, 1987)

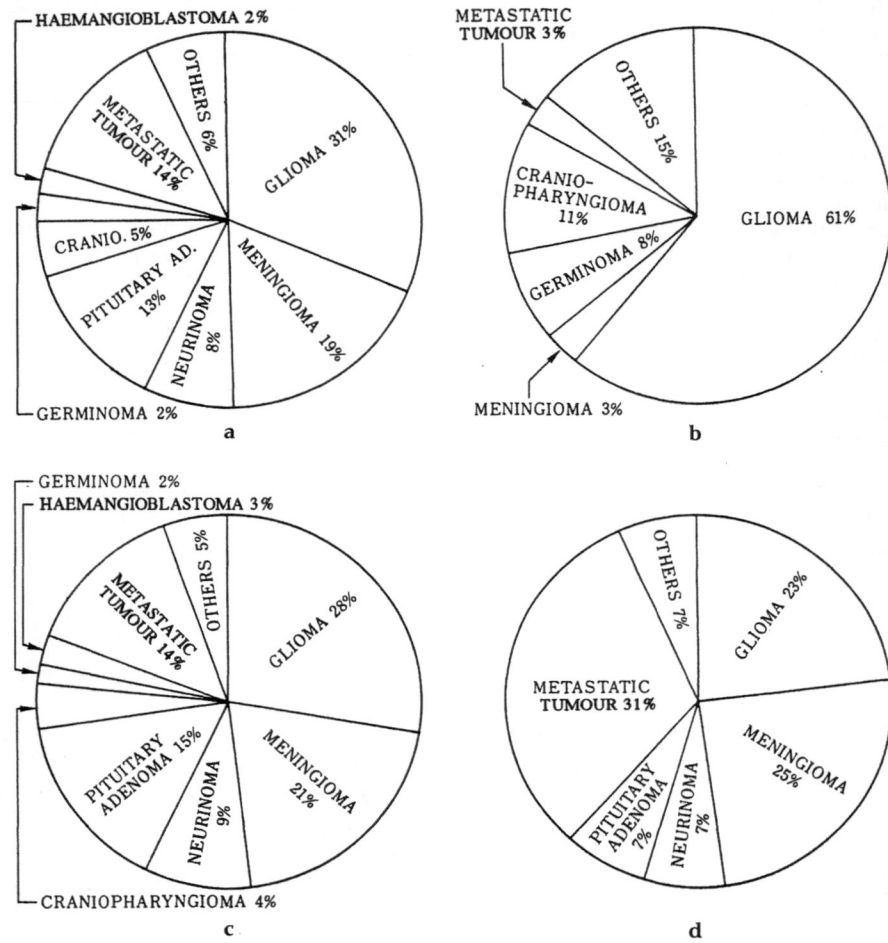

Fig. 9.2 Frequency of metastatic and major primary brain tumours. **a** All ages (26 527 cases). **b** Children (0–14 years old) (3243 cases). **c** Adult (15–64 years old) (21 062 cases). **d** Aged (over 65 years old) (2231 cases) (From The Committee of Brain Tumor Registry in Japan, 1987)

There is a small peak in childhood. The peak incidence of metastatic brain tumours in children was found in those under 9 years of age. The main intracranial metastatic tumours in children are leukaemia, retinoblastoma, neuroblastoma and various sarcomas. The period between 15 and 19 years is the most resistant to or healthy with respect to cancer. The increased number of metastatic brain tumours in people over 65 has been clearly observed in recent years (Fig. 9.2).

Regarding sex difference, 56.8% of 3709 cases of metastatic brain tumours were male, and 43.3% female. In cases of lung cancer, 77% were male and 23% female. The age distribution of lung cancer metastasizing to the brain ranged from 25 to 78 years, and its peak incidence was between the ages of 55 and 57 years in male and between 45 and 49 years in female patients. In cases of oesophageal, rectal cancers and various sarcomas, brain metastases were five to six times more frequently encountered in men than in women, while all the metastatic breast cancers were found in females.

Signs and Symptoms

Neurological Symptoms

Neurological signs and symptoms appearing in the initial or early stages of metastatic brain tumours are summarized in Table 9.5. Headache is the most frequent symptom and it was reported by 58% of the patients. The headache does not, however, always reflect the severity of increased intracranial pressure. Meningeal involvement by the tumour intensifies the headache. It is often recognized that a headache may subside partially or completely even though the intracranial mass grows. This may be due to disturbance of consciousness or depression of pain-perceptive mechanisms. Nausea, unilateral motor weakness, mental and behavioural changes are common symptoms and signs. Epileptic seizure is an especially important initial symptom, and when once complained of by a patient with a history of cancer, the diagnosis of metastatic brain tumour is most likely. Focal seizures, especially Jacksonian seizures, are characteristic in patients with metastatic brain tumours. Post-ictal unilateral weakness usually subsides in a few hours or a few days after the seizure. When seizures repeat or continue, the motor and sensory deficits may persist. It might be introduced by focal histological change of the brain due to severe cerebral oedema, ischaemia or haemorrhage adjacent to the tumour. Any seizure should be

Table 9.5   Neurological signs and symptoms of metastatic brain tumours

Signs and symptoms	Per cent of cases
Headache	58
Nausea	40
Vomiting	31
Unilateral motor weakness	39
Mental and behavioural changes	22
Seizure	19
Visual disturbance	12
Diplopia	8
Aphasia	10
Dizziness or vertigo	5
Ataxia	5
Visual field loss	2
Impairment of cranial nerves	
III	4
V	5
VI	7
VII	4
VIII	3
Papilloedema	31

treated promptly by administration of anticonvulsants, glucocorticoids and mannitol or glycerol. Impairment of cranial nerve function, especially visual disturbance or diplopia, is often noted in brain secondaries.

In cases of metastatic leptomeningeal tumours, the most common initial symptom is headache, which appears in two-thirds of patients. The headache is generally very severe and accompanied by nausea and vomiting. Since the tumour cells often involve the cranial and spinal leptomeninges at the same time, pain in the extremities or lumbago is also a frequent initial symptom. Mental deterioration, such as confusion, lethargy, memory disturbance and disorientation, is not a common initial symptom, but is generally encountered in patients in more advanced stages. Epileptic seizures are less common compared to the brain parenchymal metastasis. In cases of cranial nerve involvement, diplopia is the most common symptom. Papilloedema is noted in one-third of the patients.

Time of Onset

Metastatic brain tumours are generally found after the diagnosis and treatment of the primary tumour. The interval from the diagnosis of the primary tumour to that of the brain metastasis varies widely and the statistical data are shown in Table 9.6. The metastatic tumours were found in 28% of cases within 6 months of the diagnosis of the primary tumors, 14% from 6 to 12 months, 15% from 1 to 2 years and 15% between 2 and 5 years. Some of the metastatic brain tumours were found after an even longer interval from the primary diagnosis. The longest duration in the author's cases of breast cancer was 16 years after mastectomy.

Intracranial metastases appear most rapidly in cases of lung cancer because of the anatomical structure of the vascular system. The average duration between the diagnosis of primary lung cancer and the appearance of neurological symptoms due to intracranial metastasis was 7.2 months. The longest duration for lung cancer in the author's cases was $7\frac{1}{2}$ years. On the other

Table 9.6 Interval from the diagnosis of primary neoplasm to that of metastatic brain tumours

Interval	No. of patients
<1 month*	506 (19.9%)
1–3 months	483 (19.0%)
4–6	233 (9.1%)
6–12	372 (14.6%)
1–2 years	375 (14.8%)
2–3	199 (7.8%)
3–5	178 (6.9%)
5–8	115 (4.5%)
>8	87 (3.4%)
Total	2548 (100.0)

* Those metastatic brain tumours were found at the same time or prior to the detection of primary neoplasms.

hand, in the cases of breast cancer, the average duration was 42.4 months, considerably longer than that of lung cancer. Constans (1974) reported that the longest interval in breast cancer was 23 years after mastectomy.

There are two forms of onset of neurological symptoms due to intracranial metastasis. In the acute type, neurological symptoms develop rapidly, with epileptic seizures or the sudden onset of headaches or disturbances of consciousness due to intracranial haemorrhage. In the progressive type, which is the more common for the development of metastatic brain tumours, various symptoms including headache and nausea develop gradually. In the author's series, more than 60% of the cases were of the progressive type.

In 20% of all the cases, metastatic brain tumours were detected before or at the same time as the diagnosis of the primary tumour. In neurosurgical clinics, the differential diagnosis of metastatic and primary brain tumours is sometimes difficult before histological confirmation of the intracranial tumour. The precise examination of a chest X-ray before a brain tumour operation is very important for the diagnosis of metastatic brain tumours.

Investigations

Computed Tomography Scan

Computed tomography (CT) scan is the essential examination in metastatic brain tumours. Contrast enhancement is important for the accurate diagnosis of the tumour. The smallest tumour nodule detectable by CT scan is approximately 2–3 mm in diameter. It does not contain necrotic tissue inside when its size is less than 1 cm in diameter. The tumour nodule grows further in a spheroid or irregular nodulous shape. A degenerative change starts in the centre of the tumour tissue as the tumour grows and the blood supply becomes insufficient at the centre of the nodule. The necrotic portion appears on CT scans as a low-density area surrounded by a thin or thick high-density rim. It often looks like a ring or a doughnut in shape. The ring formation is one of the common appearances of metastatic brain tumours on CT scans. The homogeneous or heterogeneous irregular features of the interior portion of the tumour nodule are dependent on the liquefaction of the necrotic tissue. It contains solid necrotic tissue in some cases and viscous pus-like fluid or xanthochromic serous fluid in other cases. When haemorrhage occurs in the tumour, it contains haematoma or bloody fluid. Haemorrhage occurs quite often in cases of hepatoma and melanoma.

In cases of metastatic brain tumours, the extensive brain oedema surrounding the tumour is a typical characteristic. The brain oedema appears as a low-density area in the CT scans. The swelling of the brain and the shift and deformity of ventricular structures are also noted in most cases. The invasion by metastatic tumours of the skull base or calvarium is demonstrated by plain X-ray films or CT scans (Fig. 9.5). The recent three-dimensional display of CT scans can demonstrate well the structure of bone defects in the skull base. Typical examples of CT scans of metastatic brain tumours are demonstrated in Figs 9.3–9.5.

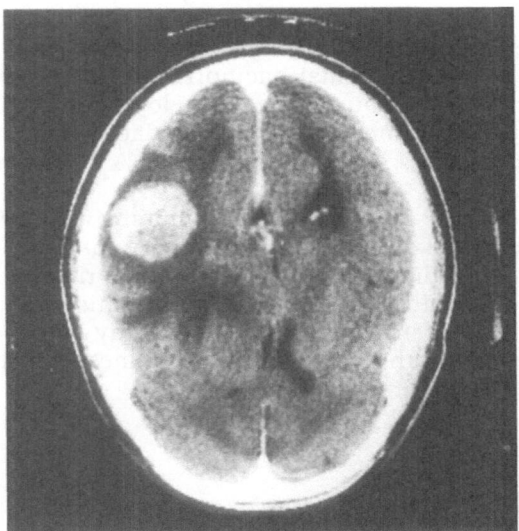

Fig. 9.3   CT scan (contrast enhancement) of metastatic solid tumour from lung cancer

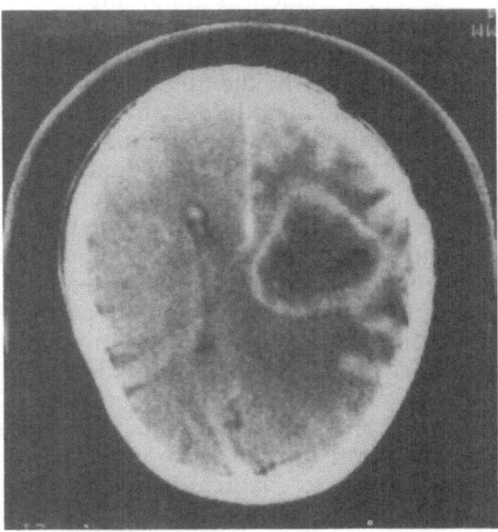

Fig. 9.4   CT scan (contrast enhancement) of metastatic cystic tumour from lung cancer

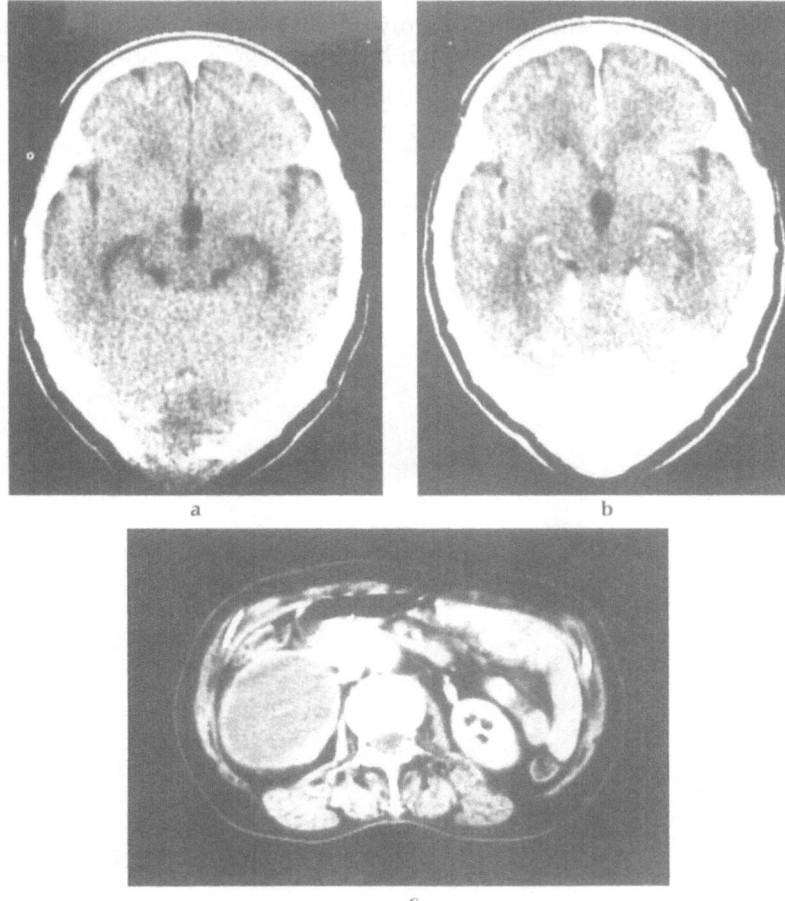

Fig. 9.5   CT scans of metastatic tumour from renal clear-cell carcinoma. **a** Non-enhanced CT scan. **b** Contrast-enhanced CT scan. **c** CT scan of the abdomen. Tumours are shown by arrows

Magnetic Resonance Imaging

Magnetic resonance imaging (MRI) can demonstrate the three-dimensional structure of the tumour better than CT scans (Figs 9.6–9.9). Metastatic tumour generally appears as a low signal intensity mass in T_1-weighted images and high signal intensity mass in T_2-weighted images. Gadolinium DTPA (Gd) enhancement can demonstrate more definitely the shape of the tumour, especially in tiny metastatic nodules. The area and extension of brain oedema is demonstrated well in T_2-weighted images, much better than in CT scans. When haemorrhage is present, it appears as

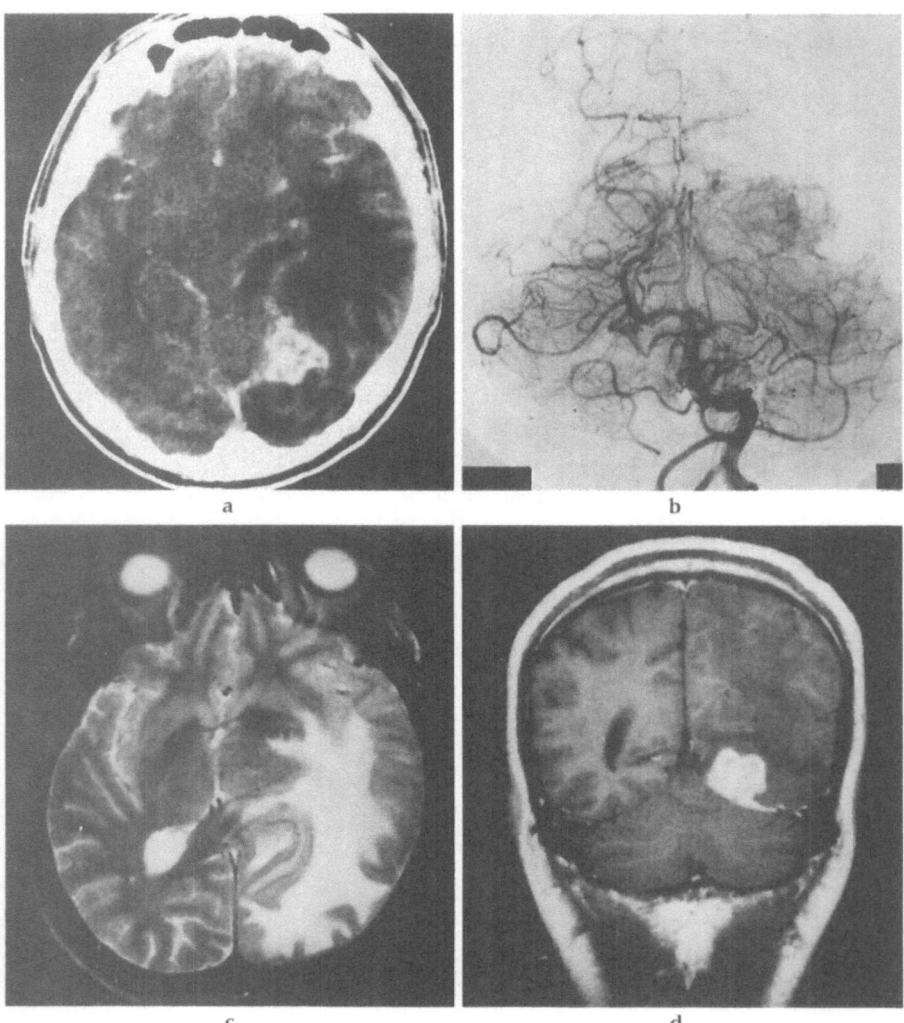

a

b

c

d

Fig. 9.6   Metastatic brain tumour from lung cancer (44-year-old male) in the left occipital lobe. **a** Contrast-enhanced CT scan. **b** Vertebral angiography. **c** T_2-weighted MRI. **d** Gd-DTPA enhanced T_1-weighted MRI

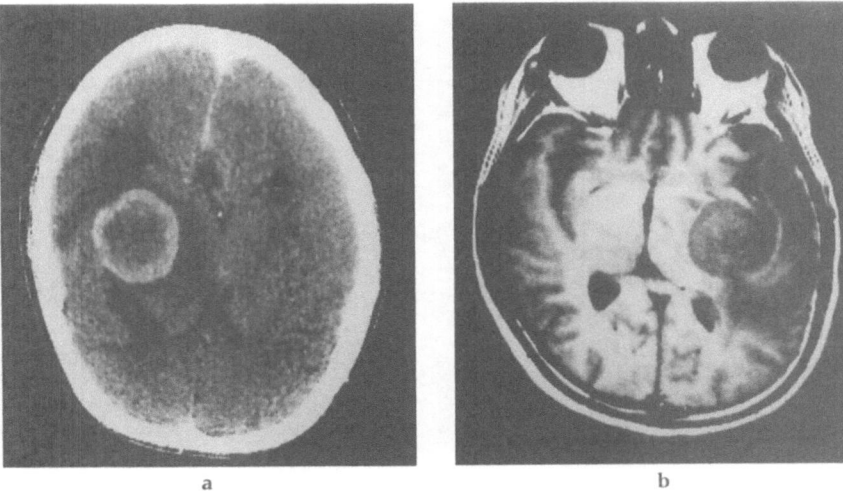

Fig. 9.7 Metastatic brain tumour from sigmoid cancer (38-year-old male) in the right parietal lobe. **a** Contrast-enhanced CT scan. **b** T_1-weighted (TR 600 ms, TE 17 ms) MRI

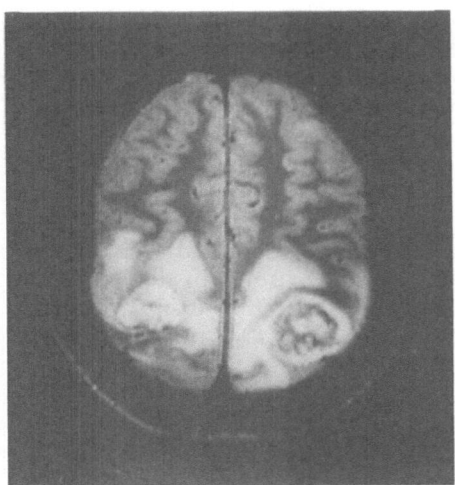

Fig. 9.8 Double metastatic brain tumours with haemorrhage in both parietal lobes (55-year-old male) from hepatoma. T_2-weighted (TR 3000 ms, TE 28 ms) MRI

a high signal intensity area both in T_1 and T_2 images in its subacute or chronic stages.

The invasion of metastatic tumours into the leptomeninges is clearly demonstrated by Gd-enhanced MRI.

Angiography

For the diagnosis of metastatic brain tumours, angiography is no longer necessary. Angiography may be necessary in some cases for safe management of

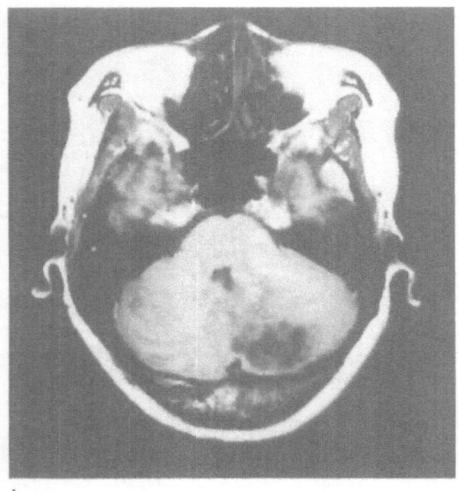

Fig. 9.9 Metastatic brain tumour in the left cerebellar hemisphere from breast cancer (47-year-old female). T_1 weighted (TR 500 ms, TE 17 ms) MRI

the surgical removal of the tumour. It is advisable to use digital angiography to minimize the risks to patients, as they are generally debilitated.

Tumour vascularization varies according to the histological type of the tumour. Extensive tumour vascular stain in the angiogram is commonly seen in cases of renal clear-cell carcinoma, melanoma, hepatoma, choriocarcinoma and several types of sarcomas. These tumours have a tendency of haemorrhage into the tumour, forming a haematoma. However, lung and breast secondaries do not generally show extensive hypervascularization.

Electroencephalography

Patients with metastatic brain tumours often demonstrate epileptic seizures. Electroencephalography (EEG) may be useful in detecting and controlling such seizures. Proper management under EEG examination can help to prevent seizures by the administration of anticonvulsants with monitoring of serum levels of the drug.

Cytological Examination of Cerebrospinal Fluid

Cytological examination of the cerebrospinal fluid (CSF) is only required when leptomeningeal involvement by tumour cells, such as in cases of leukaemia, is suspected. In general, spinal tap is contraindicated in patients whose intracranial pressure is increasing. In cases of meningeal involvement by leukaemia, the CSF examination may conclusively demonstrate tumour cells and the spinal tap may also be used to administer anti-cancer drugs, such as methotrexate intrathecally. To increase the efficacy of this as a diagnostic method, cell culture examination is recommended in addition to simple cytological examination.

Tumour Markers

Substances clinically important as tumour markers are various fetal proteins, hormones and some enzymes. Although they are not specific to the tumour, the increase of such tumour marker proteins or polypeptides in the serum or CSF gives definite information about the diagnosis. The progression, the recurrence of the tumour and the therapeutic effect can be well monitored.

Some examples of tumour markers are human chorionic gonadotrophin (HCG) in cases of choriocarcinoma, alpha-fetoprotein (AFP) in hepatoma, carcinoembryonic antigen (CEA) in colon cancer and CA-19-9 in pancreatic cancer. Lung or ovarian cancers sometimes produce various ectopic hormones (functional tumours).

Treatment

Since brain metastases are a manifestation of systemic cancer, the treatment requires not only neurosurgical management to reduce the tumour mass and intracranial pressure but also general control of the primary cancer. A multi-modality treatment including surgery, radiotherapy, chemotherapy and use of biological response modifiers is especially important for the control of the metastatic tumour. Once clinical manifestations of metastatic brain tumour appear, the fatal symptoms progress very rapidly. Without treatment, the patient cannot expect to survive more than a few months. Quick action in starting an appropriate therapy for each patient is necessary.

Surgery

Indication

The best indication for surgical treatment is a single and solitary brain metastasis located in a region accessible to surgical removal in the patient under complete control of the primary cancer. Even if a tumour is located in the subcortical area near the Sylvian fissure, the tumour can usually be removed from either the anterior or posterior side of the Sylvian fissure without any major postoperative damage to the motor or sensory function. The patient generally demonstrates good recovery from the pre-existing neurological signs and symptoms. On the other hand, when the tumour is located in the brain stem, there in no surgical indication. In cases of multiple metastases, if the major tumour nodules can be removed easily, the surgical removal of these will help further treatment with radiotherapy and chemotherapy.

Regarding the primary tumour and general condition of the patient, the indications for surgical treatment are summarized as follows:

1. The primary tumour has been cured or is likely to be cured. It can at least be properly controlled for a long period.
2. Good general condition of the patient, in terms of pulmonary, liver and renal functions.
3. No metastasis to other vital organs, in particular no metastasis to the liver or presence of uncontrollable tumours in the lung.

Preoperative Management

It is always necessary to reduce the increased intracranial pressure due to brain oedema associated with metastatic brain tumour. The administration of a steroid hormone (glucocorticoid) is essential. A suggested initial dosage of steroid hormone administration is 40–60 mg per day of prednisolone or 8–16 mg per day of dexamethasone. Dosage will be reduced depending on the patient's condition. Mannitol or glycerol intravenous infusion (200 ml at time), twice or three times a day, is also recommended. A dramatic improvement in the patient's condition is generally obtained. Anticonvulsants should be given orally. In severe cases, oxygen inhalation will help to reduce the increased intracranial pressure.

Surgical Technique

The metastatic tumour should be removed totally. The partial removal of the tumour will increase the brain oedema postoperatively and will not help the patient. Once the tumour border is recognized, the tumour will easily be separated from brain tissue by suction of the soft gliotic tissue surrounding the tumour. A cotton pledget is inserted between the tumour and brain tissue. The tumour is usually removed en bloc. When the tumour contains cyst, aspiration of the cystic fluid before tumour removal will help further manoeuvres.

Even when there are multiple metastatic tumours, it is sometimes worth while to remove all major tumour nodules, if possible.

Postoperative Care

An adequate oxygen supply and steroid hormone administration are necessary to reduce postoperative brain oedema. Mannitol or glycerol intravenous administration is recommended for a few days. Complications generally do not occur if the tumour is macroscopically totally extirpated.

Radiotherapy

Lenz and Freid first reported in 1931 the results of radiotherapy in five patients suffering from metastatic brain tumours originating from carcinoma of the breast (Lenz and Freid, 1931). Three of them showed good palliative improvement of neurological symptoms up to 20 months. Although radiotherapy is definitely effective in metastatic brain tumours, this efficacy depends on the biological characteristics of the tumour. Malignant lymphoma, leukaemia and some sarcomas are quite sensitive to radiotherapy. However, adenocarcinoma or squamous cell carcinoma are not so sensitive to simple radiotherapy. A linear accelerator is used in most cases. Combined chemotherapy enhances the effect of radiotherapy in many cases. In our clinic, ACNU and vincristine combined chemotherapy is given concomitantly with radiotherapy and tumours showed regression during radiotherapy in most cases. A total dose of 30–40 Gy is irradiated to the whole brain, and 20–30 Gy are added to the local site of metastatic tumours.

Chemotherapy

The selection of anti-cancer drugs depends on the sensitivity of the primary tumour. The drugs most often used are vincristine, vinblastine, cisplatinum, adriamycin, ACNU, BCNU, VM-16 and so on. Leucocytopenia is cured by the administration of granulocyte colony stimulating factor (G-CSF). If liver dysfunction appears, the chemotherapy should be stopped. In cases of geriatric patients, there should be more caution in the use of anti-cancer drugs, since those patients frequently develop liver dysfunction, thrombocytopenia and other side effects.

Biological Response Modifiers

The effect of biological response modifiers (BRM) on metastatic brain tumours has not been fully evaluated. The enhancement of cellular immunity is, however, expected to suppress the tumour growth and to prolong the survival time of the patients (Takakura et al., 1982). Some examples of such BRM are BCG, Krestin (plant polysaccharide), Picibanil (detoxicated *Streptococcus haemolyticus* SV strain), interferons and interleukin II.

Outcome of Treatment

The overall outlook for the treatment of metastatic brain tumours is still quite pessimistic. According to the Japanese Brain Tumor Registry, the average survival rates of 573 cases of metastatic brain tumours are 32.5% at 1 year, 17.2% at 2 years, 10.4% at 5 years and 7.1% at 10 years. The best survival rate was recorded in cases of metastatic brain tumours from uterine carcinoma, where the 5-year survival rate was 23.8% (93 cases). The 5-year survival rates of metastatic brain tumours were 4.6% for lung cancer (1229 cases), 7.3% for breast cancer (280 cases), 7.6% for gastrointestinal cancer, 12.9% for renal carcinoma and 19.1% for head and neck cancers. The factors influencing the prognosis of metastatic brain tumours are (a) primary tumour, organs, histology, (b) metastasis to another organ, (c) solitary or multiple metastases, (d) clinical grading, (e) extent of surgical removal, (f) radiotherapy, (g) chemotherapy and (h) therapy with BRM.

In the author's series, the longest survival time is more than 17 years after craniotomy for metastatic brain tumour from uterine carcinoma. Patients of choriocarcinoma and renal clear cell carcinoma have survived more than 16 and 13 years, respectively. Since metastatic brain tumour is one of the pathological conditions of disseminated cancer in the advanced stage, the outcome of treatment is still very poor compared with primary brain tumours. The diagnostic method and multimodality treatment with surgery, radiotherapy and chemotherapy have, however, greatly progressed recently and the outcome is gradually improving. The treatment of metastatic brain tumours will be more and more important in the coming years, as part of the strategy of cancer treatment in general.

References

Beerman WF (1912) Meningeal carcinomatosis. J Am Med Ass 58:1437–1439

Chu FCH, Hilaris BB (1961) Value of radiation therapy in the management of intracranial metastasis. Cancer 14: 577–581

Constans JP, Cioloca C (1974) Chimiotherapie des metastases cérébrales. Neurochirurgie (Paris) 20(Suppl 2): 212–219

Earle KM (1954) Metastatic and primary intracranial tumors of the adult male. J Neuropathol Exp Neurol 13: 448–454

Eberth CJ (1870) Zur Entwicklung des Epithelioms (Cholesteatomas) der Pia und der Lunge. Virchows Arch 49: 51–63

Lenz M, Freid JR (1971) Metastases to the skeleton, brain and spinal cord from cancer of the breast and the effect of radiotherapy. Ann Surg 93: 278–293

Lesse S, Netsky MG (1954) Metastasis of neoplasms to the central nervous system and meninges. Arch Neurol Psychiat 72: 133–53

Nisce LZ, Hilaris BS, Chu FCH (1971) A review of experience with irradiation of brain metastasis. Am J Roentgenol Radium Ther Nucl Med 3: 329–333

Perese DM (1959) Prognosis in metastatic tumors of the brain and the skull: an analysis of 16 operative and 162 autopsied cases. Cancer 12: 609–613

Richard R, McKissock W (1963) Intracranial metastases. Br Med J 1: 15–18

Simionescu MD (1960) Metastatic tumors of the brain. A follow-up study of 195 patients with neurosurgical considerations. J Neurosurg 17: 361–373

Störtebecker TP (1954) Metastatic tumors of the brain from a neurosurgical point of view. A follow-up study of 158 cases. J Neurosurg 11: 84–11

Takakura K, Sano K, Hojo S, Hirano A (1982) Metastatic Tumors of the Central Nervous System. Igaku Shoin, Tokyo

The Committee of Brain Tumor Registry in Japan (1987) Brain Tumor Registry in Japan, Vol. 6. National Cancer Center

10 Diagnostic Imaging of Brain Tumours

Brian E. Kendall and Jill V. Hunter

Introduction

The management of patients with suspected or confirmed intracranial tumour may be considered in several stages. In practice these are integrated and combined and include:

1. Detection or confirmation that a structural abnormality is present.
2. Localization and assessment of the extent of any abnormality.
3. Characterization of the abnormality and, in particular, the distinction of tumours from non-tumour processes.
4. Assessment of the nature of any tumour.
5. Facilitation of additional diagnostic procedures, and planning for surgery or other types of therapy.
6. Monitoring of response to therapy:
   a) assessment as to whether or not therapy is adequate,
   b) indications for the additional therapy, and
   c) documentation of complications and their response to appropriate treatment.

Imaging may be indicated at any of these stages and usually has a significant role in each of them.

Detection of an Abnormality

Computed Tomography and Magnetic Resonance Imaging

Computed X-ray tomography (CT) as a screening method for the demonstration of a supratentorial abnormality, supplemented as necessary by further CT after intravenous injection of contrast material, is a very accurate and freely available diagnostic procedure. For this reason, and because it often gives sufficiently specific information for management and is only minimally invasive, it is considered as the basic radiological study. However, bone-induced artefacts and partial volume effects impose limitations on the accuracy of CT in lesions close to the skull base and in particular in the posterior fossa. If a tumour in these situations is suspected, magnetic resonance imaging (MRI) is a more appropriate primary investigation (Brant-Zawadzki *et al.*, 1984a; Holland *et al.*, 1985a) – if availability permits. Otherwise, MRI is used as a supplementary study when CT is equivocal or negative in cases with high clinical suspicion (Brant-Zawadzki *et al.*, 1984b).

These imaging methods may show an abnormality in two fundamentally different ways: (a) revealing attenuation on CT or signal return on MRI which is outside the range for normal tissues; (b) showing deformity and/or displacement of normal anatomy, including the visible grey structures and white matter tracts within the brain substance and the intracranial spaces filled with cerebrospinal fluid (CSF). Intracranial tumours usually exhibit a combination of these findings.

Abnormal attenuation or signal change may be due to the tumour itself or to secondary changes within it, to reactive oedema or other changes within the cerebral substance induced by the tumour, and to abnormal contrast enhancement within the tumour.

The physical properties of various tumour tissues may be similar to those of brain substance or they may be markedly different. The physical properties of tissues determine their X-ray absorption characteristics and hence the differential attenuation of the X-ray beam in tumour and brain substance. In general, but by no means invariably, compact cell masses increase attenuation and appear dense, whereas sparsely cellular tumours tend to be of lower density than brain substance.

A variety of MRI sequences is available in which the influence of proton density, T_1, T_2, susceptibility effects and blood flow may be emphasized. In general, intracerebral tumours tend to have prolongation of both T_1 and T_2, due in large part to their high free-water content and surrounding oedema. Astrocytomas, neurinomas and metastatic tumours show longer T_1 values than meningiomas. Lipomas have the shortest T_1 values and together with melanoma and subacute haemorrhage demonstrate shortening, producing high signal on T_1-weighted (T_1W) sequences (Araki *et al.*, 1984).

While the two investigative methods may be complementary, MRI does offer advantages over CT, including its sensitivity in the demonstration of oedema. This sensitivity of MRI leads to the earlier detection of tumours, such as glioma, and the more accurate definition of the extent of surrounding oedema – it is well recognized that even tumours demonstrating clear marginal definition have microscopic infiltration into the surrounding oedema. Magnetic resonance is able to offer superior depiction of anatomical relationships, for example to vessels and the ventricles, especially in the posterior fossa where there is little bony artefact compared with CT. This is aided by the ease of imaging in any plane without radiation hazard. Magnetic resonance imaging has been shown to offer a more accurate and rapid method for planning stereotaxic surgery (Le Bas *et al.*, 1987) and radiotherapy. However, the use of MRI is rarely crucial in the detection of *supratentorial* tumours.

Fluid

Fluid in cystic and necrotic regions is usually but not always of diminished attenuation on CT. On MRI the signal return depends on the composition of the fluid, with increasing T_1 and T_2 as it more closely approaches that of

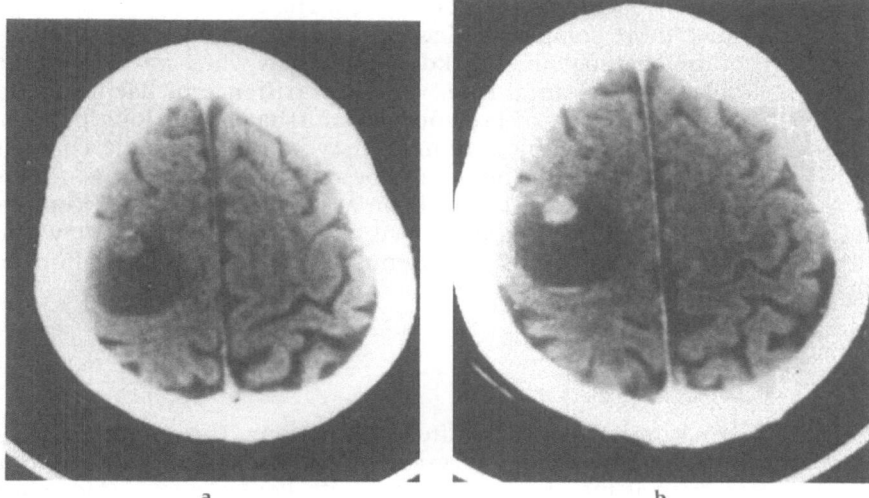

a   b

Fig. 10.1 A pair of CT axial images before (**a**) and after (**b**) intravenous contrast medium. **a** A low-density high right parietal lesion containing a nodule in its anterior wall with a fleck of calcium. Following intravenous contrast, **b** demonstrates enhancement of the nodule with layering of contrast confirming the presence of a cystic cavity. This was proved to be a cystic glioma

CSF. On both CT and MRI, fluid generally appears homogeneous, and the presence of a fluid level is the only definitive indicator of a necrotic or cystic cavity. Such a level is usually between cyst fluid and debris which may be limy, but there may be layering of fat, blood, air or, occasionally, contrast medium on enhanced scans (Fig. 10.1). As a general rule, the attenuation of, or signal returned from, fluid is unchanged between plain and enhanced scans.

Calcification

Calcification, depending on its concentration, causes increased CT attenuation, varying from the upper normal range for soft tissue, 45 Hounsfield units (HU), to more then 200 HU. On MRI, calcium returns no signal (Holland *et al.*, 1985b). However, where tissues are only partly calcified, there is often sufficient signal returned from protons in the uncalcified element for partial volume effects to mask the calcification (Oot *et al.*, 1986). When signal void is apparent, factors other than calcification, such as the presence of iron giving a susceptibility effect, may need to be considered, and though in general this can be resolved, interpretation is not usually as straightforward as with CT.

Haemorrhage

Complicating features may affect the appearance of the tumour; recent haemorrhage may increase CT attenuation up to almost 100 HU, decreasing with time as the haemoglobin is denatured. The effects of haemorrhage on MRI are much more complex, reflecting the susceptibility and paramagnetic effects of the haemoglobin breakdown products which vary with time (Fig. 10.2; Table 10.1) (Gomori et al., 1985). Persistence of haemosiderin may testify to the presence of previous haemorrhage over long periods. Occlusion of blood vessels by a tumour may cause ischaemic changes within the tumour or adjacent brain, and be reflected in local low attenuation on CT or prolongation of T_1 and T_2 on MRI. Variations in the composition of tumour tissue itself, and the effects of any complications, may lead to the abnormality having mixed CT attenuation and/or local variations in MRI signal returned.

Oedema

The brain surrounding a tumour is often oedematous. The amount of perifocal oedema is usually relatively greater with malignant intracerebral tumours. However, it may be pronounced with benign and extracerebral tumours; this occurs, for example, in about 10% of meningiomas. The increased water content of oedematous brain is reflected in diminished CT attenuation and relatively high signal on T_2W and low signal on T_1W sequences. The oedema fluid seeps through the extracellular spaces and increases the interstitial pressure within them. This tends to cause distension, which varies in different regions, depending at least partly on the size of the spaces and the compliance of the tissue. Peritumoral or vasogenic oedema tends to have sharp margins and to spread along certain white fibre tracts. The grey matter is involved to a much lesser extent, and vasogenic oedema differs in this respect from ischaemic oedema, in which both grey and white matter are involved in the affected region (Fig. 10.3). The increased sensitivity of MRI shows the oedema to be more extensive, but the involvement of grey matter is more apparent, and the distinction between vasogenic and ischaemic oedema is not so pronounced. Several studies have demonstrated difficulty, with the use of MRI, in distinguishing between tumour and oedema, despite the use of heavily T_1W or T_2W sequences, and even with the use of gadolinium enhancement (Araki et al., 1984; Rinck et al., 1985).

Contrast Enhancement

The complex compounds of iodine used for enhancement in CT and the gadolinium chelates used in MRI do not pass across the normal blood–brain barrier (BBB). Many neoplasms, including most malignant tumours, allow interstitial passage of contrast media across the BBB. The BBB may be damaged by many other pathologies, including infarction, infection and contusion. The degree of enhancement on CT is linearly proportional to the concentration of iodine within the tissue, each milligram causing an increase in density equivalent to 26 HU. The blood volume in

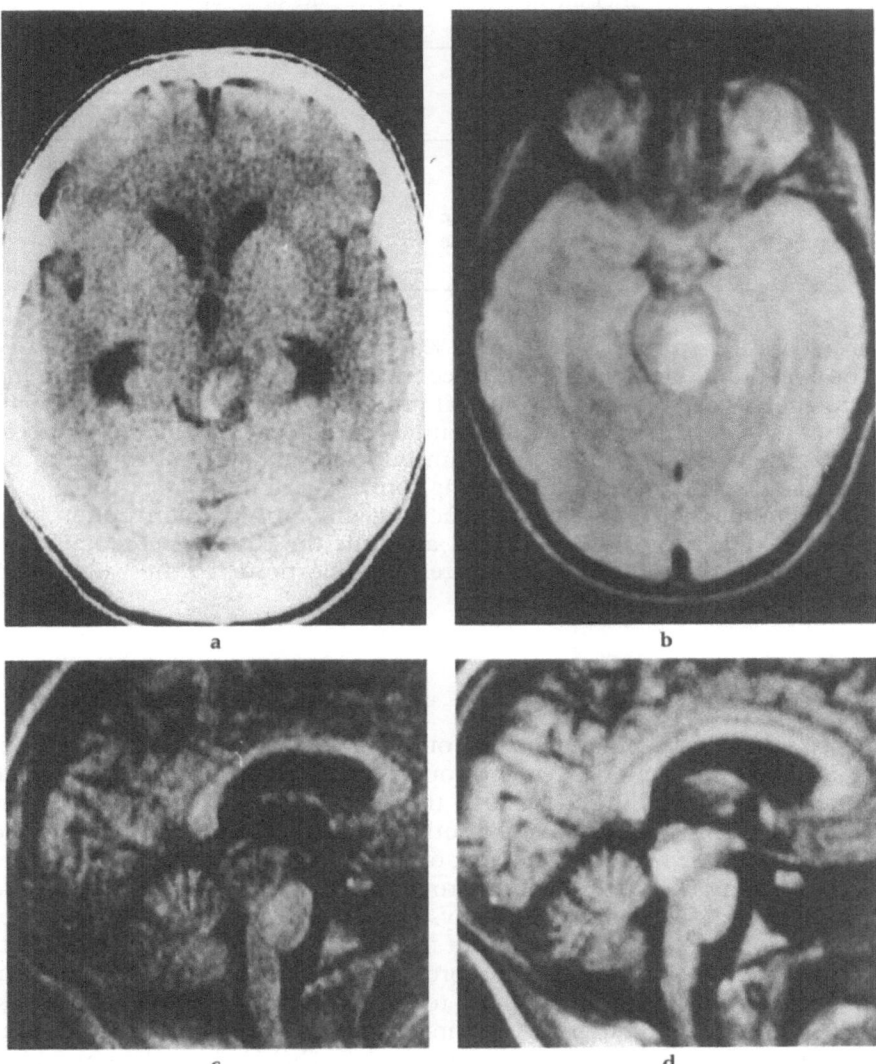

Fig. 10.2 A series of images demonstrating the changing signal characteristics on MRI of haemorrhage. Image **a** within 24 h of onset demonstrates high density to the left of the midline in the region of the quadrigeminal plate with surrounding low density causing mass effect and obstruction to the posterior end of the third ventricle with mild hydrocephalus. Image **b** axial 2000/80 sequence (acute within 48 h) reveals high signal due to partial volume from vasogenic oedema with some central lower signal. Image **c** sagittal IR 2100/500 sequence (acute) demonstrates deoxyhaemoglobin in red blood cells (isointense with brain) with reactive oedema returning lower signal then brain (longer T_1). Image **d** sagittal IR 2100/500 (subacute) demonstrates increasing signal from the lession reflecting T_1 shortening, due to the presence of paramagnetic methaemoglobin released by the lysis of red blood cells

Table 10.1 The evolution of the MRI signal characteristics of intracerebral haemorrhage over time

	T_1-weighted sequences		T_2-weighted sequences	
	Early	Late	Early	Late
Centre	isointense	rising	low	rising
Periphery	isointense	rising	isointense	rising
Rim	isointense	isointense	isointense	falling
Adjacent white matter	isointense	isointense	high	falling

normal brain tissue varies between 2% and 5%, thus the increased density of normal brain due to perfusion by contrast material is very small, ranging up to about 4 HU. Many pathological processes, including malignant neoplasms, show considerable differential enhancement, mainly due to extravasation of the contrast medium but also, in some cases, due to an increased volume of blood, containing contrast medium.

Gadolinium chelates are paramagnetic agents, causing reduction in both T_1 and T_2 relaxation times. In small amounts the T_1 effect predominates; the signal increase on T_1W sequences is proportional to the gadolinium concentration (Stack *et al.*, 1988).

Mass Effect of Tumours

Part of an intrinsic tumour may be outlined within CSF, either because it arises on the surface or extends through the cerebral substance to invade or metastasize along the margins of the ventricles or subarachnoid spaces (Fig. 10.4). This may cause a smooth or irregular local or more diffuse deformity. It may result in widening of the septum pellucidum, or enlargement of emerging nerves or nerve roots. Such changes strongly suggest a malignant tumour. More commonly, the expanding tumour causes local compression and/or displacement of the adjacent brain substance, distorting the internal anatomy and compressing the CSF spaces. The neoplasm may cause hydrocephalus due to obstruction of fluid pathways. Hemispheric swelling expands the ipsilateral gyri against the dura, compressing and eventually effacing the sulci. The ipsilateral ventricle is compressed, and the medial surface of the hemisphere is displaced contralaterally and eventually herniated beneath the falx, which causes depression of the ventricular roof, so that on axial sections it is low relative to the normal side.

Temporal lobe herniation encroaches into and compresses the chiasmatic cistern and tends to displace the brain stem, with widening of the ipsilateral ambient cistern. Large herniae of the hippocampus and uncus cause lateral compression and sagittal elongation of the brain stem. Eventually, in both lateral and axial herniations, the chiasmatic and interpeduncular cisterns are obliterated. Herniation may cause ischaemic change due to com-

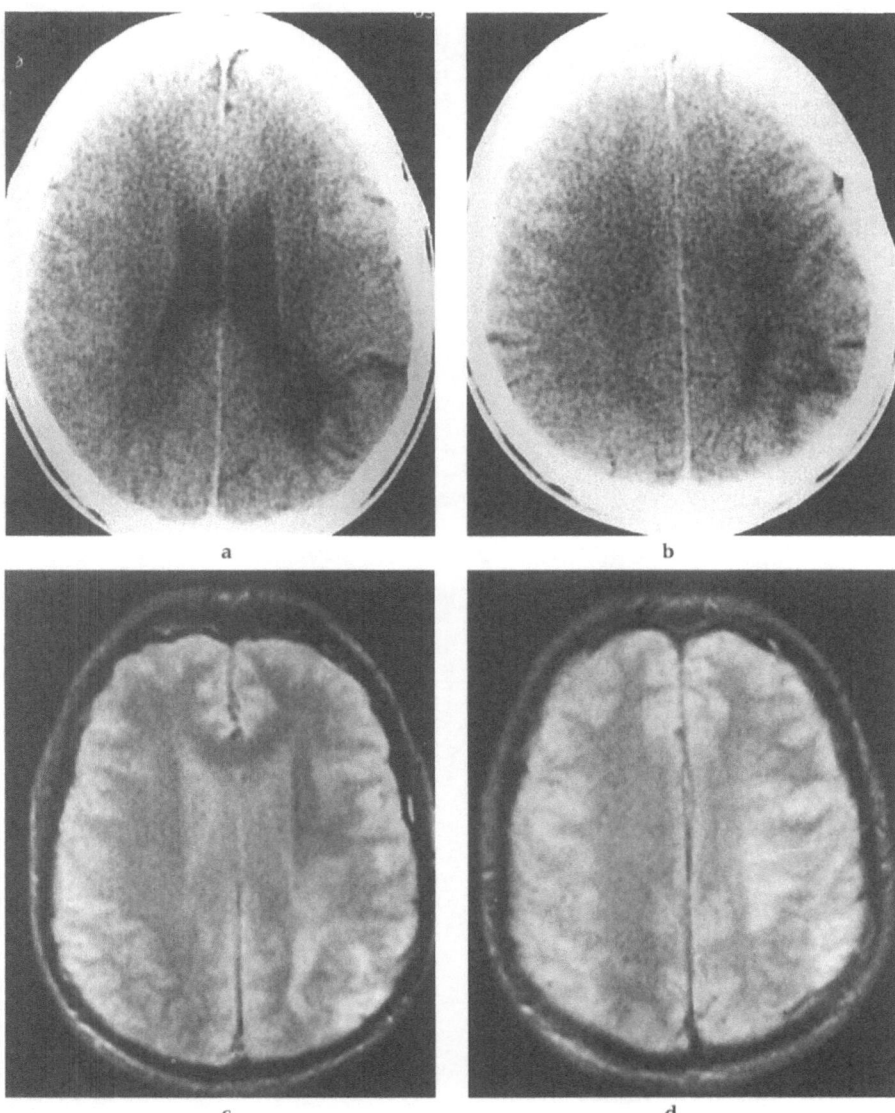

Fig. 10.3   A pair of enhanced axial CT slices (**a, b**) and the corresponding pair of axial T$_2$ weighted MRI scans (**c, d**). The examinations were performed within 24 h of each other in a 42-year-old male presenting with an acute onset of right-sided weakness. There are features of oedema with some mass effect in the involvement of both grey and white matter. Histology confirmed infarction in the left middle cerebral artery territory

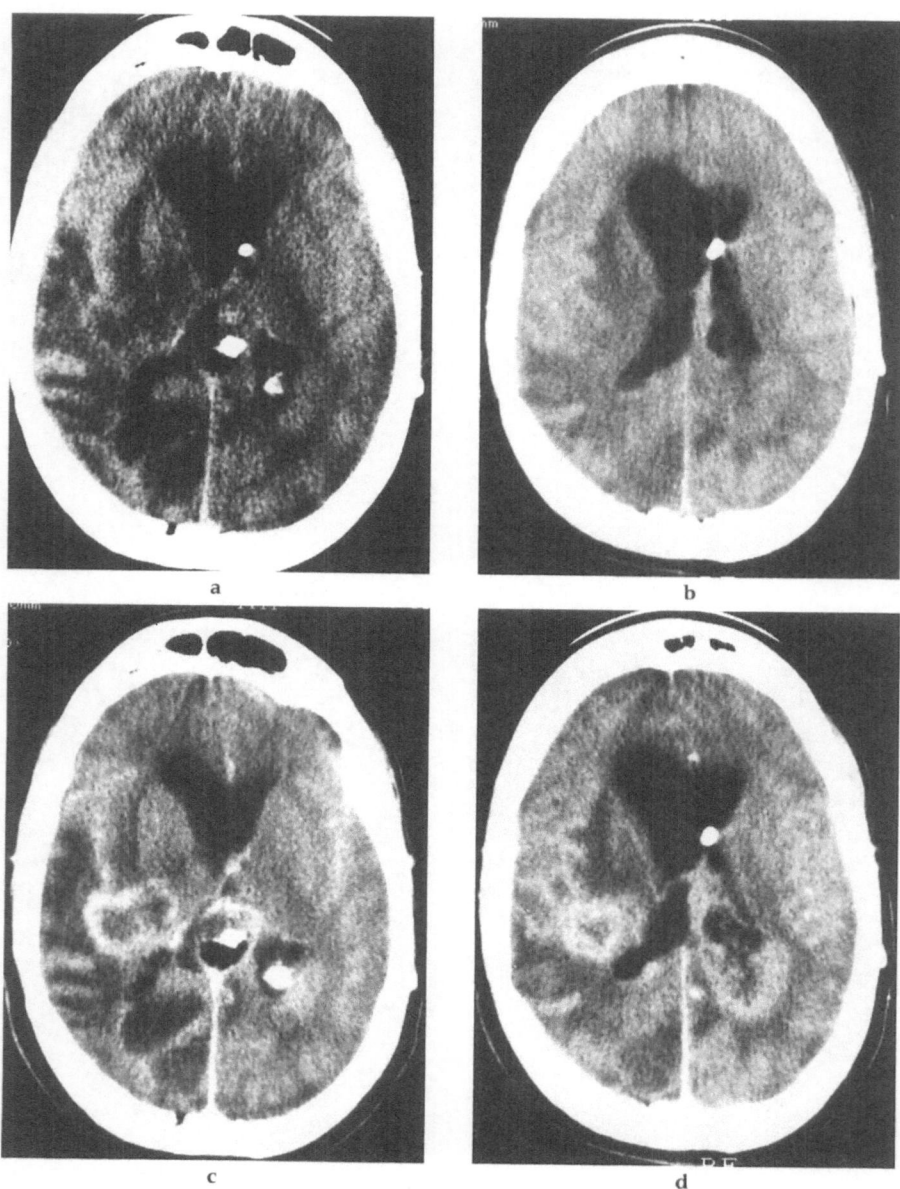

Fig. 10.4 Two pairs of axial CT images at the same levels obtained before (**a, b**) and after (**c, d**) intravenous contrast medium. Follow-up examination in a child shunted for hydrocephalus with removal of pineal teratoma. There is tumour recurrence in the pineal region and also in the right thalamus and posterior temporal regions. Periventricular spread of tumour around the trigones and posterior parts of the bodies of the lateral ventricles is shown

pression of branches of the posterior or anterior cerebral artery against the free edge of the tentorium or falx, respectively, resulting in infarction in corresponding brain substance.

An abnormality is revealed by CT and/or MRI at the time of presentation in almost all cases of intracerebral tumour (Fig. 10.5). This is a far higher detection rate than with any other imaging method. The overriding advantage of CT or MRI in the investigation of brain tumours, however, lies in the ability – in negative cases – to show that the brain anatomy is either normal or that non-space-occupying or atrophic processes are present which may clinically simulate tumours. No other procedure is capable of detecting or excluding a similar range of conditions, and no other study should be used in the routine screening for brain tumours. In particular, skull radiography are normal in over 85% of brain tumours and gamma scintigraphy in up to 50% overall. Moreover, these methods are nonspecific and are unable to demonstrate many of the conditions which enter into differential diagnosis and which *are* revealed by CT and/or MRI. A plain radiograph of the chest should, however, be made whenever metastases are likely. Gamma scintigraphy is a reasonable test for brain metastases in patients with known primary malignancy in those units without computed tomography and with free access to nuclear medicine facilities.

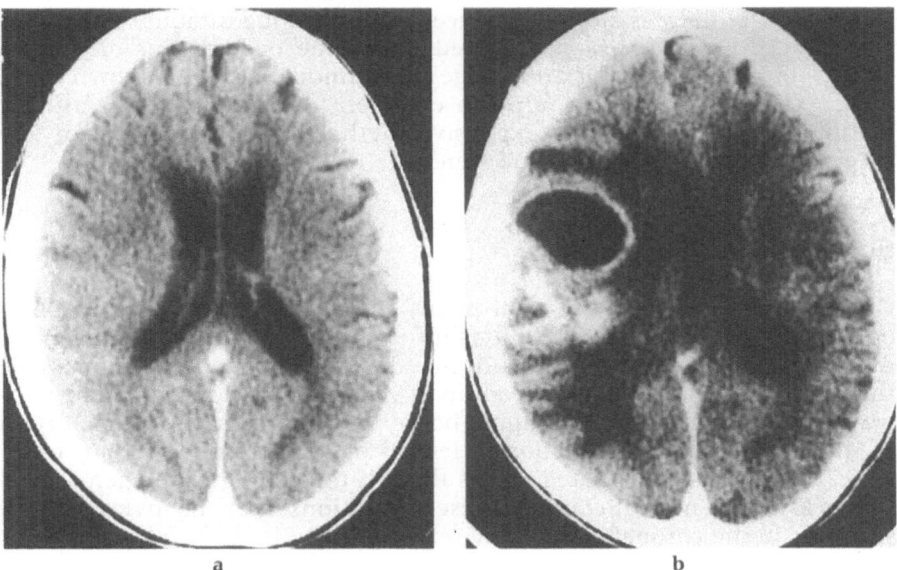

a   b

Fig. 10.5 Enhanced CT scans performed three months apart following initial presentation of a middle-aged female with a single fit. Minimal mass effect with effacement of right parietal sulci is noted on the original examination (**a**). The follow-up study (**b**) demonstrates the rapid progression of glioblastoma multiforme. There is now a partly diffuse and partly ring-enhancing mass with surrounding oedema compressing the right lateral ventricle and causing midline shift to the left

Localization and Assessment of Extent of Abnormality

In general, direct visualization of lesions on a computed tomogram allows accurate localization in relation to cerebral anatomy. The multiplanar imaging capacity of MRI is particularly useful for showing the topography and relationships of tumours, and for displaying extra-axial tumours or exophytic components of intra-axial masses.

On most machines it is now possible to indicate electronically on a scout view the position of a tumour identified on the axial sections. The facility of MRI to produce images in any plane allows exact positioning of a tumour. Using three-dimensional acquisition and free perspective display it is possible to show the tumour, the overlying brain surface (including the local distortion and expansion of overlying gyri), compressing of the adjacent gyri and their relationship to overlying bone. This information is invaluable to the surgeon and renders obsolete other localizing methods used prior to surgery (Le Bas *et al.*, 1987).

A relatively small percentage of intrinsic tumours cannot be distinguished from surrounding oedema on either CT or MRI and, when these fail to enhance, the precise position of the primary pathology cannot be determined by either method. Even where the area of primary abnormality is well shown, there is biopsy evidence that infiltrating tumours sometimes extend well beyond the apparent edge into the oedematous or normal brain substance. The infiltrating edge of a tumour has been shown to be more exactly defined with positron emission tomography (PET) using a methionine tracer, but this expensive method is only available in few specialized centres in Europe and America.

Supratentorial Masses

On scans performed in the axial plane, the temporal horn is sometimes poorly visualized and the relationship of a laterally placed hemispheric mass to a deformed Sylvian fissure may be unclear, rendering impossible the reliable distinction between an invasive temporal and frontal lobe neoplasm. Also, deformity of the upper border of a lateral ventricle, important in the diagnosis and localization of lesions in and above the corpus callosum, may not be appreciated unless there is thickening of the genu, splenium or septum pellucidum. These limitations may be overcome by scanning in the coronal plane or by reformatting.

Cerebellar Masses

Cerebellar tumours often occlude the fourth ventricle and cause hydrocephalus. The mass is recognized directly by attenuation or signal changes, but hydrocephalus with lack of visualization of the fourth ventricle due to a mass indistinguishable from brain substance may equally well be caused by a primary fourth ventricular tumour. When the forth ventricle is visible, unilateral cerebellar masses tend to compress it to a sagittal slit and cause contralateral displacement. Vermian or bihemispheric tumours reduce the height of the fourth ventricle and cause anterior displacement. This deformity is difficult to recognize on axial sections, until gross, because the appearance of the fourth ventricle changes with flexion and extension of the neck. By using sagittal MRI supplemented with coronal sections if necessary, these masses can generally be well defined and localized. The considerable normal variation in the pontine and quadrigeminal cisterns makes difficult the assessment of narrowing of these structures secondary to transmitted mass effect; such an observation should only be considered as a subsidiary sign.

Brain Stem Tumours

Brain stem gliomas are usually centred in the pons, and spread to the midbrain or medulla or along the peduncles into the cerebellum. The tumours may be symmetrical or markedly asymmetrical, displacing the floor of the fourth ventricle posteriorly, with the asymmetrical lesions also causing tilting. The ventricle is displaced posteriorly in relation to Twining's point and the height of the ventricle is diminished. The belly of the pons expands anteriorly and superiorly, narrowing the adjacent parts of the pontine and chiasmatic cisterns. Part of the mass may extrude from the pons, the exophytic component displacing the rest of the pons posteriorly and widening the surrounding parts of the pontine cistern. The basilar artery may be embedded in tissue bulging forward along its lateral aspects. The expanded midbrain may be outlined by the ambient and quadrigeminal cisterns, or obliterate them.

Characterization of a Lesion

Glial Tumours

The great majority of glial tumours are shown on CT and MRI as masses of abnormal density or signal which is frequently heterogeneous and, particularly in the more benign tumours, may not be markedly different from normal brain. There is usually perifocal, well-demarcated oedema in the white matter. Enhancement tends to be irregular and often considerable, though commonly less than that occurring in meningiomas and many metastases. Calcification occurs in about 18%, being more common in the more benign tumours and in oligodendrogliomas. In the clinical context, an accurate diagnosis of glioma can be made from plain CT in about 75% of cases, increasing to about 90% after contrast enhancement.

Enhancement is particularly valuable in showing the different components of a tumour. Circumscribed, totally unenhancing homogeneous regions, usually of low attenuation on CT and low signal on T_1W images, suggest cysts or necrosis (Fig. 10.6). Occasionally debris, or contrast medium precipitating into the dependent part and causing a fluid level, confirms the presence of cavity. Homogeneous enhancement, which occurs in 15%–20% of glial tumours, is indicative of a solid mass. Enhancement also frequently defines the proximal macroscopic limit of the tumour, which

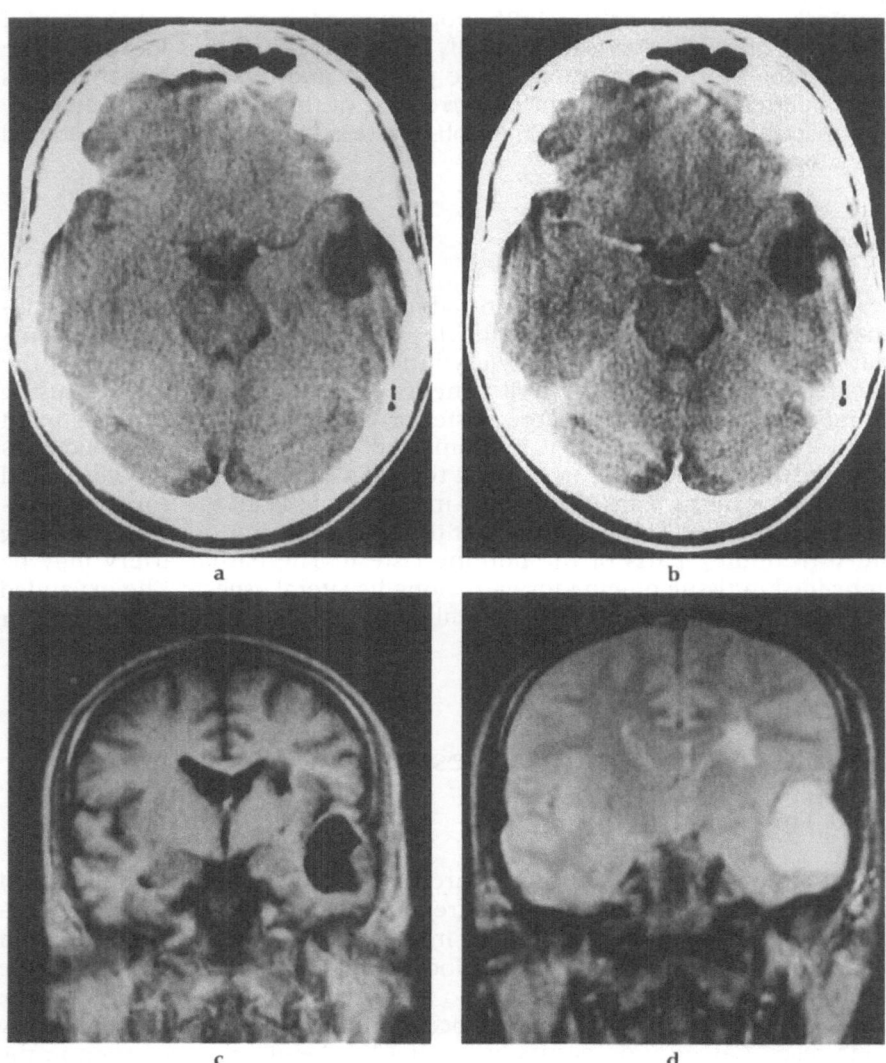

Fig. 10.6   Images of a 29-year-old with long-standing TLE worsening over the last three years with an abnormal EEG focus. There is a low-density left temporal mass on CT (**a**) demonstrating an enhancing nodule on its lateral aspect (**b**). Coronal T_1- and T_2-weighted MRI (**c, d**) scans confirm the cystic nature of the lesion and demonstrate, in addition, left lentiform signal change due to a small incidental infarct. Histology of the temporal mass revealed a malignant glioma. Note that the T_1 sequence demonstrates anatomical features to better advantage

appears similar to the surrounding normal or oedematous brain on plain scans.

In many cases, the diagnosis after CT is as certain as can be achieved by any combination of additional radiological studies; about 10% of intra-axial malignant tumours are indistinguishable from other pathologies, and a similar number of benign lesions simulate malignant intracerebral tumours. There is no doubt that in a proportion of these cases MRI increases diagnostic accuracy (Figs 10.3, 10.6, 10.7, 10.12 and 10.14).

Infarction

About 20% of infarcts show obvious mass effect in the acute phase. They present as a diffusely abnormal region involving both grey and white matter (Fig. 10.7). Oedema tends to be confined to the ischaemic region and not to produce the well-defined perifocal involvement of white matter which occurs around tumours. Breakdown of the BBB may occur very early, and enhancement with MRI has been shown within hours of onset of the event. Later, more immediate and prominent enhancement tends to occur with neocapillary vascularization. The enhancement may involve any part of the ischaemic region which is not totally deprived of blood supply. Enhancement very close to the margin of the plain scan abnormality is more suggestive of infarction because of the usual occurrence of oedema, which does not enhance, around malignant tumours.

A typical clinical presentation suggests the diagnosis; in cases where the history is not definitive or is misleading, the differential diagnosis can be achieved by observing improvement or deterioration on follow-up study. Where an immediate diagnosis is more imperative, angiography can be definitive if a major peripheral cerebral artery supplying the abnormal region is occluded or if tumour circulation is shown. Often, however, distinction by angiography is more complex or impossible. On the one hand, blockage of vessels proximal to the circle of Willis is not uncommonly asymptomatic and may be unrelated to the presenting pathology. One the other hand, lysis of a recent embolus causing infarction may occur rapidly, so that the arteries supplying the abnormal region are often patent by the time that angiography is performed in cases with a subacute or ingravescent onset of symptoms. Delay of filling of peripheral vessels may be secondary to local or diffuse increase in intracranial pressure alone. Narrowing or occlusion of arteries between herniating brain substance and dura or bone may also be responsible. Local involvement of vessels by a neoplasm may occasionally cause superadded infarction.

Intracerebral Haemorrhage

A resolving intracerebral haemorrhage can appear on CT made three to five weeks after onset as an isodense mass with ring enhancement (Fig. 10.8). Magnetic resonance imaging will generally show specific features of a resolving haematoma (Gomori et al., 1985) and sometimes a causative

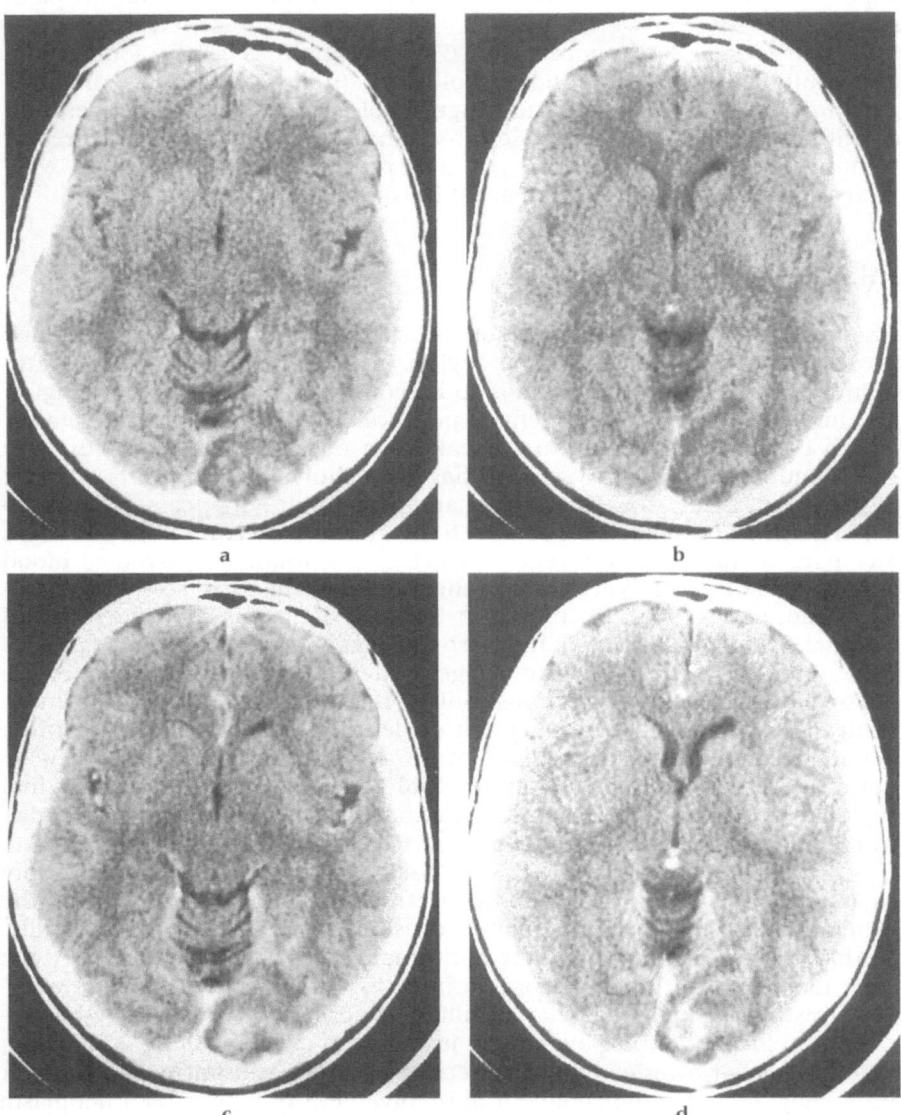

Fig. 10.7 CT before (**a, b**) and after (**c, d**) intravenous contrast with axial T$_2$-weighted (**e**) and coronal T$_1$-weighted (**f**) MRI scans. A young man presenting with right-sided visual inattention and headache. CT demonstrates serpiginous rim enhancement of a mixed mainly low-density lesion in the left occipital lobe. Such appearances could be caused by infarction, tumour or abscess. MRI performed within 24 h of CT demarcates the lesion and shows a well-defined rim of even thickness with central low signal (coronal image), suggesting cavitation. Subsequent biopsy and follow-up confirmed the presence of an abscess

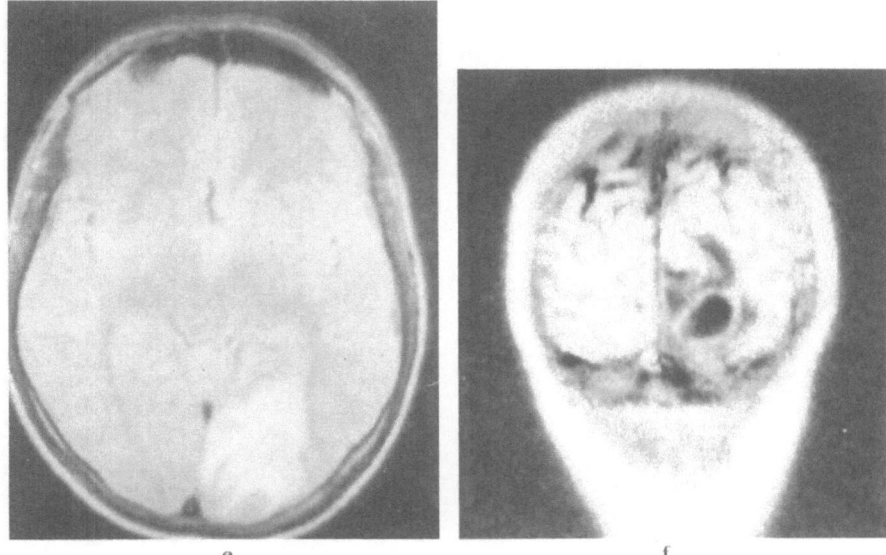

e   f

Fig. 10.7 *Continued*

aneurysm may be revealed when the vascular anatomy is well shown by
flow void or on flow sensitive sequences prior to angiographic confirmation (Young *et al.*, 1983).

Cerebral Abscess

Most cerebral abscesses are diagnosed clinically and CT or MRI confirms
the diagnosis by revealing the abscess capsule as a smoothly curved
enhancing ring of almost even thickness with central low attenuation on
CT, or low signal on T_1W sequences, due to pus with surrounding oedema.
In such cases, the only further radiological study recommended prior to
treatment is a chest radiograph. However, some abscess capsules give an
atypical uneven, irregular or thick-walled ring enhancement, or an early
abscess may present as a solid enhancing nodule. Such abscesses may be
indolent, with relatively little surrounding oedema, and due to an unusual
organism, particularly in patients with immune suppression. They may
simulate a necrotic or cystic glioma or a metastasis.

Granuloma

Tuberculomata may appear as homogeneous enhancing nodules, or may
show central necrosis, with irregular or ring-like enhancement. Oedema is

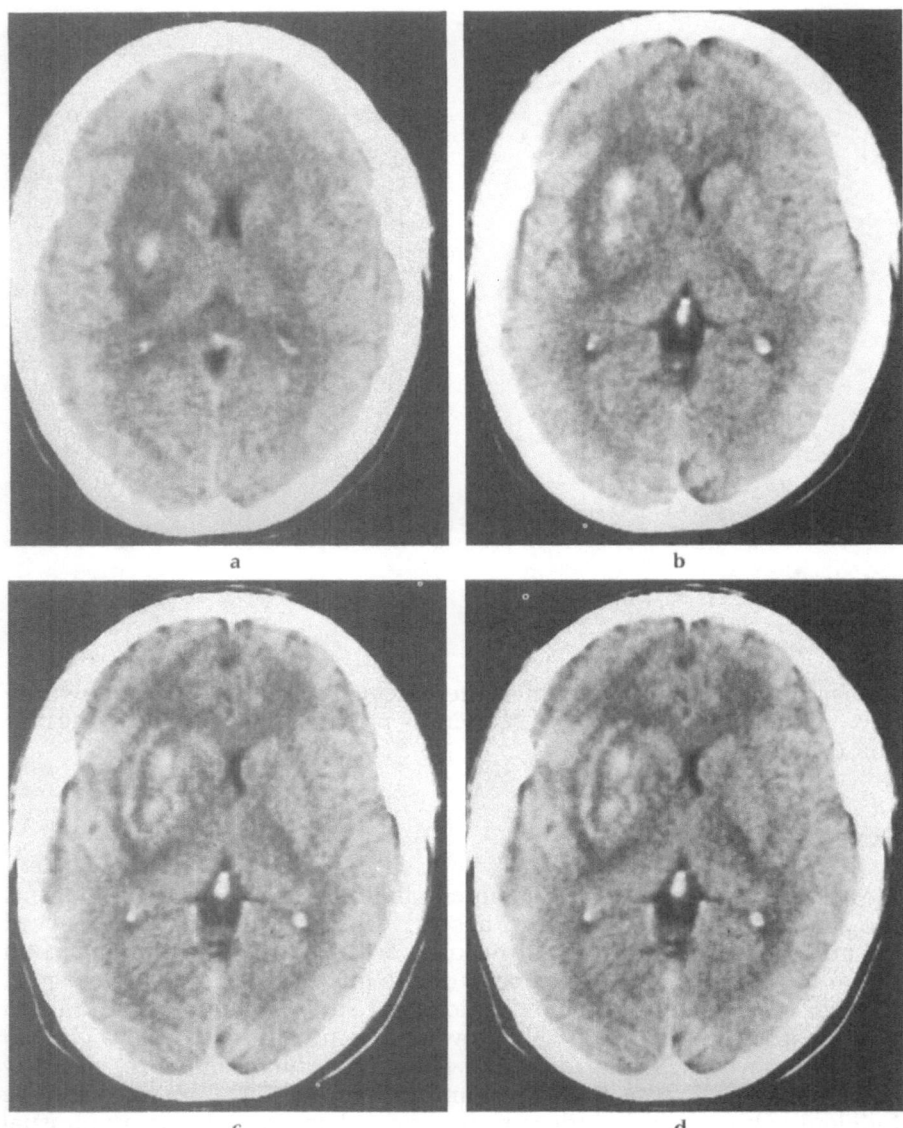

Fig. 10.8 Axial CT of a young man with resolving spontaneous intracerebral haematoma, before (**a, b**) and after (**c, d**) contrast. Note the high-density lesion with surrounding vasogenic oedema and mass effect with rim enhancement producing a "target" appearance

generally less marked than with simple abscesses, and mass effect tends to be less pronounced. Calcification is present in a small percentage of tuberculomata and may be useful in suggesting the probability of a granuloma rather than multiple metastases. Calcification is not helpful in distinguishing a solitary tuberculous lesion from a glioma.

Other granulomas may cause similar appearances, and in any of them meningeal involvement may occur and be recognized by enhancement of the meninges. In sarcoidosis, involvement of the brain substance tends to occur by extension from the meninges, and is often of brain density, homogeneously enhancing after intravenous contrast medium. On MRI, deep white matter lesions, causing little mass effect and closely simulating multiple sclerosis, are often revealed. Where inflammatory disease is suspected, examination of the whole brain with MRI or with CT after enhancement may reveal lesions which were not evident on plain CT scanning.

Opportunistic Infection

In immune-compromised patients, indolent abscesses – most frequently due to toxoplasma in AIDS patients – are commonly revealed as regions of low or brain density on CT and high signal on T_2W MRI with variable surrounding oedema. Magnetic resonance imaging is particularly valuable for showing multiple lesions if CT shows only a single focus. The appearances are often indistinguishable from lymphoma, which should be suspected when any lesion fails to respond to appropriate anti-infection treatment.

Fungi, yeasts and tuberculosis may also give rise to lesions simulating those of toxoplasma in AIDS patients. HIV encephalopathy may be associated with extensive, rather diffuse white matter low density on CT and signal change on MRI. These appearances are non-specific and may also be present with complicating cytomegalovirus or cryptococcal infection. Both these organisms may, however, cause a superimposed ependymitis and/or meningitis and necrotizing encephalitis with enhancement after intravenous contrast medium.

Progressive Multifocal Leucoencephalopathy

White matter involvement by progressive multifocal leucoencephalopathy (PML) should rarely be confused with neoplasm. There are circumscribed foci of low density on CT and of altered signal on MRI typically confined to the white matter, though the grey matter may also be involved, particularly in AIDS patients. These changes are most common in the parieto-occipital regions, though they may occur in any location, and in AIDS patients the cerebellar white matter alone is involved in about 10% of cases of PML. Many of the foci enhance after intravenous contrast medium.

Herpes Simplex Virus Encephalitis

Of the acute encephalitic processes, only herpes simplex is likely to be confused with tumour. The mass effect is frequently most pronounced in one or both temporal lobes and may be associated with haemorrhagic changes. Careful observation reveals multiple foci in the majority of cases, particularly on MRI. The differential diagnosis is usually from infarction, and only when mass effect is particularly pronounced and unicentric is tumour likely to be considered in the differential diagnosis.

Parasitic Infections

Hydatidosis

The most common sites of hydatid cysts are the liver and lung: under 5% occur in brain substance and these are rarely isolated lesions. they are relatively more common in children. They are almost always intracerebral and they are more frequent in the parietal lobes, though both extracerebral and intraventricular cysts have been recorded. The cysts are round, up to several centimetres in diameter, contain fluid resembling water in imaging characteristics and have a clearly defined thin wall. Typically there is no inflammatory response in the adjacent brain, but this can occur and then a thin rim of enhancement may be present. Calcification may occur when degeneration affects the capsule and it may occasionally be extensive in dead cysts. The appearances are unlikely to be confused with other pathologies.

Cysticercosis

Neurocysticercosis is a rare cause of intracranial abnormality in the UK, but it is common in Mexico, Central and South America, the Indian subcontinent, Africa and the Far East. It should be considered particularly when unexplained multifocal lesions are found in patients from these regions.

The clinical and neurological abnormalities depend on the stage of the infection. Early in the acute phase there is multifocal oedema in which small foci of enhancement appear. Later the cysts are recognized as round or oval well-defined regions of water density with low signal on T_1W and high signal on T_2W sequences. The scolix may be recognized within the cyst as a small region of high signal on T_1W and low on T_2W image. This provides conclusive evidence of diagnosis. Marginal enhancement and pericystic oedema occur as the cyst degenerates and this is associated with reversal of the signal from the scolix. Degeneration and death of the parasites may be followed by disappearance or by evidence of calcification or fibrosis on CT or MRI.

In most cases with multiple intraparenchymal parasites, there is little difficulty with diagnosis. Cysts within the ventricles or subarachnoid spaces may also have a characteristic configuration, but racemose cysts obstructing the narrower parts of the ventricular system or distending regions of the subarachnoid space may simulate arachnoid or epidermoid cysts.

Paragonimiasis, Schistosomiasis

The lung fluke *Paragonimus westermani* may infect the central nervous system by entering the cranial cavity through the basal foramina, particularly the jugular foramina. It tends to involve the temporal, parietal and occipital lobes. In the acute state, a cyst resembling a thin-walled abscess develops around the worm. The eggs of the parasite induce a granulomatous fibrotic reaction causing a solid mass. The lesions tend to calcify heavily, causing a soap bubble appearance associated with evidence of focal brain damage and atrophy.

Schistosoma, particularly S. *japonicum*, may also cause intracerebral granuloma formation without specific features.

Extrinsic Tumours

Extrinsic tumours can usually be diagnosed by the displacement of the surface of the brain, well-defined margins and clinical presentation, but some peripheral intrinsic tumours may closely resemble them. Changes in bone are frequently shown on CT with meningiomas as well as with other extrinsic tumours which are much less likely to cause confusion. These include well-defined pressure erosion by pituitary and parasellar tumours, neurinomas and paraganglionomas and by epidermoid and cholesterol granulomas. Ill-defined permeative destruction and/or reactive sclerosis occurs in metastasis to the skull and with direct invasion of bone from basal malignant tumours and chordomas, chondromas and chondrosarcomas.

Extrinsic masses tend to widen the subarachnoid spaces around their margins, whereas intrinsic masses tend to compress them. However, exophytic parts of gliomas may cause similar widening, and extrinsic masses may burrow into the brain substance from a narrow pedicle of attachment, and therefore simulate an intracerebral mass. Magnetic resonance imaging with an elective plane of section designed to demonstrate the region of attachment of a tumour, will generally elucidate in difficult cases (Haughton et al., 1988).

Meningioma

Meningiomas generally have characteristic features on CT. They are typically of greater density than brain and are homogeneous, with well-defined edges accentuated by a varying amount of oedema in the adjacent brain, which is extensive in about 10%. Calcification is present in 15%–20%, and cystic or degenerative change in a further 10%, causing regions of low density; a minority of meningiomas are isodense with, or of lower density than, brain.

On MRI, many meningiomas return similar signal to the adjacent brain substance on both T_1W and T_2W sequences, though the majority return lower signal than oedematous brain and cerebrospinal fluid on heavily T_2W sequences. Almost all meningiomas are highlighted on CT and MRI following intravenous contrast medium. The meningeal base is usually broad and, frequently, enhancement is evident extending along and thickening the adjacent meninges on MRI. This feature is indicative of extension of the meningioma beyond the obvious macroscopic site of attachment, and it can be of diagnostic importance in meningiomas "en plaque". Hyperostosis or erosion of bone and enlargement of meningeal channels should be sought with appropriate windowing on CT, and tumour extension into bone can usually also be appreciated on MRI. Angiography is occasionally necessary for specific diagnosis in cases with atypical features on computed imaging.

Classical angiographic features include peripheral enlargement of the meningeal trunks entering the tumour at its primary dural attachment and fanning throughout the major part of the tumour. Though other extra-axial tumours, or intra-axial tumours which have invaded or become adherent to the meninges, may also obtain a meningeal supply, this is generally less rich and extensive and not associated with marked hypertrophy of meningeal trunks. A meningioma has characteristically a diffuse capillary circulation which persists after the cerebral circulation has cleared. However, about one-third of meningiomas show large tumour vessels or arteriovenous shunting which may simulate those of malignant tumour. Gliomas frequently show transcerebral drainage connecting to the subependymal venous system, whereas meningiomas only acquire such drainage if they have involved venous sinuses and precipitated occlusion. Angiography is required much more frequently to show the relationship of the tumour to a major superficial vessel, and is then incidentally useful in confirming the diagnosis.

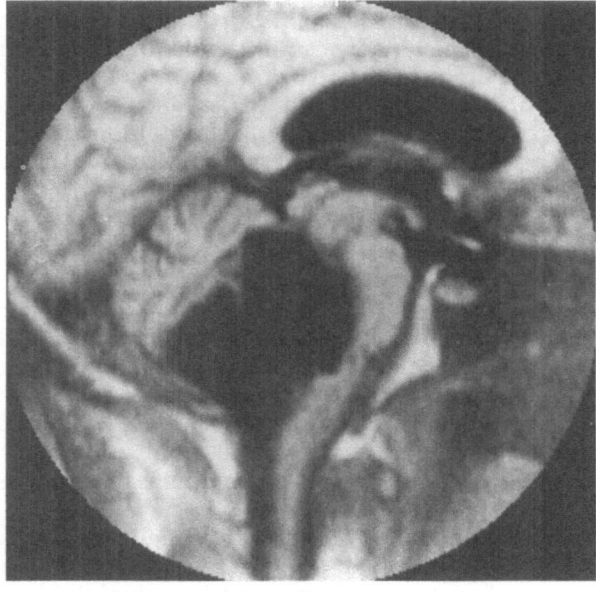

a

Fig. 10.9   Sagittal T_1-weighted (**a**) and composite of contiguous axial T_2-weighted (**b**) MRI slices. A typical epidermoid situated in and enlarging the fourth ventricle and cisterna magna. It returns low signal on T_1- and high signal on T_2-weighted sequences and demonstrates fronding of the tumour edge

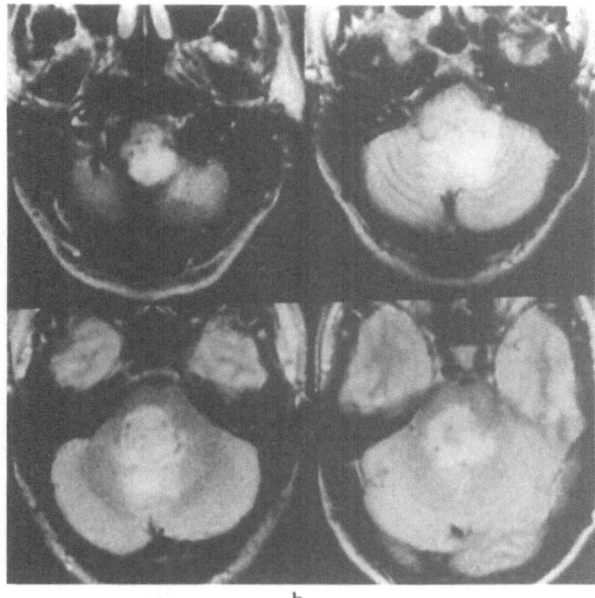

b

Fig. 10.9   *Continued*

Epidermoid

Epidermoids are well-defined, often lobulated, masses with no surrounding oedema. They are easily recognized on CT when their density is lower than water, but it is commonly between that of water and brain; there may be some calcification in the capsule. On MRI, it should be possible to distinguish the lesion from both brain and CSF by choosing appropriate imaging sequences, epidermoids classically demonstrating uniformly low signal on T_1W images with strikingly high signal on the T_2W images (Fig. 10.9). The tumour itself does not enhance, but there may be enhancement in the capsule. Only those epidermoids which lack specific features and are completely enclosed by cerebral substance, such as those occupying the Sylvian or choroidal fissures, could be mistaken for a glioma (Fig. 10.10). Once the diagnosis has been considered, multiplanar MRI should elucidate.

Cholesterol Granuloma

The appearance of this tumour on CT is similar to that of epidermoid; on MRI it returns high signal on both T_1W and T_2W sequences related to the products of recurrent haemorrhage. The temporal bone is typically involved, the classical site being at the petrous apex.

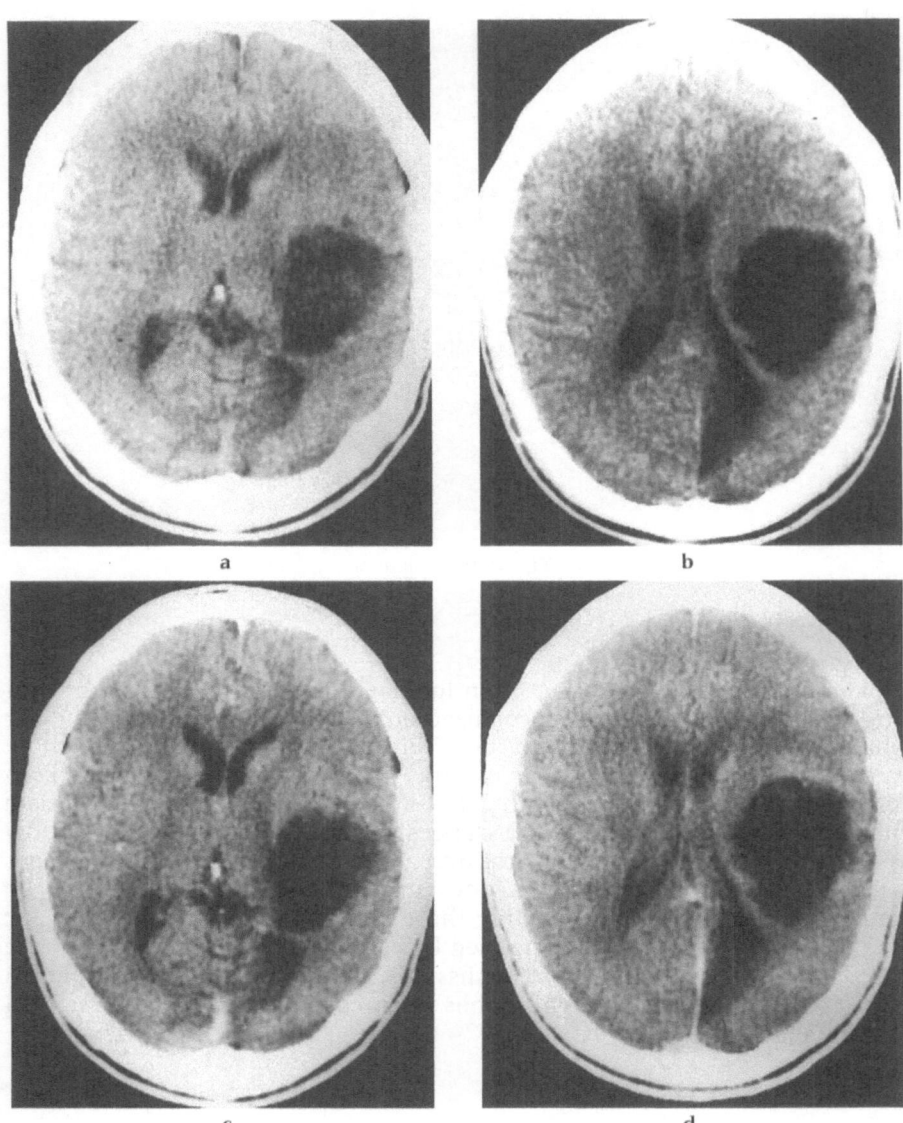

Fig. 10.10   Composite of pre- (**a, b**) and post-contrast (**c, d**) axial CT scans demonstrating a well-defined, low-density, non-enhancing left temporoparietal tumour. Close inspection reveals evidence of some inhomogeneity and structure within the lesion and a rim of grey matter density. It proved to be an epidermoid arising within and expanding the Sylvian fissure. Note incidental left occipital infarct

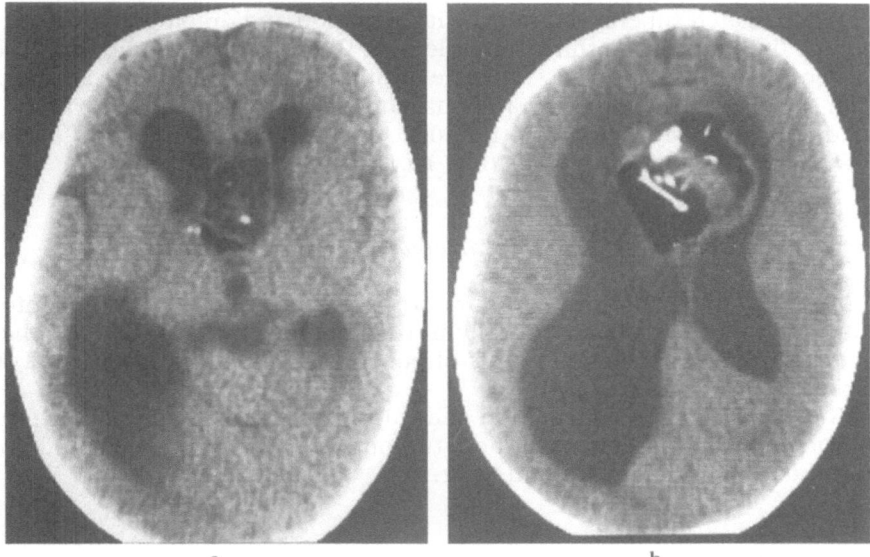

a   b

Fig. 10.11 Unenhanced axial CT images (**a, b**) of a neonate presenting with a large head and calcification identified on plain skull radiograph. CT reveals hydrocephalus with a mixed density mass containing fat (lower density than CSF), calcific areas (similar density to bone) and soft-tissue attenuation. A long bone is identifiable (**b**). The appearances are pathognomonic of a dermoid

Dermoid

Dermoids generally have components which contain neutral fat. This is lower than CSF in density, and has a short T_1 which is reflected in high signal in T_1W sequences on MRI. The fat is liquid at body temperature; should the dermoid rupture, globules may be shown in the ventricular system and subarachnoid spaces, sometimes associated with hydrocephalus precipitated by secondary inflammatory reaction. The dermoid may also contain other tissue components such as muscle, bone and teeth. The combination is virtually pathognomonic and, even when the tumour is enclosed within brain substance, it could scarcely be mistaken for a glioma (Fig. 10.11).

Parasellar Tumours (Intrinsic and Extrinsic)

Hypothalamic and chiasmal gliomas tend to cause ill-defined masses within and sometime occluding the basal cisterns. The chiasmal glioma may extend along the optic nerves or tracts in a characteristic anatomical distribution. The CT density is variable, but often close to or lower than that of brain substance. They are usually of high signal on T_2W sequences. In general, sagittal and coronal MRI scans allow easy distinction from suprasellar extension of the common extrinsic masses which tend to have well-demarcated superior margins, with elevation and draping of the structures in the floor of the third ventricle over them. Intrinsic tumours cause thickening of the floor of the ventricle or of the chiasm and of the adjacent brain substance. In addition, craniopharyngiomas are frequently partly calcified and cystic, and pituitary tumours tend to cause enlargement of the sella and destruction of its walls (Fig. 10.12).

Chordoma, Chondroma, Chondrosarcoma

Chordomas are extradural tumours arising from remnants of the notochord within the skull base, usually in the region of the body of the sphenoid. They cause varying amounts of bone destruction and may be partly calcified.

More extensive bone formation and calcification should suggest a chondroma or chondrosarcoma. All these tumours are extradural and could not be confused with an intrinsic tumour.

Neurinoma

The majority of neurinomas tend to enlarge the basal foramina and present with cranial nerve deficits. Exceptionally an exophytic pontine glioma can simulate an acoustic or trigeminal neurinoma. Careful analysis, particularly by MRI sections profiling the extrinsic tumour, will almost always demonstrate an intrinsic component of such gliomas.

Paraganglionoma

Paraganglionomas are variously known as glomus tumours, chemodectomas and carotid body tumours. The cranial tumours are derived from chemoreceptor tissue which is mainly within the adventitia of the jugular bulb and the ganglia of the vagus nerve in the jugular fossa and middle ear.

Glomus jugulare tumours are highly vascular masses which cause irregular destruction of the margins of a jugular fossa, occluding the jugular vein. They may extend through the middle ear, the cerebellopontine angle and posterior fossa, along the carotid canal towards the cavernous sinus or downwards into the cervical region along the jugular vein. The temporal bone involvement is best shown by high resolution CT, and their typical clinical presentation serves to distinguish them from tumours arising within the cranial cavity. Coronal MRI scans demonstrate well the extent of the tumour and frequently indicate that it is highly vascularized.

A glomus tympanicum usually presents with tinnitus and impaired hearing at a time when it is small. High-resolution CT scanning is capable of showing the smallest of lesions when still confined to the hypotympanum. Larger tumours may expand the tympanic cavity and erode into the jugular foramen, but they rarely extend intracranially and are not likely to be confused with a tumour arising there.

Extrinsic Malignancies

Extrinsic malignancies may cause extensive intracerebral oedema. These may be primary tumours arising within the paranasal sinuses or pharynx, or metastatic, but they almost always cause bone destruction visible on CT and MRI.

Sphenoid Mucocele

This is a retention cyst of the sphenoid sinus which characteristically returns high signal on T_2W images, outlining the sphenoid loculi. Mucoceles characteristically expand the affected sinus, but occasionally there is destruction of part of the wall and an extra axial mass may extend into the posterior or middle fossa.

Subdural and Extradural Empyema

These infections most commonly occur in children and teenagers. The diagnosis is primarily clinical in a febrile patient, with focal neurological signs and middle ear or paranasal sinus infection or previous meningitis. In the early stages the amount of extracerebral fluid which can be demonstrated by CT or MRI scanning may be minimal, and unless the index of suspicion is high it could be overlooked. The diagnosis can generally be confirmed by demonstrating enhancement on the cerebral aspect of the inflammatory process on the scans made after iodide or gadolinium injection. The not uncommon complication of cortical thrombophlebitis may be associated with cerebral oedema and infarction with brain swelling.

Meningeal Infiltration

This can be inflammatory, usually granulomatous, but fungi or yeasts may be responsible especially in immune compromised or diabetic patients. It can also be neoplastic, primary extension from the apparent edge of a meningioma or more commonly metastatic from primary CNS malignancies or systemic tumours, of which bronchial and breast carcinoma are the commonest. Gadolinium-enhanced MRI may demonstrate thickened meninges when enhanced CT is negative.

Intracerebral Metastasis

Intracerebral metastases grow by expansion and typically cause well-defined mass lesions. The majority of gliomas are infiltrating, with ill-defined edges, but a minority may grow by expansion, and these can be indistinguishable from metastases, particularly if they are multifocal and

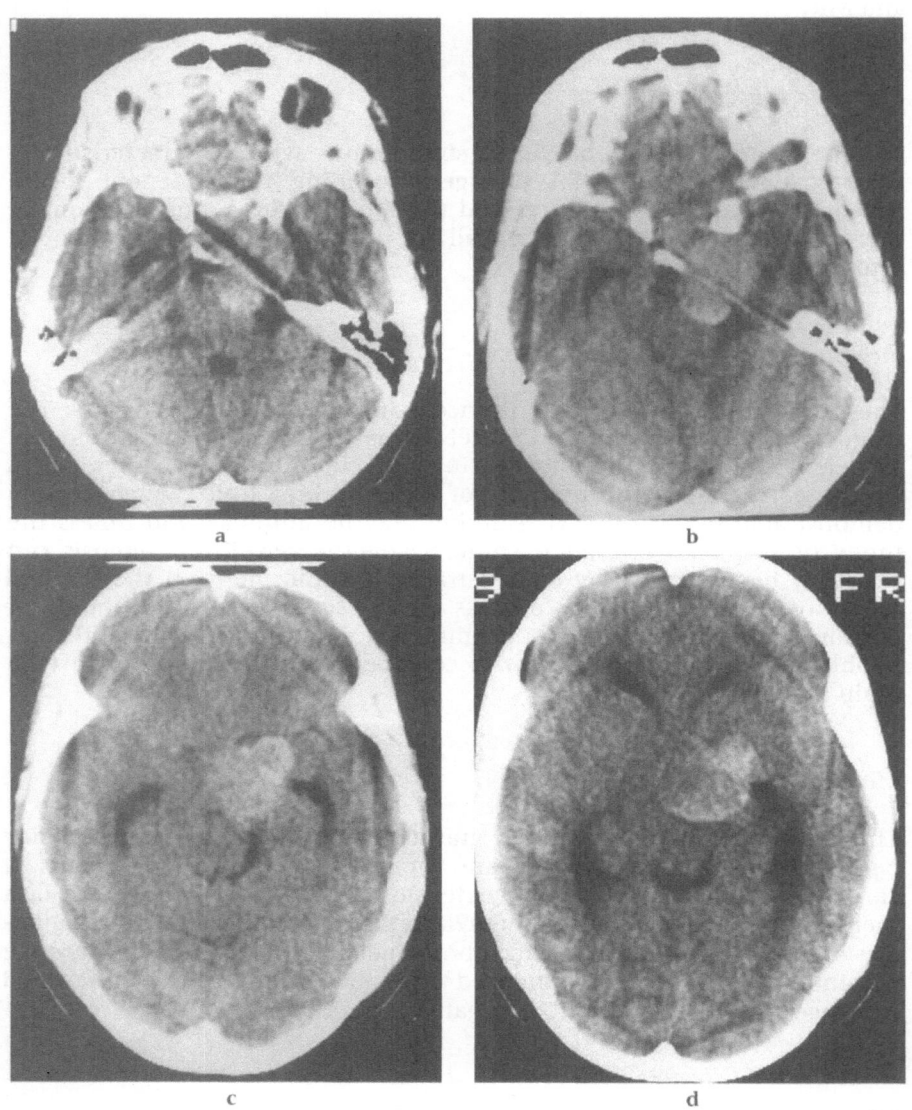

Fig. 10.12   a–d.
Caption on page 248

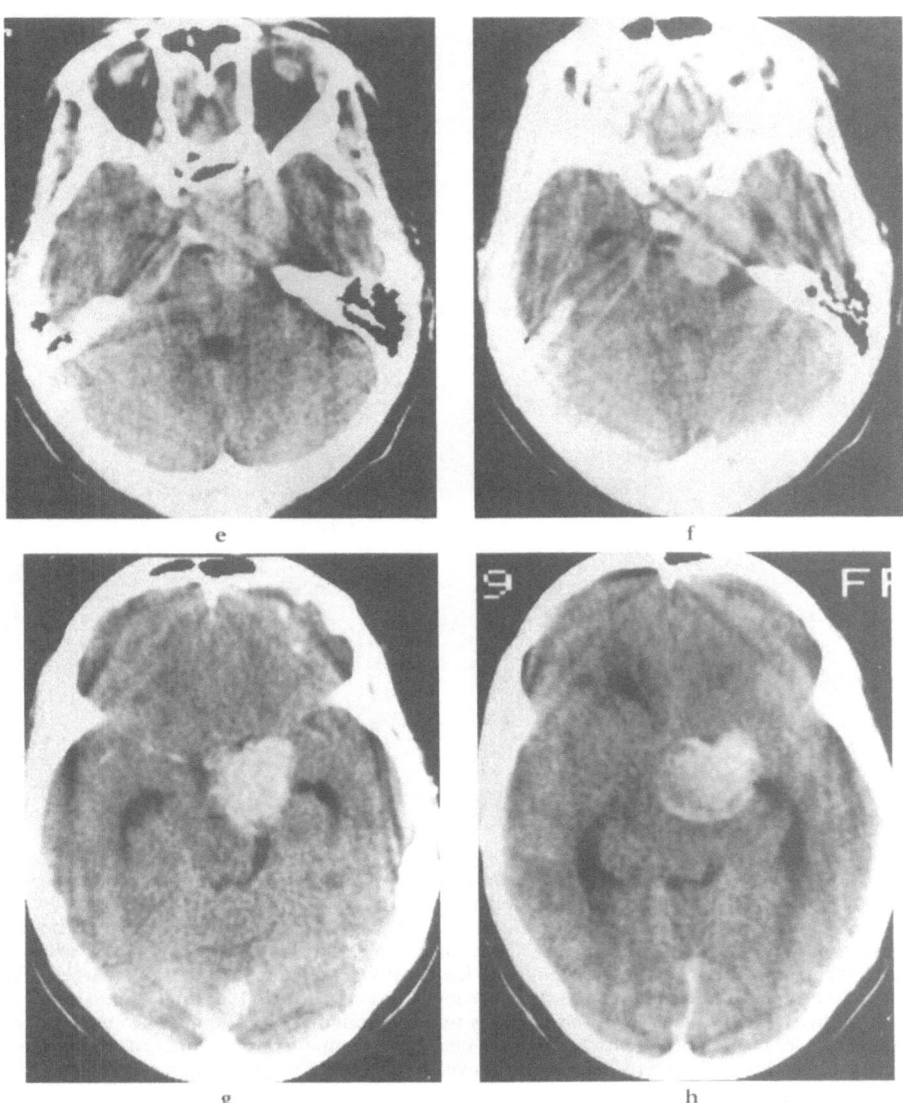

Fig. 10.12   e–h.
Caption on page 248

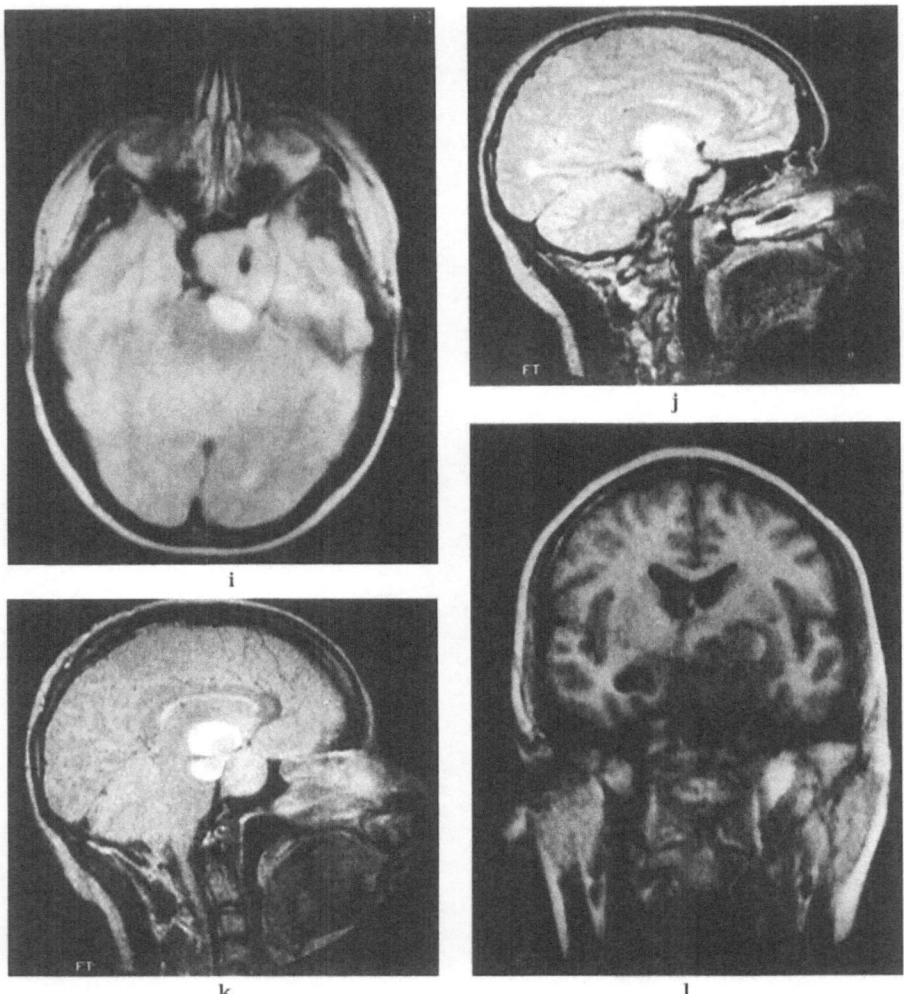

Fig. 10.12 Axial CT scans before (**a–b**) and after (**e–h**) intravenous contrast medium. Axial (**i**) and sagittal (**j, k**) 2000/80 T$_2$-weighted and coronal (**l**) IR T$_1$-weighted sequences. On CT there is a predominantly hyperdense, enhancing sellar and parasellar mass involving the left cavernous sinus, with hydrocephalus. CT is marred by beam hardening and streak artefact projected across the sella and middle cranial fossa. MRI clearly demonstrates an enlarged sella and encasement of the siphon and intracavernous portions of the left internal carotid artery – shown by signal flow void. MRI also reveals two components to the tumour – a more solid-appearing lesion isointense with brain and a peripheral component with longer T$_2$ signal characteristic (white on **j, k** and **l**). The lesion proved to be a prolactinoma with a capping cyst and has responded well to bromocriptine

contain no calcification. Metastases are most frequent in the subcortical white mater in the posterior halves of the cerebral hemispheres. They are multiple in about one-third of patients and there is evidence of a concurrent or previous systemic neoplasm in two-thirds. There is nearly always some surrounding oedema, and it is marked and disproportionate to the size of

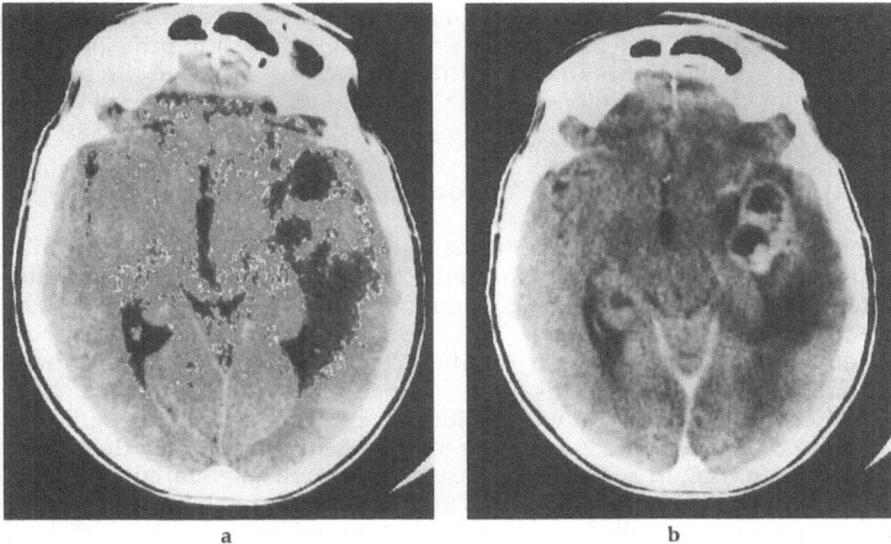

a   b

Fig. 10.13 a, b Pre- and post-enhanced axial CT images of an elderly female smoker presenting with confusion. **a** A mixed-density lesion with surrounding low density in the left
temporoparietal region. **b** Some irregular rim enhancement and confirmation of the presence
of a second lesion medial to the right trigone. Higher cuts demonstrated further lesions, confirming the diagnosis of multiple metastases

the tumour in over 90% (Fig. 10.13). About three-quarters of metastases
are of higher attenuation than brain and have increased T_1 and T_2 relaxation
times. Exceptions include melanoma metastases which are frequently
haemorrhagic or have a short T_2 due to the paramagnetic properties of
melanin itself (Atlas *et al.*, 1989). Metastases of mucoid-secreting carcinomas and osteogenic sarcomas are not infrequently calcified or ossified.
Almost all metastases enhance and, if the diagnosis is suspected, further
lesions may be revealed after intravenous contrast medium. Cavitation
occurs in about 20% of metastases, causing thick-walled neoplastic cysts.

Angiography is rarely necessary for diagnosis, but a typical metastasis
has a well-defined rounded or lobulated tumour circulation supplied
through a single cortical artery and draining through a single vein. The circulation may be an irregular network of tortuous vessels or a homogeneous, diffuse capillary blush, frequently associated with arteriovenous
shunting. Transcerebral venous drainage is a feature of anaplastic gliomas
but not of metastases.

Imaging Characteristics of Intrinsic Tumours

The abnormalities shown on CT and MRI vary between the different histological types and grades of tumour, but overlap is considerable. The previous clinical course and progress on follow-up studies may be helpful.

However, in most cases an absolute diagnosis is necessary for management decisions and histological evidence is required, so that the principal value of the imaging studies is to define the tumour and to decide the best site for open or stereotactic biopsy using CT or MRI guidance.

Anaplastic or Malignant Astrocytoma – Grades III and IV

Anaplastic astrocytoma (glioblastoma multiforme) is the commonest brain tumour of adults, with a peak incidence at about 50 years, being more frequent in males, and presenting, commonly, with a short history, usually under six months (Kieffer et al., 1982). The tumour is generally centred in a temporal or frontal lobe or in the corpus callosum. It is rare in the infratentorial regions.

The tumour grows rapidly by infiltration and expansion. It is highly vascular, and the vessel walls are weak, or may be compressed by the tumour growth, resulting in regions of necrosis, haemorrhagic infarction or frank haemorrhage within the tumour. Tumour vessels are abnormally permeable and there is usually considerable peritumoral oedema. The corpus callosum and ventricular system are commonly invaded. About 60% of the tumours are of mixed density on CT, the rest being of increased density (15%), isodense with brain (15%) or of low density (10%). Calcification occurs in less than 10%, and it may suggest that the tumour has arisen from one of lower grade.

On MRI the tumour usually returns mixed signal, reflecting the macroscopic features (Destian et al., 1988, 1989). It can, however, be almost homogeneous, returning high signal on T_2W and low signal on T_1W sequences. Contrast enhancement of varying degrees occurs in 98% (Fig. 10.14). Necrotic and cystic components, which are common, and haemorrhagic regions also, do not enhance. The overall pattern of enhancement is typically inhomogeneous; it forms a ring which is thick and markedly irregular in about 50% of cases. A thick but even ring is less frequent, a pattern equally common in metastases (see Figs 10.5 and 10.13). A thin-walled ring occurs less frequently, and may simulate a typical pyogenic abscess (Laster et al., 1984).

About 20% of malignant astrocytomas show homogeneous nodular enhancement and, when peripheral, may simulate a meningioma – bone involvement and a dominant meningeal vascular supply may help to distinguish. Multifocal enhancement may simulate metastases, though it is usually evident that the abnormal tissue extends in continuity between the enhancing regions. Nodular enhancement is more frequent in medium-grade astrocytomas, though they tend to be associated with less oedema. It also occurs in granulomas and indolent abscesses; the ultimate differential is by histology (Earnest et al., 1988).

A combination of nodular and ring patterns in a single mass occurs in 30% of malignant astrocytomas; this pattern is much less frequent in medium-grade astrocytomas and metastases. An ill-defined, intrinsic, non-enhancing mass is more typical of benign astrocytoma, but it occurs in about 2% of malignant astrocytomas. It should be noted that patterns of enhancement with gadolinium have, so far, proved unsuccessful in consistently distinguishing high- from low-grade gliomas (Vaghi *et al.*, 1986; Price *et al.*, 1986).

Angiography is rarely indicated for diagnosis, but may be required for surgical planning. Pathognomonic features include increased vascularity with irregular calibre vessels, forming a haphazard pattern. There are large draining veins, including dilated long medullary veins, connecting to the subependymal venous system. There may be avascular regions corresponding to necrosis, cyst formation or haemorrhage.

Only about a third of malignant astrocytomas have this typical pattern. A further third show slight or moderately increased vascularity, with smaller vessels of more even calibre and early-filling veins: a similar vascular pattern occurs in benign gliomas and metastases. Other malignant astrocytomas present an even or ring-like capillary blush, simulating mengingioma or metastasis. Over 10% of malignant gliomas are angiographically avascular masses.

Benign Astrocytoma – Grades I and II

Benign astrocytoma of the cerebral hemispheres occurs in adults, with a peak around 30 years. Brain stem, optic chiasm and cystic cerebellar astrocytomas occur in childhood; solid cerebellar astrocytomas are more frequent in adults. The tumour may be widely infiltrating and diffuse, with little mass effect, or it may grow by expansion as well as infiltration, forming a focal mass with variable but often large, cystic components. On CT, most grade I and about 40% of grade II tumours are of decreased density. A small proportion of grade I and about 30% of grade II are of similar density to brain substance, and about 30% of grade II are of mixed density. Calcification occurs in about a third of grade I and less often in grade II tumours. Oedema is uncommon in grade I, but oedema of minor or moderate degree is present in about two-thirds of grade II tumours. This is reflected in the MRI appearances with variable prolongation of T_1 and T_2 depending on the degree of oedema and amount of calcification, in addition to any mass effect. Enhancement after intravenous contrast medium is minimal or absent in grade I, but usual in grade II in which it tends to be inhomogeneous, but in about a third it forms a homogeneous blush or ring of fairly even thickness.

On angiography, scanty abnormal vessels or tumour staining extending into the venous phase occurs in about 20% and is occasionally accompanied by early filling veins. Most benign astrocytomas appear as avascular masses.

Subependymal Giant Cell Astrocytoma

Neoplasms occur in about 2% of cases of tuberous sclerosis. The commonest type is subependymal giant cell astrocytoma, which almost always involves the region of the head of the caudate nucleus and presents with hydrocephalus due to obstruction of the foramen of Monro. Diagnosis is usually evident on CT, which shows a mass in the region of the foramen of Monro, usually of similar or lesser density than brain and enhancing after

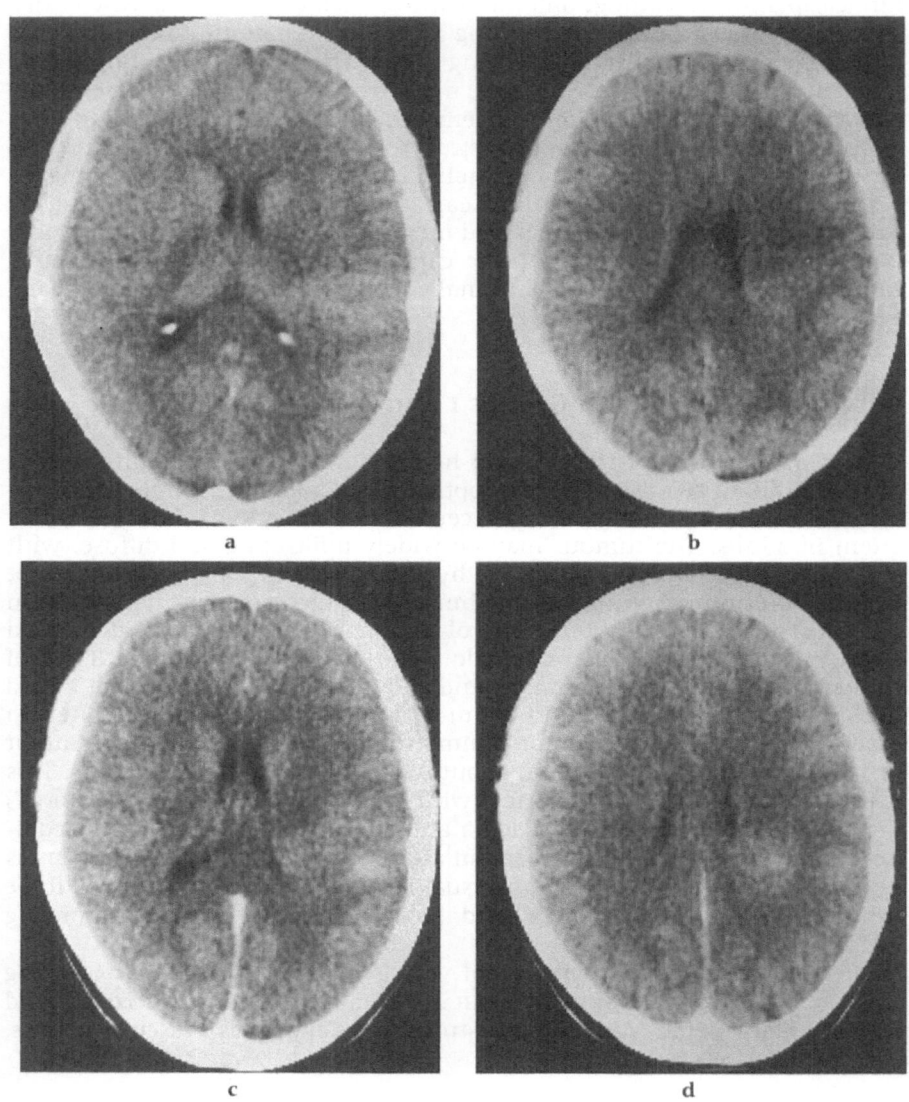

Fig. 10.14   a–d.

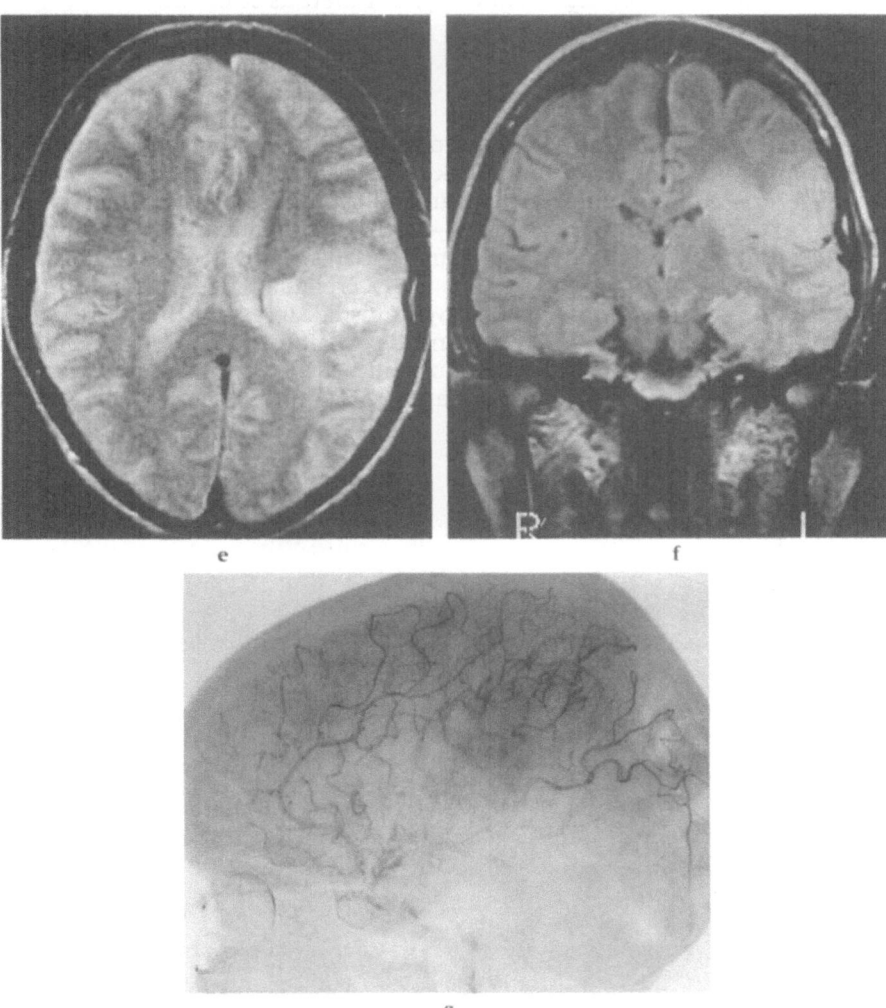

e

f

g

Fig. 10.14 Axial CT pre- (**a, b**) and post-contrast (**c, d**); axial (**e**) and coronal (**f**) MRI; lateral internal carotid angiogram (**g**). Left parietal enhancing high signal area with surrounding low density and mass effect. There is compression of the body of the left lateral ventricle. MRI better delineates the lesion identifying a deeper high signal component and a more peripheral, less intense abnormal signal extending out to the cortex. The lateral angiogram shows a tumour blush in the left parietal lobe with an early filling vein draining towards the cortex. Histology confirmed a malignant glioma grade 2–3

intraveneous contrast medium. Tubers are usually evident, showing as calcified nodules in the subependymal region or as low- or high-density nodules within the brain substance. Both ganglioglioma and ependymoma may arise, though much less frequently, in tuberous sclerosis.

Lhermitte–Duclos Disease

This condition – also known as dysplastic gangliocytoma, granular cell hypertrophy, gangliomatosis and purkinjioma – is confined to the cerebellum. There is regional enlargement and thickening of the cerebellar folia with an external layer of nerve fibres – both myelinated and unmyelinated – and an inner layer of abnormal ganglion cells, with decreased granules and absent Purkinje cells. It presents as a cerebellar mass lesion, and CT and/or MRI reveal the mass, usually with displacement of the fourth ventricle anteriorly and contralaterally and depression or apparent ectopia of the cerebellar tonsils (Fig. 10.15). In some cases the tumour simulates exaggerated cerebellar folia on images, an appearance which should suggest the correct diagnosis.

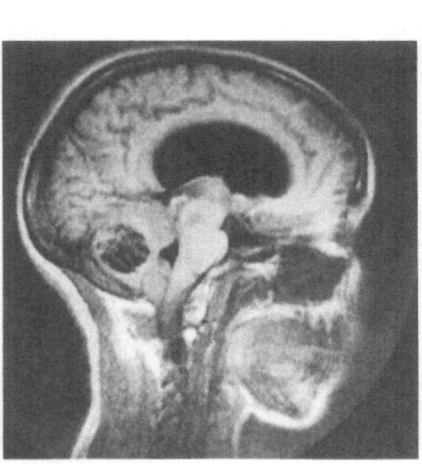 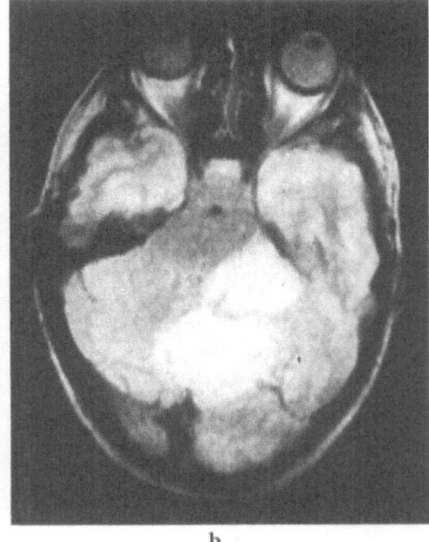

a   b

Fig. 10.15   Sagittal IR (**a**) and axial 2000/80 (**b**) MRI sequences demonstrate two discrete areas of tumour involving the vermis and left cerebellar hemisphere. The appearance of the hemispheric lesion suggests reduplication of cerebellar folia. The cerebellar tonsils are depressed. The fourth ventricle is displayed forwards and the lateral ventricles are enlarged. Histology confirmed the diagnosis of Lhermitte–Duclos disease

Ganglioglioma

Gangliogliomas are very rare, slow-growing, intrinsic tumours which may occur anywhere within the central nervous system, but are most commonly supratentorial in location. They may occur at any age but are most frequent in the first three decades of life. They are composed of mature, dysmorphic nerve cells and glial tissue. Typically, they form masses isodense with brain, which induce little oedema but enhance moderately. As with many other brain pathologies, they tend to show prolongation of both T_1 and T_2 relaxation times. Almost half contain calcification or cystic elements. On angiography they are avascular and only the larger tumours are identified by non-specific mass effect.

Hamartoma

Hamartomas are heterotopic, congenital masses of tissue, usually composed of glial cells, which can occur anywhere within the central nervous system. They are most frequent in the supratentorial region, particularly in the hypothalamus, close to the tuber cinereum – where they present with diabetes insipidus or precocious puberty – and the temporal lobe, presenting with epilepsy. On computed imaging they are generally shown as mass lesions, usually isodense and isointense with brain substance, but may contain calcification. They rarely enhance or induce oedema in the adjacent brain substance.

Ependymoma

Ependymomas are intrinsic tumours occurring throughout the central nervous system but usually arising close to an ependymal surface. They form only 5% of glial tumours and are most frequent in the first three decades. They are most commonly located in the cerebellum, adjacent to the floor of the fourth ventricle, and extending along the lateral recesses of the ventricle into the cerebellopontine angles. They tend to obstruct the fourth ventricle and present with hydrocephalus. The supratentorial tumours are most frequent adjacent to the lateral ventricular trigones, forming intraparenchymal masses. These are frequently lobulated, of higher than brain density on CT but often isointense on MRI. Calcification is present in about one-third. Enhancement tends to be moderate and homogeneous, though there may be non-enhancing cystic components. Peritumoral oedema is present in 50%.

The more malignant tumours are of higher density on CT and higher intensity on T_2W MRI. They tend to show intense enhancement, and calcification is less frequent. Nodular metastases along the CSF pathways are frequent.

On angiograms, some abnormal vascularity is usual. It is generally supplied by choroidal arteries and may therefore simulate primary intraventricular tumour. However, there is generally angiographic evidence of an intraparenchymal mass and the tumour vessels tend to be more irregular.

Oligodendroglioma

These tumours arise in the white matter, usually in the cerebral hemispheres and particularly in the frontal lobes. They mainly involve adults, and both sexes are affected equally. They tend to form masses expanding beneath the cortex or ependyma, and not uncommonly produce exophytic extra-axial or intraventricular components.

The tumours are of low density on CT with calcification occurring in 70% of the more benign and 50% of the higher grade tumours. Enhancement occurs in 90% of benign and all of the more malignant oligodendrogliomas; it is usually inhomogeneous, but is homogeneous in 30% of the more malignant lesions. Only about half the tumours incite surrounding oedema and it is frequently not extensive. Angiograms usually reveal an avascular mass; sometimes there is faint, diffuse staining, and exophytic tumours not uncommonly acquire a supply from dural or choroidal vessels. In such cases the primary presentation may be with an extra-axial effusion or haemorrhage.

Malignant Lymphoma

Included in this category are reticulum cell sarcoma, lymphosarcoma and microglioma. This group of conditions can cause widely varying appearances which may simulate many disease processes (Holtas et al., 1984).

The brain may be involved during the course of known systemic lymphoma, in patients with diseases or on drugs causing immune suppression, or as a primary tumour without known precipitating cause. The radiological appearances in all these subgroups may be similar, though the differential diagnosis varies with the clinical context. However, extracerebral involvement is more common in the presence of systemic lymphoma. The deposits may be flat or rounded, with bone destruction or sclerosis. There is usually considerable enhancement and there may be intracerebral oedema. They may simulate meningioma, but the presence of other systemic or CNS deposits is helpful in differential diagnosis. Systemic lymphoma may also present with meningeal involvement, resulting in communicating hydrocephalus: there may be no other manifestation on imaging at a time when CSF cytology is positive.

Parenchymal involvement by malignant lymphoma may present as a diffuse infiltrating process, or as focal mass lesions which are multiple in 40%–50% of cases. The most typical manifestation of the former is a mass of increased density, diffusely enhancing after contrast medium, extending around the margins of the lateral ventricles and often thickening the septum pellucidum, associated with moderate or marked oedema. The focal masses most typically involve the cortex or deep grey matter and form high-density, usually diffusely enhancing foci, with little surrounding oedema. Calcification is absent. The lesions, however, may show irregular ring enhancement, and when they involve white matter they may be of low or similar density to brain substance and occasionally fail to enhance, simulating encephalitis or infarction. Temporary, spontaneous regression may occur and be mistakenly interpreted to support the latter diagnoses.

The association of primary neoplasms, particularly B-cell lymphoma, with immune suppression has achieved particular importance in relationship to AIDS, in which 6%–8% of patients are affected. The tumours tend to be multicentric and vary in appearance from heterogeneous density, with diffuse enhancement, to low density, with ring enhancement and surrounding oedema. In such patients, failure of one or more lesions to respond to a trial of anti-toxoplasma treatment should suggest the possibility of lymphoma and the necessity for biopsy.

On angiography, lymphomas most frequently cause avascular masses, though there may be a tumour stain. Important features include a tendency to arterial encasement and dilatation of the deep medullary veins, which are more frequent in primary reticulum cell sarcoma. Involvement of vessels is also relatively more frequent in the rare variant of intracerebral lymphomatoid granulomatosis. This may result in infarction or haemorrhage within the tumour masses. The haemorrhage gives rise to the characteristic features shown at various stages of development on MRI, and the disease may be suggested by the presence of multifocal haemorrhages or haemosiderin deposit in the absence of a systematic cause, particularly with an intra-axial tumour involving the posterior fossa structures.

Primitive Neuroectodermal Tumour

Classical Medulloblastoma

These malignant parenchymal tumours usually occur in the first decade, the most typical manifestation being the medulloblastoma of the inferior cerebellar vermis. It most frequently presents as a homogeneous mass of altered signal on MRI or high density on CT, showing considerable enhancement, often well defined, but infiltrating the adjacent brain substance which shows moderate surrounding oedema. Atypical features are often present, including eccentric location, calcification, cystic or necrotic changes and lack of enhancement (Zee *et al.*, 1982). Though angiography reveals typical features, it is no longer indicated for diagnosis or treatment.

Central nervous system metastases should be sought as soon as the diagnosis has been established. Within the cranial cavity they most frequently form nodular masses posteriorly within the third ventricle, or around the basal meninges, and are well shown with enhanced CT and/or MRI scans. In the spinal subarachnoid space they are frequently shown by gadolinium-enhanced MRI. Where MRI is not freely available, myelography will reveal spinal metastases, which show as nodules on the surface of the spinal cord or meninges lining the spinal canal, the commonest site being the terminal theca. Silent spinal metastases are present in 50% of cases at the time of initial presentation: the extent of radiotherapy may be modified by their identification.

Adult Medulloblastoma

Medulloblastomas occurring in adults tend to lie more laterally, involving a cerebellar hemisphere and often infiltrating through the leptomeninges. They tend to form well-defined, high-density masses, and have a better prognosis, with less tendency to metastasize through the CSF spaces, than the childhood tumours.

Congenital Medulloblastoma

Congenital cerebellar medulloblastoma is a rare cause of hydrocephalus, occurring $2\frac{1}{2}$ more frequently in girls than in boys (Kucharczyk *et al.*, 1985; Kim *et al.*, 1985).

Supratentorial Neuroectodermal Tumour

Primitive ectodermal tumours occurring above the tentorium are frequently large, partly calcified and cystic masses, showing irregular enhancement after intravenous contrast medium. They tend to invade the ventricular system and cause cranial enlargement in younger children (Sauerbrei and Cooperberg, 1983).

Colloid Cyst

This forms a typical, well-demarcated mass within the third ventricle, adjacent to and often encroaching through the foramina of Monro, and causing hydrocephalus of the lateral ventricles and usually of the anterior extremity of the third ventricle. On CT, it is typically of high density, but occasionally isodense with brain. Calcification never occurs and enhancement is very infrequent. Magnetic resonance imaging exquisitely displays the anatomy, particularly that of the adjacent veins, on sagittal images. The cyst itself usually returns high signal on both T_1W and T_2W sequences.

Choroid Plexus Papilloma

This uncommon tumour presents as an intraventricular mass, most frequently with manifestations of hydrocephalus. In children, it usually involves the trigone of a lateral ventricle or, occasionally, the third ventricle. In adults, it most frequently involves the fourth ventricle, often extending through and sometimes confined to a lateral recess.

On CT it forms a mass of high density, frequently with mottled calcification and a well-defined rounded edge in the position of and expanding the affected ventricle. Enhancement is usually marked, but calcified and cystic components do not enhance, and some tumours have a marked central, fibrous stroma in which enhancement may be slight or absent. The tumour becomes adherent to and may transgress the ependyma, and then there may be marked oedema in the white matter, whether or not malignant changes are present. The former may be diagnosed with certainty only if nodular implants are shown on the walls of the ventricles beyond the tumour edge.

On MRI, the tumour usually returns low signal on T_1W sequences. On T_2W sequences, high signal from the tumour parenchyma is frequently mixed with low signal from the calcified and fibrous components, and from flow void in the tumour vessels when, as is usual, the tumour is markedly vascular.

On angiograms, the tumour is typically supplied by hypertrophied choroidal arteries, and shows a large number of small, tortuous blood vessels showing a homogeneous blush, which may be associated with early venous filling without malignant change being present. It should be noted that in some tumours vascularity is slight or absent, and that this finding does not exclude the diagnosis (Morrison and Sobel, 1984).

Intraventricular Meningioma

Intraventricular meningioma occurs in a similar situation to plexus papilloma, but most often arises in the trigone of the lateral ventricle and obstructs the temporal horn. On CT, it frequently has a regular outline and higher density than choroid plexus papilloma. Its MRI and angiographic features are very similar to the latter tumour (Kendall et al., 1983).

Pineal Region Tumour

Pineal region tumours tend to extend anteriorly through the pineal recess of the third ventricle to form a mass which, as it enlarges, occludes the ventricle and the aqueduct, causing three-ventricular hydrocephalus. They also grow inferiorly and posteriorly to impress the quadrigeminal plate and superior vermis, and superiorly to elevate the vein of Galen and splenium of the corpus callosum. Though the tumours are well shown by CT, the anatomy of the region and the differential diagnosis from falco-tentorial meningiomas and vein of Galen aneurysms are best revealed by MRI, particularly in the sagittal plane. The meningiomas tend to extend along the meninges, to involve the straight sinus and depress all the vessels within the tentorial hiatus. Aneurysm of the vein of Galen and any related angiomatous malformation will be distinguished by the characteristic appearances of flowing blood in various sequences or, if clotting has occurred, from the signal characteristics of methaemoglobin or haemosiderin. Curvilinear calcification is frequent in the wall of an aneurysmal vein.

Germinoma

Germinomas are more frequent in boys. They form high-density masses, often associated with heavy calcification of the pineal gland. They usually enhance homogeneously and may metastasize along the walls of the ventricular system, particularly to the hypothalamic region.

Pineal Teratoma

Pineal teratomas cause cystic masses, with enhancing capsules. They usually contain calcified or ossified components as well as fatty substance. Occasionally there are formed elements simulating a fetus.

Pineocytoma

Pineocytomas form rounded, high-density, well-circumscribed masses, showing considerable enhancement. Calcification is frequent but cyst formation uncommon, and there is little or no surrounding oedema.

Pineal Astrocytoma

Astrocytomas frequently cause smaller infiltrating masses, with ill-defined edges extending along the third ventricle or thickening the quadrigeminal plate. Such lesions may simulate aqueduct stenosis on CT, but are usually distinguished on MRI by altered signal around the aqueduct and third ventricle, often associated with thickening of the tissues.

Haemangioblastoma

Haemangioblastoma usually presents as a cerebellar tumour, most frequently in the hemisphere of a young adult. The symptomatic tumours are large and generally easily shown by CT. They may be solid, diffusely enhancing masses of slightly higher or of similar density to brain substance. Central cystic or necrotic and non-enhancing regions are frequent. Many haemangioblastomas are associated with large, intra-axial cysts, and the enhancing nodular component of the tumour may be inconspicuous. Haemangioblastomas are multiple in 15%–20% of cases, particularly in the von Hippel–Lindau syndrome. The asymptomatic tumours are frequently small and tend to lie adjacent to the pia over the cerebellum, medulla – particularly its posterior surface – and the spinal cord. Occasionally they are supratentorial. Their peripheral location may make detection with CT difficult, and they are best shown by MRI after gadolinium enhancement. When MRI is not available, they may be well demonstrated by magnification angiography. They are highly vascular, with fine tumour vessels filled in the arterial phase and a dense blush persisting through the capillary and venous phases. There are frequently large veins which may drain early.

Multiple Primary Brain Tumours

Multiple tumours should be suspected in patients (a) with genetic predisposition, as occurs in neurofibromatosis, von Hippel–Lindau syndrome and tuberous sclerosis; (b) who have undergone cranial irradiation; and (c) who have immune suppression. Otherwise, multiple brain tumours of different cell types are occasionally detected during imaging of a symptomatic tumour. The unsuspected tumour is most frequently one of the more common benign lesions, such as a pituitary tumour or meningioma. Occasionally the two tumours are in close proximity, and in certain circumstances the second tumour may only be recognized as a separate mass following an unsatisfactory response to surgery. In such cases it may be detected in retrospect on reviewing the preoperative images.

Cavernous Haemangioma

The typical appearance on MRI is of a single or occasionally multiple rounded zone(s) of decreased intensity on T_1W images, with a rather more extensive abnormality, also of low signal, when using a more T_2W sequence. Evidence of past haemorrhage may be present in the centre of the lesion, but significant mass effect is uncommon. Imaging with the use of flow-sensitive sequences does not show evidence of increased flow within these malformations (Lee *et al.*, 1985).

Neuroblastoma

Extracranial neuroblastoma is one of the commonest tumours of early childhood. Metastases to the orbits and cranial sutures causing extracerebral mass lesions and raised intracranial pressure are common, but intracerebral metastases are relatively rare. Primary cerebral neuroblastoma is a very rare, highly malignant tumour occurring anywhere within the brain substance. It grows rapidly and usually undergoes extensive necrosis, haemorrhagic and cystic degeneration; it metastasizes along the CSF pathways (Just *et al.*, 1989).

The CT and MRI changes reflect the non-specific macroscopic appearances, and diagnosis depends on histology and immunocytology.

Facilitation of Treatment and Further Surgical Planning

After detection and localization of a mass, the surgeon may require further information prior to operative intervention. This will include:

1. If possible, an indication of the consistency of the tumour, whether it is cystic, necrotic, solid, and if solid whether it may be hard or soft.
2. The relationship of blood vessels to the operative approach and whether they may be involved by the tumour.

The major part of this information can now be provided by MRI, and it is not usually necessary to perform angiography prior to surgery on many intra-axial tumours. With extra-axial tumours, however, particularly meningiomas, involvement of major arteries and, especially, involvement of the large venous sinuses may be crucial in determining the extent of resection which can safely be performed.

3. Following partial surgical resection or biopsy, and prior to radiotherapy, imaging should be repeated to allow maximum accuracy for the placement of radiation fields.
4. In tumours which commonly disseminate throughout the central nervous system, particularly through the subarachnoid space, MRI or myelography may be necessary to determine the extent of the treatment fields.

Monitoring of Therapy and Outcome of Treatment

Surgery

Craniotomy defects, metal clips and low-attenuation intracerebral lesions, due to the unavoidable contusion caused by cutting into brain substance, are the inevitable result of surgical procedures. In addition, fluid, sometimes haemorrhagic, may accumulate within cavities, or gelatin foam may be placed within them to absorb such fluid. Reactive inflammatory changes occur within a few days and may be the cause of enhancement following intravenous contrast medium. If it is considered that follow-up of a partially resected tumour will be facilitated by obtaining a control scan, this is best preformed within a day or two of the surgery, before the reactive enhancement occurs. Otherwise, follow-up scanning is probably indicated only when it will affect management, and is therefore essentially dictated by the state of the patient.

Either CT or MRI scanning is appropriate for detection of the cause of any unexpected deterioration following surgery or other therapy. Intracranial haemorrhage, cerebral infarction, hydrocephalus due to arachnoidal adhesions or obstruction of Pacchionian granulations, and reactive oedema following radiation therapy, are all well shown.

Radiotherapy

The effects of radiation therapy on tumour volume are more difficult to assess. Early radiation reactions are due to parenchymal cell injury, and if the therapy has been directed correctly it is generally assumed that a central, non-enhancing region indicates tumour necrosis or cystic change, especially if it has appeared or increased following the treatment. Such a region may, however, contain viable tumour, or be partly due to brain softening. Peripheral and peritumoral non-enhancing low density, or region of increased T_2, is usually attributed to oedema, but necrosis, infarction or tumour infiltration could also be present (Graeb et al., 1982).

The damaging effects of radiation therapy on the vascular endothelial cells usually commence after about two months and continue for up to two years. They cause secondary ischaemic changes in the white matter, with demyelination or radionecrosis. Such changes may cause swelling and enhancement indistinguishable from recurrent or residual tumour. Hence, disappearance of enhancement generally indicates regression of tumour, but recurrent enhancement in the same region may be due to recurrent tumour, granuloma formation or radionecrosis (Hopewell, 1974).

Symptomatic improvement with tumour regression does not usually occur until after completion of radiotherapy. The regression generally continues for a very variable period, sometimes as long as nine months. Progression, as indicated by increase in mass effect, in recurrence or extension of an enhancing region can be recognized in some cases before any clinical deterioration, but rarely precedes it by more than a month or two. Further treatment is usually determined by the patient's clinical state, and it is generally better to use imaging as a method for assessing the cause of deterioration rather than as a means of anticipating it.

References

Araki T, Inouye T, Suzuki H, Machida T, Masahiro I (1984) Magnetic resonance imaging of brain tumors: measurements of T_1. Radiology 150: 95–98

Atlas SW, Grossman RI, Gomori JM et al. (1989) MR imaging of intracranial metastatic melanoma. J Comput Assist Tomogr 11: 577–582

Brant-Zawadzki M, Norman D, Newton TH et al. (1984a) Magnetic resonance of the brain: the optimal screening technique. Radiology 152: 71–77

Brant-Zawadzki M, Badami JP, Mills CM, Norman P, Newton TH (1984b) Primary Intracranial tumor imaging: a comparison of magnetic resonance and CT. Radiology 150: 435–440

Destian S, Sze G, Krol G, Zimmerman RD, Deck MD (1988) MR imaging of hemorrhagic intracranial neoplasms. Am J Neurorad 9: 1115–1122

Destian S, Sze G, Krol G, Zimmerman RD, Deck MD (1989) MR imaging of hemorrhagic intracranial neoplasms. Am J Roentgenol 152: 137–144

Earnest F, Kelly PJ, Scheithauer BW et al. (1988) Cerebral astrocytomas: histopathological correlation of MR and CT contrast enhancement with stereotactic biopsy. Radiology 166: 823–827

Gomori JM, Grossman RJ, Goldberg HI, Zimmerman RA, Bilaniuk LT (1985) Intracranial hematomas: imaging by high-field MR. Radiology 157: 87–93

Graeb DA, Steinbok P, Robertson WD (1982) Transient early computed tomographic changes mimicking tumor progression after brain tumor irradiation. Radiology 144: 813–817

Haughton VM, Rimm AA, Czervionke LF et al. (1988) Sensitivity of Gd–DTPA-enhanced MR imaging of benign extraaxial tumors. Radiology 166: 829–833

Holland BA, Brant-Zawadzki M, Norman D, Newton TH (1985a) Magnetic resonance imaging of primary intracranial tumors: a review. Int J Radiat Oncol Biol Phys 11: 315–321

Holland BA, Kucharcyzk W, Brant-Zawadzki M et al. (1985b) MR imaging of calcified intracranial lesions. Radiology 157: 353–356

Holtas S, Nyman U, Cronqvist S (1984) Computed tomography of malignant lymphoma of the brain. Neuroradiology 26: 33–38

Hopewell JW (1974) The late vascular effects of radiation. Br J Radiol 47: 157–158

Just M, Goebel HH, Bohl J, Schwarz M, Thelen M (1989) Magnetic resonance imaging in primary cerebral neuroblastoma. Neuroradiology 31: 108

Kendall B, Reider-Grosswasser I, Valentine A (1983) Diagnosis of masses presenting within the ventricles on computed tomography. Neuroradiology 25: 11–22

Kieffer SA, Salibi NA, Kim RC *et al.* (1982) Multifocal glioblastoma: diagnostic implications. Radiology 143: 709–710

Kim JH, Duncan C, Manuelidis EE (1985) Congenital cerebellar medulloblastoma. Surg Neurol 23: 75–81

Kucharczyk W, Brant-Zawadzki M, Sobel D *et al.* (1985) Central nervous system tumors in children: detection by magnetic resonance imaging. Radiology 155: 131–136

Laster DW, Ball MR, Moody DM, Witcofski RL, Kelly DL (1984) Results of nuclear magnetic resonance with cerebral glioma. Comparison with computed tomography. Surg Neurol 22: 113–122

La Bas JF, Benabid AL, Camuset JP *et al.* (1987) MRI in the prestereotactic evaluation of intracerebral tumours. Comparison with CT. J Neuroradiol 14: 203–221

Lee BC, Herzberg L, Zimmerman RD, Deck MD (1985) MR imaging of cerebral vascular malformations. Am J Neurorad 6: 863–870

Morrison G, Sobel DF (1984) Intraventricular mass lesions. Radiology 153: 435–442

Oot RF, New PFJ, Pille-Spellman J *et al.* (1986) The detection of intracranial calcifications by MR. Am J Neurorad 7: 801–809

Price AC, Runge VM, Allen JH, Partain CL, James AE (1986) Primary glioma: diagnosis with MRI. J Comput Tomogr 10(4): 325–334

Rinck PA, Meindl S, Higer HP, Bieler ES, Pfannenstiel P (1985) Brain tumours: detection and typing by use of CPMG sequences and in vivo T_2 measurements. Radiology 157: 103–106

Sauerbrei EE, Cooperberg PL (1983) Cystic tumors of the fetal and neonatal cerebrum: ultrasound and computed tomographic evaluation. Radiology 147: 689–692

Stack JP, Antoun NM, Jenkins JPR, Metcalfe R, Isherwood I (1988) Gadolinium–DTPA as a contrast agent in magnetic resonance imaging of the brain. Neuroradiology 30: 145–154

Vaghi M, Visciani A, Passerini A, Longone V, Broggi G (1986) La risonanza magnetica nello studio dei gliomi cerebrali. Radiol Med (Torino) 72: 431–438 (Eng Abstr)

Young IR, Bydder, GM, Hall AS *et al.* (1983) NMR imaging in the diagnosis and management of intracranial angiomas. Am J Neurorad 4: 837–838

Zee C-S, Segall HD, Miller C *et al.* (1982) Less common CT features of medulloblastoma. Radiology 144: 97–102

11 Surgery for Primary Malignant Brain Tumours

J. Ashraf and R. Bradford

The development of new methods of diagnostic imaging, improvement in surgical techniques and perioperative care and advances in adjuvant therapy in recent years have failed to improve the overall prognosis for gliomas. Although malignant in behaviour, these tumours rarely, if ever, metastasize outside the central nervous system (CNS) and by the time of diagnosis the majority are still macroscopically confined to their site of origin. Hence, surgical resection would appear to be the logical method of treatment. However, in most cases complete resection is not possible because of the infiltrative nature of the tumour and proximity of important areas of brain. For these reasons it appears unlikely that surgery will ever be the definitive treatment for this disease. However, in current clinical practice it is of established value in providing a pathological diagnosis, safe and rapid palliation and prolongation of symptom-free survival and, by reduction of tumour volume, assisting adjuvant therapy.

Historical Introduction

Although considerable knowledge regarding brain tumours had existed in the second half of the nineteenth century, it was not until 1884 that the first operation for a malignant glioma was performed by Sir Rickman Godlee (Bennett and Godlee, 1884). These and subsequent early attempts at surgery by Horsley and Cushing preceded radiological evaluation and relied solely on the principles of cerebral function localization introduced earlier by Jackson and Ferrier (Bradford and Thomas, 1990) and on the findings at exploration. The overall results of surgery were poor and mortality was high. Hence, attempts at curative resection lost popularity and only palliative external decompression was performed, often accompanied by high surgical mortality, major neurological deficits and an unsightly extracranial mass (Jelsma and Bucy, 1967).

With the introduction of diagnostic studies like ventriculography, pneumoencephalography and cerebral angiography, it became possible to confirm the clinical diagnosis by radiological means, albeit indirectly, and hence purely exploratory craniotomy was no longer necessary. Also during the same period the general principles of neurosurgery were being defined by such neurosurgical giants as Cushing and Dandy and these, together with advances in anaesthesia, surgical technique, operative haemostasis and better understanding of intracranial physiology, improved the results of surgery, but a similar improvement in the overall survival figures could not be achieved. As, during this early period, the infiltrative nature of these tumours was not recognized and it was thought that they could be extirpated by radical resection and the overall survival period could be prolonged (Dandy, 1921; McKenzie 1936). In the process frontal, temporal and occipital lobectomies were developed and new approaches to the supratentorial and infratentorial tumours were described. However, it soon became obvious that it was the biological behaviour of the tumour rather than the extent of surgical resection that dictated the overall prognosis. Cushing and Bailey introduced the first histological classification of gliomas (Bailey and Cushing, 1926) and advocated that the extent of resection should be dictated by pathological type and location of the tumour. The policy of radical resection was more successful with cerebellar astrocytomas, which on the whole present earlier with hydrocephalus, are smaller at the time of presentation and are less aggressive in behaviour and therefore the long-term survival is good. The same results were not achieved with supratentorial gliomas, where, whatever form of surgery was performed, the overall prognosis remained poor. Although with improvement in surgical conditions the operative mortality had steadily been reduced, there was no significant prolongation in survival and gradually the role of surgery was reduced to that of establishing a diagnosis by means of a burr-hole biopsy or of providing palliation with internal or external decompression. The attempts to achieve cure through surgery came to a standstill.

The advent of radiation as adjuvant therapy, in the early 1940s, renewed enthusiasm in a curative role for surgery. Surgical resection of the tumour followed by radiation treatment of adjacent brain became an established practice. Although the tumour could not be eradicated completely, survival was significantly prolonged. Despite its morbidity, radiotherapy has remained, to date, the only form of adjuvant therapy that has consistently improved survival (Walker *et al.*, 1978).

Like other areas of medicine, the use of antibiotics, blood transfusion and attention to fluid replacement also made a significant impact on the surgical results in neurosurgery. However, it was the introduction of steroids in the early 1960s that produced a dramatic improvement in postoperative survival figures. This, together with further advances in neuroanaesthesia and in surgical techniques reduced the operative mortality figures from around 20%–25% to as low as 2%–3% in some series (Jelsma and Bucy, 1967).

The greatest revolution to date in neurosurgery occurred with the advent of computed tomography (CT scanning) in the early 1970s. With this new imaging method it has become possible to visualize not only the anatomical location and extent of the tumour, but also to predict its nature with some degree of certainty. Deep small lesions which could not be visualized in the past can be seen and biopsied. The results of surgery, extent of resection, postoperative complications and effects of radiotherapy and the development of recurrent tumour can be diagnosed by a non-invasive method. A natural extension of this new imaging technique is its application in surgical planning. CT-directed stereotaxic biopsy has made it possible to obtain tissue from small deep lesions, from heterogeneous areas within the tumour and from the adjacent brain. Magnetic resonance (MR) has added a further dimension not only to imaging for diagnosis and treatment (Bydder et al., 1982), but also MR spectroscopy may prove a highly useful non-invasive method for pathological diagnosis.

More recently, the application of the principles and instrumentation of microsurgery for tumour resection, the use of the operative microscope and lasers in some situations have refined surgical techniques even further.

Principles of Surgery

Shapiro (1982) has outlined five objectives for surgical treatment of malignant gliomas. First, it enables the neuropathologist to establish a pathological diagnosis; secondly, the bulk of the tumour is removed; thirdly, distressing symptoms are relieved; and fourthly tumour volume reduction permits time for adjuvant therapy. Finally, removal of tumour bulk alters tumour kinetics and induces quiescent cells to resume an active phase of growth, which increases their susceptibility to irradiation or to chemotherapy.

Establishment of a Diagnosis

Although the clinical diagnosis of an intracranial mass lesion can be confirmed by CT and MR scan, these highly sensitive methods of neuro-imaging do not provide enough data to enable a firm pathological diagnosis to be made. This continues to depend on histological examination of tissue samples (Thomas et al., 1988). The accuracy of diagnosis of gliomas on the basis of enhanced CT scan, considered in combination with clinical presentation, is 90% (Kendall, 1980). This leaves a 10% margin of error in diagnosis if imaging methods alone are considered. The CT and MR appearances of some types of lesions may be specific enough to enable an accurate diagnosis to be made on radiological grounds alone (Salcman et al., 1981). However, there are well-known exceptions, such as the solitary metastatic lesion, an area of inflammation or infarction, an abscess, a meningioma or a granuloma, among others; all of these can masquerade as glioma. It is therefore essential to obtain a tissue diagnosis.

Although there are exceptional situations where there is significant morbidity associated with surgery, e.g. in tumours of the brain stem and pineal region, in which a case can be made for treating these lesions on empirical grounds (Hide 1975), in most other cases, since radiotherapy and chemotherapy carry such a significant morbidity, it is essential to establish tissue diagnosis prior to subjecting a patient to either of these forms of

treatment. The untreated course of these tumours is relentlessly progressive, and establishment of a firm diagnosis is also necessary if treatment is
contraindicated for some reason (Garfield, 1980). It is also important to
classify a tumour according to its histological grade, before planning treatment and predicting prognosis. As the histological behaviour of glioma is
known to change with radiotherapy and in some cases spontaneously
(Roth and Elvidge, 1960; Jelsma and Bucy, 1967; Pool, 1967), histological
grading at the time of presentation is essential.

Cytoreduction and Alteration of Tumour Kinetics

Although surgery cannot be regarded as curative for malignant glioma, it is
a most useful component of adjuvant therapy (Bradford and Thomas,
1990). By reducing the bulk of the tumour volume it allows time for adjuvant therapy to act. With cytoreduction, there is a smaller number of cells
that require adjuvant treatment and hence its dosage can suitably be tail-
ored. Also, reduction of the tumour volume induces cells which are not in
proliferative cycle to resume an active phase of growth and this change in
tumour cell kinetics increases their susceptibility to radiotherapy and
chemotherapy. Moreover, by surgical resection the tumour cells are not
only destroyed but also removed, thereby sparing the body the necessity of
disposal of dead cells (Shapiro, 1982). It is well known that most gliomas
are made of a heterogeneous population of cells with differing sensitivity to
various forms of radiotherapy or chemotherapy. By surgical resection a
large bulk of the insensitive population of cells can be removed, thereby
further assisting adjuvant therapy (Salcman, 1982b).

Relief of Symptoms and Improvement in Survival

By immediate relief of elevated intracranial pressure and the treatment of
hydrocephalus, survival figures are improved and, when the lesion is of a
lower grade, a long duration of good-quality survival can be achieved.
Although in various series the median prolongation of survival has been
directly correlated with the extent of resection (Frankel and German, 1958;
Roth and Elvidge, 1960; Jelsma and Bucy, 1967), this improvement in survival has been recorded only in the intermediate period, and the duration
of the overall survival has been found to be independent of the method of
resection. Nevertheless, this improvement in survival in the intermediate
period provides sufficient time for adjuvant therapy.

The influence of surgery on the improvement in the quality of survival is
even more difficult to ascertain. Where the symptoms of raised intracranial
pressure, like headache and recurrent vomiting, are due to small cerebellar
or pineal lesions, relief of hydrocephalus may provide good palliation
during the period of survival. Similarly, if the focal deficit in an alert
patient is due to compression of an eloquent area by a tumour in an adjacent silent area, resection of this may relieve the symptoms. Similarly,
surgical resection of the tumour can arrest the otherwise inexorable

progression of the neurological deficit and preserve the functional quality of existence to some extent. Such a favourable situation is, however, rare and the local symptoms are more likely to be due to direct tumour infiltration of an eloquent area, so that surgery is less likely to relieve the symptoms. With the improvement in perioperative care and reduction in overall operative risks, a stronger case can be made for surgery. However, prolongation of a poor-quality survival should be discouraged. Deterioration in the level of consciousness, and the impairment of intellect due to raised intracranial pressure associated with extensive high-grade gliomas, may provide natural relief and there may be little justification in partially resecting these lesions where prolongation of such a quality of survival is inadvisable.

Principles of Perioperative Care

The presence of a mass lesion affects intracranial dynamics in several ways (see Chapter 5). There is the added volume of the mass and surrounding zone of vasogenic oedema. The autoregulation of cerebral blood flow in the tumour and in the adjacent brain is impaired and the cerebral blood flow and hence the blood volume in this region often changes in parallel with the mean arterial pressure. With the displacement of the adjacent normal brain and distortion of blood vessels, a zone of ischaemia and disturbed autoregulation is produced even in areas remote from the tumour. As the tumour grows in size, this zone of disturbed autoregulation extends. In some cases obstruction of the CSF pathways can produce generalized and focal hydrocephalus. With the increase in intracranial volume, there is a slow progressive increase in intracranial pressure and although in a slow-growing tumour the pressure–volume relationship may still be critically compensated at the time of presentation due to the loss of compliance, even the slightest increase in intracranial volume can cause a precipitous increase in intracranial pressure. As this state of critical compensation exists in all but very small tumours, the absence of symptoms and signs of raised intracranial pressure does not necessarily mean that the pressure–volume relationship is in its normal range. Optimizing the intracranial physiology before surgery and keeping it optimized in the perioperative period is essential.

The use of steroids (dexamethasone) in the perioperative period, the smooth induction, maintenance and reversal of anaesthesia with agents having minimal effects on cerebral blood flow and intracranial pressure, the prevention of hypoxia and hypercapnia during anaesthesia, the maintenance of the circulatory status within physiological limits, the observation of surgical principles of proper positioning and good access, minimal manipulation, retraction and tissue damage, meticulous haemostasis and careful monitoring in the perioperative period are essential prerequisites for good operative results.

Surgical Methods

The main objectives of surgical management are first to establish the diagnosis and secondly to reduce tumour volume sufficiently to provide palliation of distressing symptoms and allow time for adjuvant therapy. Whenever possible, it is preferable to perform complete resection of the tumour with minimal disturbance of adjacent normal brain and without producing any further major neurological deficit. In current clinical practice the surgical procedures include:

1. Biopsy.
2. Craniotomy with either partial or complete tumour resection without lobectomy.
3. Lobectomy – removal of the tumour with adjacent normal brain, constituting an anatomical lobe of the brain.
4. Additional surgical procedures such as insertion of shunts and reservoirs.

Methods of Establishing Tissue Diagnosis

Where resection of some form is undertaken as a primary procedure, tissue diagnosis is established after examination of the resected specimen. In all other cases, as the definitive diagnosis of glioma is established only after tissue analysis, biopsy of the lesion is the minimum necessary procedure.

Freehand Biopsy

Large superficial lesions, especially in the non-dominant hemisphere, where there is little doubt regarding clinical and radiological diagnosis, can be biopsied freehand, although stereotaxic techniques are preferred. Tumour cysts can be drained at the same time. A prior CT scan with a marker on the scalp assists localization and positioning of the burr hole. An immediate frozen section examination is essential to confirm that pathological material has been obtained. Diagnosis of malignant glioma is definitive. However, where the tissue diagnosis suggests a lower grade of malignancy, doubt often remains regarding other more malignant areas of the tumour, as a small specimen may include only the well-differentiated portions of an astrocytic tumour while missing the area of malignancy (Shetter et al., 1977).

Although freehand biopsy provides a simple method of diagnosis, several passes through the brain are usually necessary before a diagnostic specimen is obtained (Bradford and Thomas, 1990). The morbidity/mortality of the procedure relates directly to the number of passes required for a diagnostic specimen. Hence, increasing the accuracy necessarily increases the surgical risks. In an early comparison of biopsy versus resection, Hitchcock and Sato (1964) reported a 27% mortality associated with burr-hole biopsy in contrast to 4% associated with definitive removal of tumour volume. Similar results were reported by other workers (Frankel and German, 1958; Cheek and Taveras, 1966). This high mortality may in part reflect selection bias, as needle biopsy alone is normally reserved for patients in poor condition. After the introduction of steroids – an improvement in pre-operative management – the mortality rate was reduced to approximately 5% (Marshall *et al.*, 1974), but tissue diagnosis could only be made in 88% of the patients. In another series (Shetter *et al.*, 1977), increasing the diagnostic yield to 95% required increased number of passes and resulted in a permanent complication rate of 10% and transient deterioration in 20% of the patients.

Image-Guided Stereotaxic Biopsy

Use of CT guidance for freehand biopsy was the first step towards image-guided stereotaxic biopsy. The integration of CT guidance with stereotaxic biopsy has improved the accuracy of diagnosis and reduced the morbidity and mortality. In conjunction with computer tomography and MR scanning, stereotaxic biopsy is safe and the error associated with sampling has almost been eliminated. It is possible to obtain multiple biopsies from different areas of the tumour and adjacent brain with a small-diameter needle using a safe trajectory, and hence a correct histological diagnosis is possible with minimal disturbance to the surrounding tissue. (Fig. 11.1). The accepted accuracy of most stereotaxic systems is within 1 mm of the target. Biopsies are obtained at exact and reproducible sites with high diagnostic accuracy and low surgical risk (Ostertag *et al.*, 1980). The difficulty of making a neuropathological diagnosis on a small tissue sample is to some extent offset by the ease and safety with which multiple tissue samples can be obtained with precision from various specific areas of interest on the CT scan (Bradford and Thomas, 1990). For instance, there is often a clear difference between the tumour centre and its edge as defined on the CT scan. In 46% of the biopsies of malignant glioma, the enhancing edge has been found to be tumour free (Thomas *et al.*, 1985), suggesting that in some selected cases radical local treatment may be warranted for tumour control. Image-guided stereotaxic methods allow safe and direct access to most deep small lesions in such important areas of the brain as thalamus, basal ganglia, pineal region and the brain stem.

The first use of MRI-directed tumour biopsy in clinical practice was reported by Thomas *et al.* (1986). Subsequent experience with MRI-guided biopsy (Lunsford *et al.*, 1986; Thomas *et al.*, 1987; Bradford *et al.*, 1987) has shown that it is a useful adjunct to CT in the management of tumours in difficult situations where CT-controlled biopsy would not be possible.

CT- and MR-directed stereotaxic procedures carry a low operative risk and a high diagnostic yield. In a series of 302 stereotaxic procedures, Ostertag *et al.*, (1980) reported a mortality of 2.3% and a transient morbidity of 3% and were able to make a diagnosis in all the cases, while Thomas and Nouby (1989) were able to establish a histological diagnosis in more than 92% of their cases and had an overall complication rate of 6% and

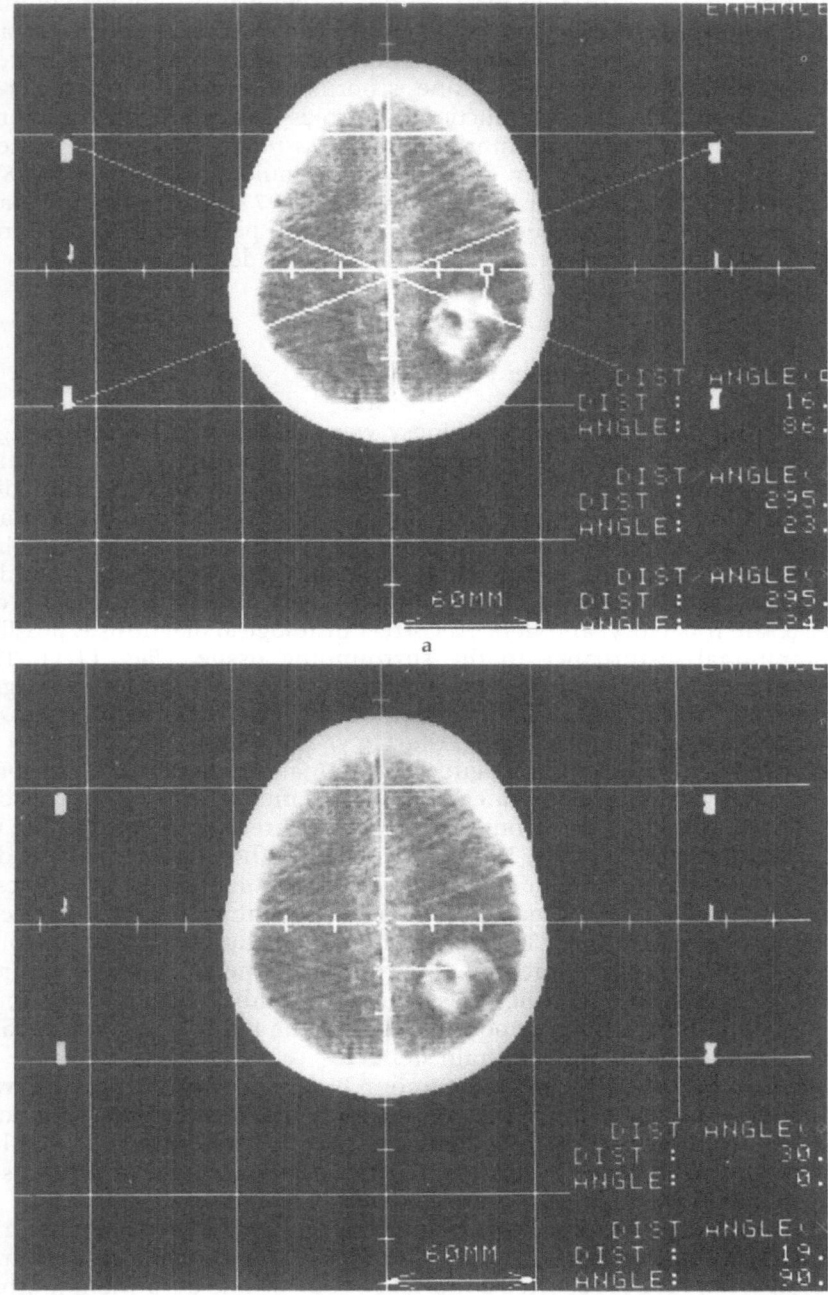

a

b

0.3% operative mortality. In another series of 240 procedures (Coffey and Lunsford, 1988), diagnostic specimens were obtained in 95.8% of the cases. There was no operative mortality and less than 5% morbidity resulted from the procedure. These and similar results from other workers (Coffey and Lunsford, 1985; Hood *et al.*, 1986; Apuzzo *et al.*, 1987) confirm the superiority of stereotaxic procedures over other methods of establishing tissue diagnosis.

Ultrasound-Guided Biopsy

Intraoperative ultrasonography provides a simple method of localization of subcortical lesions and thereby assists surgical intervention. In order to perform an ultrasound-assisted biopsy, a craniotomy is usually necessary, although recently special probes have been designed which can be used through a suitably placed burr hole (Kane *et al.*, 1988). Prior CT scan with a skin marker assists in correctly positioning the burr hole. An ultrasound-guided stereotaxic biopsy device, consisting of a swivelling pivot into which an ultrasonic probe is placed, is then screwed into the burr hole. Once the lesion is localized with the probe, the pivot is locked into position and the transducer is replaced by a needle guide. The depth of the required needle placement is calculated from the data obtained by the ultrasound and the needle is passed directly into the target. Cysts can be aspirated successfully. This method is suitable for larger lesions with distinct difference in density from surrounding tissue and has been used successfully to biopsy deep intracranial lesions.

Methods of Tumour Resection

The detailed steps involved in a craniotomy and various lobectomies are described in standard operative texts (Walters and Schmidek, 1988; McCabe 1989). Only the important principles applicable to glioma surgery are described here.

◀ **Fig. 11.1** Image-guided stereotaxic biopsy. Scanning is performed in a CT/MR compatible stereotaxic head-holder with a localizer system. Each image displays the lesion with reference markers. An axial scan through the centre of the frame is the reference image. The target areas are selected. The *z* coordinate of the target is the axial distance of the target scan from the reference scan. By diagonally connecting the reference markers, the centre of the stereotaxic frame is defined on the target scan (**a**) and the *x* and *y* coordinates calculated from the anteroposterior and lateral distance of the target from the centre (**b**). The trajectory and depth of the needle placement, with reference to the burr hole, are calculated from this information

Principles of Craniotomy

The general principles of perioperative care have been discussed above. All operations are performed under general anaesthesia with endotracheal intubation.

Positioning and Exposure. Most procedures are performed in the supine position, but posteriorly placed tumours require the semiprone, three-quarters prone or sitting position. The head end is elevated 15°–20°, but extreme rotation and forward flexion is avoided in order to maximize venous drainage. A wide exposure is planned to necessitate minimal brain retraction, to provide for perioperative cerebral swelling and to visualize the important operative landmarks. Care is taken to base the scalp flap on an adequate blood supply, as future adjuvant therapy is likely to compromise healing. An osteoplastic skull flap is preferred. If, despite the preoperative administration of steroids and hyperventilation, the dura is under tension, mannitol is administered as an infusion of 1 g kg^{-1}. Once the intracranial tension is controlled and the dura is slack, it is opened rapidly as a flap with the base towards the major cortical venous drainage.

Identification of Tumour and Resection. If the tumour extends to the surface, there may be obvious cortical discoloration, abnormal vascularity and the presence of arterialized veins on the surface. A widened gyrus is another guide to its location. Alteration in texture and consistency, as palpated on the cortical surface, may be fallacious. The brain can be probed with a brain cannula, but the difference in consistency between the tumour and the adjacent brain may not be marked enough to allow distinction. Cystic gliomas are easier to locate by this method. When difficulty in locating the tumour is anticipated, a preoperative scan with a surface marker helps to identify the position of the tumour relative to the surface. Intraoperative ultrasonography, through an intact dura, also assists in localization.

Tumours located close to the surface are approached directly through the overlying cortex. Eloquent areas are identified and protected at all times. If the tumour lies under an eloquent area, it is best approached indirectly through an adjacent silent area. A small 2–3 cm cortical incision is made along the long axis of the gyrus and is deepened to the tumour brain interface. If the tumour is cystic, the cyst is evacuated first and the adjacent tumour tissue excised later. The soft and necrotic centre of the tumour is removed with gentle suction, but the more solid portions of the tumour may require ultrasonic aspiration. If resection proceeds from within outwards till the tumour–brain interface is reached, it may be possible to achieve complete resection of identifiable tumour tissue without incurring any further neurological deficit. If there is doubt regarding the brain–tumour interface, it is best to be conservative in resection close to an eloquent area. However, if radical resection is likely to damage a relatively silent area, a more complete resection is preferred.

After scrupulous haemostasis and a watertight dural closure, the bone flap is secured with stitches and the scalp is repaired in two layers.

Lobectomy as a Method of Resection

The general principles involved in a craniotomy for a lobectomy are similar to those for a simple tumour resection. During planning of exposure it is essential to keep the important landmarks in consideration and to identify and protect these before commencing resection (Fig. 11.2a).

Frontal Lobectomy. A frontal lobectomy can be performed safely on either side, provided that the motor strip and, on the dominant side, Broca's area are identified and protected. Cortical incision commences at the supero-medial margin 7 cm behind the frontal pole and extends across the supero-lateral surface to the lesser wing of the sphenoid, but avoids the posterior half of the inferior frontal gyrus on the dominant side (Fig. 11.2a). On the medial surface, the posterior limit of resection is at the level where the two hemispheres separate anterior to the corpus callosum (Fig. 11.2b). The anterior cerebral arteries are protected in their course around the genu of the corpus callosum and the frontopolar arteries divided on the cortical surface. The gyrus rectus is left in situ to prevent damage to the cribriform plate. The inferior line of resection joins the lower ends of the medial and superolateral incisions and roughly corresponds to the lesser wing of the sphenoid (Fig. 11.2c). Any residual tumour in the margin of resection is removed separately.

If, on the dominant side, the motor strip and the speech areas have been protected during resection, the residual neurological deficit involves only minor impairment in higher cerebral function associated with some slowing of intellect, lack of initiative, apathy and some impairment of recent memory of voluntary and serial performance tasks. With a similar resection of the non-dominant frontal lobe, the postoperative deficit is minimal.

Temporal Lobectomy. A safe temporal lobectomy involves resection of the anterior 5–6 cm of the temporal lobe. On the dominant side the first and second temporal convolutions posterior to the intersection of the Sylvian and Rolandic fissures (angular and supramarginal gyri) are best left undisturbed, although on the non-dominant side these can be included in the resection if necessary. It is also important to leave a thin layer of superior temporal gyrus adjacent to the Sylvian fissure in order to avoid the Sylvian vessels. The superior margin of cortical resection extends along the superior temporal gyrus from its anterior edge to the junction of the Sylvian and Rolandic fissures posteriorly (Fig. 11.2a). From here it extends inferiorly across the middle and inferior temporal gyri to the floor of the middle cranial fossa, where the posterior limit corresponds to the vein of Labbé, which is left intact. The plane of resection is deepened through the white matter of the temporal lobe till the pia-arachnoid covering the insula and the Sylvian vessels is reached. Inferiorly the pia arachnoid of the inferior surface of the temporal lobe is divided as the middle cranial fossa floor is reached. If the tumour extends towards the uncus, it is best to resect its medial aspect under the magnification of an operating microscope, in order to protect the optic tract, the oculomotor nerve and the posterior cerebral and choroidal arteries. Any residual tumour at the margin of resection is removed with suction or aspiration. If the landmarks have been identified correctly, the likelihood of causing damage to Wernicke's area is small.

Occipital Lobectomy. In the dominant hemisphere, safe occipital lobectomy involves resection of the occipital lobe 3.5 cm from the occipital tip, thereby avoiding the angular and supramarginal gyri. The extent of safe resection can be increased to 7.0 cm from the occipital tip on the non-dominant side (Fig.11.2a). The general principles of resection are similar; in the depth of the Calcarine fissure, the calcarine artery is encountered and divided close to the brain surface. The veins draining into the torcular and the straight sinus are also divided close to the brain surface. Occipital lobectomy results in contralateral homonymous hemianopia, a functionally acceptable deficit. If during a temporal or an occipital lobectomy

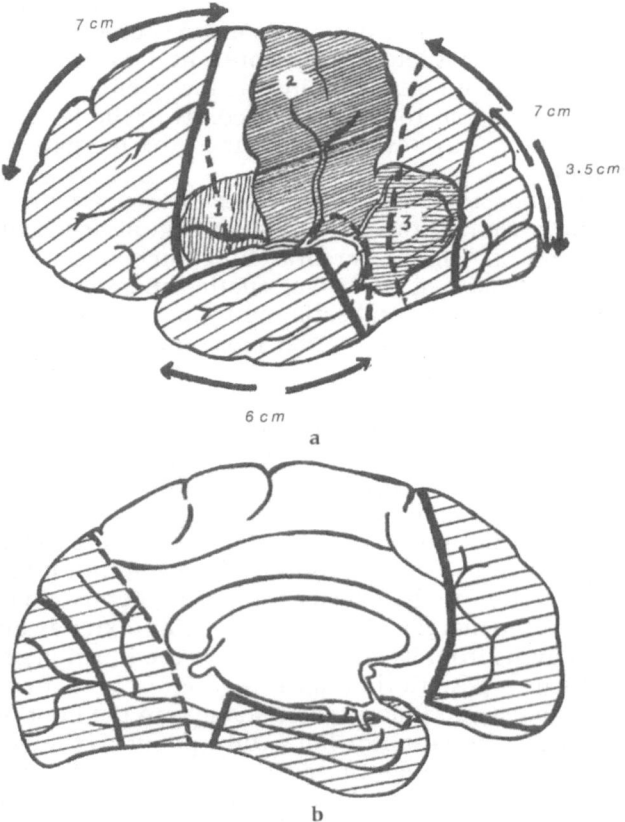

Fig. 11.2 Extent of safe resection for various lobectomies in the dominant (bold outline) and the non-dominant (broken outline) hemispheres. Superolateral (**a**), medial (**b**) and inferior (**c**) surfaces of the hemisphere are shown with important cortical areas. Broca's motor speech area (1), motor strip and somatosensory areas (2) and Wernicke's receptive speech area (3)

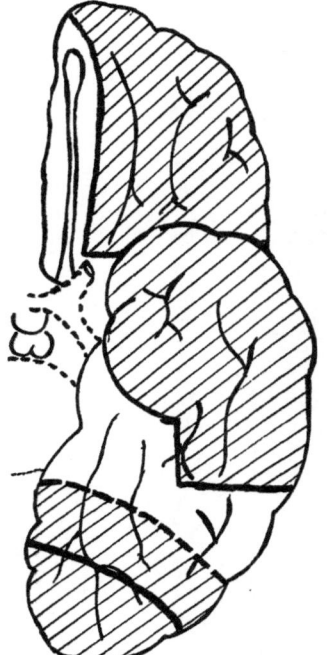

c

Fig. 11.2   *Continued*

Wernicke's area is inadvertently damaged, the residual neurological deficit can be profound with severe receptive aphasia. Minor damage can produce impairment of visual and auditory comprehension and abnormality of communication.

Additional Surgical Measures

Tumours close to the CSF pathway can present with obstructive hydro-cephalus. This requires insertion of a ventriculoperitoneal or ventriculo-atrial shunt, often as a preliminary measure, before definitive surgery is undertaken, e.g. in some cerebellar astrocytomas. However, not infre-quently surgical resection of the tumour causing hydrocephalus is not possible and in this case CSF diversion provides prolonged palliation and time for adjuvant therapy.

Recurrent astrocytomas are often cystic in nature. These cysts can be drained and an Omayya reservoir placed for future percutaneous aspiration (Fig. 11.3). Recurrent tumours often present with meningeal gliomatosis (Yung *et al.*, 1980). Intrathecal chemotherapy can be administered through a similar device implanted under the scalp and connected to a ventricular catheter.

There is no place for primary external decompression for gliomas in current surgical practice.

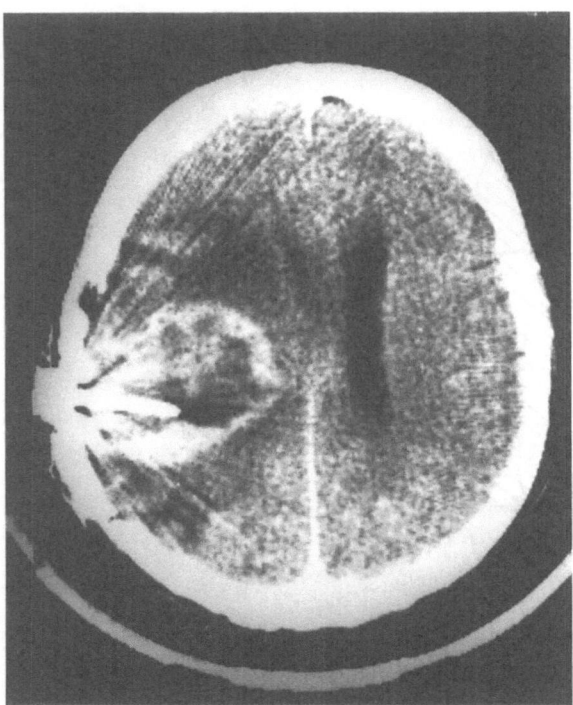

Fig. 11.3 A cystic recurrent astrocytoma with an Ommaya reservoir, for percutaneous aspiration of the recurrent cyst. Percutaneous cyst aspiration can be repeated several times and a small procedure can provide good palliation

Surgical Treatment of Gliomas Within a Specific Site

As the objective of surgical management is first to establish the diagnosis and secondly to achieve maximum tumour extirpation without producing any major deficit, an ideal surgical procedure would involve total resection of macroscopically identifiable tumour without sacrificing the adjacent unaffected brain. In practice, however, it may not be possible to completely resect the tumour and while planning surgery it may be necessary to take into account the histological grade of the tumour, its rate of growth, anatomical location and extent and the neurological and general status of the patient at the time of presentation.

For example, lower grade astrocytomas and oligodendrogliomas often present as diffuse lesions with no anatomical distinction between the tumour and the surrounding brain. As such, total macroscopic resection cannot be performed (Fig. 11.4). Moreover these tumours, unless they obstruct CSF pathways, often do not present with acutely raised intracranial pressure and so radical surgery for palliation is not essential. However, various studies (Laws *et al.*, 1984; Roth and Elvidge, 1960) suggest that more extensive resection of these lesions is associated with longer duration

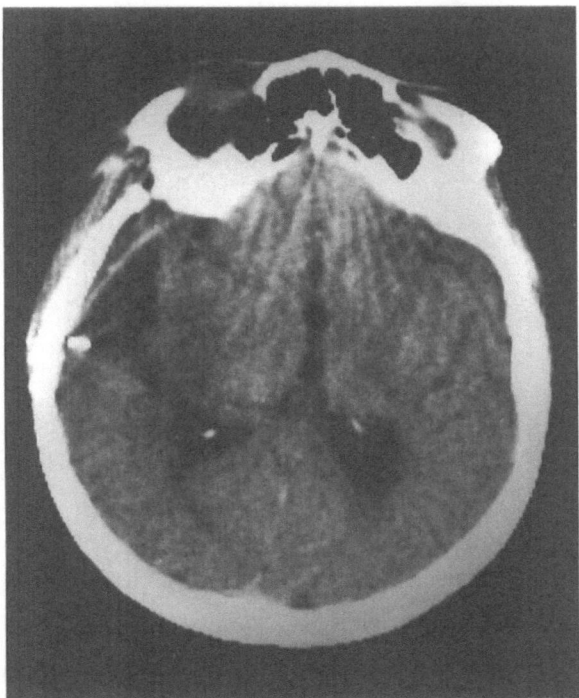

Fig. 11.4 Low-grade astrocytomas often present as a diffuse area of low attenuation on the CT scan, merging imperceptibly with the surrounding brain. This lesion, a low-grade right temporal astrocytoma, has been partly resected with a temporal lobectomy. The internal decompression provides good palliation while adjuvant therapy is being administered

of survival. Since adjuvant radiotherapy is often successful in prolonging survival for several years, while resecting this tumour emphasis is laid on preservation of existing neurological function. Higher grade gliomas, on the other hand, often present as discrete lesions with more obvious distinction between the tumour and surrounding brain. These lesions are often associated with symptoms of raised intracranial pressure and here a stronger case can be made for a radical resection.

Tumours macroscopically confined to the frontal, temporal and occipital poles can be resected by a formal lobectomy beyond the apparent tumour margin. In theory, a complete removal of identifiable tumour and part of the infiltrative edge is combined with the advantage of a generous internal decompression. However, this favourable situation, where the tumour can be resected with a margin of adjacent macroscopically unaffected brain, is only rarely encountered and lobectomies achieve little more than providing good decompression. Tumours in the parietal lobes (Fig. 11.5) and those near the eloquent areas are suitable for simple resection without a lobectomy.

Deep hemispheric gliomas, i.e. tumours of thalamus, basal ganglia and hypothalamus, usually present as diffuse lesions (Fig. 11.6a), often with major neurological deficits, and cannot be resected by conventional surgi-

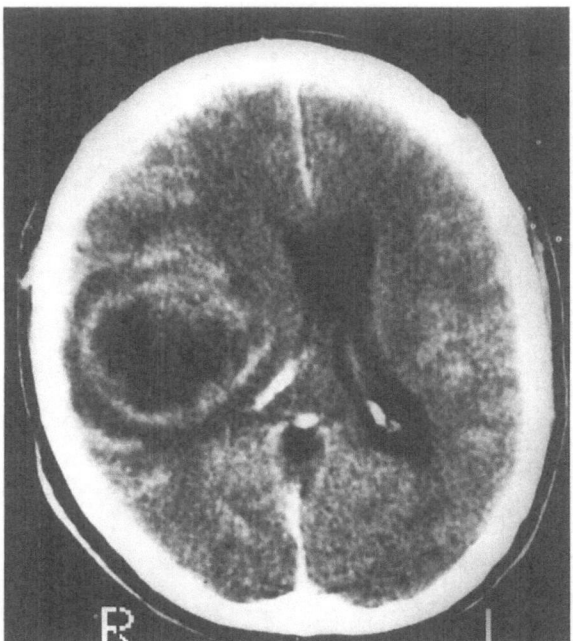

Fig. 11.5 A parietal lobe glioblastoma multiforme. Tumours in this situation are suitable for simple resection, as lobectomy cannot be performed

cal methods (McKissock and Paine, 1958; Cheek and Taveras, 1966). It is best to manage these tumours with the establishment of the histological diagnosis, by a CT or MR stereotaxic biopsy, treatment of obstruction of CSF pathways and early adjuvant therapy. By following this policy, slower growing thalamic tumours may be controlled with radiotherapy for a prolonged period (Cheek and Taveras, 1966). More recently, Kelly and Alker (1981) have attempted to resect these lesions by stereotaxic methods (described below) and have been successful in achieving results comparable to those of conventional surgery for superficial hemispheric lesions. However, this method of resection is not suitable for the diffuse type of tumour where there is no distinction of the tumour – brain interface. Cystic gliomas (Fig. 11.6B) can further be treated by stereotaxic aspiration of the cyst.

The intrinsic tumours of the pineal region include tumours originating in the pineal, from the posterior part of the third ventricle and those arising from the quadrigeminal plate. Obstruction of the cerebral aqueduct predominates in the early presentation and hydrocephalus has to be relieved as the first priority. Since direct surgery for establishment of diagnosis has been associated with high morbidity and mortality, a strong case can be made for irradiation without tissue diagnosis and follow-up imaging to monitor progress, keeping surgery in reserve for the cases where doubt

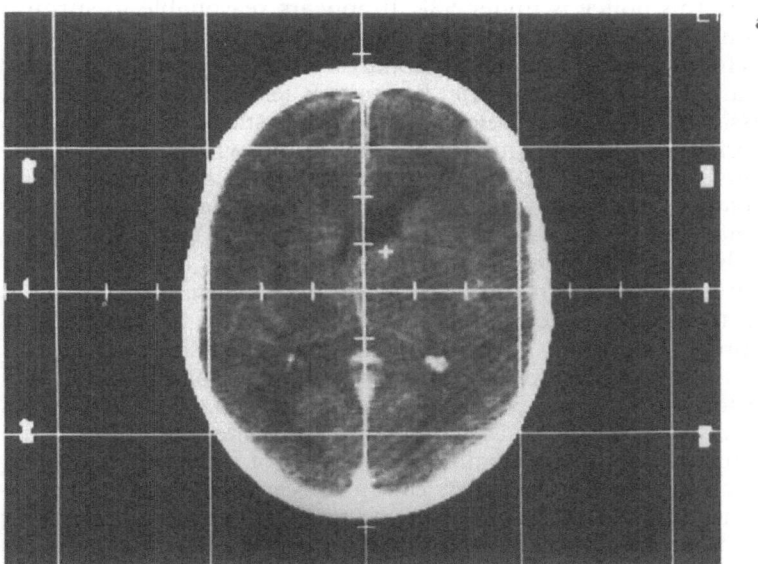

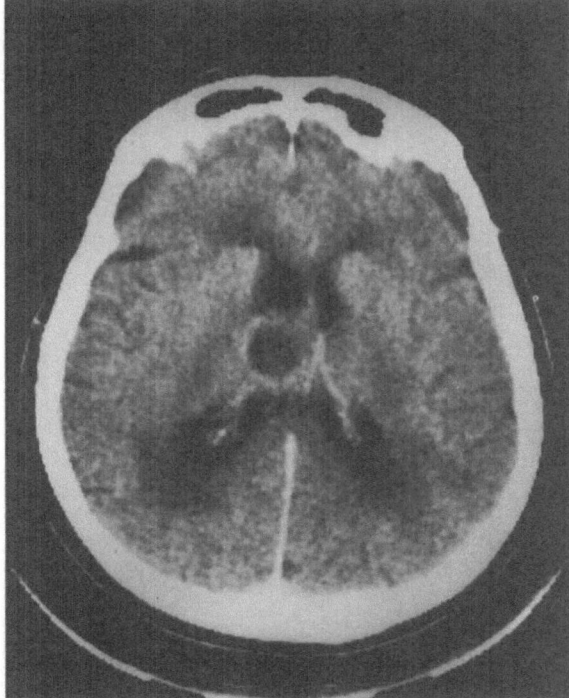

Fig. 11.6 Thalamic and basal ganglionic gliomas often present as a diffuse infiltrative lesion with no radiological or anatomical distinction of brain–tumour interface (**a**). Diagnosis can be established by the stereotaxic method. Cystic gliomas can be managed by stereotaxic aspiration of the cyst (**b**), establishment of diagnosis and irradiation

regarding diagnosis remains (Hide, 1975). The mortality rate of management following this policy is under 5%. It appears reasonable to surmise
that where a diagnosis of germinoma is suspected on clinical, radiological
and CSF biochemical and cytological findings, irradiation without tissue
diagnosis is safe. However, when the diagnosis is less certain, exploration
may be necessary to exclude a benign lesion like a cystic dermoid, a granuloma, a tentorial edge meningioma or for any lesion which shows only a
limited response to irradiation. Although tumours of this region have been
successfully biopsied by stereotaxic methods with high diagnostic accuracy
and limited morbidity (Conway, 1973; Sugita et al., 1975), because of the
proximity of important neurological structures the risks associated with
such procedures are high. On the other hand, with the improvement in
microsurgical techniques, open exploration of this region is possible by a
variety of approaches (Fig. 11.7), with modest operative risk.

The gliomas of the brain stem are primarily tumours of childhood and
may account for almost as many as 25% of all intracranial tumours at this
age (Bilaniuk et al., 1980). Pons is the most frequent site. Tumour initially
presents as uniform enlargement and hypertrophy of the brain stem (Fig.
11.8). The growth may be asymmetrical and occasionally partially exophytic. These tumours are eminently unresectable. However, in rare cases
where the tumour consists of a large intra-axial cyst, decompression by
open exploration improves the short-term outlook (Fig. 11.9) (Lassiter
et al., 1971). Similarly, resection of the exophytic extension into the extra-
axial posterior fossa or into the fourth ventricle (Hoffman et al., 1980) may
provide short-term benefit. The open explorations of the brain stem
tumours for diagnosis represent major surgical procedures, and although

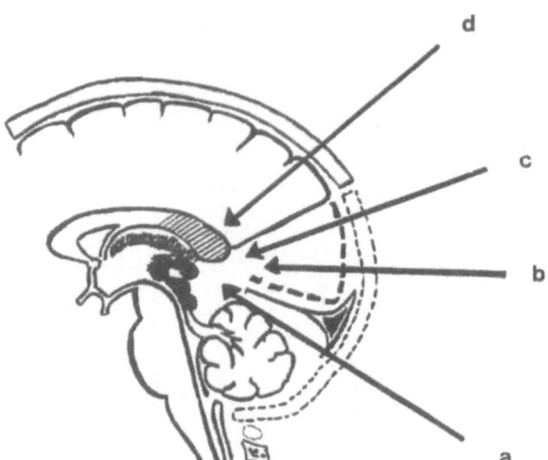

Fig. 11.7 Various approaches for open exploration of tumours in the pineal area:
a infratentorial-supracerebellar (Horsley); *b* supratentorial with resection of wedge of
tentorium cerebelli (Poppen); *c* supratentorial transcallosal with occipital lobectomy (Horrax);
d the same approach without occipital lobectomy (Dandy)

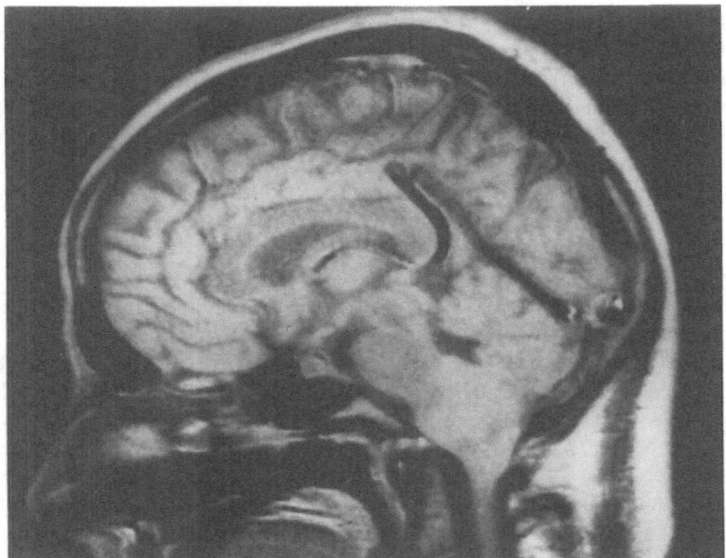

Fig. 11.8 A sagittal MRI scan demonstrating diffuse enlargement of the brain stem from a low-grade medullary astrocytoma in a 20-year-old male. Such lesions are best managed by radiotherapy after establishment of diagnosis by the stereotaxic method

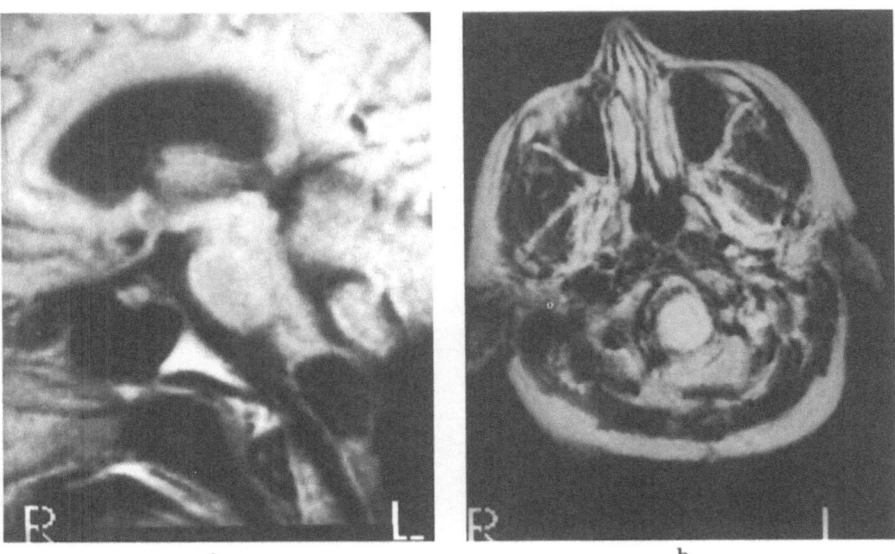

a   b

Fig. 11.9 This 52-year-old female presented with left hemiparesis due to a low-grade astrocytoma of the brain stem with a large intra-axial cyst (**a**). This was successfully decompressed at open exploration and she has remained functionally well for two years (**b**)

recent series of open biopsy procedures reported a low morbidity and mortality there is a significant rate of non-diagnostic tissue obtained (Lassiter et al., 1977). This may partly be due to the histological heterogeneity of the tumour. An apparently well-differentiated tumour may harbour a less differentiated area which will determine the ultimate prognosis. It is difficult to obtain representative tissue by open procedures and small biopsy specimens often do not portray the whole diagnostic picture. With the improvement in diagnostic accuracy of CT- and MR-guided stereotaxic biopsy, specific areas of interest within the brain stem tumour can be approached by various routes (Fig. 11.10), and a high yield of positive histological diagnosis can be achieved with low surgical risk (Coffey and Lunsford, 1985; Hood et al., 1986; Thomas et al., 1988). Hence, the case for empirical therapy of these tumours has become much less strong.

Cerebellar astrocytomas are predominantly tumours of childhood and are only infrequently encountered in adult life. They are usually benign, cystic and indolent and following resection have generally a good prognosis (Fulchiero et al., 1977). Total removal should be attempted by a posterior fossa craniectomy. It is often possible to define a plane between the tumours and the adjacent cerebellum (McLone, 1982), although part of the cerebellar hemisphere can be resected in order to perform a radical tumour excision. If the tumour is partly cystic, the cyst is drained first, after which the mural nodule is excised and the cyst wall is removed (Fig. 11.11). Solid astrocytic tumours are often malignant and infiltrate into the cerebellar peduncles. Complete resection is therefore not possible and

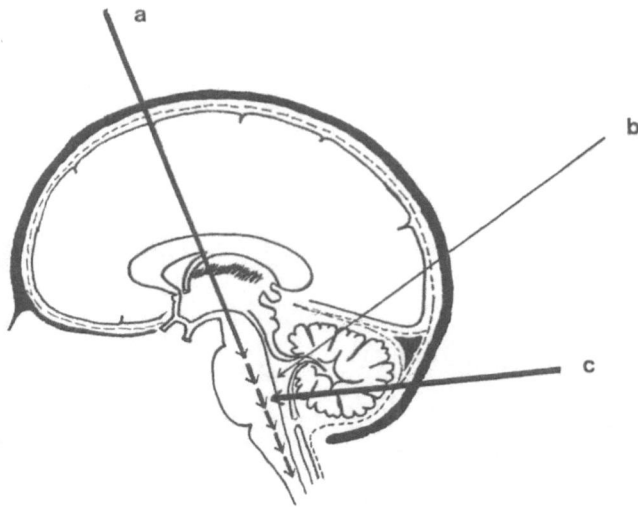

Fig. 11.10 Various stereotaxic approaches to the tumours of the brain stem: *a* transfrontal – lesions in the midbrain and pons above the level of the middle cerebellar peduncle can be approached by this route, although this approach has successfully been used for lesions in all parts of the brain stem (From Thomas *et al.*, 1988, with permission); *b* transtentorial; *c* transcerebellar – for the lesions of the lower brain stem and pons

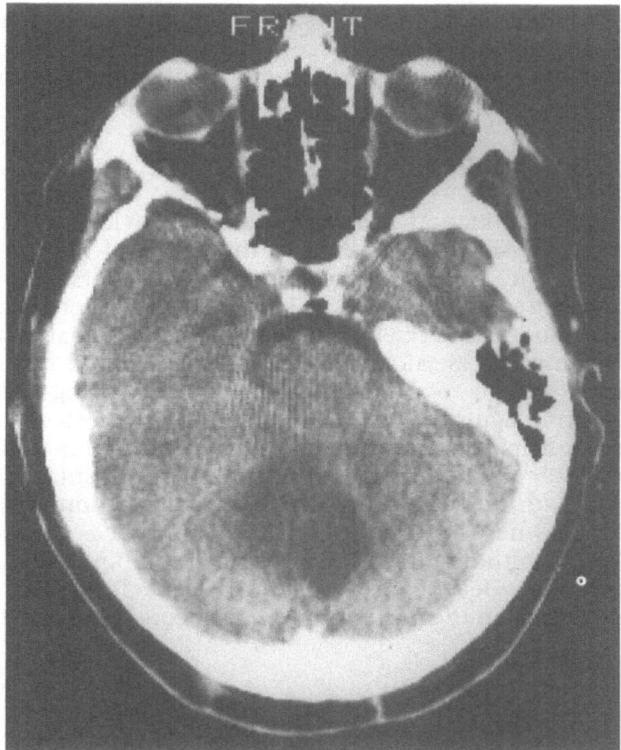

Fig. 11.11   A low-grade cystic cerebellar astrocytoma. The cyst was drained and the mural nodule excised. The patient remained free of recurrence for 10 years

surgery is restricted to relief of hydrocephalus, partial decompression and establishment of the histological diagnosis. The neuroradiological appearance of many of the adult onset gliomas may be difficult to differentiate from metastatic tumours, haemangioblastomas and cerebellar abscesses, and histological diagnosis is required.

Recent Advances in Surgical Methods

In order to achieve total removal of the tumour by conventional surgical methods, the neurosurgeon is faced with serious limitations of access, exposure and visibility in the depth of tumour bed. Hence, there is uncertainty regarding demarcation of the tumour and normal brain, as well as about the optimum methods of resecting the tumour without disturbing adjacent tissue, and the degree of acceptable safe manipulation and retraction. In recent years the introduction of new neurosurgical instrumentation, namely the operative microscopes, lasers and ultrasonic aspirators, and the use of stereotaxic techniques and integration with computer tomography and MR scanning, have allowed improved access and maximum tumour resection with minimal damage and manipulation of the surrounding normal brain.

Operative Microscope

With the coaxial illumination and magnification provided by the operative microscope, it is possible to visualize in the depth of the operating field through a small cortical incision. Large subcortical tumours up to about 8 cm in diameter can be resected through 2–3 cm cortical incisions (Salcman *et al.* 1982a). Since tumour tissue can be visualized and differentiated from normal brain, a more complete resection is possible. The use of self-retaining retractors minimizes damage to the surrounding brain during exposure, and bipolar diathermy coagulation improves operative haemostasis. The applications of these principles of microsurgery have been found to be of importance in such difficult situations as resection of recurrent gliomas, where routine use of the operative microscope in conjunction with lasers for tumour resection has been influential in improved operative results (Salcman, 1982a). Although for most superficial tumours visualization may not be a problem, selective use of a microscope for resection of deep tumours, for example in the posterior fossa, is of considerable assistance.

Lasers

Normal cerebral tissue is uniquely susceptible to trauma during manipulation. As gliomas vary in consistency and often, as in recurrent cases, are adherent to the adjacent normal brain and areas of radiation necrosis, removal of these tumours is potentially hazardous. In order to debulk them safely it is necessary to resort to the minimal possible manipulation. With the recent addition of lasers to the neurosurgical armamentarium, it has become possible to vaporize tumour tissue progressively with extreme precision.

Of the various types of surgical laser the two which have found application in neurosurgery are the carbon dioxide laser and the neodymium : YAG laser (Moser, 1985). The CO_2 laser is more commonly employed. The energy generated from this is absorbed by the tissue water in the superficial layers which are vaporized in the process. As 90% of the energy is absorbed at a depth of 0.03 mm, tissue penetration is limited and only the surface tissue is destroyed. As the laser beam is moved across the tumour, its volume is steadily reduced. It is possible to remove the tumour with precision from adjacent vital structures with functional preservation. However, because of limited tissue penetration, inadequate haemostasis is provided and the technique is time consuming.

The neodymium : YAG laser achieves greater tissue penetration and therefore provides haemostasis during resection. The advantage of a faster removal of tumour volume is lost in the lack of precision of this system, which has been emphasized by recent reports suggesting its potential hazards (Jain, 1985). More recently, the advantages of lasers have been successfully integrated with CT- and MR-guided stereotaxic surgery for resection of deep lesions (Kelly and Alker, 1981; Kelly *et al.*, 1983). When used in conjunction with other methods of tumour resection, laser resection is a safe and precise technique and has found increased application in surgery for gliomas of the posterior fossa and the brain stem.

Ultrasonic Surgical Aspirator

High-frequency sound waves in the ultrasonic range can fragment tumour tissue. When this is combined with a system of irrigation and suction, tumour tissue can sequentially be fragmented and removed. This principle has been utilized in the ultrasonic surgical aspirator. The high-frequency vibration of the tip of the aspirator fragments tumour tissues within 2 mm and simultaneous irrigation and suction evacuates it from the operative area. As the intensity of vibration of the tip, the flow rate of irrigation and magnitude of suction are adjustable, a high degree of precision can be obtained through a wide range. Since a small volume of tumour is removed at a time without retracting adjacent unaffected areas, safe and precise tumour resection is possible. The ultrasonic aspirator does not provide any haemostasis and hence surgical resection has to be combined with bipolar coagulation for haemostasis. Although most gliomas are soft in consistency and can be removed by other methods, the firmer areas which often separate from the adjacent brain with difficulty may safely be removed with ultrasonic aspirator.

Computer-Interactive Stereotaxic Tumour Excision

The introduction of CT and MR scanning has made it possible to define the anatomical limits of the tumour by non-invasive methods. However, the tumour–brain interface is less obvious at surgery; hence complete resection of the image-defined tumour has often not been possible. In an attempt to achieve this, a system has been designed wherein the data obtained from a CT and MR scan can be placed precisely in a three-dimensional stereotaxic coordinate system (Kelly and Alker, 1981; Kelly *et al.*, 1983) and, with specially designed instruments, the image-defined tumour can be resected completely.

For this purpose the data are acquired by imaging in a CT-compatible stereotaxic head-holder and a localizer system, which produces images of the lesion with localizing markers as artefacts (Fig. 11.12a). These data are transformed within a three-dimensional image matrix in a space related to a stereotaxic coordinate system. Each CT slice images the lesion with nine fiducial markers which are digitized by an intensity sweep program, automatically marking their coordinates. The margins of the tumour are simultaneously digitized and the X and the Y coordinates of these points are connected by computer program and shown as a continuous outline (Fig. 11.12b). The centre point of the lesion is chosen on the slice with the largest cross-section area of the tumour and the entire image is then reconstructed around this point in the image matrix in which the three-dimensional image is created by a program which interpolates intermediate

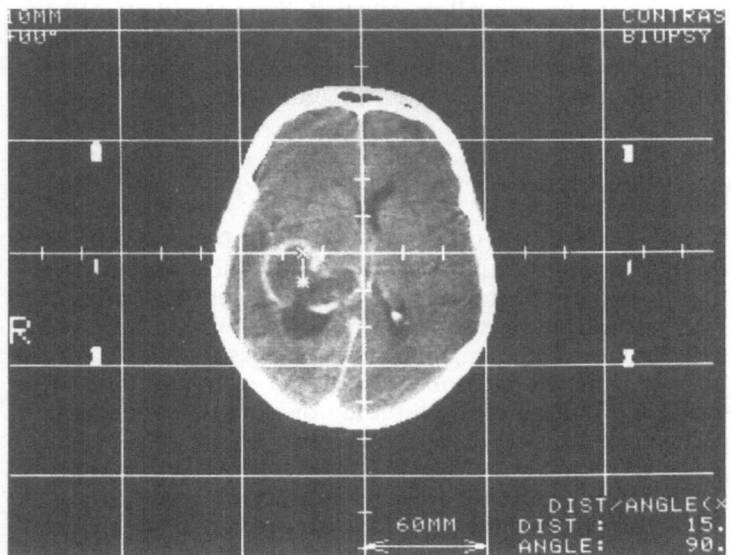

a

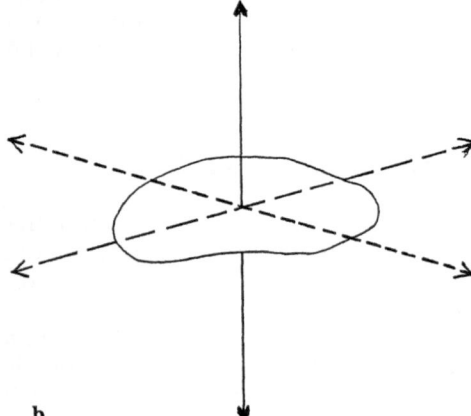

b

Fig. 11.12 Data acquisition and interpolation in a three dimensional image
matrix. CT scan with the lesion and
fiducial markers (**a**). The margins of the
lesion are digitized in the image matrix
(**b**). Successive slices are sequentially
transformed in the image matrix and,
by interpolation of intermediate slices
(**c**), the three dimensional outline of
the tumour is reconstructed (**d**)

slices at 1 mm intervals (Fig. 11.12c, d). Reformatting of images in different
planes is possible and a safe surgical trajectory can be planned accordingly.
The reformatted slices can be displayed on the operation theatre video
monitor. The computer also selects mechanical adjustments on the stereotaxic frame to position the selected target point at the focal point of the arc
quadrant system.

Surgical resection is performed in a specially designed computer interac-
tive arc quadrant stereotaxic system in which the patient's head can be
positioned into the focal point of the internal arc quadrant. An operating
microscope and a laser manipulator apparatus is suspended from the car-

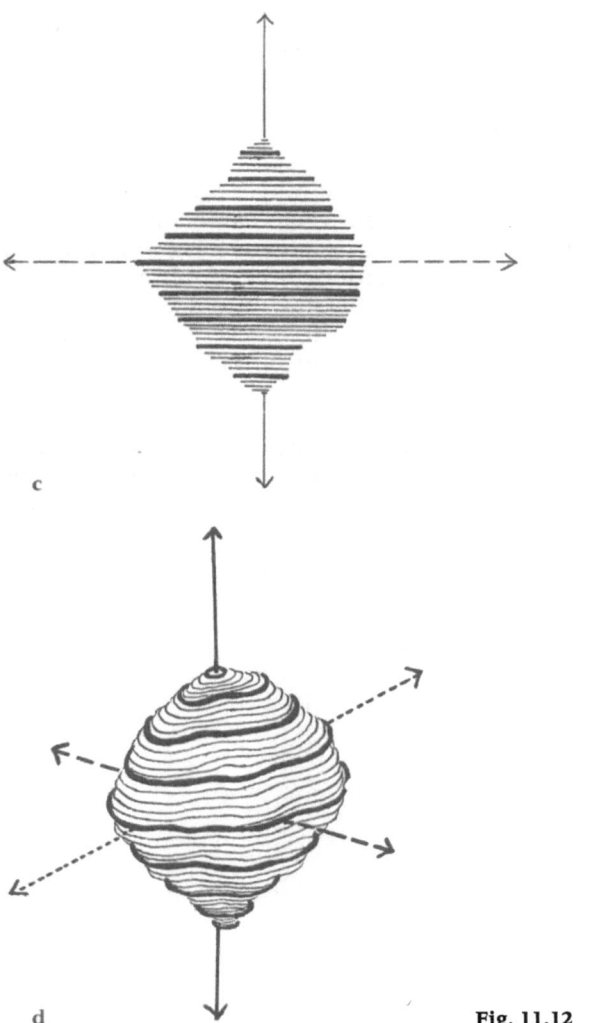

c

d

Fig. 11.12 *Continued*

riage that runs perpendicular to the arc quadrant. The angle of safe trajectory is determined from the data already acquired and a small craniotomy is performed at the chosen site of entry. The computer displays the relationship of the trephine to the reformatted tumour outline. A specially designed stereotaxic retractor is directed towards the focal point. The superficial tumour slices are then displayed on the monitor and a plane is created around the tumour with a stereotaxically guided laser beam, manipulated by a remote joystick using computer-generated images as a guide. The attached operating microscope allows direct vision to monitor the position of the laser beam and surgical resection. The tumour is thus removed slice by slice, proceeding from the most superficial slices to the deepest until complete resection of image-defined tumour has been performed.

Intraoperative Ultrasonography

High-frequency sound waves, in the ultrasonic range, are reflected from surfaces of differing densities. These reflected waves can be displayed as a real-time ultrasonographic image. The potential advantages of this method of imaging include simplicity of interpretation, precise delineation and localization of the lesion and its characterization into solid and cystic components, and dynamic display of the pathological anatomy during the course of an operation, a benefit which no other imaging method provides. Hence, surgical intervention can be better planned with minimal tissue damage. For ultrasound-guided procedures, a craniotomy site is selected after a marker CT scan has been performed. The appropriate ultrasound probe is covered in a sterile sheath after a scanning gel has

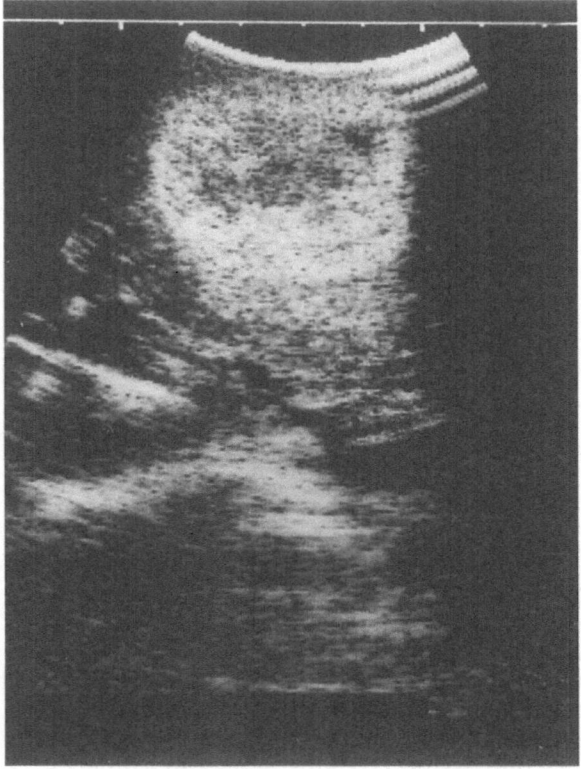

Fig. 11.13 An ultrasonic image of a malignant glioma showing hyperechoic (solid) margin and hypoechoic (cystic) centre. Part of the lateral ventricle is seen adjacent to the tumour

been smeared on the probe surface for acoustic coupling. Scanning through the intact dura produces good-quality images (Fig. 11.13). The size, the position and the distance of the lesion from the surface can be calculated, and its consistency defined. Most solid gliomas are hyperechoic in relation to the surrounding normal or oedematous brain tissue. Increased echogenicity is also present where there is calcification in the lesion. Hypoechoic and anechoic areas represent cyst formation, necrosis, haemorrhage or degeneration (Kane *et al.*, 1988). From the ultrasonographic data, surgical intervention is planned, resection monitored and completeness of resection assessed intraoperatively.

Smaller probes have recently been designed which allow imaging and assist biopsy through a burr hole (see above).

Results of Surgery

The natural course of malignant gliomas is relentlessly progressive and without any form of treatment the average life expectancy is two months (Hitchcock and Sato, 1964; Roth and Elvidge, 1960). The introduction of various forms of treatment has succeeded only in prolongation of survival without being able to eradicate the disease in a significant proportion of the treated population. Whatever form of treatment is undertaken, only 10%–20% of the patients will survive beyond two years (Lieberman *et al.*, 1982). Since there is significant morbidity and mortality associated with all methods of treatment, it is necessary to take this fully into consideration when analysing the results of any form of treatment.

In recent years, surgery alone has rarely been used as a treatment modality. It is therefore difficult to assess the results of different forms of surgery uninfluenced by the effects of adjuvant therapy. The existing data are retrospective and suffer from selection bias, as age, general and neurological status of the patient, nature, site and size of the lesion and the philosophy of the surgeon have all been influential in the choice of treatment. Moreover, there have been significant advances in diagnostic imaging with early diagnosis and resection of smaller lesions, correct estimation of completeness of resection, early detection of postoperative complications, introduction of steroids and improvement in perioperative care, neuroanaesthesia and surgical technique, all of which have been influential in reducing operative complications. Hence, the operative results of the previous series may not necessarily be valid in the present context.

The dramatic reduction in operative mortality after the introduction of steroids has already been mentioned. The availability of non-invasive methods of early diagnosis of operative complications and their prompt treatment have also contributed to this sustained improvement in the results of surgery. When the various surgical factors are taken into account, one feature which has consistently been associated with a favourable outcome is the extent of resection; more radical extirpation of tumour and stereotaxic procedures are associated with reduced operative mortality and morbidity compared with limited resection at an open craniotomy (Hitchcock and Sato, 1964; Jelsma and Bucy, 1967; Shapiro, 1982) and are more likely to produce improvement in the immediate postoperative period (Taveras *et al.*, 1962), whereas partial resection and external decompression make the patients worse (Jelsma and Bucy, 1967).

The influence of surgery in producing improvement in the quality of survival remains unsettled. Since the operative morbidity necessarily affects the subsequent quality of life, prospective analyses of the effect of surgery on the preoperative symptoms is essential. In one such study, Shapiro examined a group of 82 patients for evidence of speech, personality, visual, sensory and motor function and noted 191 abnormalities in 410 functional areas examined (Shapiro, 1982). Postoperatively, while 151 of these were either improved or unchanged, 40 had become worse. In 125 instances the potential abnormality was not present either before or after surgery, while there were 55 instances of an abnormality first appearing postoperatively. It was concluded that surgical resection was more likely to alleviate existing symptoms than to produce additional ones. In general, around 70% of the survivors have either an excellent or good functional quality of existence, with minimal or no neurological deficit, and are independent and self-caring. About 30% of the survivors have severe deficit and require assistance (Roth and Elvidge, 1960; Hitchcock and Sato, 1964). The fact that these latter figures are from the pre-steroid series, and that the introduction of steroids and non-invasive diagnostic imaging has improved the surgical outcome even further, when considered in the context of the relentlessly progressive untreated course of the disease, substantiates Shapiro's conclusion that overall patients are helped rather than hurt by surgical procedures (Shapiro, 1982).

Various studies have indicated that surgery for malignant gliomas prolongs the median life span of the patients by approximately 14 weeks if no further anticancer treatment is utilized (Walker *et al.*, 1978). On the whole, more extensive resection of tumour is associated with higher median survival. Although the survival rates are higher with extensive resection in the first one to two years, after this period the influence of resection progressively lessens and by five years the survival figures for groups with partial and extensive resection are similar (Frankel and German, 1958; Roth and Elvidge, 1960; Jelsma and Bucy, 1967). The extent of surgical resection has also been reported to be of importance for lower grade astrocytomas (grades I and II), with higher 5-year survival in the group treated with total excision of the tumour compared with those who underwent radical subtotal excision or biopsy alone (Laws *et al.*, 1984)

This influence of the extent of resection on survival for glioblastomas is not substantiated by the study of Hitchcock and Sato (1964), who found no difference in survival between patients with glioblastomas undergoing partial removal followed by radiotherapy, and those who underwent "complete" removal and radiotherapy. Weir (1973) also came to a similar conclusion, and found that for glioblastomas the extent of tumour resection was of negligible importance in prolonging survival compared with the influence of the patient's age and postoperative radiation therapy.

The value of postoperative irradiation is undeniable and is discussed in greater detail in Chapter 12. In brief, postoperative radiation therapy positively influences survival in all grades and forms of primary malignant brain tumours (Leibel *et al.*, 1975; Walker *et al.*, 1978; Scanlon and Taylor, 1979; Chen *et al.*, 1980; Walker *et al.*, 1980).

It appears reasonable to infer that, on the whole, surgery for gliomas prolongs survival. More extensive resection is associated with longer duration of good-quality survival and lower operative morbidity and mortality, and the addition of postoperative irradiation improves survival in all grades of tumours.

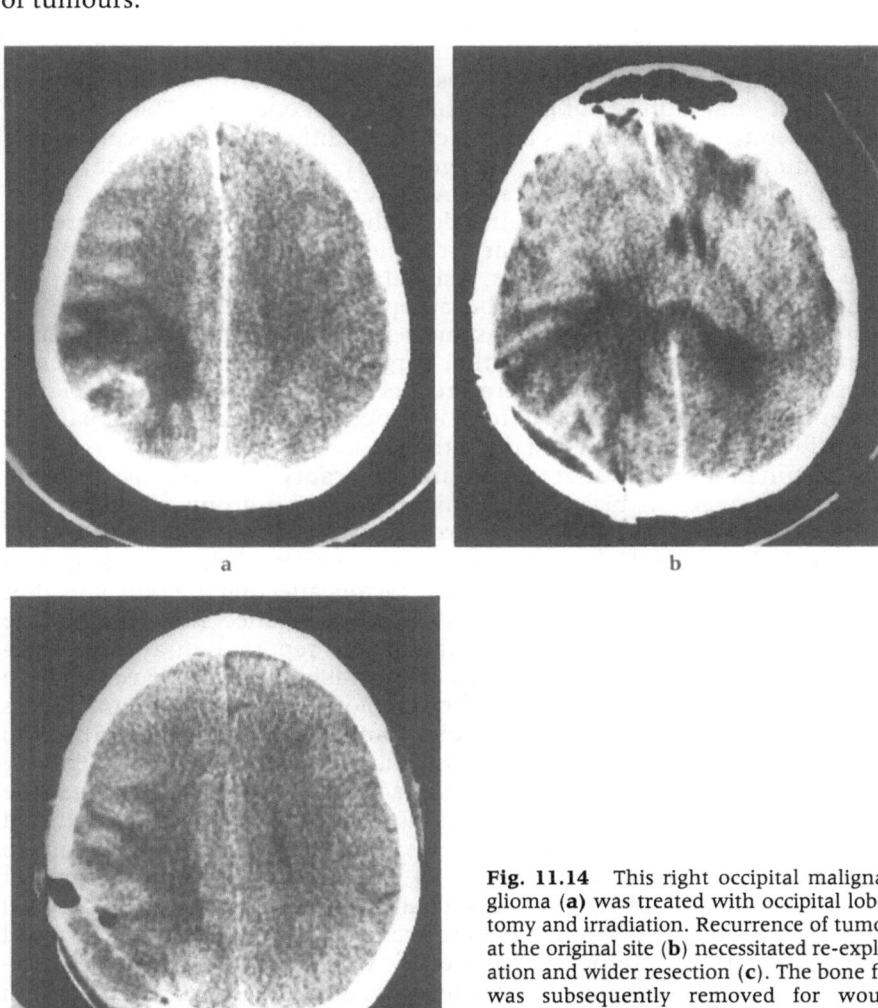

Fig. 11.14 This right occipital malignant glioma (**a**) was treated with occipital lobectomy and irradiation. Recurrence of tumour at the original site (**b**) necessitated re-exploration and wider resection (**c**). The bone flap was subsequently removed for wound infection. The patient has residual, functionally acceptable contralateral hemianopia and some cortical sensory loss and has remained free of recurrence for eight months

The Case for Reoperation for Gliomas

The case for a second operation for resection of recurrent tumours, after failure of primary treatment, is even more debatable. Since for most recurrent gliomas radiation therapy alone is likely to be of little value (Pool, 1967) and since the rationale for the second operation is similar to that for the first operation, i.e. maximal tumour extirpation to improve survival and arrest the progression of neurological deficit, one expects reoperation rates to be higher than the published rates of between 0% and 10% (Salcman et al., 1982a). However, the morbidity and mortality associated with a second procedure are high and the prolongation of survival has been, at best, similar to surgical excision alone without radiotherapy, so that only a small number of patients have been reoperated. The reason for a high operative mortality lies in the technical difficulties of a second operation, impaired tissue healing and generally poor physical state of the patient following adjuvant therapy. Another important reason for lower reoperation rate is the negative attitude of the clinicians, who regard recurrence of tumour following primary treatment as a sign of failure and, in view of the overall poor prognosis, a signal to discontinue aggressive treatment. While this attitude may be acceptable in some cases, with the improvement in surgical conditions and reduction in operative morbidity a stronger case can be made for further surgery in a selected group of cases whose tumours recur. It is therefore necessary to analyse the relative influence of various selection criteria in order to define a group of patients who will benefit from a second operation.

Unfortunately, the available data are too scanty and retrospective and suffer from selection bias. Only a highly selective group of patients of younger age with more benign, surgically accessible tumours, in good neurological status with longer recurrence-free intervals have undergone a second operation, and in many series the factors affecting the outcome after a second operation have not been analysed. For instance, Roth and Elvidge (1960) reoperated on 3% of their patients and reported an average additional survival of 4.5 months, a figure comparable to median postoperative survival following first operation without radiotherapy. They did not discuss the criteria affecting outcome, nor did Frankel and German (1958) who reoperated on 26 of their 183 patients. Young et al. (1981) retrospectively studied 24 reoperated cases of glioma, grades 3 and 4, and found that the median survival period of 14 weeks after second operation correlated significantly with the performance status (Karnofsky's rating, KR) prior to second operation. Patients with a KR of 60+ had a mean survival time of 26.8 ± 6.29 weeks and patients with a KR less than 60 had a mean survival time of only 11.2 ± 2.7 weeks. There was a weak correlation of survival with the length of interoperative interval, with better results if the interoperative interval was more than 6 months, and no significant correlation between the length of survival and age, sex and location of the tumour. There was no definite relationship between surgeons' impression of completeness of resection and length of survival after the second operation.

In a somewhat similar study (Harsh et al., 1985), on 39 reoperated patients with glioblastoma multiforme and 31 patients with anaplastic astrocytoma, the median survival of 35 weeks for patients with glioblastoma was statistically significantly related to both age and preoperative Karnofsky rating in high-quality survivors, but not with survival independent of quality. For anaplastic astrocytomas, only age was a significant correlate. On the other hand, in a prospective series of 40 patients reported by Salcman *et al.* (1982a), the median survival time after reoperation of 37 weeks was unrelated to the patients' Karnofsky rating, sex, tumour grade, interoperative interval or age. In the 60 reoperations there were no operative deaths and overall morbidity was 8.3%. Even after correction for early reoperation for residual rather than recurrent tumour, there was sufficient prolongation of median survival after a reoperation to suggest its benefit in conjunction with other modalities of treatment.

Although the principles of surgery governing the second operation are similar to the primary procedure, there are certain important technical differences. As most of these patients have undergone radiotherapy, the scalp is devitalized and therefore at a higher risk of wound infection and poor healing. Whenever extension of the original incision is planned, the vascularity of the flap should be kept in consideration. The dura is often adherent to the bone and cerebral cortex and separates with difficulty. Since recurrence usually takes place in the bed of resected tumour, the original bone flap may provide the necessary access (Fig. 11.14). The tumour is not usually difficult to identify, but often cannot be differentiated from the associated radiation changes which make the tissue friable and firmly adherent to the brain. Gentle separation with minimal manipulation of normal brain is essential and may necessitate the use of lasers and ultrasonic aspirator. Total macroscopic removal often requires a lobectomy or a wide resection. With improvement in imaging method and neurosurgical techniques, it has become possible to detect the recurrent tumours at an earlier stage and resect them more completely with minimal risk. Further prospective studies are therefore necessary to identify the group of patients who would benefit from a second operation.

Conclusion

With the improvement in diagnostic imaging, surgical techniques and perioperative care, the overall risks associated with surgery have reduced and it is finding increasing application in the management of primary malignant brain tumours. However, the infiltrative nature of gliomas, with microscopic extension of tumour cells into adjacent, functionally intact neurological structures, makes it appear unlikely that surgery will ever become the definitive treatment of this disease. It is, however, conceivable that with the development of safe and more effective adjuvant treatment directed at the peripheral infiltrative edge, complete resection of image-defined tumour will become a part of standard management. The future advances may be in further improvement in surgical techniques and methods of complete resection of image-defined tumour with minimal disturbance of adjacent normal structures. In the foreseeable future, however, surgery will retain its place in providing tissue diagnosis, prolongation of improved quality of survival with relief of some distressing effects of tumour and as an adjunct to current adjuvant treatment.

References

Apuzzo MLJ, Chandrasoma PT, Cohen D et al. (1987) Computer imaging stereotaxy; experience and perspective related to 500 procedures applied to brain masses. Neurosurgery 20: 930–937

Bailey P, Cushing H (1926) A Classification of the Tumours of the Glioma Group on a Histogenic Basis with a Correlated Study of Prognosis. Lippincott, Philadelphia

Bennett AH, Godlee RJ (1884) Excision of a tumour from the brain. Lancet 2: 1090

Bilaniuk LT, Zimmerman RA, Littman P et al. (1980) Computed tomography of brain stem gliomas in children. Radiology 134: 89–95

Bradford R, Bydder GM, Thomas DGT (1987) MRI directed stereotactic biopsy of cerebral lesions. Acta Neurochir Suppl 39: 25–27

Bradford R, Thomas DGT (1990) Advances in surgery for malignant brain tumours. In: Thomas DGT (ed) Neuro-oncology. Edward Arnold, London, pp 148–163

Bydder GM, Steiner RE, Yound IR et al. (1982) Clinical NMR imaging of the brain: 140 cases. Am J Radiol 139: 215–236

Cheek WR, Taveras JM (1966) Thalamic tumours. J Neurosurg 24: 505–513

Chen HW, Hanel JJ, Kim TH, Webster JH (1980) Oligodendrogliomas: 1. A clinical study of cerebral oligodendrogliomas. Cancer 45: 1458–1466

Coffey RJ, Lunsford LD (1985). Stereotactic surgery for mass lesions of the midbrain and pons. Neurosurg 17: 12–18

Coffey RJ, Lunsford LD (1988) Localization and biopsy of intracranial lesions with computed tomography and magnetic resonance imaging. In: Schmidek HH, Sweet WH (eds) Operative Neurosurgical Techniques. Indications, Methods and Results, Vol.1. Grune and Stratton, Orlando, Fl, pp 463–474

Conway LW (1973) Stereotactic diagnosis and treatment of intracranial tumours including an initial experience with cryosurgery for pinealomas. J Neurosurg 38: 453–460

Dandy WE (1921). The treatment of brain tumours. J Am Med Ass: 1853

Frankel SA, German WJ (1958) Glioblastoma multiforme; review of 219 cases with regard to natural history, pathology, diagnostic methods and treatment. J Neurosurg 15: 489–503

Fulchiero A, Winston K, Leviton A et al. (1977) Secular trends of cerebellar gliomas in children. J Nat Cancer Inst 58: 839–843

Garfield J (1980) Surgery for cerebral gliomas. In: Thomas DGT, Graham DI (eds) Brain Tumours, Scientific Basis Clinical Investigations and Current Therapy. Butterworths, London

Harsh IV, Levin GR, Gutin VA et al. (1985) Reoperation for recurrent glioblastoma and anaplastic astrocytomas. Presented at the American Association of Neurological Surgeons, Atlanta

Hide TAH (1975) A rational approach to treatment in pineal tumours (Abstract). Fifth Congress of European Association of Neurosurgical Societies, p 90

Hitchcock E, Sato F (1964) Treatment of malignant glioma. J Neurosurg 21: 497–505

Hoffman HJ, Becker L, Gaven MA (1980) A clinically and pathologically distinct group of benign brain stem gliomas. Neurosurgery 7: 243–247

Hood TW, Gebanski SS, McKeever PE, Venes JL (1986) Stereotaxic biopsy of intrinsic lesions of brain stem. J Neurosurg 65: 172–176

Jain KK (1985) Complications of use of the neodymium yag : Yttrium-aluminium-garnet laser in neurosurgery. Neurosurgery 16: 759

Jelsma R, Bucy PC (1967) The treatment of glioblastoma multiforme of the brain. J Neurosurg 27: 388–400

Kane RA, O'Leary DH, Matthews ES (1988) Intraoperative ultrasonography in neurosurgery. In: Schmidek HH, Sweet WH (eds) Operative Neurosurgical Techniques. Indications, Methods and Results, Vol. 1. Grune and Stratton, Orlando, Fl, pp 213–222

Kelly PJ, Alker GJ (1981) A stereotactic approach to deep seated central nervous system neoplasms using the carbon dioxide laser. Surg Neurol 15: 331–337

Kelly PJ, Kall BA, Goerss SJ et al. (1983) Precision resection of intraaxial CNS lesions by CT-based stereotactic craniotomy and computer-monitored CO_2 laser. Acta Neurochirurg 68: 1

Kendall B (1980) Neuroradiology. In: Thomas DGT, Graham DI (eds) Brain Tumours, Scientific Basis, Clinical Investigations and Current Therapy. Butterworths, London, pp 231–267

Lassiter KRL, Alexander E Jr, Davis CH Jr, Kelly DL Jr (1971) Surgical treatment of brain stem gliomas. J Neurosurg 34: 719–725

Laws ER, Taylor WF, Clifton MB et al. (1984) Neurosurgical management of low-grade astrocytomas of the cerebral hemispheres. J Neurosurg 61: 665

Leibel SA, Sheline GE, Wara WM, Boldrey EB, Nielsen SL (1975) The role of radiation therapy in the treatment of astrocytomas. Cancer 35: 1551–1557

Lieberman AN, Foo SH, Ransohoff J et al. (1982) Long term survival among patients with malignant brain tumours. Neurosurgery 10: 450–453

Lunsford LD, Martinez J, Latchaw RE (1986) Stereotaxic surgery with a magnetic resonance and computerised tomography compatible system. J Neurosurg 64: 872–878

Marshall LF, Jennett B, Langfitt TW (1974) Needle biopsy for the diagnosis of malignant glioma. J Am Med Ass 228: 1417–1418

McCabe JJ (1989) Surgery for gliomas. In: Symon L, Thomas DGT, Clark K (eds) Rob and Smith's Operative Surgery, 4th edn, Neurosurgery. Butterworths, London, pp 231–240

McKenzie KG (1936) Glioblastoma. A point of view concerning treatment. Arch Neurol Psych 36: 542–546

McKissock W, Paine KWE (1958) Primary tumours of the thalamus. Brain 81: 41

McLone DG (1982) Cerebellar astrocytomas. In Wilkins RH, Rengachary SS (eds) Neurosurgery, Vol. 1. McGraw-Hill, New York, pp 754–757

Moser RP (1982) Tumours, Tools and Technology – The Role of the Neurosurgeon. Prog Exp Tumour Res, Vol. 29. Karger, Basel, pp 256–268

Ostertag CB, Mennel HD, Kiessling M (1980) Stereotactic biopsy of brain tumours. Surg Neurol 14: 275–283

Pool JL (1967) The management of recurrent gliomas. Clin Neurourg 15: 265–285

Roth JG, Elvidge AR (1960) Glioblastoma multiforme. A clinical survey, J Neurosurg 17: 736–750

Salcman M, Levine H, Rao K (1981) Value of sequential computed tomography in the multimodality treatment of glioblastoma multiforme. Neurosurgery 8: 15

Salcman M, Kaplan RS, Ducher TB (1982a). Effect of age and reoperation on survival in combined modality treatment of malignant astrocytomas. Neurosurgery 10: 454

Salcman M, Kaplan RS, Samaras GM et al. (1982b) Aggressive multimodality therapy based on a multicompartmental model of glioblastoma. Surgery 92: 250–259

Scanlon PW, Taylor WF (1979) Radiotherapy of intracranial astrocytomas: analysis of 417 cases treated from 1960 through 1969. Neurosurgery 5: 301–308

Shapiro WR (1982) Treatment of neuroectodermal brain tumours. Ann Neurol 12: 231–237

Shetter AG, Thomas VB, Altman HW (1977) Closed needle biopsy in the diagnosis of intracranial mass lesions. Surg Neurol 8: 341–345

Sugita K, Matsuga N, Takaoka Y, Shibuya M, Aoi T (1975) Stereotaxic exploration of parathyroid ventricle tumour. Confin Neurol 37: 156–162

Taveras JM, Hartwell GT, Poole JL (1962) Should we treat glioblastoma multiforme? Am J Roentgenol 87: 473–479

Thomas DGT, Powell MP, Bradford R et al. (1985) Correlation of CT directed target site with histology and cell culture in cerebral glioma. Appl Neurophysiol 48: 460–462

Thomas DGT, Davis CH, Ingram S et al. (1986) Stereotaxic biopsy of the brain under MR imaging control. Am J Neurorad 7: 161–163

Thomas DGT, Bradford R, Bydder GM (1987) Magnetic resonance directed stereotactic brain biopsy. J Neurol Neurosurg Psychiat 50: 645

Thomas DGT, Bradford R, Gill S, Davis CH (1988) Computer-directed stereotactic biopsy of intrinsic brain lesions. Br J Neurosurg 2: 235–240

Thomas DGT, Nouby RM (1989) Experience in 300 cases of CT directed stereotactic surgery for lesion biopsy and aspiration of haematoma. Br J Neurosurg 3: 321–326

Walker MD, Alexander E Jr, Hunt WE et al. (1978) Evaluation of BCNU and/or radiotherapy in the treatment of anaplastic gliomas – a cooperative clinical trial. J Neurosurg 49: 333–343

Walker MD, Green SB, Byar DP et al. (1980) Randomised comparison of radiotherapy and nitrosoureas for the treatment of malignant glioma after surgery. N Engl J Med 303:

1323–1329

Walters CL, Schmidek HH (1988) Surgical management of intracranial gliomas. In: Schmidek HH, Sweet WH (eds) Operative Neurosurgical Techniques. Indications, Methods and Results, Vol. 1. Grune and Stratton, Orlando Fl, pp 431–450

Weir B (1973) The relative significance of factors affecting postoperative survival in astrocytomas grade 3 and 4. J Neurosurg 38: 448

Young B, Oldfield EH, Markesbery WR *et al.* (1981) Reoperation for glioblastoma. J Neurosurg 55: 917–921

Yung WA, Horten BC, Shapiro WR (1980) Meningeal gliomatosis; a review of 12 cases. Ann Neurol 8: 605–608

12 Radiotherapy for Malignant Brain Tumours

M. Brada and L. Vanuytsel

The development of radiotherapy in the management of brain tumours has been largely empirical, but since its introduction it has developed into one of the most effective treatment modalities in the management of patients with intracranial tumours. In some rare malignant brain tumours such as germinoma, radiotherapy is curative. The excellent disease control following conservative surgery and radiotherapy in the more benign tumours such as optic nerve glioma, pituitary adenoma and craniopharyngioma has established radiotherapy as the essential component of treatment. In high-grade gliomas, the commonest of malignant brain tumours, radiotherapy achieves prolongation of disease-free survival and survival. The value of radiotherapy has been demonstrated in prospective randomized studies in high-grade gliomas. In other brain tumours the role of radiotherapy is not proven beyond doubt and we have to rely for proof of its effectiveness on selected series of patients referred for radiotherapy largely following incomplete tumour excision.

The technique of radiotherapy has evolved from wide-field irradiation to high-precision stereotaxic radiotherapy which in the form of single fraction radiosurgery is considered equivalent to a non-invasive neurosurgical procedure. Stereotaxic neurosurgical technology is also employed in the implantation of radioactive sources directly into tumours (interstitial radiotherapy). Such new developments in radiotherapy require close collaboration between neurosurgeons and radiation therapists and are likely to lead to improved care of brain tumour patients.

Biological Basis of Radiotherapy

Radiation in the treatment of brain tumours is most commonly employed in the form of high-energy X-rays generated by a linear accelerator or gamma-rays from a cobalt-60 source. Electrons are rarely used and the role of other particle radiations (protons, neutrons, alpha particles) is largely experimental and limited. In clinical practice, radiotherapy (RT) is given as a series of daily treatments five times a week with daily doses of 1.6–2.0 Gy to total doses of 45–60 Gy. The overall treatment times range from four to seven weeks.

Given the high proportion of water in the mammalian cells, most of the energy is absorbed in the aqueous component of irradiated cells. The radiation-induced dissociation of water results in the formation of free radicals, which damage the cell by reacting with vital biological molecules such as nuclear DNA, cytoplasmic organelles and cell membranes. Free radicals react with oxygen to produce peroxy-radicals, leading to fixation of the damage. This irreversible reaction competes with repair in which hydrogen atoms are transferred from endogenous hydrogen-donating molecules. Under hypoxic conditions, less damage is fixed and cells are therefore more resistant to X-rays or gamma-rays.

From a mechanistic point of view, it is proposed that at the cellular level part of the inflicted radiation damage is irreparable, leading to mitotic cell death, while part can be repaired if the cell is allowed enough time. This latter phenomenon is described as repair of sublethal and potentially lethal radiation damage. When the number of sublethal lesions exceeds a critical limit (e.g. after a high single radiation dose) or when the time between successive treatments is too short to allow for repair, the lesions can accumulate and become lethal.

Lethally damaged cells usually die only after their reproductive integrity is tested by one or more attempts at mitotic division. The rate at which radiation injury develops in a tissue is therefore related to the proliferative activity of its component cells. Rapidly proliferating tissues, such as skin and mucosa, develop injury early, within weeks after the start of irradiation. Non-proliferating or slowly proliferating tissues, such as the central nervous system, will express damage late, months to years after irradiation. Once the tissue has recognized the radiation-inflicted damage, it will attempt to restore itself by compensatory repopulation of the surviving stem cells. Early reacting tissues have an extensive regenerative capacity and are rarely dose-limiting in modern conventional RT. Repopulation starts during therapy and increases at an exponential rate. Shortening the overall treatment time will increase acute reactions. Late-reacting tissues exhibit only limited regenerative potential and lost cells are not usually replaced, leading to late irreversible damage. The radiation tolerance of these tissues is thus a limiting factor which is important in determining current RT practice. Although some tumours proliferate very slowly, most exhibit rapid proliferation kinetics and are considered to react similarly to acutely responding normal tissues.

Clonogenic cell survival curves for mammalian cells exposed to X-rays or gamma-rays are usually plotted as logarithm of surviving fraction of cells (log S) against dose (D). They show an initial shoulder region followed by a continuous bending part. Different mathematical models have been developed to fit the actual cell survival data. Widely used to describe the cell survival relationship is the linear quadratic (LQ) model (Chadwick and Leenhouts, 1973), where

$$\log S = -\alpha D - \beta D^2$$

The linear term with constant α would represent the single hit component of damage and determines the initial slope of the survival curve; the quadratic term with constant β would represent two-hit damage, sublethal damage accumulating at high doses, and determines the final slope of the curve. The ratio α/β gives the dose at which the lethal effect of both components is equal. Early-reacting tissues have a large α/β ratio and late-reacting tissues a small ratio.

With increasing dose, cell killing from accumulated sublethal injury (β component), relative to killing from direct single hit events (α component), increases more rapidly in late-reacting target cells. As a consequence, late-responding tissues will be more spared by fractionation than acutely reacting tissues (Fig. 12.1).

The LQ model can be used to calculate equivalent doses when other than conventional radiation treatment schedules are applied:

$$D_{new} = D_{ref} \times \frac{\alpha/\beta + d_{ref}}{\alpha/\beta + d_{new}}$$

where D is total dose and d is dose per fraction (of *new* versus *reference* regimen). The application of this and also other cell survival models is restricted by two assumptions: the interfraction interval is long enough to allow complete repair of sublethal damage, and overall treatment time is kept constant or short enough to rule out repopulation. If interfraction intervals are insufficient to allow full repair, survival will be less than for longer intervals and more elaborate models are necessary (Thames, 1985).

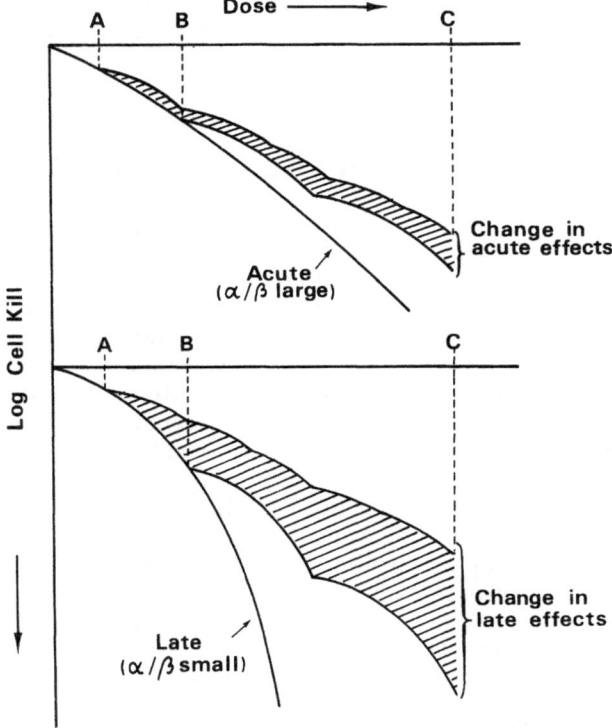

Fig. 12.1 Hypothetical single and multifraction dose survival curves for target cells in acute and late-responding tissues. Changing the fraction size from fraction A to B results in little change in acute effects but a large change results in late effects (From Fowler, 1985, with permission)

Present radiation practice, giving one fraction of 1.6–2.0 Gy per day, five days a week, is largely empirical, and different alternative radiation schedules are being investigated in an attempt to improve the therapeutic ratio (the balance between tumour control and normal tissue complications). Hyperfractionation and accelerated treatment are the most frequently employed modifications. In hyperfractionated schedules, smaller doses per fraction are used, while the overall treatment time is unchanged. This is achieved by giving two or more fractions per day separated by a time interval that should allow for complete repair of sublethal damage of late-responding tissues. The rationale is to exploit the sparing effect of fractionation for late-responding tissues relative to acutely-responding tissues and tumours. This allows for higher total doses causing more pronounced acute reactions, but unchanged late damage. Accelerated treatment is based on a shortening of the overall treatment time, giving the same dose per fraction. The interval between fractions should also be kept as long as possible. The rationale is to minimize the potential for tumour growth or regeneration, which has its greatest effect later in the treatment. The limiting factor is acute tissue reaction.

The total dose, dose per fraction, time interval between fractions and total treatment time are all important factors in determining the final outcome of treatment. This is assessed as the probability of tumour control and probability of late damage and sequelae. Modification of any of the parameters has to be done with caution, based on clinically and experimentally derived data (see below).

Radiation Tolerance of the Central Nervous System

Radiation-induced CNS damage is characterized by a latent period which corresponds to the depletion of target cell population. In the CNS, the most likely target cells are slowly proliferating oligodendroglia and vascular endothelial cells and the resultant damage is a mixture of demyelination and necrosis which is usually confined to the white matter. Necrosis within vital areas of the CNS, such as brain stem or spinal cord, produces severe disability which does not recover and is not acceptable even at a low frequency.

Adverse effects of radiation in the CNS have historically been divided according to the time of appearance. Acute reactions become manifest during irradiation; they are usually transient and are difficult to distinguish from tumour progression. Delayed damaged is described as either early delayed reaction, appearing a few weeks to a few months after RT, or late delayed reaction which starts months to years later (Sheline, 1986). In addition to the classical late radiation reactions which have been correlated with histological changes, a less severe late deficit may follow lower doses of radiation. This has been recognized, particularly in children, as cognitive impairment. Such late radiation effects are also suspected in adults. Because of the difficulty in distinguishing the effects of radiation from those of the tumour itself and because of the poor long-term survival in many of the tumours treated, the relationship to radiation parameters and the true incidence are not clear. The radiation sensitivity of normal CNS is also altered by the addition of chemotherapeutic agents. The extreme example is the simultaneous administration of systemic methotrexate with radiation causing leucoencephalopathy (Bleyer, 1981).

Spinal Cord Tolerance

Early delayed radiation reaction in the spinal cord is expressed as Lhermitte's phenomenon. It occurs between two weeks and seven months following even low doses of radiation (20 Gy or more using standard 1.6–2.0 Gy per fraction) and is more common with higher doses. In experimental animal models it is the result of transient demyelination due to damage to oligodendroglia (Mastaglia et al., 1976).

A late delayed radiation damage to the spinal cord results in chronic progressive myelopathy with features of pyramidal or spinothalamic tract impairment and can also include the signs of lower motor neuron deficit. The pathogenesis is the combination of vascular and demyelinating lesions and damage to the anterior horn cells and their axons (Reinhold et al., 1984). Although it may occur anywhere between six months and five years following radiation, there are two peaks of incidence corresponding to two latent periods. This is in keeping with the hypothesis that white matter parenchymal lesions have a shorter latent period than vascular lesions (Schultheiss et al., 1988).

Delayed radiation changes in human spinal cord have been correlated with experimental radiation myelopathy observed in rodents. Although the radiation tolerance of rat spinal cord is different from the clinically observed radiation sensitivity, the experimental data have helped to derive time/dose/fractionation factors. The dose–response curves for spinal cord damage are steep, suggesting that even small increments of dose beyond radiation tolerance will significantly increase the risk of myelopathy. Recent experiments, using clinically relevant fraction sizes between 1.2 and 2 Gy, reveal a bi-exponential repair mechanism for sublethal damage with repair half-times of 0.7 and 4 h (Ang et al., 1992). As a consequence of the long half-time of the slow component of repair, shortening the interfraction interval from 24 to 8 h decreases the spinal cord tolerance by about 15% (Ang et al., 1992; Ruifrok et al., 1992a). This suggests that when the CNS is the dose-limiting organ, improvement in therapeutic ratio by hyperfractionation will be small if two fractions per day are given in less than an 8–10 h interval.

Spinal cord radiation tolerance studies in primates are closer to the clinical setting (Schultheiss et al., 1990). Using 2.2 Gy fractions the estimated doses for 1% and 0.1% myelopathy are 59 Gy and 52 Gy, respectively, which is close to clinical experience. Field size effects in primates are small (Gutenberger et al., 1992).

Studies on the developing rat spinal cord would indicate that the immature cord is less sensitive to the sparing effect of fractionation with an

α/β value of 4.5 Gy (Ruifrok et al., 1992b), and that the capacity for long-term recovery in the months after radiation treatment is less in the immature than in the adult cord (Ruifrok et al., 1992c).

The dose limits of spinal cord tolerance in clinical RT have initially been derived by Boden (1948). He recognized that tolerance is not only related to the number of fractions and total dose but also to the length of cord treated. Other data on cord sensitivity have been derived from irradiation of the neck, thoracic or upper abdominal structures where the cord is included in the target volume. Published clinical data are inadequate for valid statistical dose–response analyses and can only serve as a guide for "best estimate". Evaluating more than 1100 patients, Marcus and Million (1990) concluded that the risk of permanent neurological damage is very low for fractions of less than 2 Gy given once a day or 1.2 Gy given twice a day to a total dose of 55 Gy or less. Reviewing the literature, Schultheiss (1990) estimated the incidence of myelopathy at 45 Gy in 1.8–2 Gy fractions as most likely below 0.2%. In conventional fraction schedules, the dose causing a 5% incidence of myelopathy (TD_5) was estimated to be in the 57–61 Gy range and the dose for 50% (TD_{50}) to be in the 68–73 Gy range. Emami et al. (1991), as a result of clinical consensus, proposed a somewhat lower TD_5 of 50 Gy for a treated length of 5–10 cm and 47 Gy for a length of 20 cm.

Brain Tolerance

Following even low doses of radiation to the whole brain, patients experience a transient period of drowsiness and other non-specific symptoms which starts 2–6 weeks after completing treatment. This is described as somnolence syndrome and represents an early delayed radiation reaction. As in the spinal cord it is most likely due to transient demyelination and axonal swelling and does not result in permanent damage. Transient somnolence syndrome has been described, particularly after whole-brain irradiation in children with acute lymphocytic leukaemia (ALL) (CNS prophylaxis). It develops in up to 80% of children receiving doses of 24 Gy and is less common with lower doses. It is not clear if the high incidence is due to the concomitant use of chemotherapy or to the more detailed observation in a closely scrutinized group of children who are neurologically normal prior to RT. The frequency of somnolence syndrome following RT for brain tumours in children and adults is not known, but it may be associated with irradiation of hypothalamic areas.

The end result of late radiation damage to the brain is necrosis. It may develop either at the site of tumour where it is difficult to distinguish from tumour recurrence, or at a distant site. Clinically derived limits of radiation tolerance are based on reports of documented cases of radiation necrosis (Sheline et al., 1980). In the majority of studies the population of patients at risk is not defined and the frequency of damage at particular dose levels is therefore not clear. Nevertheless a summary of the reported experience serves as a guide to the tolerance doses in clinical practice. The tolerance limits are largely derived from conventional daily fractionation. It is possible to obtain limits of radiation tolerance for unconventional fractionation by extrapolating from the available data, but this has to be done with considerable caution and the safety of such derived limits cannot be guaranteed.

The dose-fractionation schemes reported as threshold for brain necrosis are approximately 35 Gy in 10 fractions, 60 Gy in 35 fractions and 70 Gy in 60 fractions (Sheline *et al.*, 1980) or 60 Gy in 31 fractions (Marks *et al.*, 1981) and 62 Gy in 30 fractions (Pezner and Achambeau, 1981). Although these limits provide overall guidelines, they have severe limitations because they do not take into account the volume of brain irradiated. Emami *et al.* (1991) estimated, on the basis of clinical consensus, the TD_5 at 1.8–2 Gy fractions to be 60 Gy, 50 Gy and 45 Gy when the irradiated volume was one-third, two-thirds or the whole brain, respectively. It is also assumed that different parts of the CNS have equivalent radiation tolerance. There is little data to substantiate it, and while in principle it may be correct it is the acceptability of radiation-induced necrosis in different parts of the brain which may specify different clinical tolerance limits. A detailed analysis of patients treated with protons for skull base tumours showed that the complication rate for brain stem and cranial nerves is 1% at 60 cobalt Gy equivalent (CGE) and 5% at 70 CGE (Urie *et al.*, 1992).

Lower doses of radiation which do not cause necrosis may result in neurological disturbance without an identifiable structural defect. Children receiving brain irradiation as CNS prophylaxis or as part of therapy for primary brain tumours may suffer intellectual impairment. It has been clearly demonstrated in children with ALL (Duffner *et al.*, 1985), and in children with primary brain tumours (Packer *et al.*, 1989), although it is occasionally difficult to distinguish the deficit caused by tumour and raised intracranial pressure from the effects of radiation. The degree of cognitive function impairment in children is related to age. The younger the child at the time of RT, the more severe the deficit. Radiotherapy to the pituitary–hypothalamic axis also leads to the deficiency of pituitary hormones. Children receiving doses over 30 Gy develop growth hormone (GH) deficiency (Shalet, 1986) and higher doses may lead to failure of ACTH, TSH and gonadotrophin secretion. Growth hormone deficiency is increasingly reported in adults following cranial irradiation, but the clinical significance is not yet clear.

Adults surviving radical RT for brain tumours have also been reported to develop neuropsychological impairment (Hochberg and Slotnick, 1980), although this is difficult to distinguish from damage caused by tumour. Cognitive function impairment has also been reported in patients with small cell carcinoma of the lung following relatively low doses of prophylactic whole-brain irradiation together with chemotherapy (Twijnstra *et al.*, 1987). Such deficit may only be detected on detailed neuropsychological tests and may be due to significant doses of radiation delivered to critical parts of normal brain. The information on sensitive structures and the doses responsible for functional impairment is not available.

Radiotherapy Technique

Photon irradiation is delivered to the brain either as external beam RT, or directly by brachytherapy where radioactive sources are implanted into the tumour. Conventional external beam RT can be given to the tumour, whole brain or whole craniospinal axis. Most glial tumours are treated by local irradiation, with craniospinal axis radiotherapy reserved for tumours with known potential for cerebrospinal fluid (CSF) spread. Whole-brain irradiation is most commonly employed in the treatment of metastatic disease and as cranial prophylaxis to prevent meningeal spread of ALL and some solid tumours.

Practical steps in the RT of brain tumours are listed in Table 12.1. The correct delivery of irradiation requires accurate localization of tumour on computed tomography (CT) or magnetic resonance imaging (MRI) scans, and transfer of this information for RT treatment planning. The intended volume to be treated to the prescribed dose is defined as target volume. It includes the visualized tumour, a margin of suspected microscopic tumour extension and a safety margin to allow for technical inaccuracy such as movement of patient during treatment. Treatment planning defines the ideal arrangement of radiation portals to deliver the prescribed dose to the target volume with maximum sparing of the surrounding normal brain. Individual techniques include treatment by parallel opposed pair of beams for large tumours crossing the midline and for whole-brain irradiation and more localized field arrangements of three to four fields for smaller or unilateral tumours. The intended treatment plan is checked on a simulator, a dedicated radiographic unit, before treatment delivery. Both treatment planning and therapy are carried out in an immobilization device to allow for precise and repeatable treatment. Depending on the position of tumour, the patient is treated either in a supine or prone immobilization shell. Lateral position is used rarely because of the discomfort for the patient and low accuracy of the set-up.

Traditional two-dimensional treatment planning is defined in the central treatment plane and the homogeneity of the dose within the target volume (less than 10% variation) is only visualized in that plane. With three-dimensional computer planning it is possible to reconstruct target volume

Table 12.1 Practical steps in radiotherapy of CNS tumours

Tumour localization
  CT, MRI, other
  Definition of target volume

Treatment planning

  Immobilization device
  Computer planning of dose distribution
  Verification of treatment plan on simulator

Treatment delivery

in space based on CT images. The homogeneity within the target volume and the dose distribution in normal tissues can be assessed more objectively by the use of dose–volume histograms (Chen, 1988), which give a more accurate representation of the dose of radiation received by tumour and normal tissue.

New Radiotherapy Techniques

The need for high-dose focal irradiation has led to the introduction of more sophisticated and precise techniques of external beam RT. Dedicated multiheaded cobalt units containing over 200 focused cobalt sources (gamma unit) can deliver radiation to 0.5–1.8 cm diameter volumes with little dose to surrounding normal brain. Similar dose distribution can be obtained with a linear accelerator using multiple non-coplanar arcs of rotation or multiple non-coplanar fixed beams around the target volume centre (Fig. 12.2) (Colombo *et al.*, 1985; Hartmann *et al.*, 1985; Lutz *et al.*, 1988). The technique is described as stereotaxic external beam RT or radiosurgery when delivered in a single fraction. It requires precise patient fixation which is achieved with neurosurgical-type stereotaxic frames and accurate delineation of treatment volumes in three-dimensional space defined by stereotaxic coordinates relative to the appropriate frame.

Single fraction radiosurgery is most commonly employed for the treatment of arteriovenous malformations and in patients with solitary brain metastases. It has also been used in the treatment of acoustic neuromas (Leksell, 1987), recurrent gliomas and a variety of other tumours. Fractionated stereotaxic external beam RT using linear accelerator may be

Fig. 12.2 Arcs of rotation of stereotaxic external beam radiotherapy using linear accelerator with patient in supine position

considered as a high-precision conformal external beam irradiation which is likely to have wide application in neuro-oncology where minimizing the volume of normal brain irradiated is important.

It is also possible to deliver high-dose radiation to small tumours by implanting radioactive sources directly into the tumour (Gutin *et al.*, 1984). The most commonly used isotopes in the interstitial RT of brain tumours are iridium-192 or iodine-125. Iridium has a half-life of 74 days and is used as a removable source. Iodine seeds (half-life 60 days) are implanted individually or as linear sources and, depending on their activity, can be employed either as temporary or permanent implants. The technical details of implantation vary between different centres. The initial step involves tumour localization with CT or MRI using stereotaxic neurosurgical equipment. Radioactive sources are placed via catheters along defined coordinates and trajectories and this procedure can be carried out either under general or local anaesthetic. The loading of radioactive isotope can be done immediately at the time of implantation or as after-loading into catheters. Ideally a sufficient number of sources should be inserted to achieve a homogeneous dose within the intended treatment volume. In practice, the number of sources implanted is a compromise between the ideal dose homogeneity and the safety of insertion of multiple catheters. The precise role of interstitial RT is currently under investigation both as adjuvant therapy in high-grade astrocytoma and as treatment for recurrent glioma.

High linear energy transfer radiation such as neutrons or protons have been used in selected centres. Protons preferentially deposit energy at the end of their path through tissues. This region (Bragg peak) is defined for each proton energy and can be modulated to coincide with a target volume. Proton irradiation has been used for the treatment of pituitary tumours, arteriovenous malformations and other selected lesions, such as chordomas. Neutron therapy has been exploited largely to overcome the putative problem of hypoxia where poorly oxygenated tumour cells may be resistant to photon irradiation. The potential advantages of increased tumour cell kill and effectiveness under hypoxic conditions are outweighed by increased normal tissue damage. So far there is no evidence that neutron therapy improves the therapeutic ratio in the treatment of brain tumours. In addition, proton and neutron therapy require specialized equipment (cyclotron), which is not available for widespread use.

Radiotherapy in Specific Tumours

High-grade Astrocytomas

High-grade (HG) astrocytomas, classified as anaplastic astrocytoma and glioblastoma multiforme or according to the Kernohan system as grades III and IV, comprise the single largest group of primary malignant brain tumours. After surgery alone, even when excision is apparently complete, HG astrocytomas recur following a short disease-free interval. Addition of radiotherapy has been shown to prolong survival in three randomized studies, although the overall results of combined surgery and radiotherapy are disappointing with few long-term survivors (Walker *et al.*, 1979; Kristiansen *et al.*, 1981). Nevertheless, radiotherapy is the mainstay of treatment, with the aim of prolonging survival and improving quality of life both by improving functional status and by maintaining neurological function for the duration of the disease-free interval.

The overall median survival of patients with HG astrocytoma treated by surgery and radiotherapy is 40–50 weeks, with few patients surviving beyond two years. Many well-conducted multicentre randomized trials have identified age, performance status and tumour histology as the most important determinants of survival. The median survival of patients less than 40, 40–60 and greater than 60 years of age have been reported as 30, 10 and 6 months, respectively (Chang *et al.*, 1983). Functional status has been assessed either by Karnofsky performance scale (KPS) or MRC neurological performance status (NPS) scale. Patients with a KPS score of 70–100 had a median survival of 12 months compared to 6 months in patients with a score of 40–60 (Chang *et al.*, 1983). Other factors such as length of history, presence of seizures and the extent of excision may also be important, but in the majority of studies they are of lesser prognostic significance (Shapiro, 1986). Based on predictive factors, patients can therefore be divided into prognostic categories. The survival advantage gained by RT in each prognostic group has not been defined.

High-grade astrocytomas were traditionally treated by whole-brain irradiation. This was based on pre-CT scanning experience of spread of tumour beyond the suspected margin visualized surgically or by radioisotope imaging. Currently, RT is confined to the CT- or MRI-visualized tumour and a margin to encompass the suspected extent of microscopic tumour spread. Although there is no randomized study comparing wide-field and localized irradiation, the reported results of localized radiotherapy are comparable to the survival results of wide-field irradiation (for review, see Leibel and Sheline, 1987).

Although the majority of HG gliomas recur within 2 cm of the tumour margin (Hochberg and Pruitt, 1980), comparison of CT and detailed postmortem studies suggests that tumour invasion is more variable and may exceed 3 cm in up to 13% of cases (Burger *et al.*, 1988). Nevertheless the main cause of death is the progression of tumour at the primary site with only 5%–9% developing multiple lesions (Choucair *et al.*, 1986; Barnard and Geddes, 1987) and our current practice is therefore to treat enhancing tumours to a target volume which is defined at 3 cm beyond the rim of CT enhancement.

Initial studies of the Brain Tumour Study Group (BTSG) indicated a dose–response relationship with increasing median survival, with total doses increasing from 45 to 50 and to 60 Gy at 2 Gy per fraction (Walker *et al.*, 1979). Prolongation of survival with increasing radiation dose has been confirmed in a randomized MRC study comparing 45 Gy in 20 fractions with 60 Gy in 30 fractions (Bleehen and Stenning, 1991). A randomized study comparing 60 Gy whole-brain irradiation with 60 Gy plus boost of 10 Gy failed to show further benefit of the extra dose (Chang *et al.*, 1983). It is therefore considered that the current optimum dose to large volumes using conventional RT technique is 55–60 Gy at 1.8–2.0 Gy per fraction.

New Approaches in High-grade Gliomas

The relative radioresistance of glial tumours seen clinically, coupled with limited radiation tolerance of the CNS, results in a poor therapeutic ratio with poor tumour control. A number of studies attempted to modify radiation response and improve the therapeutic ratio. The use of radiosensitizers (MRC Working Party, 1983), hyperbaric oxygen (Chang, 1977) and neutron beam therapy (Catterall *et al.*, 1980; Duncan *et al.*, 1986), to overcome the presumed problem of hypoxia, have not shown a significant survival advantage. The high proliferation of HG gliomas suggested that accelerated fractionation, which shortens the overall treatment time, may further inhibit the regeneration of tumour cells. By giving more than one fraction per day, tumour repopulation during treatment may be reduced with a potential for a better control for a given dose level, provided that there is no increase in late normal tissue injury. Early studies of accelerated RT, often combined with hyperfractionation, have not demonstrated a survival advantage (Douglas and Worth, 1982; Payne *et al.*, 1982; Keim *et al.*, 1987). Nevertheless, treatment can be completed in 3–4 weeks with little adverse effect and this avoids protracted six weeks of therapy which is often unacceptable in patients with limited prognosis. Median survival of over 100 patients treated with accelerated regimen of 55 Gy in 34 fractions given twice daily (minimum 6-hour gap) at the Royal Marsden Hospital, Surrey, UK, was 10 months which is comparable to a matched cohort treated with conventional fractionation. Accelerated fractionation carries the potential risk of increased late normal tissue damage and the low toxicity demonstrated in these studies cannot be considered as proof of safety of these regimens, particularly as the survival of the patients studied is often too short for the full expression of late CNS injury.

Hyperfractionation, the use of small dose fractions in the same overall treatment time, exploits the differences in repair capacity between tumour and late-responding normal tissues. In the CNS it may allow for a higher total dose and an increased tumour cell kill. Small randomized studies of hyperfractionation have yielded conflicting results, but overall there is little survival advantage (Fulton *et al.*, 1984; Packer *et al.*, 1987; Ludgate *et al.*, 1988). The optimum dose level may not have been reached and increasing the total dose using small doses per fraction is being tested further, but this will have to be done with caution so as not to precipitate late normal tissue damage.

Attempts have been made to sensitize tumour cells by administration of halogenated pyrimidine analogues. BUDR and IUDR are taken up in the DNA of cycling cells and may act as selective radiosensitizers of proliferating tissues. They have been administered throughout the course of RT in glioma patients, but a survival benefit with this approach has not yet been demonstrated (Kinsella *et al.*, 1988; Greenberg *et al.*, 1988).

Accurate high-dose localized irradiation is best achieved with interstitial or stereotaxic external beam RT but these techniques are currently limited to small volumes and to tumours within less functionally important regions of the brain. High radiation doses employed are well beyond known radiation tolerance, although the radiation dose outside the confines of the tumour is minimized. Interstitial RT with iodine-125 sources has been employed for the treatment of recurrent supratentorial HG gliomas. Tumour doses range from 50–120 Gy (retreatment). Although the median survival from recurrence is beyond 50 weeks in patients with glioblastoma multiforme, and beyond 80 weeks in patients with anaplastic astrocytoma, the treatment is accompanied by a high incidence of necrosis within the irradiated region which requires reoperation (Leibel et al., 1989). The selection of patients, which includes many favourable prognostic characteristics, makes it difficult to ascribe the excellent survival results to the procedure alone (Florell et al., 1992). Patients who undergo resection of necrotic tissue have the most favourable outcome which may be the result of repeat surgery rather than interstitial RT. The role of localized radiation boost after conventional treatment is currently being tested in randomized studies.

Stereotaxic external beam RT achieves a similar dose distribution compared to brachytherapy and is suitable for localized RT of brain tumours. Single fraction radiosurgery (Alexander and Loeffler, 1992) and fractionated stereotaxic RT (Laing et al., 1993) in patients with recurrent HG glioma result in survival comparable to brachytherapy and provide a non-invasive alternative to interstitial RT.

The use of adjuvant chemotherapy following surgery and RT for HG astrocytomas has demonstrated a limited but significant survival advantage in randomized studies (Stenning et al., 1987). The details are described in Chapter 13. Although used in some centres, chemotherapy is not commonly employed as adjuvant therapy in the UK because the survival advantage achieved is considered to be of questionable clinical significance.

Low-grade Astrocytomas

Low-grade (LG) tumours are defined as grade I or II (Kernohan and Sayre, 1972) or as astrocytomas on a three-tiered grading system (Rubinstein, 1972; Burger and Vogel, 1982). This recognizes the difficulty in differentiating between Kernohan grades I and II and the failure of most reports to demonstrate longer survival for the grade I tumours. The revision of WHO classification classifies pilocytic astrocytoma and subependymal giant cell astrocytoma (ventricular tumour of tuberous sclerosis) as separate LG tumours.

Astrocytomas in adults occur primarily in the cerebral hemispheres. Pilocytic astrocytomas, which present in late childhood and adolescence, are probably the supratentorial counterpart of the cerebellar pilocytic astrocytomas of childhood. They tend to be localized to the temporal lobes, basal ganglia, thalamus and hypothalamus (Shaw et al., 1989a) and tend to be more localized with more indolent behaviour than other LG astrocytomas.

Although LG astrocytic tumours are often considered as benign, based on their histological appearances and relatively slow growth (Hoshino, 1984), the long-term survival rates and treatment results are poor. Similar to high-grade astrocytomas, histology, age and performance status have been considered the most important prognostic factors. The reported 5- and 10-year survival rates of patients with pilocytic astrocytoma are 85% and

79%, respectively. The 5- and 10-year survival results for ordinary astrocytomas and mixed oligo-astrocytomas range from 40% to 50% and from 10% to 20%, respectively. The independent significance of age has been more difficult to define, largely because of the failure to take into account the different histological subtypes. Pilocytic astrocytomas preferentially occur in a younger age group and represent the better demarcated, less infiltrative lesions, more amenable to surgical excision (Shaw *et al.*, 1989a) and this may account for better results in younger patients. However, other reported results and the Royal Marsden Hospital experience of ordinary astrocytomas suggests that older patients with LG tumours have worse prognosis.

The role of RT in the treatment of LG astrocytomas is controversial, and general agreement about indications, timing and total dose is lacking. Randomized prospective studies are not available and we have to rely on historical data where the results are not stratified by age or histology. Leibel *et al.*, (1975) reported the results of 108 patients with incompletely excised lesions treated between 1942 and 1967, at the University of California, San Francisco. Seventy-one patients who received postoperative RT had 5- and 10-year survival rates of 46% and 35%, compared to 19% and 11% for unirradiated patients. Following complete resection, the reported 5- and 10-year survival rates were 100%, but this group of only 14 patients included 9 children or young adults and 11 cerebellar tumours, usually of the cystic pilocytic type. A multi-institutional retrospective study of 68 patients spanning the years 1958–1974 reported a 5-year survival rate of 54% for irradiated and 32% for non-irradiated patients (Fazekas, 1977). This apparent survival advantage for RT was demonstrated despite an age and complete excision bias in favour of the unirradiated group. When only incompletely resected patients were considered, the 5-year survivals were 41% with and 13% without radiotherapy.

A retrospective analysis of the Mayo Clinic experience of the modern RT era (1960–1982) included only patients with supratentorial LG gliomas. The results demonstrated a survival advantage for patients receiving RT, and suggested an improved survival for patients receiving doses above 53 Gy. The extent of surgery was of no prognostic significance (Shaw *et al.*, 1989a). The interpretation of the results is, however, hampered by the inclusion of astrocytomas and mixed oligo-astrocytomas in the same histological group; the same authors reported 5-year survival of 83% for 20 patients with oligodendrogliomas and mixed oligo-astrocytomas and only 40% for 24 patients with astrocytomas (Shaw *et al.*, 1989b).

Patients with pilocytic astrocytoma have excellent prognosis after complete excision, reaching 100% survival irrespective of adjuvant postoperative treatment. The reported 5- and 10-year survival rates of incompletely excised tumour are 85% and 79% respectively. After incomplete resection, improvement of survival with postoperative RT has been claimed, both in supratentorial (Shaw *et al.*, 1989a) and infratentorial tumours (Griffin *et al.*, 1979), but the number of patients is small and the exact role of RT is not clear.

Until the results of ongoing prospective randomized trials of adjuvant RT in LG gliomas conducted by the EORTC (European Organisation For Research and Treatment of Cancer), and BTCG (Brain Tumour Cooperative

Group) become available, treatment recommendations have to be based on the results of retrospectively analysed patient groups. The available data favour the use of postoperative radiation of incompletely resected LG gliomas. After gross total removal, adjuvant RT is not indicated in patients with pilocytic astrocytoma. In cystic pilocytic tumours, complete resection of the mural tumour and immediate surrounding cyst wall may be sufficient, without the need to excise the remaining cyst wall (Palma and Guidetti, 1985). The value of adjuvant RT following complete resection of other histological subtypes is unclear, but the results of retrospective studies suggest a benefit for RT. In addition, a successful control of LG glioma by RT is accompanied by functional improvement which usually lasts the duration of disease-free interval.

The recommended radiation dose is 55–60 Gy in 1.6–1.8 Gy fractions and the treatment should be restricted to the tumour and a surrounding margin of normal tissue. There is no survival advantage demonstrated for whole-brain irradiation, particularly as treatment failures tend to occur at the primary site within the irradiated volume (Shaw et al., 1989b). Despite the poor long-term tumour control, a number of patients survive beyond five or 10 years. New treatment strategies employing radiation schemes beyond the recognized limits of radiation tolerance may result in unacceptable radiation damage prior to tumour recurrence. They have to be used with considerable caution and such strategies can only be evaluated after more than 5-year follow-up.

Oligodendroglioma

Oligodendroglioma is an uncommon tumour of neuroectodermal origin, usually arising in frontal and, less frequently, temporal and parietal lobes. It occurs in all age groups, with peak incidence between 40-60 years. Although it is a relatively indolent tumour, it usually recurs at the primary site. The rare anaplastic form has a tendency for distant metastasis (Macdonald et al., 1989).

Based on cytological and tissue features, oligodendrogliomas have been classified into grades A–D, but this is of marginal prognostic significance (Ludwig et al., 1986) and the grading system is not universally applied. Burger et al. (1987) were only able to identify necrosis as an independent prognostic variable. A large proportion of oligodendrogliomas contain elements of other glial tumours, particularly astrocytoma. Based on small retrospective series, the prognosis of pure oligodendrogliomas and mixed oligo-astrocytomas is similar (Wallner et al., 1988).

Treatment of oligodendrogliomas has been largely surgical, and the role of RT is uncertain. There are no prospective randomized trials and the available information is based on small uncontrolled retrospective studies. These usually span long periods of time, during which the criteria for diagnosis and the surgical and RT techniques have changed. Bullard et al. (1987) reported their experience of 71 patients treated over a period of 43 years. Thirty-seven patients had postoperative RT of varying techniques and doses and the median survival was not significantly different compared to unirradiated patients (5.2 versus 4.5 years). Other reports (Neumann *et al.*, 1987) also failed to demonstrate a survival benefit for RT. The treatment results of 170 patients with oligodendroglioma treated in Norway over a 24-year period demonstrated a marginal survival advantage for RT (Lindegaard *et al.*, 1987). One hundred and eight patients who received postoperative RT had a median survival of 38 months compared to 27 months after surgery alone. The long-term survival was poor in both groups, with no advantage for patients treated with postoperative RT; the 8-year survival rate for irradiated patients was 17% compared to 14% in unirradiated patients. Following complete excision, which is relatively uncommon in oligodendrogliomas, RT did not confer a survival advantage. Sheline *et al.*, (1964) reported on 32 patients, half of whom received postoperative RT. Survival rates at five years were significantly better with combined treatment (85% versus 55%), but the survival benefit disappeared by 10 years.

It is difficult to interpret such conflicting results. At present there is no convincing evidence that adjuvant RT prolongs survival. However, oligodendrogliomas are infiltrating tumours which can rarely be completely excised. Most patients have residual tumour and recur at the primary sites; consequently the long-term results are unsatisfactory. Although RT does not result in a higher cure rate, it may postpone tumour progression after subtotal resection. In patients with unresectable progressive tumours, RT frequently produces tumour responses with clinical improvement. It is, therefore, reasonable to advocate postoperative RT for incompletely excised tumours and primary RT for unresectable progressive disease. The recommended doses and technique are the same as for low-grade astrocytoma with treatment localized to the tumour and a surrounding margin of 2–3 cm.

The role of chemotherapy is not fully defined. A subgroup of anaplastic oligodendrogliomas appear to have a chemoresponsiveness higher than other glial tumours (Cairncross *et al.*, 1992), but a survival benefit for combined chemotherapy and RT needs to be demonstrated. Preliminary reports also suggest effectiveness of chemotherapy in other oligodendrogliomas (Glass *et al.*, 1992).

Ependymoma

Ependymomas occur most frequently in children and young adults; they constitute 10% of all intracranial neoplasms in childhood, but only 1%–4% of tumours when all age groups are considered. Intracranial ependymomas project from ependymal surfaces, most commonly the floor of the fourth ventricle, and less frequently, lateral and third ventricles. Tumours may also arise within the brain parenchyma without apparent ependymal attachment. The majority of tumours are infratentorial and they account for 60%–70% of all intracranial ependymomas.

There is a lack of consensus regarding the nomenclature and grading of ependymomas. Kernohan and Sayre (1952) proposed a four-grade classification, with grades I and II considered as low-grade and grades III and IV as high-grade tumours. WHO classification distinguishes between LG ependymoma, which includes myxopapillary and papillary ependymoma and subependymoma and HG anaplastic ependymoma (Zulch, 1980). The term ependymoblastoma should be reserved for highly cellular embryonal neoplasms with primitive cellular features and ependymal rosettes (Rubenstein, 1970). This tumour should be distinguished from the anaplastic or malignant ependymoma, and included in the "primitive neuroectodermal tumour" (PNET) category (Rorke et al., 1985). Unfortunately, most reports divide ependymomas into low and high grade without reference to a specific grading system and the terms high-grade ependymoma, anaplastic ependymoma and ependymoblastoma are used as synonyms.

In a number of studies, tumour differentiation is the most important prognostic factor (Bloom, 1982; Salazar et al., 1983; Shaw et al., 1987; Vanuytsel et al., 1992), but there is no uniform agreement on this point (Mørk and Løken, 1977; Ross and Rubinstein, 1989). With more intensive treatment the significance of histological grade may be less important (Salazar et al., 1983).

The pattern of growth and spread varies between supratentorial and infratentorial tumours. Hemisphere tumours invade the white matter, while paraventricular tumours may penetrate through the ventricular walls. Tumours in the region of the fourth ventricle typically fill the ventricular space, and subsequently extend either through the cervical canal or via the subarachnoid space, down to and below the level of the foramen magnum. Ependymomas also have a tendency for CSF seeding. The incidence is not known, and the reported rates vary from 0% to 60% (mean 9%). The difference is largely accounted for by the different methods of assessment of seeding, although other factors may be important. Initial experience suggested high risk of seeding in HG compared to LG tumours and infratentorial compared to supratentorial ependymomas (Salazar et al., 1983; Bloom, 1986). As the majority of ependymomas fail at the primary site, the high incidence of seeding may be a reflection of an uncontrolled disease at the primary site, and the true incidence of isolated seeding is not known.

Despite the lack of formal comparison of surgery with and without RT, the use of postoperative irradiation in patients with ependymomas has become accepted as standard treatment. This is based on retrospective studies which demonstrate good long-term control following partial tumour excision and RT (Salazar et al., 1983; Shaw et al., 1987; Vanuytsel et al., 1992). The reported survival rates in adults following treatment with surgery and RT range from 50% to 70% at 5 years and from 25% to 60% at 10 years (Shaw et al., 1987; Vanuytsel et al., 1992). These are comparable with the survival results reported in children (Pierre-Kahn et al., 1983; Shaw et al., 1987), with the exception of infants, aged ≤ 2 years, who have a bleak prognosis (Nazar et al., 1990; Healey et al., 1991). Tumour grade is a significant prognostic factor, with reported 5-year survival rate of 70% for LG and 20%–30% for HG lesions in both paediatric and adult ependymomas (Kim and Fayos, 1977; Shaw et al., 1987; Vanuytsel et al., 1992). Complete tumour resection improves disease-free survival in LG but not in HG tumours (Healey et al., 1991; Vanuytsel et al., 1992).

Historically, recommendations on the extent of RT have been based on the apparent pattern of spread. Craniospinal irradiation was considered appropriate for the treatment of HG tumours, irrespective of site (Bloom,

1982; Salazar, 1983; Wallner *et al.*, 1986). The recommended treatment for LG supratentorial tumours was either whole-brain irradiation (Salazar *et al.*, 1983) or local irradiation with generous margins (Bloom, 1986; Wallner *et al.*, 1986; Shaw *et al.*, 1987). Craniospinal irradiation has been advocated for LG infratentorial tumours by Bloom (1986). Most authors were, however, more conservative, recommending either whole-brain irradiation with cervical spine extension (Salazar *et al.*, 1983) or local irradiation with generous margins based on operative findings and modern imaging (Wallner *et al.*, 1986; Shaw *et al.*, 1987).

An extensive review of the recent literature (Vanuytsel and Brada, 1991) and a retrospective analysis of the patients treated at the Royal Marsden Hospital (Vanuytsel *et al.*, 1992) suggest that progression/recurrence of tumour at the primary site remains the major problem in the treatment of ependymoma, even though tumour grade, tumour localization and tumour control at the primary site are all factors influencing the incidence of spinal seeding. In addition, there is no evidence that prophylactic spinal irradiation prevents spinal seeding. Our recommendation (Vanuytsel and Brada, 1991; Vanuytsel *et al.*, 1992) supported by others (Nazar *et al.*, 1990; Healey *et al.*, 1991; Lyons and Kelly, 1991) is that irrespective of site and grade, post-operative RT should be confined to the tumour site with appropriate margins, guided by the preoperative radiological extent of the tumour, especially on MRI scan in infratentorial tumours (Griffin *et al.*, 1988). Craniospinal irradiation should be reserved for ependymoblastomas and for patients with HG tumours presenting with positive CSF cytology or with evidence of spinal seeding on spinal MRI or myelography. The recommended dose to primary site is 55–60 Gy. If wide-field irradiation is considered appropriate, our practice is to treat cranial fields at fraction sizes of 1.6–1.8 Gy to 30–40 Gy and spinal fields at 1.25–1.5 Gy per fraction to 30–35 Gy. Appropriate dose reductions should be applied in children (see Chapter 7).

Medulloblastoma

Medulloblastoma is a primitive tumour of neuroectodermal origin and is considered as part of the spectrum of PNETs. While in children these tumours comprise 15%–20% of intracranial neoplasms and usually arise in the region of the cerebellar vermis, medulloblastomas in adults represent less than 2% of intracranial tumours and are more frequently located in the cerebellar hemispheres. Adult tumours most likely represent the upper limit of age incidence with its peak in childhood. Consequently, the biological behaviour and treatment approach in adults is similar to that defined for medulloblastomas in childhood.

The results of surgery alone are unsatisfactory, with survival ranging from 6 to 13 months. Even with improved surgical techniques and apparently complete tumour excision, the local tumour control is poor. This is accompanied by high incidence of spread largely confined to the CNS. Since the introduction of radiotherapy the survival rates in children and adults have markedly improved, and RT has become the essential component of therapy regardless of the extent of surgery.

Medulloblastomas, in common with other PNETs, have a propensity for CSF spread with frequent seeding throughout the subarachnoid space, particularly to the spine. This can be detected by myelography or high-resolution gadolinium-enhanced spinal MRI and has been demonstrated in up to 30% of patients. CSF cytology is routinely perfomed as a staging investigation, but is a poor predictor of spinal seeding. Patients with definite spinal metastases on myelography may have negative CSF cytology, and the presence of tumour cells does not correlate with subsequent spinal seeding (Berry et al., 1981), although it is considered as a poor prognostic factor. Approximately 5% of patients with medulloblastoma develop distant metastases (Kleinman et al., 1981). Although ventricular-systemic shunts are a possible contributing factor, particularly in patients with locally controlled tumours (Berry et al., 1981), systemic disease may occur without shunting. The tendency for seeding through the subarachnoid space, and the known radiosensitivity of medulloblastomas, led to the introduction of postoperative craniospinal axis irradiation. It is a complex high-precision technique which must cover the entire subarachnoid space, as geographical misses increase the number of recurrences (Jereb et al., 1982).

Reported survival rates in adults are comparable to those obtained in children, with 5-year survivals ranging from 50% to 60%. The extent of surgical resection and radiation dose to the posterior fossa are the most important prognostic factors (Bloom and Bessell, 1990). A dose of 50 Gy or more to the posterior fossa is associated with greater local control and improved survival. Most relapses occur within the first three years after treatment, although late recurrences up to and beyond 10 years have recently been recognized, especially after complete resection and high radiation doses (Bloom and Bessell, 1990). This suggests RT cure in some patients and a prolonged growth delay in others. The quality of life of long-term survivors is good. Eighty-four per cent of the surviving patients were able to lead normal active lives and retained normal fertility, with nine healthy children born to six patients (Bloom and Bessell, 1990).

The recommended treatment technique of craniospinal axis irradiation is described by Bloom (1986). The posterior fossa is treated to a dose of 50–55 Gy in daily fractions of 1.6–1.8 Gy. To control subclinical disease, recommended doses vary between 35 and 45 Gy for the whole cranium, and from 30 to 35 Gy to the spine. Attempts at reducing craniospinal axis dose to 25 Gy to reduce late radiation morbidity has resulted in worse control rates in early analysis of POG studies and such lower doses are no longer recommended. Our present policy is to treat whole brain to 30–35 Gy in 18–21 fractions and spine to 30–35 Gy in 20–24 fractions.

The role of chemotherapy in the treatment of adult patients with medulloblastoma is not established. Bloom and Bessell (1990) reported improved survival results after adjuvant chemotherapy, but the introduction of chemotherapy coincided with the use of higher radiation dose to the posterior fossa, and the subgroup of patients receiving chemotherapy included more patients with completely or subtotally resected tumours. As the prognosis of medulloblastoma in adults and children is similar, the definitive guidance as to the value and use of chemotherapy will have to come from the ongoing international multicentre trials for childhood medulloblastoma.

Pineal Tumours

Germinomas are histologically indistinguishable from testicular seminomas and comprise the single largest group of tumours in the pineal region, although they may also arise in the suprasellar area. Malignant teratomas are less frequent at either site. Pineal tumours may also be of pineal parenchymal origin (pineocytoma or pineoblastoma) or gliomas (Bloom, 1983).

The pineal region has not been easily accessible to surgery but histological confirmation by stereotaxic or endoscopic biopsy is now more frequently available. In the absence of histology, the diagnosis of germinoma has been defined by radiosensitivity, although pineoblastomas may show similar response to low-dose irradiation (Bloom, 1983).

Following initial surgery, which may include biopsy, attempted excision or shunting, patients should have assessment of the extent of disease by measurement of serum and CSF alphafetoprotein (AFP) and beta chorionic gonadotrophin (β-HCG) and by CSF cytology. The role of myelography or MRI to detect spinal seeding is not clear, although high-resolution gadolinium-enhanced spinal MRI is recommended in the initial staging. CSF cytology and myelography are not reliable in predicting the risk of spinal seeding, but positive cytology is accepted as a high risk factor.

Radiotherapy has been and remains the treatment of choice in localized pineal and suprasellar germinoma. The extent of RT and the dose to primary site are open to some debate. The known radiation sensitivity of seminoma would suggest that 30 Gy may be sufficient to control intracranial germinomas. Sung et al., (1978) reported a higher control rate of pineal tumours with doses over 50 Gy, but this retrospective study includes a variety of pineal tumours and the minimum required dose for the control of germinomas is not known. A dose of 50 Gy localized to small pineal volume carries low risk of late normal tissue damage and has been the recommended total dose to the pineal region with daily fractionation of 1.6–1.8 Gy. Increasing numbers of reports suggest an excellent tumour control with doses of 40–45 Gy to the primary site.

The unresolved issue is the incidence of dissemination of germinomas along the CSF pathways. Surgical intervention has been reported to increase the incidence of seeding (Leibel and Sheline, 1987), but in retrospective analyses the numbers reported may reflect the true incidence of spinal seeding in histologically verified germinomas (Bloom, 1983). The summary of the reported incidence of spinal seeding is 13% in histologically verified germinomas treated by whole-brain irradiation and only 5% in patients treated by craniospinal RT (Brada and Rajan, 1989). The current reasonable and safe practice therefore includes craniospinal irradiation to a dose of 30 Gy to the whole brain and 30 Gy to the spinal cord and a further boost to the primary site to a total dose of 40–50 Gy. In children with incomplete skeletal growth and in young women where ovaries are in the exit beam of the spinal field, the decision on the inclusion of spinal field is more debatable and has to be made individually.

Unverified pineal region tumours are becoming less common. They should initially be investigated with serum and CSF markers (β-HCG and AFP). If the nature of the tumour remains in doubt and histological confirmation is considered inadvisable, the pineal region should receive a therapeutic trial of RT of 20 Gy in 12 fractions with pre- and post-treatment CT scan. A significant reduction in tumour size is accepted as evidence for germinoma and RT continues to the whole brain and spine to 30 Gy each, reaching a total dose of 50 Gy to the pineal region. Results of RT alone are excellent, with cure rates of 70%–80%. With histologically verified germinomas the figure may be even higher, with cure rate up to 100% (Dearnaley et al., 1990). Suprasellar germinomas are more often his-tologically confirmed. The results of treatment have been reported together with pineal region tumours and there is little evidence to suggest that their behaviour is different. They retain the same responsiveness to RT and the propensity for CSF spread. Radiotherapy to craniospinal axis and boost to the primary site is therefore also the recommended treatment.

Cisplatin- or carboplatin-based chemotherapy is effective in recurrent germ cell tumours by producing radiological and clinical responses. It is increasingly employed as first-line treatment, particularly in patients with extensive tumours at presentation, and is highly effective in producing tumour responses, although the use of systemic chemotherapy has not been proven in preventing seeding through the CSF. Routine use of chemotherapy in small localized intracranial germinomas is at present not justified, but in future studies combined regimens may allow for a reduc-tion in the radiation dose (Allen et al., 1987).

The results of RT for pineal teratomas (non-seminomatous germ cell tumours) are disappointing, with 5-year cause-specific survival of 10%–30% (Bloom, 1983; Dearnley et al., 1990). At present, the same RT regimens are employed as for germinoma, but there is a need to exploit new treatment strategies based on international cooperative groups. Primary chemotherapy is effective in producing radiological responses and has been used either alone or combined with RT. Because of the rarity of these tumours, results of large series are not yet available but it is likely that chemotherapy may become the treatment of choice as in tumours of testicular origin. Residual masses may remain in the pineal region even after successful chemotherapy of teratoma, and attempts at resection or radiosurgery/stereotactic RT boost to the residual pineal tumour after com-pletion of treatment may be indicated.

Pineoblastomas are histologically indistinguishable from medulloblas-tomas and they retain similar radiosensitivity and tendency for CSF spread. They are treated with RT as a PNET/medulloblastoma of the pineal region. Pineocytomas do not share the same radioresponsiveness, but following local radiotherapy to a dose of 55–60 Gy prolonged survival and occasional cures have been reported (Disclafani et al., 1989).

Primary Cerebral Lymphoma

Primary cerebral lymphoma (PCL) is a clonal B-cell neoplasm, phenotypi-cally and genotypically identical to systemic non-Hodgkin's lymphoma (NHL) (Smith et al., 1988), although in its distribution it is confined to the CNS. The majority are intermediate and high grade histology NHL (International Working Formulation), particularly diffuse large cell and diffuse small non-cleaved lymphoma (Hochberg and Miller, 1988). Primary cerebral lymphoma is associated with immune deficiency states such as following organ transplantation, but the recent increase in incidence has been ascribed to the association with acquired immune deficiency syndrome. It has also been suggested that the incidence of PCL not associated with immune deficiency is also rising (Eby et al., 1988).

Patients with PCL present with either solitary or multiple lesions with occasional subependymal and CSF involvement and, rarely, disease in the eye. It is an aggressive disease with median survival of untreated patients between 1 and 3 months and only minimal prolongation following surgical intervention (Hochberg and Miller, 1988). Radiotherapy has been the treatment of choice, but chemotherapy is being increasingly employed, with the suggestion of improved results.

Radiation induces rapid and complete resolution of lesions on CT or MRI scan and, in most patients, this is accompanied by functional improvement. Despite the excellent initial response to RT, the median survival is only 10–18 months (Brada et al., 1990; Nelson et al., 1992). In the majority of patients the disease recurs within the CNS outside the primary site in the brain parenchyma, but relapses also occur in the CSF and occasionally outside the CNS (Hochberg and Miller, 1988).

The published studies report a variety of RT techniques, including involved field, whole-brain or craniospinal axis irradiation. Studies comparing survival results following different treatment volumes have failed to demonstrate advantage of a specific technique. Nevertheless, the frequent recurrences outside the primary site within the brain would argue for the use of whole-brain irradiation with a boost to the ·site of initial involvement. The role of spinal irradiation is not clear and a reasonable practice is the use of craniospinal RT in patients with positive CSF cytology after the initial eradication of CSF disease with intrathecal chemotherapy.

A review of published retrospective series reported better tumour control with higher local dose (>50 Gy) (Murray et al., 1986), but this is not confirmed in other studies and in more restricted reviews (Berry et al., 1981; Leibel and Sheline, 1987). Dose–response data of localized intermediate and high grade lymphoma outside the CNS suggest improved local control with doses ≥ 35 Gy (Sutcliffe et al., 1985). The poor disease control of PCL would therefore argue for maximum tolerated dose to whole brain and the usual recommended policy is whole-brain irradiation to a dose of 40 Gy followed by a boost to the site of initial disease to a total dose of 55–60 Gy (Nelson et al., 1992). The role of spinal irradiation is debatable.

The effective use of chemotherapy in the management of advanced systemic lymphoma has led to the use of chemotherapy either as initial or adjuvant treatment in PCL. Despite the theoretical objection of the blood–brain barrier (BBB), which may prevent penetration of drugs to the tumour, the blood tumour barrier appears to be permeable and initial chemotherapy has demonstrated excellent radiological responses. The early results of series of patients with PCL treated with initial chemotherapy followed by craniospinal axis irradiation have so far failed to show an improvement in survival (Brada et al., 1989; De Angelis et al., 1992). However, the combination of chemotherapy and RT may result in better disease-free survival (De Angelis et al., 1992), and a proportion of long-term survivors. In a combined series of 31 patients from the Royal Marsden Hospital and the Hogarth Oncology Centre, Nottingham, UK, treated with initial chemotherapy, the 3- and 4-year survival is 30% (Bessel and Brada, unpublished) and this is comparable to the results of advanced systemic aggressive NHL treated with conventional chemotherapy. Other new approaches include the use of combination chemotherapy in association with osmotic BBB disruption (Neuwelt et al., 1991) without the use of RT. Although the role of chemotherapy remains to be proven, the poor overall results of radiation alone and the promising early results of a policy of primary chemotherapy followed by RT would argue for further exploration of the combined modality approach.

Radiotherapy in Other Tumours

Optic Nerve Glioma

Optic gliomas are usually low-grade astrocytomas most commonly present-ing in childhood, although they also occur in adults. The natural history of slow progression has led to controversy regarding the optimum manage-ment. Some tumours undoubtedly display aggressive behaviour, with involvement of the optic chiasm and local extension along the optic pathway and into the third ventricle and adjacent brain, particularly hypo-thalamus. Complete excision of optic glioma is difficult, and surgical inter-vention, particularly of chiasmal lesions, is usually confined to biopsy to avoid the risk of visual impairment.

There are no randomized studies comparing surgery with irradiation. The results of RT, usually in patients with progressive inoperable disease, suggest that radiation may prevent progression and improve or stabilize visual deficit. Horwich and Bloom (1985) reported the results of 30 patients with optic nerve glioma treated with RT at the Royal Marsden Hospital. The 5-year survival was 100% and 10- and 15-year survivals 93%, with 90% progression-free survival. Visual acuity improved in 43% of patients, remained stable in 48% and deteriorated only in 9%. Similar results have been reported by others (Flickinger et al., 1988; Jenkin et al., 1993). The recommended RT dose is 50 Gy in 30–35 fractions given in 6–7 weeks, with appropriate dose reduction in young children. The adverse effects of radiation in children, particularly under 2 years of age, have led to the use of initial chemotherapy, which may produce tumour response and temporary stabilization of the visual deficit (Rosenstock et al., 1985).

Meningioma

Histologically benign meningiomas are managed surgically. Radiotherapy has been employed in patients with incompletely excised tumours, which are usually confined to poorly accessible sites. Retrospective studies from a number of centres suggest that the addition of RT improves local tumour control (Barbaro et al., 1987; Taylor et al., 1988; Salazar 1988). The 10-year cause-specific survival of patients with incompletely excised meningioma treated with RT is 67% (Glaholm et al., 1990). The results of surgery alone in patients with malignant meningioma are disappointing. Local control is poor and the median survival is only seven months (Chan and Thompson, 1984). Radiotherapy has been recommended as adjuvant treatment, particularly in incompletely excised malignant meningiomas. Its effectiveness is not proven, although a number of small studies suggest improvement in tumour control (Salazar, 1988).

Meningiomas recurrent after conventional surgery and RT and not considered accessible to further radical surgery may be treated with stereotaxic RT/radiosurgery, provided that they are of appropriate size. The role of stereotaxic RT in the primary treatment of benign meningiomas is unproven and single fraction radiosurgery is potentially hazardous (Engerhart et al., 1990; Kondziolka et al., 1991). However, stereotaxically guided tumour localization and treatment delivery improve the accuracy of treatment with potential sparing of normal tissue and are likely to be adopted in future fractionated RT of meningiomas.

Craniopharyngioma

Despite the apparent benign histological nature of craniopharyngioma, the natural history of untreated craniopharyngioma varies from benign to rapidly progressive and locally invasive tumour causing damage to optic nerve and chiasma, pituitary and hypothalamus. Total surgical resection is curative treatment in selected patients, but this procedure carries high morbidity and mortality (reported at 16% and 17%, respectively (Yassargil et al., 1990), largely because of poor accessibility and close proximity of vital structures and a significant recurrence rate. After incomplete resection, the local recurrence rate ranges from 33% to 100% (mean 73%) (Rajan et al., 1993). The addition of local RT markedly improves the local control, with recurrence rate ranging from 0% to 47% (mean 18%) (Bloom, 1982; Wen et al., 1989; Rajan et al., 1993). The progression-free survival of a cohort of 173 patients treated with conservative surgery and RT at the Royal Marsden Hospital was 83% at 10 years and 79% at 20 years (Rajan et al., 1993). Despite the lack of randomized studies, the excellent local control and survival suggest that combined partial excision and RT is the treatment of choice.

Craniopharyngioma is frequently cystic and despite external beam RT and partial excision, the cyst may refill, requiring repeat aspiration. The instillation of radiolabelled (P^{32}) colloidal chromic phosphate may prevent the reaccumulation of fluid.

Brain Metastases

Ten to fifteen per cent of patients with malignancy develop intracranial metastases during the course of their illness. Brain metastases occur most frequently with cancer of the lung and breast, and less frequently in patients with renal and gastrointestinal tumours. However, most malignant tumours have been associated with the development of brain metastases and the risk has been considered sufficiently high in some, such as small cell carcinoma of the lung, to advocate prophylactic cranial irradiation as part of initial therapy.

The treatment of brain metastases is essentially palliative. The initial use of corticosteroids will improve symptoms in up to 60% of patients. This may be sufficient treatment in the face of extensive metastatic disease at other sites. Radiotherapy provides effective palliation and can produce neurological improvement in 35%–70% of patients (Borgelt et al., 1980). Despite the temporary tumour and symptom control, the long-term results of brain irradiation are disappointing, with a median survival of four months and a 1-year survival of less than 20% (Borgelt et al., 1980). Several large randomized studies of the RTOG group addressed the problem of the optimum radiation schedule. They compared 20 Gy in one week, 30 Gy in two or three weeks, 40 Gy in three or four weeks and 50 Gy in four weeks, given as five daily fractions a week. There was no difference in the efficacy of the different schedules in terms of survival, improvement of neurological deficit, time to progression of neurological symptoms and the frequency of intracranial failure, which is the cause of death in up to a half of the patients. Performance status, control of primary tumour, age and the extent of metastatic disease were identified as independent prognostic factors for survival (Diener-West et al., 1989) and this enables the separation of patients into prognostic subgroups. No benefit has been demonstrated for any of the treatment schedules when patients were stratified into the prognostic groups (Borgelt et al., 1980; Gelber et al., 1981; Kurtz et al., 1981). Brain metastases from melanoma are often considered as a separate category, but the survival results of the RTOG study were no different compared to other tumour types (Carella et al., 1980). The use of unconventional fractionation, such as large doses per fraction which on theoretical grounds may be considered beneficial, has not shown a survival benefit. Considering the convenience of short treatments, particularly in patients with poor overall prognosis, and the lack of demonstrable advantage of large doses of radiation, we employ the short treatment schedules of 20 Gy in five fractions, or 30 Gy in 10 fractions given to the whole brain. Shorter regimens of two or three fractions are being exploited in randomized studies.

Patients undergoing excision of solitary metastases before radiation have better tumour control and survival compared to patients treated with RT alone. The addition of postoperative RT is considered to reduce the risk of intracranial relapse, although survival benefit has not been demonstrated. Despite improvement in local control with surgery and RT, patients with solitary brain metastases have poor overall prognosis and it is reasonable to consider a non-invasive alternative to surgery. One or two fraction radiosurgery/stereotaxic RT achieves excellent tumour control at the treated site, with overall survival and local progression-free survival equivalent to tumour resection (Laing et al., in press). Although randomized studies are not available, stereotaxic RT is now considered as an alternative to surgery in the treatment of solitary metastases.

Conclusion

Radiotherapy is an effective treatment modality in the management of patients with malignant brain tumours, although its usefulness is limited by poor radiation tolerance of the CNS and relative radioresistance of some CNS tumours. Recent technical advances in RT and the increasing knowledge of radiobiological parameters may help in optimizing treatment. Any new therapeutic approaches will have to be evaluated, not only in terms of improved tumour control, but also in terms of quality of life.

References

Alexander E III, Loeffler JS (1992) Radiosurgery using a modified linear accelerator. Neurosurg Clin N Am 3: 167–190

Allen JC, Kim JH, Packer RJ (1987) Neoadjuvant chemotheraphy for newly diagnosed germ-cell tumours of the central nervous system. J Neurosurg 67: 65–70

Ang KK, Jiang GL, Guttenberg R et al. (1992) Impact of spinal cord repair kinetics on the practice of altered fractionation schedules. Radiother Oncol 25: 287–294

Barbaro NM, Gutin PH, Wilson CB et al. (1987) Radiation therapy in the treatment of partially resected meningiomas. Neurosurgery 20: 525–528

Barnard RO, Geddes JF (1987) The incidence of multifocal cerebral gliomas. A histologic study of large hemisphere sections. Cancer 60: 1519–1531

Berry MP, Jenkin DT, Keen CW, Nair BD, Simpson WJ (1981) Radiation treatment for medulloblastoma. A 21 year review. J Neurosurg 55: 43–51

Berry MP, Simpson WJ (1981) Radiation therapy is the management of primary malignant lymphoma of the brain. Int J Rad Oncol Biol Phys 7: 55–59

Bleehen NM, Stenning SP (1991) A Medical Research Council trial of two radiotherapy doses in the treatment of grades 3 and 4 astrocytoma. Br J Cancer 64: 769–774

Bleyer WA (1981) Neurologic sequelae of methotrexate and ionising radiation: A new classification. Cancer Treat Rep 65: 89–98

Bloom HJG (1982) Intracranial tumors: response and resistance to therapeutic endeavours, 1970-1980. Int J Radiat Oncol Biol Phys 8: 1083–1113

Bloom HJG (1983) Primary intracranial germ cell tumours. Clin Oncol 2: 233–257

Bloom HJG (1986) Tumours of the central nervous system. In: Voute PA, Barret A, Bloom HJG, Lemerle J, Neidardt MK (eds) Cancer in Children – Clinical Management. Springer-Verlag, Berlin, pp 197–222

Bloom HJG, Bessel EM (1990) Medulloblastoma in adults: a review of 97 patients treated between 1956 and 1981. Int J Radiat Oncol Biol Phys 18: 763–773

Boden G (1948) Radiation myelitis of the cervical spinal cord. Br J Radiol 21: 464–469

Borgelt B, Gelber R, Kramer S et al. (1980) The palliation of brain metastases: final results of the first two studies by the Radiation Therapy Oncology Group. Int J Radiol Oncol Biol Phys 6: 1–9

Brada M, Dearnley D, Horwich A, Bloom HJG (1990) Management of primary cerebral lymphoma with initial chemotherapy: preliminary results and comparison with patients treated with radiotherapy alone. Int J Radiol Oncol Biol Phys 18(4): 787–792

Brada M, Rajan B (1990) Spinal seeding in cranial germinoma. Br J Cancer 61: 339–340

Bullard DE, Rawlings CE, Phillips B et al. (1987) Oligodendroglioma. An analysis of the value of radiation therapy. Cancer 60: 2179–2188

Burger PC, Vogel FS (eds) (1982) Surgical Pathology of the Nervous System and Its Coverings. John Wiley, New York

Burger PC, Rawlings CE, Cox EB et al. (1987) Clinicopathologic correlations in the oligodendroglioma. Cancer 59: 1345–1352

Burger PC, Hemz ER, Shibata T et al. (1988) Topographic anatomy and CT correlations in the untreated glioblastoma multiforme. J Neurosurg 68: 698

Cairncross J, Macdonald DR, Ramsay DA (1992) Aggressive oligodendroglioma: a chemosensitive tumour. Neurosurgery 31(1): 78–81

Carella RJ, Gelber R, Hendrickson F et al. (1980) Value of radiation therapy in the management of patients with cerebral metastases from malignant melonoma. Radiation Therapy Oncology Group Brain Metastases Study I and II. Cancer 45: 679–683

Catterall M, Bloom HJG, Ash DV et al. (1980) Fast neutrons compared with megavoltage x-rays in the treatment of patients with supratentorial glioblastoma: a controlled pilot study. Int J Radiat Oncol Biol Phys 6: 261

Chadwick KH, Leenhouts HP (1973) A molecular theory of cell survival. Phys Med Biol 18: 78–87

Chan RC, Thompson GB (1984) Morbidity, mortality and quality of life following surgery for intracranial meningiomas – a retrospective study in 257 cases. J Neurosurg 60: 52–60

Chang CH (1977) Hyperbaric oxygen and radiation therapy in the management of glioblastoma. In Modern Concepts in Brain Tumour Therapy: Laboratory and Clinical Investigations. National Cancer Institute Monograph No. 46. US Govt Printing Office, Washington DC, p 163

Chang CH, Horton J, Schoenfeld D et al. (1983) Comparison of postoperative radiotherapy and combined postoperative radiotherapy and chemotherapy in the multidisciplinary management of malignant gliomas. Cancer 52: 997

Chen GTY (1988) Dose volume histograms in treatment planning. Int J Radiat Oncol Biol Phys 14: 1319–1320

Choucair Ak, Levin VA, Gutin PH et al. (1986) Development of multiple lesions during radiation therapy and chemotherapy in patients with gliomas. J Neurosurg 65 : 654

Colombo F, Benedetti A, Pozza F et al. (1985) External stereotactic irradiation by linear accelerator. Neurosurgery 16: 154

DeAngelis LM, Yahalom J, Thaler HT et al. (1992) Combined modality therapy for primary CNS lymphoma. J Clin Oncol 10(4): 635–643

Dearnaley DP, A'Hearn R, Whitaker S et al. (1990) Pineal and CNS germ cell tumours: Royal Marsden Hospital experience 1962–1987. Int J Radiat Oncol Biol Phys 18(4): 7773–7781

Diener-West M, Dobbins TW, Phillips TL et al. (1989) Identification of an optimal subgroup for treatment evaluation of patients with brain metastases using RTOG study 7916. Int J Radiat Oncol Biol Phys 16: 669–673

Disclafani A, Hudgins RJ, Edwards MSB et al. (1989) Pineocytomas. Cancer 63: 302–304

Douglas BG, Worth AJ (1982) Superfractionation in glioblastoma multiforme – results of a phase II study. Int J Radiat Oncol Biol Phys 8: 1787

Duffner PK, Cohen ME, Thomas PRM et al. (1985) The long-term effects of cranial irradiation on the central nervous system. Cancer 56: 1841–1846

Duncan W, McLelland J, Jack WJL et al. (1986) Report of a randomised pilot study of the treatment of patients with supratentorial gliomas using neuron irradiation. Br J Radiol 59: 373

Eby M, Grufferman S, Flannelly CM et al. (1988) Increasing incidence of primary brain lymphoma in the US. Cancer 62: 2461–2465

Emami B, Lyman J, Brown A et al. (1991) Tolerance of normal tissue to therapeutic irradiation. Int J Radiat Oncol Biol Phys 21: 109–122

Engenhart R, Kimmig BN, Höver KH (1990) Stereotactic single high dose radiation therapy of benign intracranial meningiomas. Int J Radiat Oncol Biol Phys 19(4): 1021–1026

Fazekas JT (1977) Treatment of grades I and II brain astrocytomas. The role of radiotherapy. Int J Radiat Oncol Biol Phys 2: 661–666

Flickinger JC, Tarres C, Deutsch M (1988) Management of low-grade gliomas of the optic nerve and chiasm. Cancer 61: 635–642

Florell RC, Macdonald DR, Irish WD et al. (1992) Selection bias, survival, and brachytherapy for glioma. J Neurosurg 76: 179–183

Fowler JF (1985) In: Proc XVI Int Conf Radiology, Hawaii, pp 201–206

Fulton DS, Urtasun RC, Shin KH et al. (1984) Misonidazole combined with hyperfractionation in the management of malignant glioma. Int J Radiat Oncol Biol Phys 10: 1709

Gelber RD, Larson M, Borgelt BB et al. (1981) Equivalence of radiation schedules for the palliative treatment of brain metastases in patients with favorable prognosis. Cancer 48: 1749–1753

Glaholm J, Bloom HJG, Crow JH (1990) The role of radiotherapy in the management of intracranial meningiomas: The Royal Marsden Hospital experience with 186 patients. Int J Radiat Oncol Biol Phys 18: 755–761

Glass J, Hochberg FH, Gruber ML et al. (1992) The treatment of oligodendrogliomas and mixed oligodendroglioma–astrocytomas with PCV chemotherapy. J Neurosurg 76: 741–745

Greenberg HS, Chandler WF, Diaz RF et al. (1988) Intra-arterial bromodeoxyuridine radiosensitization and radiation in treatment of malignant astrocytomas. J Neurosurg 69: 500

Griffin BR, Shuman WP, Wisbeck W et al. (1988) Improved localization of infratentorial ependymoma by magnetic resonance imaging: implications for radiation treatment planning. J Neuro-oncol 6: 147–155

Griffin TW, Beaufait D, Blasko JC (1979) Cystic cerebellar astrocytoma in childhood. Cancer 44: 276–280

Gutin PH, Phillips TL, Wara WM et al. (1984) Brachytherapy of recurrent malignant brain tumors with removable high-activity iodine-125 sources. J Neurosurg 60: 61

Guttenberger R, Thames HD, Ang KK (1992) Is the experience with CHART compatible with experimental data? A new model of repair kinetics and computer simulations. Radiother Oncol 25: 280–286

Hartmann GH, Schlegel W, Sturm V et al. (1985) Cerebral radiation surgery using moving field irradiation at a linear accelerator facility. Int J Radiat Oncol Biol Phys 11: 1185

Healey EA, Barnes PD, Kupsky WJ et al. (1991) The prognostic significance of postoperative residual tumor in ependymoma. Neurosurg 28: 666–672

Hochberg FH, Pruitt A (1980) Assumptions in the radiotherapy of glioblastoma. Neurology 30: 907

Hochberg FH, Slotnick B (1980) Neuropsychologic impairment in astrocytoma survivors. Neurology 30: 172–177

Hochberg FH, Miller DC (1988) Primary central nervous system lymphoma. J Neurosurg 68: 835–853

Horwich A, Bloom HJG (1985) Optic gliomas: radiation therapy and prognosis. Int J Radiat Oncol Biol Phys 11: 1067–1079

Hoshino T (1984) A commentary on the biology and growth kinetics of low-grade and high-grade gliomas. J Neurosurg 61: 895–900

Jenkin D, Angyalfi S, Becker L et al. (1993) Optic glioma in children: surveillance, resection, or irradiation? Int J Radiat Oncol Biol Phys 25: 215–225

Jereb B, Reid A, Ahuja RK (1982) Patterns of failure in patients with medulloblastoma. Cancer 50: 2941–2947

Keim H, Potthoff PC, Schmidt K et al. (1987) Survival and quality of life after continuous accelerated radiotheraphy of glioblastomas. Radiother Oncol 9: 21

Kernohan JW, Sayre GP (1952) Tumors of the central nervous system. In: Atlas of Tumor Pathology; 1st Series; Section 10; fasc. 35. Armed Forces Institute of Pathology, Washington DC

Kim YH, Fayos JV (1977) Intracranial ependymomas. Radiology 124: 805–808

Kinsella TJ, Dobson PP, Mitchell JB et al. (1987) Enhancement of x-ray induced DNA damage by pre-treatment with halogenated pyrimidine analogs. Int J Radiat Oncol Biol Phys 13: 733

Kleinman GM, Hochberg FH, Richardson EP (1981) Systematic metastases from medulloblastoma: report of two cases and review of the literature. Cancer 48: 2296–2309

Kondziolka D, Lunsford LD, Coffey RJ et al. (1991) Stereotactic radiosurgery of meningiomas. J Neurosurg 74: 552–559

Kristiansen K, Hagen S, Kollevold T et al. (1981) Combined modality therapy of operated astrocytomas grade III and IV. Confirmation of the value of postoperative irradiation and lack of potentiation of bleomycin on survival time: a prospective multicenter trial of the Scandinavian Glioblastoma Study Group. Cancer 47: 649–652

Kurtz JM, Gelber R, Brady LW et al. (1981) The palliation of brain metastases in a favourable patient population: a randomized clinical trial by the radiation therapy oncology group. Int J Radiat Oncol Biol Phys 7: 891–895

Laing RW, Warrington AP, Graham J, Britton J, Hines F, Brada M (1993) Efficacy and toxicity of fractionated stereotactic radiotherapy in the treatment of recurrent gliomas (Phase I–II Study). Radiother Oncol 27: 22–29

Leibel SA, Sheline GE, Wara WM et al. (1975) The role of radiation therapy in the treatment of astrocytomas. Cancer 35: 1551–1557

Leibel SD, Sheline GE (1987) Radiation therapy for neoplasms of the brain. J Neurosurg 66: 1–22

Leibel S, Gutin P, Wara W et al. (1988) Survival and quality of life following interstitial implantation of removable high-activity iodine-125 sources for recurrent malignant gliomas (Abstract). Int J Radiat Oncol Biol Phys 15: 155

Leksell DG (1987) Stereotactic radiosurgery. Neurol Res 9: 60

Lindegaard KF, Mørk SJ, Eide GE et al. (1987) Statistical analysis of clinicopathological features, radiotherapy and survival in 170 cases of oligodendroglioma. J Neurosurg 67: 224–230

Ludgate CM, Douglas BG, Dixon PE et al. (1988) Superfractionated radiotherapy in grade III, IV intracranial glioma. Int J Radiat Oncol Biol Phys 15: 1091

Ludwig CL, Smith MT, Godfrey AD et al. (1986) A clinicopathological study of 323 patients with oligodendroglioma. Ann Neurol 19: 15–21

Lutz W, Winston KR, Maleki N (1988) A system for stereotactic radiosurgery with a linear accelerator. Int J Radiat Oncol Biol Phys 14: 373

Lyons MK, Kelly PJ (1991) Posterior fossa ependymomas: report of 30 cases and review of the literature. Neurosurg 28: 659–665

MacDonald D, O'Brien R, Gilbert JG et al. (1989) Metastatic anaplastic oligodendroglioma. Neurology 39: 1593–1596

Marcus RB, Million RR (1990) The incidence of myelitis after irradiation of the cervical spinal cord. Int J Radiat Oncol Biol Phys 19: 3–8

Marks JP, Baglan RJ, Prassad SC et al. (1981) Cerebral radionecrosis: incidence and risk in relation to dose, time fractionation and volume. Int J Radiat Oncol Biol Phys 7: 243–252

Mastaglia FL, McDonald WI, Watson JV et al. (1976) Effects of X-radiation of the spinal cord: an experimental study of the morphological changes in central nerve fibres. Brain 99: 101–122

Mørk SJ, Løken AC (1977) Ependymoma. A follow up study of 101 cases. Cancer 40: 907–915

MRC Working Party (1983) A study of the effect of misonidazole in conjunction with radiotherapy for the treatment of grades 3 and 4 astrocytomas. A report from the MRC Working Party on misonidazole in gliomas. Br J Radiol 56 : 673

Murray K, Kund L, Cox J (1986) Primary malignant lymphoma of the central nervous system. Results of treatment of 11 cases and review of the literature. J Neurosurg 65: 600

Nazar GB, Hoffman HJ, Becker LE et al. (1990) Infratentorial ependymomas in childhood: prognostic factors and treatment. J Neurosurg 72: 408–417

Nelson DF, Martz KL, Bonner H et al. (1992) Non-Hodgkin's lymphoma of the brain: can high dose, large volume radiation therapy improve survival? Report on a prospective trial by The Radiation Therapy Oncology Group (RTOG): RTOG 8315. Int J Radiat Oncol Biol Phys 23: 9–17

Neuwelt EA, Goldman DL, Dahlborg SA et al. (1991) Primary CNS lymphoma treated with osmotic blood–brain-barrier disruption: prolonged survival and preservation of cognitive function. J Clin Oncol 9(9): 1580–1590

Packer RJ, Littman PA, Sposto RM et al. (1987) Results of a pilot study of hyperfractionated radiation therapy for children with brain stem gliomas. Int J Radiat Oncol Biol Phys 13: 1647

Packer RJ, Sutton LN, Atkins TE et al. (1989) A prospective study of cognitive function in children receiving whole-brain radiotherapy and chemotherapy: 2-year results. J Neurosurg 70: 707–713

Palma L, Guidetti B (1985) Cystic pilocytic astrocytomas of the cerebral hemispheres. Surgical experience with 51 cases and long-term results. J Neurosurg 62: 811–815

Payne DG, Simpson WJ, Keen C et al. (1982) Malignant astrocytoma. Hyperfractionated and standard radiotherapy with chemotherapy in a randomized prospective clinical trial. Cancer 50: 2301–2306

Pezmer RD, Achambeau JO (1981) Brain tolerance unit: a method to estimate risk of radiation brain injury for various dose schedules. Int J Radiat Oncol Biol Phys 7: 397–402

Pierre-Kahn A, Hirsch JF, Roux FX et al. (1983) Intracranial ependymomas in childhood. Child's Brain 10: 145–156

Reinhold HS, van Putten WLJ, Hopewell JW et al. (1984) The latent period in clinical radiation myelopathy. Int J Radiat Oncol Biol Phys 10: 2385–2387

Rorke LB, Gilles FH, Davis RL et al. (1985) Revision of the World Health Organization classification of brain tumours from childhood brain tumours. Cancer 56: 1869–1886

Rosenstock JG, Packer RJ, Bilaniuk L et al. (1985) Chiasmatic optic glioma treated with chemotherapy: a preliminary report. J Neurosurg 63: 862–866

Ross GW, Rubinstein LJ (1989) Lack of histopathological correlation of malignant ependymomas with postoperative survival. J Neurosurg 70: 31–36

Rubinstein LJ (1970) The definition of the ependymoblastoma. Arch Path 90: 35–45

Rubinstein LJ (1972) Tumours of the central nervous system. Atlas of Tumour Pathology, ser 2. Armed Forces Institution of Pathology, Washington DC, p 6

Ruifrok ACC, Kleiboer BJ, van der Kogel AJ (1992a) Fractionation sensitivity of the rat spinal cord during radiation retreatment. Radiother Oncol 25: 295–300

Ruifrok ACC, Kleiboer BJ, van der Kogel AJ (1992b) Radiation tolerance and fractionation sensitivity of the developing rat cervical spinal cord. Int J Radiat Oncol Biol Phys 24: 505–510

Ruifrok ACC, Kleiboer BJ, van der Kogel AJ (1992c) Reirradiation tolerance of the immature rat spinal cord. Radiother Oncol 23: 249–256

Salazar OM (1983) A better understanding of CNS seeding and a brighter outlook for postoperatively irradiated patients with ependymomas. Int J Radiat Oncol Biol Phys 9: 1231–1234

Salazar OM, Castro-Vita M, Van Houtte P et al. (1983) Improved survival in cases of intracranial ependymoma after radiation therapy. Late report and recommendations. J Neurosurg 59: 652–659

Salazar OM (1988) Ensuring local control in meningiomas. Int J Radiat Oncol Biol Phys 15: 501–504

Schultheiss TE, Stephens LC, Moor MH (1988) Analysis of the histopathology of radiation myelopathy. Int J Radiat Oncol Biol Phys 14: 27–32

Schultheiss TE, Stephens LC, Jiang GL et al. (1990) Radiation myelopathy in primates treated with conventional fractionation. Int J Radiat Oncol Biol Phys 19: 935–940

Shalet SM (1986) Irradiation induced growth failure. Clin Endocrinol Metab 15: 591–606

Shapiro WR (1986) Therapy of adult malignant brain tumours: what have the clinical trials taught us? Semin Oncol 13: 38

Shaw EG, Evans RG, Scheithauer BW et al. (1987) Postoperative radiotherapy of intracranial ependymoma in pediatric and adult patients. Int J Radiat Oncol Biol Phys 13: 1457–1462

Shaw EG, Daumas-Duport C, Scheithauer BW et al. (1989a) Radiation therapy in the management of low-grade supratentorial astrosytomas. J Neurosurg 70: 853–861

Shaw EG, Scheithauer BW, Gilbertson DT et al. (1989b) Postoperative radiotherapy of supratentorial low-grade gliomas. Int J Radiat Oncol Biol Phys 16: 663–668

Sheline GE, Boldrey E, Karlsberg P et al. (1964) Therapeutic considerations in tumours affecting the central nervous system: oligodendroglioma. Radiology 82: 84–89

Sheline GE, Wara W, Smith V (1980) Therapeutic irradiation and brain injury. Int J Radiat Oncol Biol Phys 6: 1215–1228

Sheline GE, (1986) Normal tissue tolerance and radiation therapy of gliomas of the adult brain. In: Bleehen NM (ed) Tumours of the Brain, Springer-Verlag, New York

Smith WJ, Gasson JA, Bourne SP et al. (1988) Immunoglobulin gene rearrangement and antigenic profile confirm B cell origin of primary cerebral lymphoma and indicate a mature phenotype. J Clin Path 41: 128–132

Stenning SP, Freedman LS, Bleehen NM (1987) An overview of published results from randomized studies of nitro-sources in primary high grade malignant glioma. Br J Cancer 56: 89

Sung D II, Havisiadis L, Chang CH (1978) Midline pineal tumors and suprasellar germinomas highly curable by irradiation. Radiology 128: 745–751

Sutcliffe SB, Gospodarowicz MK, Bush RS et al. (1985) Radiation therapy in localised non-Hodgkin's lymphoma. Radiother Oncol 4: 211–223

Taylor BW, Marcus RB, Friedman WA et al. (1988) The meningioma controversy: postoperative radiation therapy. Int J Radiat Oncol Biol Phys 15: 299–304

Thames HD (1985) An "incomplete-repair" model for survival after fractionated and continuous irradiations. Int J Radiat Oncol Biol Phys 47: 319–339

Twijnstra A, Boon PJ, Lormans ACM et al. (1987) Neurotoxicity of prophylactic cranial irradiation in patients with small cell carcinoma of the lung. Eur J Cancer Clin Oncol 23: 983–986

Urie MM, Fullerton B, Tatsuzaki H et al. (1992) A dose response analysis of injury to cranial nerves and/or nuclei following proton beam radiation therapy. Int J Radiat Oncol Phys 23: 27–39

Vanuytsel L, Brada M (1991) The role of prophylactic spinal irradiation in localized intracranial ependymoma. Int J Radiat Oncol Phys 21: 825–830

Vanuytsel L, Bessel EM, Ashley SE et al. (1992) Intracranial ependymoma: long-term results of a policy of surgery and radiotherapy. Int J Radiat Oncol Phys 23: 313–319

Walker MD, Strike TA, Sheline GE (1979) An analysis of dose-effective relationship in the radiotherapy of malignant gliomas. Int J Radiat Oncol Phys 5: 1725

Wallner KE, Wara WM, Sheline GE et al. (1986) Intracranial ependymomas: results of treatment with partial or whole brain irradiation without spinal irradiation. Int J Radiat Oncol Phys 12: 1937–1941

Wallner KE, Gonzales M, Sheline GE (1988) Treatment of oligodendrogliomas with or without postoperative irradiation. J Neurosurg 68: 684–688

Wen BC, Hussey DH, Staples J et al. (1989) A comparison of the roles of surgery and radiation therapy in the management of craniopharyngiomas. Int J Radiat Oncol Phys 16: 17–24

Yassargil MG, Curcic M, Kis M, et al. (1990) Total removal of craniopharyngiomas. Approaches and long-term results in 144 patients. J Neurosurg 73: 3–11

Zulch KJ (1980) Review article: Principles of the new World Health Organization (WHO) classification of brain tumours. Neuroradiology 19: 59–66

13 Chemotherapy of Malignant Brain Tumours

P. Krauseneck and B. Müller

Summary

Over the past 20 years, chemotherapy for malignant brain tumours has found a wider application. Particularly in the malignant supratentorial gliomas of adults, large systematic randomized studies have attempted to establish the aims, value and relative risks of surgery, radiotherapy and chemotherapy. At present, response rates to chemotherapy range between 20% and 30%, adjuvant chemotherapy prolongs median survival time by eight weeks, while with the addition of BCNU monotherapy to surgery and radiotherapy, the proportion of patients surviving more than 18 months increases from 5% to 20%, with most of the patients remaining in a good general condition.

Medulloblastomas/primitive neuroectodermic tumours (PNETs) are highly sensitive to chemotherapy. However, in primary cerebral lymphomas and other malignant tumours arising from the brain itself, response rates vary. In brain metastases, particularly those from breast carcinoma and choriocarcinoma, cytotoxic treatment has achieved long-standing remissions. In carcinomatous meningitis, the local intrathecal application of anti-cancer drugs and corticosteroids is now an established palliative treatment, and regularly results in an impressive improvement in the clinical symptoms. However, although at the present time these effects do not result in the complete cure of the malignant disease, the knowledge of well-established prognostic factors and of histopathological findings form a basis on which to optimize and tailor current strategy to the individual patient, e.g. the development of risk-adapted therapeutic concepts where the effect of treatment for malignant gliomas depends on the general condition/Karnofsky performance status (KPS) of the individual patient.

Introduction

Today, the prognosis of malignant brain tumour is still unfavourable but computed tomography (CT) and magnetic resonance (MR) scanning and angiography have greatly improved preoperative diagnosis so that surgery may be planned more precisely and the indications for the different therapeutic modalities differentiated. Similarly, the risks of postoperative complications have been greatly diminished.

More precise and objective evaluation criteria have created better possibilities of systematic clinical trials. Large multicentre randomized studies can determine the value of therapeutic modalities, identify prognostic factors and broaden our knowledge of the pathology of the central nervous system (CNS) tumours. Nevertheless, therapeutic efficacy so far remains only modest and, even with the great advances in molecular biology and immunology which enable us to use cytokines and specifically immunized lymphocytes in the treatment of malignant gliomas, there is no prospect in the near future of a major breakthrough, and progress is likely only in a series of short steps.

As most experience with chemotherapy for brain tumours has been accumulated in the treatment of malignant gliomas, the most frequently occurring tumours in the brain, the authors will concentrate on the problems of this tumour entity, but will also deal with those basically different aspects of brain tumours not arising from the brain parenchyma, especially brain metastases; also, because of its growing clinical relevance, the treatment of leptomengingeal involvement will be discussed.

Curative Concepts

Of the fundamental therapeutic goals, i.e. cure, prolongation of life and palliation of symptoms, by far the most important goal in neuro-oncological chemotherapy is prolongation of life.

At the present time in primary malignant brain tumours, cures may be expected in about 50% of medulloblastoma or PNETs occurring in childhood, but in other types of tumour and in most adult tumours 5-year survival figures are very low. Nevertheless, the concept of primary multimodal treatment of brain tumours aims at a cure. Until now there has been no controlled trial to compare adjuvant postoperative chemotherapy with a palliative chemotherapy beginning only at the time of relapse in respect of overall survival. Such adjuvant therapy strictly would require a total tumour resection, which can only rarely be achieved in intrinsic brain tumours. Therefore, in brain tumour the term "adjuvant" means therapy starting immediately postoperatively irrespective of remaining tumour. Chemotherapeutic "cure" of secondary CNS tumours at present may be expected only in haematological disease and, possibly, choriocarcinoma. In these cases, aggressive chemotherapy with its higher risk of complications is justified. However, single 5-year "cures" are possible with all tumour varieties.

Realistic goals in a major percentage of patients are a longer lasting remission or at least a delay of further progression. Since about one-third of the patients do not have any measurable benefit from chemotherapy, it is always necessary to weigh possible complications, toxicity and quality of life against the possible gain in survival, particularly if adjuvant chemotherapy is considered. Because individual needs are highly important in this process of decision-making, in the authors' opinion an essential task of the physician is to counsel the patients and their next of kin very extensively to enable them to take an informed part in the decision.

Evaluation of Therapy

Tumour status is well defined by modern neuroimaging techniques, although the use of corticosteroids which reduce brain oedema and sometimes visible tumour on CT or MR has to be taken into account. Another diagnostic pitfall may be brain necrosis after radiotherapy or intensive chemotherapy, particularly intra-arterial treatment. With CT or even MR, brain necrosis may be indistinguishable from tumour growth. Positron emission tomography (PET) may be able to differentiate necrosis from tumour. Single photon emission computer tomography (SPECT) may be helpful, as well as a combination of different neuro-imaging techniques.

For Phase I (testing toxicity) and Phase II (testing efficacy of a new substance) clinical trials, the response defined by CT or MR in combination with clinical remission of symptoms is a crucial criterion. Usual categories are:

CR = complete remission = complete disappearance of tumour in the CT/MR.

OR = objective remission = regression of tumour volume or largest diameter by 50% or more.

PR = partial remission = regression of tumour volume or largest diameter by 25%–50%.

ST = stable disease = change of tumour volume or largest tumour diameter by less than 25%.

PD = progressive disease = unequivocal tumour enlargement.

These criteria are not always used in reports of clinical trials in this standardized way. Therefore, the evaluation of individual clinical trials always requires a critical review of how these criteria have been applied.

In Phase III clinical trials (comparison of a new treatment with an already established treatment), median survival time (MST) is a very important criterion.

Other important criteria are time to tumour progression (TTP = time when 50% of the patients show progression clinically or with imaging techniques) and the proportion of so-called long-term survivors (LTS) after a certain time (in malignant gliomas usually 18 or 24 months). Monitoring and evaluation of side effects of the therapy have become more important, and a good standard is given by the WHO criteria. Distinction from tumour-related complications may be difficult and needs careful consideration. In long-term survivors, late sequelae may occur, like seeding of the tumour down the CSF pathways, regional or even systemic metastases and particularly neurotoxic symptoms, e.g. slowly progressing dementia or stroke-like syndromes due to brain necrosis.

Special Features of Chemotherapy of Malignant Brain Tumours

Space Limitation

Because of the strict volume restriction in the skull, the secondary effects of the tumour, particularly brain oedema and disturbance of the CSF flow, are usually much more important for the clinical symptomatology than local infiltration or destruction by the tumour itself. Even very small tumours may lead to an acute life-threatening situation if sited in the lower brain stem or if they occlude the CSF flow. Tumours within the hemispheres usually cause symptoms only after reaching a volume of about 30 ml and frequently cause progressive serious symptoms when they have reached double that volume because they then occupy, together with the associated surrounding brain oedema, a total volume in excess of 100 ml, leading to potentially lethal increments in intracranial pressure (Shapiro, 1982). Consequently, two statements have to be made. First, before starting chemotherapy (or any other intensive treatment modality) treatment of brain oedema and possibly a shunt procedure to allow CSF flow are usually necessary. Treatment of oedema is usually highly effective in alleviation of symptoms.

Secondly, there is only a short time for the chemotherapy to have an effect unless the tumour is resected or growth is halted by radiotherapy since tumour doubling in malignant gliomas, for instance, can take place in a few weeks. Palliative surgery is, therefore, often needed.

Absence of Lymph Drainage

No lymph vessels exist in the brain. This explains in part the fact that intrinsic, even highly malignant brain tumours only extremely rarely develop regional metastases. The lack of lymph vessels is, on the other hand, one of the main reasons that necrosis of the brain may not be resorbed naturally and, therefore, often has to be operatively resected to avoid an increase in brain oedema and space occupation.

Blood–Brain Barrier

The blood–brain barrier (BBB) restricts the passage of most molecules from the vessel lumen into the brain parenchyma or CSF. It is a function of the special endothelial cells of the capillaries of the brain which are connected by "tight junctions". Essentially, only water and small, inert and lipophilic molecules up to a molecular weight of 200 dalton may passively penetrate this barrier. For some substances like glucose, specific mechanisms of transport exist. Another important factor is the concentration gradient for a given substance. Therefore, the physicochemical properties of a substance do not strictly correlate with its BBB penetration in vivo (Ausman *et al.*, 1977; Haid *et al.*, 1987) (Table 13.1). Besides, the barrier is disturbed within

Table 13.1 Penetration of blood–brain barrier by antineoplastic agents

Substance	BBB Penetration	Remarks
Nitrosourea compounds (ACNU, BCNU, CCNU, etc.)	++	
DAG	++	
DBC	++	
PCZ	++	
HU	++	
AZQ	++	
ARA-C	++	Only with prolonged infusion ≥ 2 h
5-FU	++	Equivocal
MTX	–/+	Only with high-dose administration
VP 16	–/+	Only with high-dose administration
VM 26	–/+	Equivocal
DTIC	(+)	Low penetration in animals
CPM	+/–	Only maternal substances,
IFO	+/–	but not the active metabolites
Actinomycin C	–	
ADM	–	
Bleomycin	–	
MIT	–	
CDDP	–	
Vinca-alkaloids (VCR, VBL)	–	
IFNs	–	

++ 0.2–0.3 of serum level achieved in CSF.
+ cytotoxic levels in CSF achievable.
– only very poor CSF levels achievable
 (approx. 0–0.02% of serum level).

For explanation of abbreviations see Appendix I.

the tumour area and this is the case in a highly variable mode, depending on the structure of the tumour region and also on the nature of the given tumour. Primary brain tumours are supplied by the original brain vessels and therefore the barrier is well preserved in the infiltrating zone at the border of the tumour with normal brain tissue, whereas in the centre of (at least the larger) tumours, necrosis damages the vessels and the barrier virtually does not exist. Therefore, chemotherapy drugs may reach rather easily the necrotic sector of a tumour but only poorly the active proliferating and infiltrating cells in the periphery. This explains why good clinical efficacy may not be expected from a drug which does not penetrate the BBB barrier, although such a drug may be able to induce tumour regression (Krauseneck and Dommasch, 1987; Shapiro et al., 1987).

Metastatic tumours initially develop behind an intact BBB, but with further growth (from a diameter of 0.8–1.0 cm and above) develop their own fenestrated vessels without any barrier function. This phenomenon explains why brain metastases may develop during chemotherapy which is proving effective in other sites in the body, but also the fact that many brain metastases of chemosensitive tumours are also sensitive to systemic chemotherapy and may even show a complete regression.

Thus, neither the BBB penetration of a given cytotoxic drug can precisely be predicted nor is there a good overall correlation between clinical efficacy of a cytotoxic drug with its BBB penetration; this depends greatly on the given tumour and the individual patient. Nevertheless, a categorical classification of BBB penetration of cytotoxics is helpful and is given in Table 13.1.

Neurotoxicity

Besides their general cytotoxicity on proliferating cells, some of the cytostatic drugs have specific neurotoxic effects, limiting their applicability. These side effects may become life-threatening with high-dose treatment or, though rarely, intrathecal application of chemotherapy. In the framework of standardized therapeutic modalities, it may be difficult to differentiate a spectrum of neurotoxic side effects (Table 13.2) from infections, cerebral ischaemia or – most commonly – tumour progression. This may well interfere with assessing a drug's therapeutic efficacy. Acute side effects such as fever, disorientation and mental disorders/psychosis are, in general, reversible within a few days. In this context, the acute hearing loss after application of cisplatinum, though rare, constitutes an exception because improvement is unlikely. Systemic as well as local application of interferon-α (and to a lesser extent of interferon-β) may cause acute and severe neurotoxicity which seems to be only partially reversible.

It appears from the literature that chronic neurotoxic effects are predominantly found in the peripheral nervous system, as with the application of vinca-alkaloids, cisplatinum and, rarely, of epipodophyllotoxin derivatives VM 26 and VP 16. Chronic cerebral neurotoxicity usually accompanies cerebral irradiation and is seen as cortical atrophy or toxic leucoencephalopathy with or without mental changes of all degrees of severity. Inflammatory leucoencephalopathy recently was described with 5-FU and levamisole (Hook *et al.*, 1992).

In the authors' studies acute mental changes were observed with longterm oral administration of procarbazine and also with antimetabolites (Ara-A, Ara-C, 5-FU and methotrexate) given intravenously, particularly in patients with a large tumour burden (Kaplan and Wiernik, 1984; Shapiro and Young, 1984; Cerny and Meier, 1989; Mahaley *et al.*, 1989; Forman, 1990; Merimsky *et al.*, 1990; Yamashima *et al.*, 1990; Anderson and Tandon, 1991; Baker *et al.*, 1991; Harmers *et al.*, 1991).

The particular side effects of intrathecal therapy are reviewed in the literature (Shapiro and Young, 1984; Colamaria *et al.*, 1990; Yamashima *et al.*, 1990; Meyers *et al.*, 1991). With this technique it is vitally important to ensure sufficient dilution of the chemotherapy substances with adequate volumes of solvent to avoid serious complications.

Topographical Considerations and Tumour Vascularization

At present, neuroradiological diagnosis with CT, angiography and MR indicates exactly topographical correlations of the tumour and brain and frequently gives substantial hints as to differential diagnosis. Thus topo-

Table 13.2 Neurotoxicity of antineoplastic agents commonly used in neuro-oncology

Substance	CNS-toxicity				PNS-toxicity	
	Acute/Transient		Late/Persistent			
	Grade*	Rate	Grade	Rate	Grade	Rate
HN2						
Normal dosse	–		–		–	
≥0.3 mg kg⁻¹	4	0.65	4	0.35	–	
IFO	4	0.05–0.3	–		–	
PCZ	2–4	0.3–0.08	–		2	0.1–0.2
DTIC	3–4	rare				
BCNU						
normal dose			–		–	
≥1.5 g			4	?		
i.a.	4	?	4	?		
CCNU	4	rare				
MTX						
normal dose	–		2–3	<0.02		
+ simult. RAD			>2	0.15		
HD	3	rare	4	<0.02		
5-FU	2	0–0.07	–		–	
Ara-C					?	rare
normal dose	3–4	0.1				
HD	4	0.05–0.15	4	0.16		
Vinca-alkaloids					2–4	1.0–0.2
(VCR, VBL, VDS)						
VP 16						
normal dose	–		–		–	
≥2.4 g	4	(6/8)				
VM 26	–		–		2	0.1–0.2
CDDP						
normal dose					2	<0.1
≥2.00 mg/m⁻²/cycle					4	>0.5
i.a.			3	(5/11)		
IFN-α	–		3	0.17	–	
IFN-β	–		–		–	

* "Grade" refers to WHO criteria of toxicity;
  if not otherwise specified, i.v. or oral route of drug administration.

For explanation of abbreviations see Appendix I.

graphical aspects can early be taken into account in planning the therapeutic approach. The functional importance of the affected brain regions as well as their vascular supply must be constantly kept in mind, not only with regard to surgical intervention but also with conservative or symptomatic methods of therapy. A knowledge of the vascular supply of a given tumour requires particularly knowledge of its blood supply. A close connection of the tumour to the CSF pathways or the brain surface seen in scans or during operation points to the risk of seeding tumour cells into the CSF. In these cases the authors monitor the CSF and administer cytotoxic drugs, sometimes prophylactically intrathecally, to avoid meningeal involvement (see below).

Postoperative Radiological Appearances

To differentiate residual postoperative tumour from operative changes, early contrast medium-enhanced series (up to 38 h after operation) of CT or MR are necessary. Magnetic resonance imaging is more sensitive in this regard (Albert *et al.*, 1991; Forsting *et al.*, 1992). This information provides the best basis to evaluate the subsequent effect of adjuvant therapy.

Specific Treatment

Symptomatic Therapy

Since, initially, chemotherapy may worsen the symptoms of raised intracranial pressure by increasing oedema, the application of cytotoxic agents in solid brain tumours must be accompanied by an optimized symptomatic treatment. Crises of intracranial pressure can largely be controlled through high corticosteroid doses and parenteral osmotherapy.

In the treatment of chronic raised intracranial pressure, the oral administration of glycerol in addition to corticosteroid therapy has proved to be effective and appropriate. Steroid doses can be reduced and side effects diminished. Peripheral analgesics are of less effect in symptomatic relief of headache due to raised intracranial pressure, while centrally-acting narcotic analgesics may cause dangerous complications due to the already impaired cerebral function present in the patient.

Carcinomatous meningitis, although not producing a mass effect, causes various neurological deficits by elevating the intracranial pressure, impairing CSF flow, invading brain tissue and irritating cranial nerves. Here, combined intrathecal therapy of corticosteroids and cytostatic agents is the best palliative approach. In a similar way the local installation of cytotoxic drugs via a catheter into tumour cysts may in some cases be effective in diminishing the production of malignant exudate in the cyst and the consequent mass effect.

Mode of Application

Since, for chemotherapy drugs, the therapeutic margin between efficacy and unacceptable toxicity is small, particularly in brain tumours, in principle intravenous administration is preferable to avoid the uncertainty of gastrointestinal absorption.

Systemic Administration

Intravenous Treatment. To overcome the BBB, high blood levels of the chemotherapy agents are of advantage, or even a prerequisite, to achieve drug concentrations in the tumour tissue. Therefore, for most of the cytotoxic drugs a single high-dose intravenous infusion is advisable. Some of the cytotoxics applied in neuro-oncology are exceptions because of special pharmacodynamic criteria: for cell-cycle specific drugs like 5-FU, vinca-alkaloids, VP 16 and VM 26, the more vulnerable cells can be hit if the drug level is maintained in the tumour for a long time. To achieve a usable compromise between these two requirements, it is advantageous to divide the total dose into repeated short-term intravenous infusions over several days. In the case of Ara-C, where the drug is not degraded within the CSF but rapidly in the blood, prolonged infusions over 2–4 h reach a higher drug concentration across the BBB (Krauseneck, 1990).

Oral Treatment. For substances needing microsomal activation in the liver and which are satisfactorily absorbed in the gastrointestinal tract, oral dosing may be favourable, e.g. for procarbazine. Also, with some well-absorbed cell-specific drugs like VP 16 the oral application may be indicated.

Local Application

Intra-arterial. Selective chemotherapy with intra-arterial application of cytotoxic drugs may reach high local concentrations with less systemic toxicity. Technical developments have made available catheters fine enough to reach small branches of the main cerebral arteries. In animal experiments it has been shown that the drug concentration in the tumour may be increased but, unfortunately, the same is true of the concentration in the normal brain, and therefore the toxicity rises steeply. The therapeutic index for some commonly used cytotoxic drugs may even be reversed, so that in the normal brain tissue higher concentrations are reached than in the tumour tissue, particularly with lipophilic drugs. A common complication is (transient) blindness as a consequence of retinal toxicity, but severe and permanent impairment of higher intellectual function and stroke-like syndromes have also been observed (Popovic and Popovic, 1986; Shapiro and Shapiro, 1986; Shapiro and Green, 1987; Shapiro *et al.*, 1987). Hydrophilic substances are a better choice for intra-arterial therapy, but before broad clinical application can be recommended a better understanding of the pharmacokinetics and toxicity is necessary.

Intratumoral. In the early stages of chemotherapy of brain tumours there was a large number of attempts to apply cytotoxic substances directly into the tumour or, postoperatively, into the tumour bed, and more recently by implantation of catheters connected to a reservoir. Due to poor diffusion within the brain tissue and the diffuse infiltrating growth of gliomas, results of these techniques are poor, as they have been also with cytokines, monoclonal antibodies and radioisotopes.

A new approach of this kind is to create a local reservoir within the tumour bed by binding the cytostatic drugs to liposomes or polymers (BCNU wafers) to prolong and extend the local drug delivery, but – not unexpectedly – first results are not very promising (Brem *et al.*, 1991).

Modifications of the Blood–Brain Barrier

Attempts to make the BBB more permeable by intra-arterial injection of high concentrations of mannitol, sorbitol or urea have been successful. However, the combination with cytotoxic drugs produces severe neurotoxic side effects (leucoencephalopathy) either with lipophilic or with hydrophilic drugs, since the therapeutic index tumour/brain was shifted to the negative-producing high, toxic concentrations in the normal brain and relatively low levels in the tumour (Shapiro, 1982; Shapiro and Shapiro, 1986).

Administration into the Cerebrospinal Fluid

With this form of therapy it is possible to circumvent the BBB and to reach high drug concentrations in the CSF, but not in the brain or in the tumour. Diffusion from the CSF into the brain or into the tumour is very poor and the drug concentration drops exponentially with depth of penetration. Therefore, it is not possible with solid tumours to improve drug delivery using the CSF pathway. This mode of administration is useful in treating CSF dissemination of tumour cells throughout the CSF pathway or to prevent CSF seeding by malignant tumours situated adjacent to the CSF. Methotrexate and Ara-C are well tolerated and in common use for lumbar and ventricular application (Shapiro et al., 1975; Dienst, 1985). With methotrexate, toxicity outside the CNS can be prevented by systemic (oral or intravenous) administration of citovorum factor, which does not penetrate the BBB.

Intrathecal chemotherapy is – when correctly performed – a safe procedure with a very low complication rate (<1%). The cytostatic drugs should be dissolved in CSF-isotonic and pH-neutral solutions, for instance natural or artificial CSF. Ventricular administration may be carried out only using reservoir or catheter systems, which can also be used for tumour cysts. Here the risk of infection and of local complications with the catheter is significant, as is a higher risk of neurotoxicity. Careful monitoring and good local care of the skin is necessary.

Timing of Chemotherapy

Most experience has been obtained in postoperative chemotherapy. Usually this is begun with radiotherapy after wound healing has taken place. In smaller, sometimes randomized, studies feasibility of preoperative chemotherapy was shown to have no more complications than postoperative chemotherapy, but no advantage could be demonstrated (Butti et al., 1984; and Goebel et al., 1987). Because chemotherapy should be done only following histological confirmation of the tumour, preoperative chemotherapy is an option for metastases of a known primary tumour or in those cases where a favourable response to chemotherapy would considerably ease the performance of and lower the risk of surgery. Preoperative or intraoperative chemotherapy is a rational concept to prevent tumour dissemination and experimentally there are hints of better efficacy using this method (Ochs *et al.*, 1991). There are no acute problems with simultaneous administration of chemotherapy and radiotherapy, provided that corticosteroids are used, but the risk of late sequelae (leucoencephalopathy) may be higher. This has to be accepted because of the dismal prognosis in malignant brain tumours in adults, whereas in childhood sequential protocols are preferred.

Clinical Classification of Malignant CNS Neoplasia

Both histolopathological and topographical aspects must be considered for diagnostic and therapeutic decisions. The precise histological classification is essential for prognosis and for the choice of therapeutic modalities. The topographical site is responsible for the symptoms and signs and is crucial not only for surgery but also for the whole therapeutic concept regarding timing and coordination of the single treatment modalities, optimum symptomatic therapy and mode of application of chemotherapy and of radiotherapy. The relevant histological types of tumour (see Chap. 3) are given in Table 13.3.

Table 13.3 Histological classification of tumours

1. *Malignant neuroectodermal tumours*
   Anaplastic astrocytoma (grade III)
   Anaplastic oligodendroglioma (grade III)
   Anaplastic mixed glioma (grade III)
   Glioblastoma multiforme and variants (grade IV)
   Anaplastic ependymoma (grades III/IV)
   PNET (primitive neuroectodermal tumours, medullobastoma, pineoblastoma, neuroblastoma)
   Other rare malignant brain tumours

2. *Malignant tumours of the mesenchymal CNS structures*
   Malignant meningioma (grade III)
   Malignant haemangiopericytoma (grade III)
   Sarcoma or sarcoma variants (grade IV)

3. *Neoplasms arising from the haematopoetic system*
   Primary cerebral lymphomas (non-Hodgkin's lymphomas)
   (classified according to the Kiel or Rappaport systems like non-Hodgkin's lymphomas in other sites of the body, but independent of this classification their prognosis is less favourable than non-CNS lymphomas)
   Cerebral manifestations of haematological malignancies (systemic leukaemias and lymphomas)

4. *Metastases of solid tumours (always grade IV)*

5. *Carcinomatous meningitis (grade IV)*

Proliferation

In recent years results of immunocytology and molecular biology have complemented classical histopathological findings in many respects. Thus, the proliferation marker Ki-67, bromodeoxyuridine (BdUR) and PCNA (proliferative cell nuclear antigen), may all contribute better definition of growth potential and prognosis of a given brain tumour. Clinical studies to evaluate their use prognostically are in progress.

Chemotherapy of Specific Tumour Types

Malignant Intrinsic Brain Tumours

Primary brain tumours develop in the brain parenchyma, predominantly in the white matter. The active areas of proliferation and invasion are at the periphery of the tumour in the transitional zone to normal brain, where the blood supply is provided by essentially normal brain capillaries with intact BBB. In contrast, the BBB is disrupted in the central tumour areas, so allowing for nearly unimpaired drug delivery to these regions (Long, 1970; Seiler and Zimmermann, 1979; Groothius *et al.*, 1984). This hetero-geneity explains why chemotherapy with non-BBB penetrating cytotoxics, like cisplatinum or VM 26, is effective. However, better results are gener-ally achieved with BBB–penetrating substances like the nitrosoureas. For theoretical reasons a complete remission by chemotherapy alone should only be expected by BBB penetrating drugs, whereas in clinical experience some cases of complete remission are seen also with other drugs.

Intrinsic brain tumours are the most common (40%) of all brain tumours in adults and 80% of them are malignant gliomas, mainly glio-blastomas (80%), anaplastic astrocytomas (10%–15%), anaplastic oligo-dendrogliomas (2%–8%) and a few anaplastic mixed gliomas. Therefore, a large body of data from randomized clinical studies is available for these entities, but for other rare malignant intrinsic brain tumours well-controlled data are lacking (Krauseneck, 1990).

Anaplastic Gliomas and Glioblastomas

Overview of Results of Intravenous Therapy. In the early 1970s the American Brain Tumor Study Group (BTSG) demonstrated in their second random-ized multicentre trial the relative effects of surgery, radiotherapy and BCNU chemotherapy (Walker *et al.*, 1978) (Table 13.4): BCNU alone was effective but less so than radiotherapy alone. The main effect of BCNU was not a prolongation of median survival but an increase in the rate of long-term survivors, i.e. patients surviving more than 18 months. These results were essentially confirmed by further studies of the BTSG and other groups. With better radiotherapy and more experience with chemotherapy, the median survival time (MST) of the combination of radiotherapy + BCNU reached 52 weeks and the rate of 18 months survivorship was increased to

Table 13.4 Clinical trials with adjuvant chemotherapy for malignant gliomas in combination with cranial irradiation

Drugs	n	Mean age	KPS (%)	Histology grade	MST(weeks)	Reference*
BCNU	72	57	?	90	34	Walker et al. (1978)
	147	57	70	89	50	Green et al. (1983)
	242	54	75	82	50	Krauseneck et al. (1989)
CCNU	59	62	?	41	43	EORTC Brain Tumor Group (1978)
	94	47	66	41	52	Trosanowski et al. (1989)
MeCCNU	91	54	60	82	43	Walker et al. (1978)
PCZ	153	56	70	89	47	Green et al. (1983)
ACNU	27	59	?	33	61	Shibata et al. (1987)
BCNU + VCR	17	58	?	76	44	Shapiro et al. (1976)
CCNU + PCZ + VCR + MTX	35	51	≥50	60	64	Jellinger et al. (1979)
CCNU + VM 26	61	53	60	28	58	EORTC Brain Tumor Group (1981)
BCNU + PCZ	176	56	73	79	50	Shapiro et al. (1989)
BCNU + PCZ/HU + VM 26	168	56	72	78	59	Shapiro et al. (1989)
BCNU + VM 26	259	55	75	81	53	Krauseneck et al. (1989)

* All except Reference 68 are randomized trials.
For explanation of abbreviations see Appendix I.

20%–35% (Potthoff, 1981; Lieberman et al., 1982). Five-year survival was reported only anecdotally. In the German–Austrian study (Krauseneck et al., 1989) it was 8%, with 15% for Grade III tumours and 4% for glioblastomas (DÖG). Therefore, in general, chemotherapy has a clear-cut effect in up to one-third of the patients, a modest effect in another third and no measurable effect in the remaining third.

The question of whether symptomatic treatment with corticosteroids – frequently dramatically improving signs and symptoms rapidly after its introduction into therapy – adds to the prolongation of median survival can be answered by another study of the BTSG showing no effect of routinely administered high-dose methylprednisolone on survival (Green et al., 1983). Other single agent chemotherapy was, in randomized studies, no more and often less effective than BCNU.

European studies from the EORTC Brain Tumour Study Group (EORTC Brain Tumour Group, 1978; Eagan et al., 1981; Hildebrand, 1991) and the Scandinavian and Italian groups (Paoletti et al., 1983; Hatlevoll et al., 1985) argue for slightly less effectiveness of CCNU compared with BCNU, whereas other American studies demonstrated comparable effectiveness of methyl-CCNU (Walker et al., 1980) or procarbazine (EORTC Brain Tumour Group, 1981; Green et al., 1983; Hildebrand, 1991). ACNU, used in Japan and Germany, also achieved comparable results (Voth et al., 1984; Ushio et al., 1987; Bamberg et al., 1988). In smaller clinical trials or phase II studies evaluating response, a number of other substances proved efficacious: PCNU (Green et al., 1985), another nitrosourea, was effective but no better than BCNU. Other non-BBB-penetrating alkylating agents like cyclophosphamide have not proved effective in malignant gliomas. In other intrinsic brain tumours like medulloblastomas, however, they apparently do work.

Blood–brain barrier penetrating alkylating drugs like the hexitols dianhy-drogalactitol (DAG), dibromodulcitol (DBD) and diacetyl-DAG are possibly as effective as the nitrosoureas, but are much less tested (Afra *et al.*, 1983; Eagan *et al.*, 1991). Hydroxyurea may have a role as radiosensitizer, but the data of the Levin group are not unequivocal (Levin *et al.*, 1979). Cisplatinum intravenously produced responses, but no additional effect in combination with chemotherapy was seen (Hildebrand, 1991). Aziridin-ylbenzoquinone (AZQ), a lipophilic, specially designed anthraquinone, was only moderately effective in various trials (Haid *et al.*, 1987; Taylor *et al.*, 1987). Early promising results with other anthraquinones could not be confirmed (Pouillart *et al.*, 1976; Green *et al.*, 1984).

The antimetabolities have at least limited efficacy, as shown for fluorouracil (Adelstein *et al.*, 1987), 6-mercaptopurine (Zimm *et al.*, 1985), high- but not low-dose methotrexate (Dgerassi *et al.*, 1985) and Ara-C (Hobert *et al.*, 1985). In the authors' experience, Ara-C is the most promis-ing, but has considerable haematological toxicity with some reversible neurotoxicity.

Mitotic poisons also have some therapeutic potential, e.g. vincristine in early trials but with less strict evaluation criteria (Afra, 1973), for VM 26 and VP 16 in the past few years (EORTC Brain Tumour Group, 1981; Tirelli *et al.*, 1984). In a randomized trial, VM 26 showed in addition to BCNU a marginal, but statistically significant, advantage over BCNU alone (Krauseneck *et al.*, 1989).

For a variety of other substances, remissions have been described (Krauseneck *et al.*, 1987, 1989). None of these has been more promising than the above-mentioned and therefore no succeeding trials were done or reported.

Of the biological response modifiers (BRM), interferons, α, β, in a few cases also γ and interleukin-2, with or without lymphokine activated killer (LAK) cells, have been tested in small series. No convincing data of more than modest activity against malignant gliomas have been reported (Boethius *et al.*, 1983; Nagai and Arai, 1984; Duff *et al.*, 1986; Mahaley, 1991). Interferon α and β may have a response rate of 15%, but also may have considerable toxicity in patients with brain tumours.

Although, as cited above, there exists some therapeutic potential in the different classes of cytostatic drugs, the well-known advantage of polychemotherapy for tumour entities elsewhere in the body could not be achieved in malignant gliomas. Levin *et al.* (1985) reported superiority of procarbazine/CCNU/vincristine over BCNU alone, but the study design is not free from possible flaws. In the German–Austrian trial, there was a significant (but modest) prolongation of the relapse-free interval and, only for the group with Karnofsky status of at least 70%, of the median sur-vival with BCNU + VM 26 in comparison with BCNU alone (Krauseneck *et al.*, 1989). Also the response rates are only slightly, if at all, higher with different rationally designed polychemotherapy schedules than with single drug therapy (Mahaley, 1991). It must be considered also that poly-chemotherapy may lead to even worse results, adding undue toxicity to an already rather toxic regimen with simultaneous radiotherapy and chemotherapy. This may explain the finding in the German–Austrian study that the certainly moderately toxic combination of BCNU and

VM 26 yielded slightly shorter survival in the low Karnofsky group (KPS 50 + 60%) (Krauseneck *et al.*, 1989). This must be borne in mind and makes aggressive schedules like "8 drugs in 1 day" (Rozenthal *et al.*, 1989) questionable.

Nevertheless the design of better and risk-adapted polychemotherapy protocols is one of the possibilities for achieving results in the near future, both in terms of survival and in terms of quality of life and avoidance of unjustifiable toxicity. In this context, the search for a better definition of individual prognosis and risk profile plays an important role, and one new and possibly major prognostic factor may be in vitro chemosensitivity of a given tumour (Thomas *et al.*, 1985). A systematic randomized multicentre study has been started in the UK which addresses the latter issue.

Various other approaches to improve chemotherapy of gliomas promise better results in the future, e.g. new drugs like temozolomide with better therapeutic ratio or with new modes of action like calcium antagonists, enzyme inhibitors or invasion inhibitors. Other possible agents include cytokines, growth factors and monoclonal antibodies.

As standard treatment today, adjuvant monotherapy with BCNU in addition to resection and radiotherapy is established.

BCNU, STANDARD PROTOCOL. BCNU 80 mg m^{-2} per day, days 1–3 per infusion (about 20 min), starting about 14 days after resection, simultaneously with radiotherapy.

Interval. 6 (–8) weeks, if leucocytes $\geq 4000 \times 10^9$ per litre and platelets $\geq 100\ 000 \times 10^9$ per litre.

Routine controls. Blood counts including platelets together with renal, hepatic and lung function tests. If nadir of leucocytes $< 1500 \times 10^9$ per litre or platelets $< 50\ 000 \times 10^9$ per litre the dose is reduced to 75%.

Side effects. The most critical side effect of BCNU is the development of an interstitial pneumonia/pulmonary fibrosis occurring rarely after only one single cycle and sometimes progressing to death, even after having discontinued BCNU. Even routine pulmonary function tests were not sufficient to prevent this severe side effect (Krauseneck *et al.*, 1989). During the days of administration, patients are more at risk of seizures and thrombosis. Allergic reaction to BCNU is rare.

Supportive therapy. There is no clear advantage of routine administration of anti-epileptic drugs. During treatment days, complications may be avoided by (a) use of steroids (at least 25 mg of prednisolone or 4 mg of dexamethasone) for anti-emesis and prophylaxis of brain oedema (also in patients not needing continuous steroid treatment); and (b) in some centres, low-dose heparin and the wearing of anti-embolism stockings, because of the risk of thrombosis and lung embolism.

Indications for cessation of treatment

1. Symptoms or signs of developing interstitial pneumonia, including dyspnoea at rest or on mild activity and declining performance on lung function testing (diffusion capacity of CO (DCO) $\leq 50\%$ or marked restriction with reduction of lung volume to 50%–60%).

2. Allergic reaction to BCNU.
3. Tumour relapse or progression (after at least 2 cycles of BCNU).

ALTERNATIVE PROTOCOLS. The following protocols have been widely evaluated, including side effects, and can be recommended as an alternative to BCNU as a single agent.

BCNU + VM 26. The protocol is very similar to the BCNU monotherapy (see above). In addition, VM 26 is given in a dose of 50 mg m^{-2} at days 1 and 2 over 20 min, immediately after BCNU. An increase of VM 26 dose to 100 mg m^{-2} did not show higher toxicity in the authors' series. The BCNU + VM 26 protocol has identical risks regarding pulmonary and haematological toxicity to BCNU alone. In a few cases, allergic reactions to VM 26 were seen.

Side effects of VM 26

1. Rapid "push injections" can cause an acute idiosyncratic reaction with arterial hypotension and tachycardia.
2. Anaphylactic or allergic reactions (these seem to be induced by the solvent agent rather than the drug).
3. Intensified anorexia and weight loss.
4. Rarely, mild polyneuropathy.
5. Thrombophlebitis.

ACNU monotherapy. ACNU 100 mg m^{-2}, intravenously (bolus injection) only day 1 every 6 weeks. Toxicity is similar to other nitrosourea compounds. Pulmonary fibrosis is less frequent than with BCNU, but thrombocytopenia seems to be more pronounced.

Procarbazine – oral monotherapy. 150 mg m^{-2} days 1–14 orally, repeated every 4 weeks.

Side effects: Sometimes severe haemato- and hepatotoxicity. Often a considerable nausea and, rarely, allergic skin reactions are encountered. This therapy can be conducted on an outpatient basis, but nevertheless close monitoring is mandatory.

PCV protocol – triple agent chemotherapy. This protocol have been used widely in the USA, but is less well documented than the other regimens:

110 mg m^{-2} CCNU day 1 orally

60 mg m^{-2} PCZ days 8–21 orally

1.4 mg m^{-2} VCR days 8 + 29 intravenously

every 6–8 weeks (Levin *et al.* 1985, 1989)

or

80 mg m^{-2} CCNU day 1 orally

100 mg m^{-2} PCZ days 1–10 orally

1.4 mg m^{-2} VCR day 1 intravenously (every 6 weeks for 18 months) (Thomas *et al.*, 1985)

INTRA-ARTERIAL CHEMOTHERAPY. Anecdotally impressive tumour regressions have been demonstrated by direct intra-arterial injections, mainly of the nitrosoureas BCNU, ACNU and HeCNU, but also of cisplatinum and of some other compounds. However, a large randomized trial in the USA could not prove any advantage of intra-arterial (i.a.) BCNU versus the conventional intravenous BCNU protocol. In contrast, serious neurotoxic side effects required reduction of the i.a. dose and survival data were worse with the i.a. route. Therefore, the trial was stopped early (Levin *et al.*, 1985).

Because of a possible improvement of the therapeutic index tumour/brain, i.a. application of a hydrophilic, non-BBB-penetrating agent like MTX (Shapiro, 1986) or CDDP (Mahaley *et al.*, 1989) is theoretically promising. Small studies and the preliminary results of the American BTCG trial showed high radiological response rates of about 35% for i.a. CDDP (Mahaley *et al.*, 1989).

PROGNOSTIC FACTORS. Looking for prognostic factors is another main goal in large systematic studies: age, histology and grade of malignancy as well as general condition of the patient (KPS) proved to be of prognostic value in nearly all trials. In some studies, but not others, the residual tumour volume after surgery was also of prognostic relevance. It is not yet clear whether chemosensitivity testing adds predictive power. Mainly negative results in general oncology nourish scepticism. However, in brain tumours a clinical correlation between proliferation in vitro (methionine incorporation rate) and response to CCNU or PCZ but not VCR has been demonstrated (Thomas *et al.*, 1985).

CHEMOTHERAPY OF MALIGNANT PRIMARY BRAIN TUMOURS OTHER THAN CEREBRAL GLIOMAS. These neoplasms are relatively rare in adults and therefore only small personal series exist and no proven generally effective chemotherapy schedules can be given.

The risk of CSF dissemination must be borne in mind with some of these uncommon tumours which occur adjacent to the CSF pathways, particularly in paraventricular or infratentorial locations.

Anaplastic ependymomas in adults behave like malignant gliomas, but seem to be more sensitive to radiotherapy and chemotherapy. Therefore, one of the glioma protocols can be recommended.

PNET. Histological similarities and coexistence of features of medulloblastomas, pineoblastomas, ependymoblastomas and neuroblastomas led to the concept that these tumours derive from a common, pluripotent stem cell. The term "primitive neuroectodermal tumours (PNETs)" (Chap 3) incorporates these related entities. They are radiosensitive and therefore radiotherapy is the primary therapeutic option.

The group of PNETs shows a less tight BBB and systemic, haematogenous spread is more common.

Systemic chemotherapy has not yet been evaluated in adults, with the exception of a small series of medulloblastomas (Levin *et al.*, 1983); but relapses frequently not taking their origin from the primary tumour location and extracranial seeding argue for use of adjuvant chemotherapy, and a prospective controlled trial of this is urgently needed. Studies of medulloblastomas in children showed benefit from adjuvant chemotherapy only in high-risk patients (i.e. with primary subarachnoidal spread or tumour rest).

In case of relapse or failure of radiotherapy in these tumours, a systemic chemotherapy is recommended. Protocols established for treatment of malignant gliomas can be used. Complex and toxic polychemotherapy derived from schedules for systemic mesenchymal neoplasms might theoretically be more effective and have been tested successfully in a small series from the literature (Friedman and Schold, 1985; Levin *et al.*, 1985) as well as in the authors' own experience with this sequential poly-chemotherapy protocol:

I.   1. Procarbazine 100 mg m^{-2}, days 1–14 orally
     2. Epirubicin   50 mg m^{-2}, day 1
     3. DTIC   400 mg m^{-2}, days 1 and 2
     4. MTX   20 mg m^{-2}, days 1, 8 and 15 orally

Dosage of procarbazine and MTX if leucocytes × 10^9 per microlitre are:

     3.000   100%
     2.500   75%
     2.500   50%
     <2.000 wait until >2.000 or until next cycle

II. 1. Ifosfamide (+ Mesna) 1.600 mg m^{-2}, days 29–33 (= 5 days)
     2. CDDP 50 mg m^{-2}, days 29 and 30

Repetition at day 51 (6 times in about 1 year). Semi-cycles should only be started at day 1 or 29, respectively if leucocytes ≥ 4000 × 10^9 per litre and platelets ≥ 100 000 × 10^9 per litre.

In patients after neuraxis irradiation, a considerably impaired bone marrow reserve has to be taken into consideration and often doses and number of cycles of chemotherapy have to be reduced.

Malignant Mesenchymal Neoplasms (Sarcomas)

Like primary brain tumours, neoplasms of the mesenchymal tissues of the CNS cause signs of intracranial pressure, but in contrast they grow outside the brain parenchyma. Therefore, the BBB does not play a role and extracranial, haematogenous dissemination is more common.

As for chemotherapy, a protocol established for the treatment of sarcomas elsewhere in the body may be applied successfully (CAD or CDDP/IFO). Due to the rarity of these entities there are no trial results available.

Lymphomas and Leukaemias

Although lymphomas arising primarily in the CNS have to be regarded as a disease of the haematopoietic system, they do not tend themselves to spread extracerebrally (<<10%). Therefore, in contrast to secondary CNS lymphomas local treatment is successful, albeit for a limited period of time. It is thus crucial to differentiate between primary cerebral lymphomas and CNS involvement of systemic disease.

Nevertheless, in the first approach, staging examinations should be restricted to an easily and quickly achievable programme with routine blood tests, including β-2 microglobulin, chest X-ray, abdominal ultrasonography and bone marrow aspiration, in order to avoid delay in treatment. When no other primary site is involved, early histological examination of the brain tumour must be carried out, since mental disorders in lymphoma patients develop early and may not be reversible, even after successful treatment. Therefore, a straightforward diagnostic approach is mandatory, resulting in early histological confirmation, usually by stereotaxic biopsy. Because of the frequency of multiple lesions and diffuse invasion these neoplasms are rarely resectable and the surgical procedure has often to be restricted to histological verification.

Another reason for early biopsy is that most lymphomas show a prompt but transient remission after administration of corticosteroids, and then histological verification might be impossible. Relapses may occur within weeks or months, but occasionally only after an interval of several years.

Lymphomas are highly radiosensitive and radiotherapy is recommended for basic treatment, but remission is in general limited to 1–1.5 years. Although lymphomas are in principle chemosensitive, chemotherapy has not yet been evaluated in randomized studies. From retrospective series it can be concluded that additional chemotherapy achieves longer median survival and more long-term remissions. It also appears from the literature and the authors' personal experience that prophylactic intrathecal chemotherapy with MTX or Ara-C is beneficial.

Early diagnosis is mandatory for successful management of relapses. Therefore, routine examination of CSF and prophylactic intrathecal chemotherapy should be done every 6–8 weeks for the first year, and every 3 months for the second year, in order to detect CSF dissemination as an early symptom of a relapse.

Complete remissions of primary cerebral lymphomas have been reported with the following protocols:

1. CHOP (BHOP and AHOP)
2. MTX high dose (only before radiotherapy, because of the risk of leucoencephalopathy)
3. Ara-C low/high dose.
4. HAM: high-dose Ara-C + mitoxantrone (as this regimen is very toxic, it should be used with caution)

In cases of cerebral or spinal manifestations of primary extracerebral lymphomas or leukaemias, standard haematological treatment should be accomplished by regional radiotherapy, with the addition of BBB penetrating agents to the chemotherapy regimen (Ara-C, nitrosoureas) and intrathecal chemotherapy.

Brain Metastases

Secondary neoplasms of the brain share some features independent of their histological nature: CT scans demonstrate, usually clearly, contrast-enhancing masses with marked perilesional oedema. Despite the rich pathological vascularization, with breakdown of the BBB, in larger metastases (more than 8–10 mm in diameter), micrometastasis initially develops protected by the BBB. This explains the clinically observed phenomenon that under systemic chemotherapy brain metastases may develop while extracerebral tumour shrinks.

Occurrence of brain metastases is a sign of generalized neoplastic disease. About half of these patients die of their primary disease and not of the CNS manifestation. Standard treatment is whole-brain irradiation, but the efficacy of chemotherapy in inducing response of brain metastases and prolongation of survival has been shown in series of patients with disseminated carcinoma (mainly lung and breast carcinoma). Median survival times correspond to those achieved with radiotherapy (4–6 months) and approximately 10% of the patients survive more than 1 year. Remarkable success is seen in small cell lung carcinoma, where brain metastases commonly present as the initial manifestation, as they do sometimes in cases of breast and germ cell carcinoma, as well as in bronchogenic and renal adenocarcinoma.

Chemotherapy should be guided by the sensitivity of the primary tumour type to treatment. If an effective chemotherapy for the primary tumour is available it should be applied unless progression of the disease contraindicates aggressive therapy.

Carcinomatous Meningitis

This special pattern of CNS neoplasia is well known as a cause of relapse in acute leukaemias in remission. But it may also occur as an isolated manifestion in other neoplasias, including solid tumours (4%–7%). Known brain metastases show meningeal involvement in up to 50% (Jellinger, 1983).

Regarding prognosis, there is a fundamental difference between the CNS relapse of haematological neoplasias, which may sometimes even be cured by specific treatment, and the meningeal spread of solid tumours where long-term survival is rare.

To relieve distressing symptoms, palliative intrathecal chemotherapy is indicated in all but moribund patients. Survival will be prolonged from typically 4 weeks without treatment to several months. With intensified therapy, survival of more than one year was achieved in 10%–25% (Wasserstrom et al., 1982). Repeated lumbar intrathecal injections (every 2–10 days) is recommended for standard treatment. If performed correctly this procedure only rarely (< 1%) induces marked side effects or complications.

Management of Carcinomatous Meningitis

Requirements are:

1. No intrathecal CT simultaneously with radiotherapy.
2. No use of highly concentrated solutions of the drug, but dilution with 6–10 ml of the patient's aspirated CSF or artificial CSF (Elliott's B solution without phenol red).

3. No injection if there is no free CSF flow out the cannula.
4. No more than three injections per week, not daily!
5. Dose reduction with high protein content of the CSF, because of the risk of impaired CSF flow with high local concentrations.
6. Concomitantly administered intrathecal corticosteroids to relieve symptoms and diminish side effects (Damm, 1985; Dienst, 1985). The authors use triaminoonacetonid crystal suspension and do not use acetates, because of local irritation (arachnoiditis).
7. Doses of MTX up to 50 mg, of Ara-C up to 120 mg per injection may be used by the lumbar route. The authors recommend 15 mg m^{-2} MTX and 40 mg m^{-2} Ara-C. With doses of >20 mg MTX, leucovorin rescue should be used (15 mg orally every 6 h for 3–5 days) to prevent haematological toxicity.

Treatment response is evaluated by the clinical course and number of malignant cells in the CSF, but the most sensitive and reliable parameter is the total protein content of the CSF. In a few cases, tumour markers like CEA, in the CSF, are helpful.

For patients who are otherwise fit, an intensified long-term treatment should be planned and an intraventricular catheter or reservoir system (Ommaya, Rickham) should be implanted for intraventricular administration. Meticulously sterile, low-flow injection technique and half the doses described above should then be used. Thio-TEPA (5–10 mg per injection) can also be used ventricularly, if MTX and Ara-C have failed. In special melanoma cases dacarbazine (20–30 mg, up to 80 mg per injection), which is more toxic than the other substances and not yet approved by regulatory authorities, for intrathecal use may be administered (Reuther *et al.*, 1983).

Conclusion

During the past few years the value of chemotherapy has been broadly evaluated in supratentorial high-grade gliomas in adults. A meta-analysis by Fine and colleagues (Mahaley, 1991), comprising 2400 patients in large randomized trials during the past decade, shows a prolongation of survival and a substantially higher proportion of long-term survivors (> 2 years) by multimodal treatment. Thus, with RT alone, 41% survive after 1 year, compared to a combination of radiotherapy and chemotherapy 53%; after 2 years 16% survive with radiotherapy alone, compared to 23% after combined treatment. Intensive polychemotherapy is indicated in patients with favourable prognostic factors. The small number of substances suited for neuro-oncological purposes, and their relatively high toxicity, limit the application of cytotoxic treatment. Cure of malignant glioma is not in sight. Improvement of chemotherapy might be achievable by the development of new BBB-penetrating substances, optimized combination of antineoplastic drugs already evaluated, extending the dose limitation by improved supportive therapy, e.g. growth factors, and by avoiding neurotoxicity wherever possible.

Besides exploring new ways in basic research and clinically well-controlled Phase II studies, there is a need to continue large, randomized trials.

Knowledge of confirmed prognostic criteria in the future will allow a risk-adapted therapy, avoiding on the one hand unnecessary toxicity in patients with unfavourable prognosis and accepting on the other hand a higher toxicity in patients with better prognosis. Evaluation of timing/schedule of chemotherapy, identification of additional predictive parameters, e.g. chemosensitivity testing and more detailed histological and molecular biological criteria are the next steps toward the aim of offering an optimized treatment for the individual patient. This can only be realized with intense interdisciplinary cooperation and systematic clinical research.

Appendix I
Some Abbreviations and Acronyms

ACNU	Nimustine – nitrosourea compound
ADM	Doxorubicin
Ara-C	Cytosine–arabinoside – antimetabolite
AZQ	Aziridinylbenzoquinone
BBB	Blood–brain barrier
BCNU	Carmustine – nitrosourea compound
CCNU	Lomustine
CDDP	Cisplatin
CPM	Cyclophosphamide
DAG	Dianhydrogalactitol
DBC	Dibromodulcitol
5-FU	5-Fluorouracil – antimetabolite
HU	Hydroxyurea
IFO	Ifosfamide
KPS	Karnofsky performance score
MIT	Mitoxantrone – antrachinone derivative
MTX	Mithotrexate – antimetabolite
PCZ	Procarbazine
VBL	Vinblastine – vinca-alkaloid
VCR	Vincristine – vinca-alkaloid
VDS	Vindesine

References

Adelstein DJ, Likavec MJ, Sharan VM *et al.* (1987) Intravenous 5-fluorouracil infusion and simultaneous radiotherapy in glioblastoma multiforme. ASCO Proc 6: 274

Afra D (1973) Vincristine therapy in malignant glioma recurrencies. Neurochirurgia (Stuttg) 16: 189–198

Afra D, Kocsis B, Dobay J *et al.* (1983) Combined radiotherapy and chemotherapy with dibromodulcitol and CCNU in the postoperative treatment of malignant gliomas. J Neurosurg 59: 106–110

Albert FK, Forsting M, Sartor K *et al.* (1991) Frühes postoperatives CT und MR (~/+KM) nach Exstirpation von High grade-Gliomen – Erste Ergebnisse einer prospektiven Studie,

Verhandlungen der Deutschen Gesellschaft für Neurologie (Firnhaber W, Dworschak K, Lauer K, Nichtweiss M, eds). Springer-Verlag, Berlin, pp 284–287

Anderson NR, Tandon DS (1991) Ifosfamide extrapyramidal neurotoxicity. Cancer 68: 72–75

Ausman JI, Levin VA, Brown WE et al. (1977) Brain-tumour chemotherapy – pharmacological principles derived from a monkey brain-tumour model. J Neurosurg 46: 155–164

Baker WJ, Royer GLJ, Weiss RB (1991) Cytarabine and neurologic toxicity. J Clin Oncol 9: 679–693

Bamberg M, Budach V, Stuschke M et al. (1988) Preliminary experimental results with the nitrosourea derivative ACNU in the treatment of malignant gliomas. Radiother Oncol 12: 25–29

Boethius J, Blomgren H, Collins VP et al. (1983) The effect of systemic human interferon-alpha administration to patients with glioblastoma multiforme. Acta Neurochir (Wien) 68: 239–251

Brem H, Mahaley MSJ, Vick NA et al. (1991) Interstitial chemotherapy with drug polymer implants for the treatment of recurrent gliomas. J Neurosurg 74: 441–446

Butti G, Knerich R, Tanghetti B et al. (1984) Perioperative carmustine chemotherapy for malignant brain tumours. Cancer Treat Rep 68: 1505–1506

Cerny T, Meier C (1989) [Neurotoxicity with cytostatic therapy: a review]. Schweiz Med Wochenschr 119: 1137–1147

Colamaria V, Caraballo R, Borgna Pignatti C et al. (1990) Transient focal leukoencephalopathy following intraventricular methotrexate and cytarabine. A complication of the Ommaya reservoir: case report and review of the literature. Child's Nerv Syst 6: 231–235

Damm W (1985) Intrathekale Verträglichkeit von Methotrexat, Methotrexat/Triamcinolon-acetonid, Gentamycin, Streptomycin. Würzburg Universität

Dienst P (1985) Intrathekale Verträglichkeit von Triamcinolon-acetonid, Cytosin-Arabinosid, Cytosin-Arabinosid/Triamcinolon-acetonid. Dissertation. Würzburg Universität

Djerassi I, Kim JS, Regev A (1985) Response of astrocytoma to high-dose methotrexate with citrovorum factor rescue. Cancer 55: 2741–2747

Duff TA, Bordon E, Bay J et al. (1986) Phase II trial for treatment of recurrent glioblastoma multiforme. J Neurosurg 64: 408–413

Eagan RT, Creagan ET, Bisel HF et al. (1981) Phase II studies of dianhydrogalactitol-based combination chemotherapy for recurrent malignant glioma. Oncology 38: 4–6

EORTC Brain Tumor Group (1978) Effect of CCNU on survival rate of objective remission and duration of free interval in patients with malignant brain glioma – final evaluation. Eur J Cancer 14: 851–855

EORTC Brain Tumor Group (1981) Effect of CCNU, VM-26 plus CCNU, and procarbazine in supratentorial brain gliomas. J Neurosurg 55: 27–31

Forman A (1990) Recent studies on neurotoxicity and on the epidemiology of brain and nervous system tumours. Curr Opin Oncol 2: 691–698

Forsting M, Albert FK, Kunze S et al. (1992) Residual tumor after brain tumor resection? How to cut the Gordian knot with early postoperative MRI Abstract. Cancer Res Clin Oncol 118 (Suppl): R106

Friedman HS, Schold SC (1985) Rational approaches to the chemotherapy of medulloblastoma. Neurosurg Clin 3: 843–853

Geobel WE, Trappe AE, Weinzierl FK (1987) Perioperative cytostatische Behandlung supratentorieller malinger Hirntumoren mit BCNU unter spezieller Beruecksichtigung der Tumorlage (Perioperative cytostatic treatment of malignant supratentorial brain tumors with BCNU with special reference to tumor site). Neurochirurgia (Stuttg) 30: 82–87

Green SB, Byar DP, Walker MD et al. (1983) Comparisons of carmustine, procarbazine, and high-dose methylprednisolone as additions to surgery and radiotherapy for the treatment of malignant glioma. Cancer Treat Rep 67: 123–132

Green SB, Byar DP, Strike TA et al. (1984) Randomized comparisons of BCNU, streptozotocin, radiosensitizer, and fractionation of radiotherapy in the postoperative treatment of malignant glioma (study 7702). ASCO Proc 3: 230

Green SB, Byar DP, Strike TA et al. (1985) Randomized phase II comparisons of PCNU and AZQ for treatment of brain tumor (study 8120) (BTSG Bethesda). ASCO Proc 4: 143

Groothius DR, Molnar P, Blasberg RG (1984) Regional blood flow and blood-to-tissue transport in five brain tumor models. Implications for chemotherapy. In: Brain Tumor Biology. Karger, Basel, pp. 132–153

Haid M, Kandekhar JD, Merril JM et al. (1987) Phase II AZQ therapy for recurrent of progressive glioma of the central nervous system – final report. ASCO Proc 6: 266

Harmers FP, Gispen WH, Neijt JP (1991) Neurotoxic side-effects of cisplatin. Eur J Cancer 27: 372–376

Hatlevoll R, Lindegaard K-F, Hagen S *et al.* (1985) Combined modality treatment of operated astrocytomas grade 3 and 4. Cancer 56: 41–47

Hildebrand J (1991) Chemotherapy of malignant supratentorial gliomas in adults: ten year experience of the EORTC brain tumor group. In: Voth D, Krauseneck P (eds) Chemotherapy of Gliomas. De Gruyter, Berlin, pp

Hobert U, Krauseneck P, Bogdahn U *et al.* (1985) Phase I–II study of cytosine-arabinoside in malignant brain tumors. J Neurol Suppl 7:232

Hook CC, Kimmel DW, Kvols LK *et al.* (1992) Multifocal inflammatory leukencephalopathy with 5-fluorouracil and levamisole. Ann Neurol 31: 262–267

Jellinger K, Kothbauer P, Volc D *et al.* (1979) (eds) Combination chemotherapy (COMP protocol) and radiotherapy of anaplastic supratentorial gliomas. Acta Neurochir 51: 1–13

Jellinger K (1983) Häufigkeit und Charakteristik der zerebralen Karzinommetastasen, Hirnmetastasen (Heyden HW, Krauseneck P, Zuckschwerdt-Verlag, München

Kaplan RS, Wiernik PH (1984) Neurotoxicity of antitumor agents. In: Toxicity of Chemotherapy. Grune and Stratton, Orlando, Florida, pp 365–431

Krauseneck P, Dommasch D (1987) Zytostatische Therapie in der Neurologie, Neurologische und psychiatrische Therapie (Flügel KA, ed). Perimed-Verlag, Erlangen, pp 363–378

Krauseneck P, Mertens H-G (1987) Results of chemotherapy of malignant brain tumours in adults. In: Jellinger K (ed) Therapy of Malignant Brain Tumours. Springer-Verlag, Wien, pp 349–395

Krauseneck P, Mertens H-G, Messerer D *et al.* (1989) Zwischenergebnisse der deutsch-österreichischen Studie zu den malignen supratentoriellen Gliomen des Erwachsenenalters, Verhandlungen der Deutschen gesellschaft für Neurologie (Fisher P-A, Baas H, Enzensberger W, eds). Springer-Verlag, Berlin, pp 1090–1093

Krauseneck P (1990) Gehirntumoren, Medikamentöse Therapie maligner Erkrankungen. (Huhn D, Herrmann R, Stuttgart (eds) Fischer, pp 347–373

Levin VA, Vestnys P, Edwards MS *et al.* (1983) Improvement in survival produced by sequential therapies in the treatment of medulloblastoma. Cancer 51: 1364–1370

Levin VA, Wilson CB, Davis R *et al.* (1979) Phase III comparison of BCNU, hydroxyurea and radiation therapy to BCNU and radiation therapy for treatment of primary malignant gliomas. J Neurosurg 51: 526–532

Levin VA, Wara WM, Davis RL *et al.* (1985) Phase III comparison of BCNU and the combination of procarbazine, CCNU, and vincristine administered after radiotherapy with hydroxyurea for malignant gliomas. J Neurosurg 63: 218–223

Levin VA, Silver P, Hannigan J *et al.* (1989) Superiority of postradiotherapy adjuvant chemotherapy with CCNU, procarbazine, and vincristine (PCV) over BCNU for anaplastic gliomas: NCOG 6G61 final report. Int J Radiat Oncol Biol Phys 18: 321–324

Lieberman AN, Foo SH, Ransohoff J *et al.* (1982) Long term survival among patients with malignant brain tumours. Neurosurgery 10: 450–453

Long JM (1970) Capillary ultrastructure and the blood–brain barrier in human, malignant brain tumours. J Neurosurg 39: 127–144

Mahaley MSJ, Hipp SW, Dropcho EJ *et al.* (1989) Intracarotid cisplatin chemotherapy for recurrent gliomas. J Neurosurg 70: 371–378

Mahaley MS Jr (1991) Neuro-oncology index and review (adult primary brain tumours). J Neuro-oncol 11: 85–147

Merimsky O, Reider Groswasser I, Inbar M *et al.* (1990) Interferon-related mental deterioration and behavioral changes in patients with renal cell carcinoma. Eur J Cancer 26: 596–600

Mertens H-G (1982) Postoperative Strahlen- und Chemotherapie mit BCNU and VM26 bei melignen supratentoriellen Gliomen des Erwachsenenalters. Protokoll der multizentrischen Therapiestudie der deutschen Hirntumorgruppe (unpublished)

Meyers CA, Obbens EA, Scheibel RS *et al.* (1991) Neurotoxicity of intraventricularly administered alpha-interferon for leptomeningeal disease. Cancer 68: 88–92

Nagai M, Arai T (1984) Clinical effect of interferon in malignant brain tumours. Neurosurg Rev 7: 55–64

Ochs J, Mulhern R, Fairclough D *et al.* (1991) Comparison of neuropsychologic functioning and clinical indicators of neurotoxicity in long-term survivors of childhood leukemia given cranial radiation or parenteral methotrexate: a prospective study. J Clin Oncol 9: 145–151

Paoletti P, Knerich R, Butti G *et al.* (1983) Italian cooperative study on malignant glial tumor therapy, Therapie maligner Neoplasien des Gehirns (Krauseneck P, Mertens H-G, eds). Perimed Verlag, Erlangen

Popovic P, Popovic V, Wheathers DR et al. (1986) Effect of intracarotid administration of BCNU in rats. In: Thomas DGT, Walker MD (eds) Biology of Brain Tumours. Martinus Nijhoff, Boston, pp 435–438

Potthoff PC (1981) Ergebnisse der Therapie maligner Hirntumoren, Maligne Hirntumoren (Potthoff PC, Schreml W, eds). Huber-Verlag, Berne.

Pouillart P, Mathe G, Thy Th et al. (1976) Treatment of malignant gliomas and brain metastasis in adults with a combination of adriamycin, VM26. Cancer 38: 1909–1916

Reuther P, Dommasch D, Fuhrmeister U et al. (1983) Intrathekale DTIC-Therapie bei leptomeningealer Melanommetastasierung, Therapie maligner Neoplasien des Gehirns (Krauseneck P, Mertens H-G, eds). Perimed, Erlangen, pp 135–138

Rozenthal JM, Robins HI, Finlay JL et al. (1989) Eight-drugs-in-one-day chemotherapy in postirradiated adults patients with malignant gliomas. Med Pediatr Oncol 17: 471–476

Seiler RW, Zimmermann A (1979) Preoperative radiotherapy and chemotherapy in hypervascular, high grade supratentoriell astrocytomas. Surg Neurol 12: 131–133

Shapiro WR, Young DF, Mehta BM (1975) MTX distribution in CSF after intravenous, ventricular and lumbar injections. N Engl J Med 293: 161

Shapiro WR, Young DF (1976) Treatment of malignant glioma. Arch Neurol 33: 494–500

Shapiro WR (1982) Treatment of neuroectodermal brain tumours. Ann Neurol 231–237

Shapiro WR, Young DF, (1984) Neurological complications of antineoplastic therapy. Acta Neurol Scand 70 (Suppl 100): 125–132

Shapiro WR (1986) Therapy of adult malignant brain tumours: what have the clinical trials taught us? Semin Oncol 13: 38–45

Shapiro WR, Shapiro JR (1986) Current approaches to chemotherapy, In: Walker MD, Thomas DGT (eds) Biology of Brain Tumors. Martinus Nijhoff, Boston, pp 383–392

Shapiro WR, Green SB (1987) Reevaluating efficacy of intra-arterial BCNU [letter]. J Neurosurg 66: 313–315

Shapiro WR, Green SB, Burger PC et al. (1987) A randomized comparison of intraarterial vs. intravenous BCNU for patients with malignant glioma (study 8301): interim analysis demonstrating lack of efficacy for ia BCNU. ASCO Proc 6: 268

Shapiro WR, Green SB, Burger PC et al. (1989) Randomized trial of three chemotherapy regimens and two radiotherapy regimens in postoperative treatment of malignant glioma. Brain Tumor Cooperative Group Trial 8001. J Neurosurg 71: 1–9

Shibata S, Mori K, Moriyama T et al. (1987) Randomized controlled study of the effect of adjuvant immunotherapy with Picibanil on 51 malignant gliomas. Surg Neurol 27: 259–263

Taylor S, Eyre HJ (1987) Randomized phase II trials of acivivi and fludarabine in recurrent malignant gliomas. ASCO Proc 6: 275

Thomas DGT, Darling JL, Paul EA et al. (1985) Assay of anti-cancer drugs in tissue culture: relationship of relapse free interval (RFI) and in vitro chemosensitivity in patients with malignant cerebral glioma. Br J Cancer 51: 525–532

Tirelli U, D'Incalci M, Canetta R et al. (1984) Etoposide in malignant brain tumors: a phase II study. J Clin Oncol 2: 432–437

Trojanowski T, Peszynski J, Turowski K et al. (1989) Quality of survival of patients with brain gliomas treated with postoperative CCNU and radiation therapy. J Neurosurg 70: 18–23

Ushio Y, Abe H, Suzuki J et al. (1985) [Evaluation of ACNU alone and combined with tegafur as additions to radiotherapy of the treatment of malignant gliomas – a cooperative clinical trial]. No To Shinkei 37: 999–1006

Voth D, Hüwel N, Al-Hami S et al. (1984) Mono-treatment of malignant glioma with a derivate of nitrosourea, ACNU (first results). In: Voth D, Krauseneck P (eds) Chemotherapy of Gliomas – Basic Research, Experience, Results. Walter de Gruyter, Berlin, pp 361–372

Walker MD, Alexander E Jr, Hunt WE et al. (1978) Evaluation of BCNU and/or radiotherapy in the treatment of anaplastic gliomas. J Neurosurg 49: 333–343

Walker MD, Green SB, Byar DP et al. (1980) Randomized comparisons of radiotherapy and nitrosoureas for the treatment of malignant glioma after surgery. N Engl J Med 303: 1323–1329

Wasserstrom WR, Glass JP, Posner JB (1982) Diagnosis and treatment of leptomeningeal metastases from solid tumors: Experience in 90 patients. Cancer 49: 759–772

Yamashima T, Yamashita J, Shoin K (1990) Neurotoxicity of local administration of two nitrosoureas in malignant gliomas. Neurosurgery 26: 794–799

Zimm S, Ettinger LJ, Holgenberg SJ et al. (1985) Phase I and clinical pharmacological study of mercaptopurine administered as a prolonged intravenous infusion. Cancer Res 45: 1869–1873

14 Immunotherapy of Brain Tumours

Yutaka Sawamura and Nicolas de Tribolet

Introduction

Host immunity and tumour biology are important cofactors in brain tumour immunotherapy. Much of the research for brain tumour immunobiology has centred upon evaluation of tumour cell antigenicity and the cell-mediated immune responses to the malignant tumours. Over recent years, evidence has been accumulated from monoclonal-antibody research for the presence of brain tumour associated antigens. There has also been an increasing awareness of the ways in which a brain tumour can apparently evade host immune reactions. In addition, it should be recognized that a host versus tumour reaction occurs at the site of tumour growth and that the cell-mediated immune system may have a central role for possible control of neoplastic growth. Ideas for potential therapeutic manipulation of host–glioma immune interactions will be reviewed critically.

Brain Tumour Immunology

Anatomically there is no lymphatic system in the central nervous system (CNS) as well as the anterior chamber of the eye, the thyroid and the testicle. It has, however, been suggested that microglia in the CNS might function as antigen-presenting cells. Microglia in the brain express some pan-macrophage markers and are phagocytic in culture; this evidence indicates its monocytic origin. The expression of major histocompatibility complex (MHC) class I and class II antigens has also been demonstrated on microglia; furthermore, microglia can secrete various factors such as interferon-γ, TNF-α (tumour necrosis factor-α) and interleukin-1 (IL-1). In normal brain vessels, vascular structures in combination with astrocytes constitute the blood–brain barrier (BBB) which shields the brain to a certain extent from systemic humoral and cellular immunity. Immunoglobulins, interleukin-2 (IL-2) and interferons would not be expected to permeate the BBB effectively. However, as a result of tumour-induced endothelial changes, the blood vessels lose their BBB properties and become abnormal. The large pores in the altered vascular walls provide a pathway both for extravasation of larger molecules including serum proteins and conversely for many factors produced by tumour cells. These factors then allow the interaction between the host immune response and the tumour cells. Since an adhesion molecule (intercellular adhesion molecule-1; ICAM-1) is expressed on endothelial cells and can be enhanced by interferon-γ, IL-1, or TNF-α and since T cells have the cell surface receptor (lymphocyte associated antigen-1; LFA-1) for ICAM-1, an adhesion interaction between the T cells and the brain endothelial cells can be induced. The T cells may then pass through the BBB into tumour tissue.

Surface Antigens on Brain Tumour

The central question in tumour immunology is whether tumour cells show differences from their normal cellular counterparts that can be recognized by the immune system. A number of tumour-associated antigens on brain tumour cells, which are, however, shared by histogenetically related tumours, have been found and characterized, as for example the neuroectodermal antigens (Carrel et al., 1982; de Muralt et al., 1985; Fischer et al., 1988). Although an absolutely brain-tumour specific antigen has not been found, it does not matter from an immunotherapeutic point of view if a glioma antigen, which may exhibit therapeutically functional specificity, is also expressed by other tumours, since distinction from the host's normal tissue is the essential requirement. Three significant groups of antigens have been identified on brain tumour cells using monoclonal antibodies, including tumour-associated antigens, MHC antigens and lymphoid differentiation antigens.

Glial antigens are primarily displayed on gliomas and monoclonal antibodies directed to glial antigens have some additional minor specificity for other non-neuroectodermal tissues and reactive astrocytes. Neuroectodermal antigens form a major component of the surface antigens found on glioma cells; they have been detected on gliomas, melanomas, neuroblastomas as well as fetal brain cells and endothelial cells within gliomas.

The expression of certain MHC class II antigens (particularly HLA-DR) is necessary for activation of the immune system to present new foreign antigens (MHC-restricted immune response). A putative tumour antigen can be immunogenic only if it is recognized together with the self-MHC molecule by antigen-presenting cells. In addition, MHC class I (particularly HLA-A) expression is required on the target cells for the lytic action of effector cells to take place. Although normal resting brain astrocytes do not exhibit MHC class II antigens, it is noteworthy that some activated astrocytes and glioma cells as well as hyperplastic endothelial cells can express HLA-DR (de Tribolet et al., 1984; Frank et al., 1986). HLA-DR expression can be induced and enhanced by interferon-γ on a number of glioma cell lines. Interferon-γ and TNF-α enhance the expression of HLA-ABC. In contrast, transforming growth factor-β2 (TGF-β2) which can be secreted by glioma cells produces a partial but significant decrease of constitutive and interferon-γ induced HLA-DR surface antigen expression on glioma cells.

Normal brain and tumours of neuroectodermal origin express lymphoid differentiation antigens for instance CD3 and CD10 (common acute lymphocytic leukaemia antigen; CALLA). The expression of these lymphoid differentiation antigens as well as secretion of some cytokines from glioma cells and astrocytes suggest common functional properties shared by immunocompetent cells and glial cells. Recently it has been found that CALLA is a membrane-bound enzyme identical to neutral endopeptidase which can catabolize the IL-1 molecule.

Humoral Immune Response

The presence of limited antiglioma activity was reported within the sera of a small number of patients harbouring gliomas. Some patients with malignant gliomas have immune complexes in their sera; the survival of these patients was reported to be shorter than that observed in those who did not have such complexes. Furthermore, elevated immunoglobulin levels in the cerebrospinal fluid (CSF) in certain patients with either malignant or benign gliomas have been reported. In certain glioma tissues, populations of B cells are present, but their ability to produce anti-glioma antibodies capable of participating in complement-dependent or antibody-dependent cell-mediated cytotoxic responses is doubtful. It is of interest to mention that interleukin-6 (IL-6), which promotes immunoglobulin production by activated B cells, can be secreted by glioma cells.

Cell-Mediated Immune Response

It has been repeatedly emphasized that the systemic cellular immune response in patients with a malignant brain tumour is significantly depressed according to the following evidence: (a) impaired blastogenic response of peripheral blood lymphocytes (PBL); (b) inhibitory effect of patients' sera on T-cell responsiveness to mitogens; (c) suppressed natural killer (NK) cell activity; and (d) reduced IL-2 production and IL-2 receptor expression of mitogen-stimulated T cells. Recently, however, it has been demonstrated that a reaction of the host immune system to tumour tissue occurs predominantly at the site of tumour growth. A number of reports have demonstrated that the degree of lymphocyte infiltration in brain neoplasms has some relevance to the prognosis. The mononuclear cells found in glioma tissue have been identified as being mostly T cells, with a predominance of the cytotoxic/suppressor cell subset. The infiltration is most pronounced at the tumour periphery and NK cell, macrophage and B-cell infiltrations have also been demonstrated. The maturation and activation of glioma-derived tumour-infiltrating lymphocytes (TILs) can be generated after in vitro exposure to exogenous IL-2, and expanded TILs are capable of killing autologous tumour cells in vitro (Fig. 14.1). However, lymphocyte infiltration in glial tumours is scarce and the number of TILs isolated from gliomas is extremely low in contrast to those found in metastatic brain tumours or cancers of other organs. Furthermore, T cells isolated from brain tumours have been shown to be small, non-blastic and negative for activation antigens. Therefore, it is suspected that suppression of glioma-

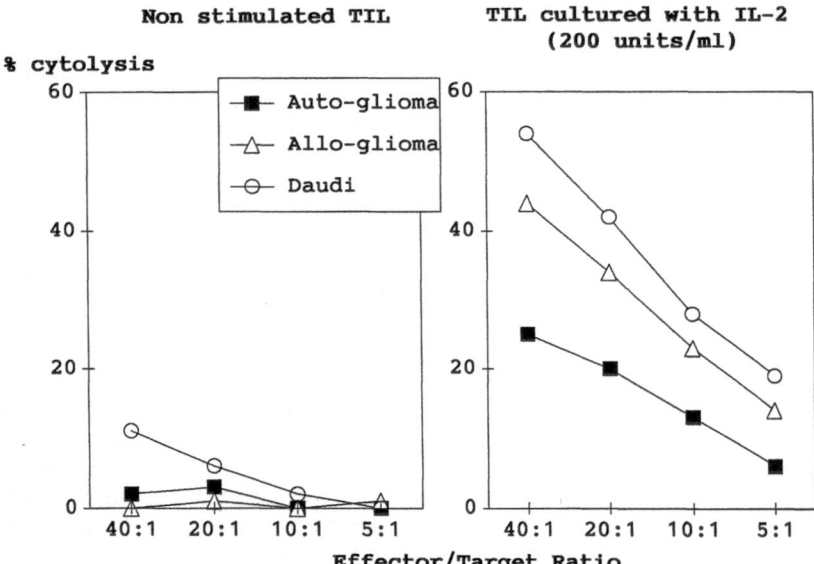

Fig. 14.1 In vitro induction of cytotoxic activity of glioma-infiltrating lymphocytes. Tumour-infiltrating lymphocytes (TILs) were isolated from a glioblastoma and cultured for 72 h in vitro with or without IL-2. The TILs were mainly composed of T cells. The number of TILs isolated was extremely low and their cytolytic activity was negative. After culture with IL-2, a cytolytic activity was induced, but no specificity of cytolysis against autologous glioma cells was observed

infiltrating precursors by the tumour cells in situ may be more profound than that of TILs in the cancers of other organs.

Modulation of the Host Immune Response by Gliomas

Augmentation of immunity requires induction of increased effector function, and also requires concomitant abatement of suppressor activities. Human brain tumours can secrete a variety of immunoregulatory (suppressive) factors, as summarized in Table 14.1, and this inhibitory role is thought to provide the means for tumour self-defence.

TGF-β2 and prostaglandin E2 (PGE2) may be responsible for the immunosuppression observed in glioma patients (Fontana *et al.*, 1982; de Martin *et al.*, 1987). TGF-β2 has a potent immunosuppressive effect on multiple immune functions such as: (a) IL-2-dependent proliferation; (b) IL-2 receptor expression on stimulated T cells and NK cells; (c) IL-1 dependent lymphocyte proliferation; and (d) immunoglobulin production and B-cell proliferation. Increasing concentrations of IL-2 can partially overcome the suppressive activity of TGF-β2 on IL-2-dependent proliferation of lymphocytes (Kuppner *et al.*, 1988). TNF may reverse inhibition of cytotoxic T-cell development by TGF-β. PGE2 also exerts profound suppressive effects on a variety of immune functions such as: (a) IL-2 production and IL-2 receptor expression on lymphocytes; (b) activation of

Table 14.1 Theoretical immunomodulation by human malignant glioma

Factors	Possible mechanism
TGF-β	Suppression of: (1) IL-2-dependent proliferation and IL-2 receptor expression on stimulated lymphocytes; (2) LAK cell generation; (3) IL-1-dependent lymphocyte proliferation; (4) immunoglobulin production by B cell
PGE-2	Suppression of: (1) IL-2 production and IL-2 receptor expression on stimulated lymphocytes; (2) CTL, NK cell, LAK cell and B-cell activation; (3) expression of MHC class II molecules and cytotoxicity of macrophage
Gangliosides	Suppression of: (1) IL-2-stimulated T-cell proliferation; (2) IL-2 binding to high affinity IL-2 receptor; (3) function of CD4 molecule; (4) accessory cell function and IL-1 production by monocytes; (5) masking of glioma-associated antigens
Mucopolysaccharides	Inhibition of direct interaction of lymphocytes against glioma cells

cytotoxic T lymphocyte (CTL), NK cell, lymphokine-activated killer (LAK) cell and macrophage-mediated cytotoxicity, (c) activation and proliferation of B cells; and (d) expression of MHC class II molecules on the cell surface of macrophages. In addition, glioblastoma cell lines secrete some plasminogen activator, plasminogen-activator inhibitor and alpha-1 antitrypsin, which were found to inhibit cytotoxic activities of lymphocytes. Certain glioma cell lines produce a mucopolysaccharide coat on their cell surface which impairs activation of lymphocytes and hinders direct interaction of glioma cells with other cells. Ganglioside compositions of human gliomas correlate with their histological grade. The degree of malignancy is associated with an increase of certain gangliosides. Gangliosides inhibit IL-2-stimulated proliferation of T cells, binding of IL-2 to the high-affinity IL-2 receptor, function of the CD4 molecule, and accessory cell function and IL-1 production by monocytes. On the other hand, glioma cells have the ability to secrete some cytokines including IL-1 and IL-6. Human astrocytoma cell lines exhibit a proliferative response and high levels of IL-6 production in the presence of IL-1. IL-3-like or interferon-α/β-like activities in glioma cell cultures have also been reported. In addition, the expression of an intercellular adhesion molecule-1 on astrocytes or glioma cells as the ligand for lymphocyte associated antigen-1 enhances the connection between effector and target cells. IL-1, interferon-γ and TNF-α should enhance the expression of intercellular adhesion molecule on glioma cells as well as endothelial cells.

Immunotherapy

During the past decade, new immunotherapeutic approaches against malignant brain tumours have become available owing to developments of genetic engineering and cell culture technology. Appropriate monoclonal antibodies may be used as carriers for chemotherapeutic and radiotherapeutic agents. The cellular immune system can be manipulated by either administration of exogenous biological response modifiers or by the transfer of activated lymphocytes.

Monoclonal Antibodies

A monoclonal antibody which is specific for one single structure on malignant brain tumours can be produced in large quantities either from ascites or in vitro culture of murine hybridoma with a high degree of purity. The possibility of using monoclonal antibodies as therapeutic agents is attractive. This concept has been considered for many years.

However brain tumour-associated antigens detected by monoclonal antibodies, which have been well characterized, are often a part of "self" antigens rather than tumour specific antigens; most monoclonal antibodies against brain tumours cross-react with some normal tissues. Monoclonal antibodies to regulatory molecules, i.e. epidermal growth factor receptor and transferrin receptor expressed on brain tumours, have been also investigated. The following monoclonal antibodies may have clinical value:

1. Monoclonal antibody 81C6 defines the glioma mesenchymal extracellular matrix (GMEM) antigen on the basement membrane of glioblastomas and is associated with proliferative endothelium of the hyperplastic vessels. Glioma mesenchymal extracellular matrix antigen is present on most glioma and fibroblast cell lines, some human cancer cell lines, fetal and adult spleen, liver, adult kidney and mesenchymal cells, but not on normal adult brain.

2. UJ13A recognizes all neuroectodermal tumours except melanomas and also reacts with fetal brain, peripheral nerves, adrenal medulla, adult thyroid epithelium, fetal kidney tubules, and primary cultures of fetal myoblasts.

3. Mel-14 reacts with an antigen found on a large number of melanoma cell lines and other neuroectodermal tumours including gliomas. Cross-reactivity is not detected with the normal neuroectodermal tumours including gliomas. Cross-reactivity is not detected with the normal adult tissues tested.

4. Monoclonal antibody 425 binds to an abnormal protein determinant on the external domain of the epidermal growth factor receptor. The epidermal growth factor receptor has been shown to be overexpressed in approximately 50% of malignant gliomas. Epidermal growth factor receptor is expressed rarely on normal brain tissue, and not all on bone marrow and peripheral blood cells.

5. Gangliosides on certain malignant tumour cells are immunogenic membrane-associated glycolipids. Monoclonal antibody Ofa-I-2 and 14.18 react with diasialoganglioside GD2. Monoclonal antibodies 11C64 and R24 are reactive with the diasialoganglioside DG3. Ofa-I-2 recognizes cell surface GD2 on glioma cells.

Some unlabelled monoclonal antibodies, i.e. not bound to drugs, toxins or isotopes, have been reported to have a direct antitumour effect. Augmentation of antibody-dependent macrophage cytotoxicity and anti-body-dependent cellular cytotoxicity are supposed to be responsible for the antitumour effect of some antibodies. Although monoclonal antibodies could potentially be used to kill tumour cells when infused alone, clinical studies using such native monoclonal antibodies have been limited to patients suffering from B-cell lymphoma or melanoma, and results were disappointing. However, gangliosides still may be relevant target antigens for monoclonal antibody-mediated immunotherapy of gliomas.

A variety of potentially toxic molecules have been linked to monoclonal antibodies (Fig. 14.2). Immunotoxins comprise a new class of cell-type specific cytotoxic heteroconjugates which consist of a monoclonal antibody linked to a protein toxin such as the A chain of ricin. Anti-epidermal growth factor receptor or anti-transferrin receptor monoclonal antibodies have been linked to protein toxins (Johnson *et al.*, 1989). These immuno-

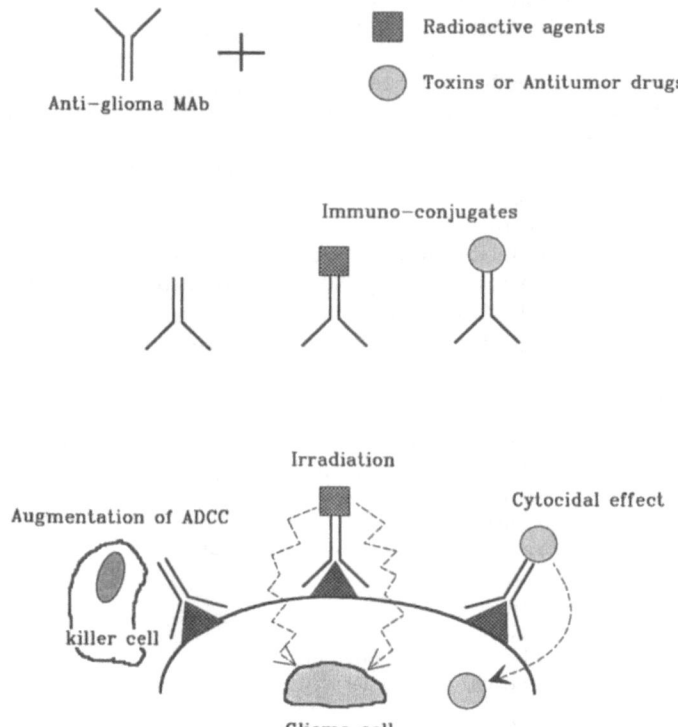

Fig. 14.2 Monoclonal antibody targeted therapy. Most monoclonal antibodies have no direct cytotoxic effects. Unlabelled antibodies may enhance antibody-dependent cell-mediated cyto-toxicity. Otherwise monoclonal antibodies can be efficiently coupled to isotopes, drugs and toxins. Radioisotope conjugates can irradiate tumour tissue beyond the antibody and drugs require internalization to destroy the target cells

toxins displayed specific in vitro toxicity for human glioma cells. Protein toxins are extremely potent since a single molecule of immunotoxin may destroy a tumour cell; however, it requires internalization to destroy the cell. Another possibility is the antibody–drug conjugate aiming at high drug delivery to the target cells and reducing the adverse effect on normal cells. In contrast to immunotoxins which are active on any cells, the effectiveness of the chemotherapeutic agent–monoclonal antibody conjugate depends on the drug sensitivity of the tumour cells as much as on the specificity of the monoclonal antibody. Monoclonal antibodies can also act as carriers for radionuclides. This method has been widely investigated and may be the most promising among monoclonal–antibody directed therapies for brain tumour patients (Wikstrand *et al.*, 1987; Lee *et al.*, 1988). As opposed to antibody-carrying toxins or chemotherapeutic agents, antibody-carrying radionuclides can destroy antigen-negative cells adjacent to antigen-positive cells binding the antibody.

Realization of these monoclonal antibodies has been, however, hampered by major obstacles, as listed in Table 14.2. All antigens recognized by the monoclonal antibodies are known to have some secondary representation on other cells. In order for monoclonal antibody-directed therapy to be safe and feasible, the antigens must be fully characterized and their expression on normal cells must be minimal. One problem in evaluation of monoclonal antibody directed therapy is the antigenic heterogeneity of glioma cells. Therefore, the use of a group of monoclonal antibodies as a "polymonoclonal" antibody may have a better chance of destroying a critical amount of glioma cells.

After intravenous administration to a patient, the monoclonal antibody is diluted into the circulating blood, where it can bind to normal cells with cross-reactivity and to pre-existing human anti-mouse antibodies (anti-globulin response). Antimouse antibodies may be present in any patient, although the level is variable between patients and to some extent depends

Table 14.2   Critical factors for monoclonal antibody therapy

1. *Monoclonal antibody*
   Reactivity and affinity against the tumour
   Cross-reactivity with normal brain
   Purity and available amount
   Technical difficulties of making immunoconjugates
   Cost

2. *Host factors*
   Anaphylactic reaction
   Neutralization by anti-mouse globulin response
   Trapping into reticulo-endothelial system
   Target antigen shed into circulation (formation of immune complexes)

3. *Tumour factors*
   Antigenic heterogeneity of cells in the tumour
   Blood–brain barrier
   Vascularity and blood flow
   Antigen modulation
   Enzymic degradation of cell-bound antibody

on the monoclonal antibody being used. After a second infusion of the monoclonal antibody, the level of circulating anti-mouse antibody is so much higher that there is an increased risk of anaphylactic reaction. Additionally, some tumour-associated antigens are shed from the tumour and circulate in the blood. These might bind the antibody and prevent tumour localization. Furthermore, in certain cases the radiolabelled antibody can be trapped rapidly into the reticuloendothelial system, especially into the liver, therefore reducing the tumour uptake.

Another point to be taken into consideration is that the blood-to-tumour tissue transport rate is influenced by the regional blood flow, the vascular permeability of the substance, the extent of vascularization, arteriovenous shunts, and the extracellular fluid circulation. In particular, the delivery of the antibody could be partially or completely blocked by a low blood flow. The regional blood flow in the tumour parenchyma is relatively high and the BBB is abnormal with an increased vascular permeability which may allow the delivery of monoclonal antibody into the brain tumour. A problem is the lower blood flow in the growing peripheral area where the BBB may be preserved. Since whole immunoglobulin crosses brain and tumour capillaries passively, the smaller size of antibody fragments may lead them to enter the tumour more efficiently. A higher specific localization in the tumour tissue by the F(ab')$_2$ fragments of monoclonal antibody and a substantially higher radiation dose to the tumour than to normal tissue have been demonstrated (Colapinto et al., 1988). Attempts at artificially altering the vascular permeability and opening the BBB have also been used as a means of enhancing monoclonal antibody access (Neuwelt, 1984).

Prior to the delivery of a therapeutic dose of isotopes conjugated with monoclonal antibody, imaging of patients for dosimetry is of critical importance (Foon, 1989). Paired-label studies in glioma patients revealed a definitive specific uptake of intravenous or intracarotid injected monoclonal antibody into glioma tissue, but with low localization indices in the tumour. In addition, it has been reported that tumour samples obtained at operation after injection of isotope-labelled monoclonal antibody contained levels of radioactivity which were clearly insufficient for therapeutic use (Behnke et al., 1988; Moseley et al., 1987). Similar low tumour uptake of monoclonal antibodies carrying radionuclides in patients with melanoma, breast, ovarian and gastrointestinal carcinomas have been reported. These findings in pre-clinical studies have led some to question the feasibility of using radiolabelled monoclonal antibodies for tumour therapy, reasoning that adequate radiation doses cannot be achieved in the tumour without unacceptable radiation exposure to normal tissues.

Despite the limitations mentioned above, several experimental clinical trials have been attempted for brain tumour patients. The efficacy of the trials is questionable. The pilot clinical studies have revealed at least the tolerability of a single infusion of murine monoclonal antibody in patients with malignant glioma (Brady et al., 1988). Another attractive way of using monoclonal antibody for antibody-guided irradiation is the intrathecal route for neoplastic meningitis. Under these conditions, the neoplastic cells are suspended within the CSF in the subarachnoid space, thus being more accessible to monoclonal antibody than cells in the solid mass. Some remissions using the intrathecal application of a radiolabelled monoclonal antibody in patients who suffered from neoplastic meningitis secondary to pineal tumours or melanomas have been reported (Coakham et al., 1988).

Biological Response Modifiers

Adjuvant immunotherapy with immunomodulating agents is an attempt to enhance or stimulate the suboptimal immune response of patients with brain malignancies. Early attempts at this type of therapy used the immunization with mycobacteria, bacille de Calmette-Guérin (BCG), Corynebacterium parvum or levamisole. Unfortunately, none of these approaches has been convincingly shown to be effective in controlled trials. Recent studies have concentrated on the use of the so-called biological response modifiers. Highly purified molecules including recombinant DNA-derived forms of interferons, interleukins and TNFs are now available. Some agents have reached the stage of clinical trials for brain tumour treatment.

Interferons

Three major types of human interferons have been identified: interferon-α interferon-β and interferon-γ. Numerous studies have shown that human interferon-α and interferon-β can inhibit the growth of human neuroectodermal tumours in murine models. In vitro studies have suggested that interferon-β exhibits a more marked direct antiproliferative activity on human glioma cell lines than either interferon-α or interferon-γ. A similar growth inhibitory effect of interferon-β has been reported on human glioma xenografts in nude mice. However, the growth inhibitory effects of interferons on experimental models have been variable. The mechanism of interferon-mediated antitumour effects in humans are unclear. In addition to a possible direct antiproliferative effect on the tumour, putative mechanisms have been postulated, such as NK and monocyte cytotoxicity, and modulation of surface antigens on the tumour cell membrane which may facilitate the recognition of tumour cells by host immunocompetent cells.

Several types of interferons, such as human lymphoblastoid interferon-α, fibroblast interferon-β and recombinant interferon-α, recombinant interferon-β and recombinant interferon-γ, are now available for clinical use (Nagai, 1988). Early works on various interferons in malignant brain tumour patients reported a 0%–40% response rate. Several large clinical studies using interferon-α or interferon-β have been performed for malignant brain tumour patients. Interferon-α or interferon-β was administered daily in doses ranging from 1.0 to 9.0×10^6 units/body locally, intravenously or intramuscularly, and was continued for approximately 4 weeks or longer (Nagai, 1988). A relative optimal responsiveness was obtained with interferon-β compared to interferon-α. These clinical trials reported 14%–28% response rates (significant tumour volume reduction) in glioma patients treated by interferon-β or interferon-α either locally or systematically. The efficacy via the local (intratumoral or intrathecal) route was similar to that of systemic administration. Long-term continuous administration of interferon-β is favourable in the suppression of tumour recurrence and growth. The overall results among a number of clinical reports have been equivocal and clearly interferons given as single agents have not given a major impact in the therapy of brain tumours. A controlled phase III study has not been yet carried out.

Transient side effects of interferon-β occurred in 60%–100% of patients, including fever, headache, lassitude, chills, convulsive seizures, hypotension, nausea and vomiting. Abnormal laboratory findings were seen in about 40% of cases, including leucopenia, thrombocytopenia, anaemia, and elevated serum GPT. Suppression of haematopoietic function and liver dysfunction were mild and generally returned to normal without withdrawal of the medication. However, in certain cases bone marrow suppression was profound enough to interrupt the therapy.

The receptor (type 2) for interferon-γ differs from the receptor (type 1) for interferon-α and interferon-β. Interferon-γ has demonstrated limited anti-glioma activity in vitro in comparison with interferon-α or interferon-β. A clinical trial of interferon-γ in glioma patients has been reported with disappointing results (Mahaley et al., 1988) and the side effects of interferon-γ are generally so severe that clinical trials of interferon-γ alone have been interrupted. In addition to their antitumour activity, interferons are potent regulators of cell gene expression, structure and function. Manipulation of the enhancement of tumour antigenicity and of host effector cytotoxicity is another potential area, especially for interferon-γ. Current trials of interferon therapy are directed to combinations with other types of immunotherapy or chemotherapeutic agents. In particular, interferon-γ at a proper low dose in order to enhance tumour immunogenicity might be utilized in combination with the other immunotherapeutical approaches.

Interleukin-2

IL-2 is a cytokine produced by activated T cells that activates a variety of lymphocyte populations and induces lymphocyte proliferation. IL-2-activated NK cells and T cells can mediate cytotoxicity against a variety of malignant brain tumours. Thus it is supposed that IL-2 should enhance tumour cell lysis by immune effector cells infiltrating brain tumour tissue. A huge dose of intravenous IL-2, approximately more than 10^6 units of IL-2 every 8 h, is required to achieve an adequate IL-2 dose (3–6 units/ml) in the CSF, which can activate LAK-precursor lymphocytes and/or maintain LAK cells in the CSF (Saris et al., 1988). Systemic IL-2 therapy as a single agent for cancer patients usually does not result in tumour regression, in spite of serious toxicity. Furthermore, the patients with intracranial metastasis have been excluded from the protocol of systemic IL-2 therapy because of the risk of brain oedema and poor results in pilot studies. At present, IL-2 has been utilized locally to augment the effect of intracranially transferred LAK cells. In preliminary clinical studies when IL-2 was injected into the peritumoral area of glioma patients either during or after surgery, no irreversible toxicity was reported (Jacobs *et al.*, 1986). In contrast, transient adverse effects were frequent including headache, moderate-grade fever, nausea, chill, mild confusion or occasional dysaesthesia in lower limbs after intratumoral or intrathecal injection of IL-2 (doses ranged from 10^3 to 2×10^4 units). In addition, an intravenous high-dose IL-2 therapy elicits mental status changes (confusion, disorientation or lethargy) in approximately one-third of patients, which may last days or weeks.

Tumour Necrosis Factor

TNF has originally been thought to have direct cytostatic and cytotoxic effects on tumour cells. Recently, two molecules have been characterized as TNF-α and TNF-β which are identical to cachectin and lymphotoxin, respectively. As opposed to previous expectations, TNF exhibits either minimal or no antiproliferative effects against glioma cell lines in vitro as assessed at escalating concentrations of 40–400 units/ml (Rutka *et al.*, 1988). Some phase I trials in advanced cancer patients reported that toxicity was present both in high and low doses, while antitumour activity was minimal. Therefore, TNF is currently being studied as an adjuvant agent in brain tumour therapy. Its immunomodulatory properties are activation of macrophages, enhancement of HLA-ABC expression and of certain tumour-associated antigens on human glioma cells (Zuber *et al.*, 1988).

Ok-432 and PS-K

The two largest selling drugs by dollar value in the cancer chemotherapy market including malignant brain tumours are the two immunomodulators Ok-432 (lyophilized powder of *Streptcoccus pyogenes* preparation) and PS-K (protein-bound polysaccharide Kureha form Basidiomycetes). It is noteworthy, however, that their use is largely limited to Japan. It has been reported that Ok-432 possesses various immunopharmacological activities, such as augmentation of cytotoxic activity and stimulation of cytokine production of various lymphocytes and macrophages. According to an early report of a cooperative study using Ok-432 with chemoradiation therapy including ACNU and vincristine, the survival of patients receiving Ok-432 was described to be significantly longer than that of controls. In contrast, another cooperative study demonstrated that survival rates of glioma patients treated with or without Ok-432 were similar, and that there was no clinical effect of Ok-432 (Shibata *et al.*, 1987).

Adoptive Cellular Therapy

Glioma patients' peripheral blood leucocytes (PBLs) can be activated by in vitro culture in the presence of IL-2. The activated PBLs, so-called LAK cells, become capable of killing autologous glioma cells and allogeneic glioma cells (Fig. 14.3). All normal brain cells are presumably LAK resist-

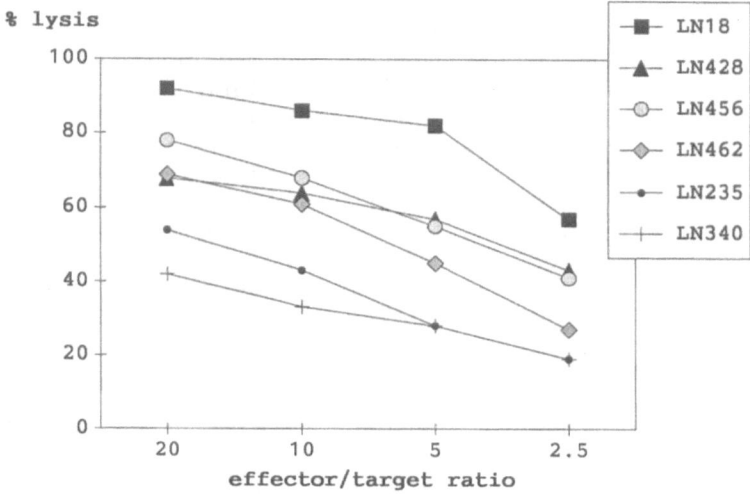

Fig. 14.3 Anti-glioma cytolytic activity of LAK cells. LAK cells were generated by seven days in vitro culture of peripheral blood lymphocytes in the presence of high concentrations of recombinant IL-2 (1000 U ml^{-1}). The cytolytic activity against six glioma cell lines (LN18, LN428, LN456, LN462, LN235, LN340) was evaluated by a 4-hour chromium release assay. LAK cells derived from one donor showed a capability of killing all allogeneic glioma cells. The susceptibility of each glioma cell was, however, significantly variable

ant. The antitumour effector cells in LAK cell culture are reported to be large granular lymphocytes. LAK cells are distinguished from NK cells only because of their IL-2 dependence. For clinical application, at least 10^9 up to 10^{11} LAK cells are required for local immunotherapy against malignant brain tumours. This therapy is usually scheduled in the following manner.

1. Precursor autologous peripheral blood mononuclear cells (10^8–10^{10} cells per one phaeresis) are prepared from patients by repetitive leuco-phaersis and enriched with PBLs by density-gradient separation.
2. PBLs are then activated in an in vitro culture system in the presence of high concentrations of IL-2 (1000–2000 U ml^{-1} medium) for approximately 3–5 days.
3. In vitro cultivation of PBLs yields non-adherent and adherent LAK cells, which can be expanded by further in vitro culture for several weeks or more. LAK cells are washed and resuspended with IL-2.
4. Adherent LAK cells have been demonstrated to be more potent in killing glioma cells than non-adherent populations.
5. LAK cells are transferred to patients by a direct injection into residual tumour tissue during a craniotomy or administered into the tumour cavity postoperatively through a subcutaneous reservoir.

Although glioma patients are known to have an impaired cell-mediated immune response, it is possible to activate PBLs derived from patients who have been treated by chemotherapy, irradiation or corticosteroid administration after surgery. Since the number of PBLs obtained from patients receiving those therapies is significantly reduced, the volume of patients' blood required to generate the same number of LAK cells was longer in glioma patients than in control subjects.

Adoptive immunotherapy, where LAK cells are transferred to patients with malignant brain tumours in combination with IL-2, has been reported in several preliminary clinical trials (Yoshida et al., 1988; Barba et al., 1989; Merchant et al., 1989). In these studies, approximately 1×10^9 to 10×10^9 (total dose) of autologous LAK cells were infused locally with 5×10^5 to 5×10^6 total units of IL-2. Improvement of clinical symptoms and tumour regression provided by LAK therapy have been demonstrated by some investigations. However, the therapy has not appeared yet to have a significant impact on patient survival and a majority of the reported remissions were transient. To improve the clinical result there has been an effort to transfer a more potent and larger number of effector cells. Although the local adoptive immunotherapy using LAK cells with IL-2 caused no irreversible toxicity, numerous reversible side effects have been observed including frequent fever, nausea, vomiting, headache, fatigability and occasional mild somnolence. In addition, an intracranial injection of large amounts of LAK cells caused cerebral oedema, increased intracranial pressure and hydrocephalus. The mechanisms of the adverse effects can, in part, be explained by endothelial injury induced by LAK cells or increased vascular permeability by IL-2.

The in situ activity of the transferred LAK cells is totally obscure. Since effector cells must physically bind to the tumour cells to kill, successful adoptive immunotherapy presumably depends on the accumulation of transferred effector cells at the site of tumour growth. Regarding the distribution pattern of LAK cells, these transferred cells remained localized at the injection site and did not appear to migrate preferentially to the peripheral site of tumour growth. It is thus supposed that one of the major limitations of therapy with LAK cells may be their inability to infiltrate actively the tumour tissue.

Another approach to adoptive immunotherapy is to expand tumour-specific T lymphocytes with IL-2 in vitro. In many animal models the therapeutic efficacy of tumour-specific CTL has been repeatedly emphasized. The existence of brain tumour-specific killer cells present in patients' PBL was suggested in early reports, and attempts to sensitize patients' PBL in vitro against autologous tumour cells have also been performed. However, the specificity of CTL clones against autologous tumour cells is doubtful. One should remember that a single human solid tumour is composed of a heterogeneous population of cells with respect to susceptibility to lysis by autologous CTL clones. This implies that a therapeutic modality must be capable of activating the overall immune response to the heterogeneous tumour cells rather than being directed to a single tumour clone.

Adoptive immunotherapy using autologous TILs in place of LAK cells has already been administered to cancer patients with the exception of brain tumour patients. The TILs have been expected to include CTL populations which can recognize the tumour-associated antigens together with self-antigens. Glioma TILs can also be isolated from a surgical specimen and expanded by an in vitro culture system with a substantial increase in cell numbers (10^{10} TILs from a glioma). The expanding glioma-derived TILs consist of approximately 90% $CD3^+$ T cells including both $CD4^+$ and $CD8^+$

subpopulations. Leu19 positive NK cells are also observed in TIL culture. In contrast to LAK cells, CD16, an NK cell marker, is expressed on a small percentage of the cells. However, the majority of TIL cultures cease to grow within six or eight weeks, losing their cytolytic activity. Therefore expanded TILs, which are quiescent or reaccumulating into the resting phase, may no longer be suitable as adoptively transferred effector cells. There is no evident specificity or selectivity for lysis of autologous glioma, inasmuch as the bulk TIL cultures are cytotoxic against both autologous and allogeneic glioma targets (Sawamura et al., 1989). On the other hand, in certain cases glioma-derived TILs obtained in microculture systems have been found to exhibit selective lytic activity against autologous tumour cells in comparison with broad LAK cell activity. At present, the definitive benefit of glioma-derived TIL over LAK cells as transferred cells is still not clear.

Conclusion

There are two obvious established and efficient modalities of malignant brain tumour treatment: surgical removal and radiation therapy. Their efficiency, however, is very limited and chemotherapy adds very little to it. Biological therapy must be considered as a fourth modality of brain tumour treatment. In the past decade, studies on brain tumour biology have shown that there is indeed a potential role for biological immune factors in therapy, although there has been only little practical success in the development of immunotherapy with interferons and even less with adoptive cellular therapy. The research in the immunology of brain tumours enabled us to gain some more insight into how we might augment the immune response to these tumours. Overall augmentation of immunity requires induction of increased effector function, in addition to concomitant abatement of the powerful suppressive activities originating from tumour itself. Since the host–tumour interaction is complex, more sophisticated combined approaches are necessary in order to establish effective immunotherapy.

References

Barba D, Saris SC, Holder RN et al. (1989) Intratumoral LAK cell and interleukin-2 therapy of human gliomas. J Neurosurg 70: 175–182

Behnke J, Coakham HB, Mach JP et al. (1988) Monoclonal antibodies in the diagnosis and therapy of brain tumors. In: Kornblith PL, Walker MD (eds) Advances in Neuro-oncology. Futura Publishing Co. Inc., Mount Kisco, NY, USA, pp 249–285

Brady LW, Woo DV, Karlsson U et al. (1988) Radioimmunotherapy of human gliomas using I-25 labeled monoclonal antibody to epidermal growth factor receptor. Proceedings of ASCO 7: p 83

Carrel S, de Tribolet N, Mach JP (1982) Expression of neuroectodermal antigens common to melanomas, gliomas and neuroblastomas. I. Identification by monoclonal anti-melanoma and anti-glioma antibodies. Acta Neuropathol 57: 158–164.

Coakham HB, Richardson RB, Davies AG et al. (1988) Neoplastic meningitis from a pineal tumor treated by antibody-guided irradiation via the intrathecal route. Br J Neurosurg 2: 299.

Colapinto EV, Humphrey PA, Zalutsky MR et al. (1988) Comparative localization of murine monoclonal antibody Mel-14 F(ab')2 fragment and whole IgG2a in human glioma xenografts. Cancer Res 48: 5701–5707

de Martin R, Haendler B, Hofer-Warbinek R et al. (1987) Complementary DNA for human glioblastoma-derived T cell suppressor factor, a novel member of the transforming growth factor-β gene family. EMBO J 6: 3673–3677

de Muralt B, de Tribolet N, Diserens AC et al. (1985) Phenotyping of 60 cultured human gliomas and 34 other neuroectodermal tumors by means of monoclonal antibodies against glioma, melanoma and HLA-DR antigen. Eur J Cancer Clin Oncol 21: 207–216

de Tribolet N, Hamou MF, Mach JP et al. (1984) Demonstration of HLA-DR antigens in normal human brain. J. Neurol Neurosurg Psychiat 47: 417–418

Fischer DK, Chen TL, Narayan RK (1988) Immunological and biochemical strategies for the identification of brain tumor-associated antigens. J. Neurosurg 68: 165–180.

Fontana A, Kristensen F, Dubs R et al. (1982) Production of prostaglandin E and an inter-leukin 1-like factor by cultured astrocytes and C6 glioma cells. J Immunol 129: 2413–2419

Foon KA (1989) Perspective in cancer research. Biological response modifiers: the new immunotherapy. Cancer Res 49: 1621–1639

Frank E, Pulver M, de Tribolet N (1986) Expression of class II major histocompability antigens on reactive astrocytes and endothelial cells within the gliosis surrounding metastasis and abscesses. J. Neuroimmunol 12: 29–36

Jacobs SK, Wilson DJ, Kornblith PL et al. (1986) Interleukin-2 or autologous lymphokine-activated killer cell treatment of malignant glioma: phase I trial. Cancer Res 46: 2101–2104

Johnson VG, Wrobel C, Wilson D et al. (1989) Improved tumor-specific immunotoxins in the treatment of CNS and leptomeningeal neoplasia. J. Neurosurg 70: 240–248

Kuppner MC, Hamou MF, Bodmer S et al. (1988) The glioblastoma-derived T-cell suppressor factor/transforming growth factor beta2 inhibits the generation of lymphokine-activated killer (LAK) cells. Int J Cancer 42: 562–567

Lee Y, Bullard DE, Humphrey PA et al. Treatment of intracranial human glioma xenografts with 131I-labeled anti-tenastin monoclonal antibody 81C6. Cancer Res 48: 2904–2910

Mahaley MS, Bertsch L, Cush S et al. (1988) Systematic gamma-interferon therapy for recurrent gliomas. J. Neurosurg 69: 826–829

Merchant RE, Merchant LH, Cook SHS et al. (1989) Intratumoral infusion of lymphokine-activated killer (LAK) cells and recombinant interleukin-2 (IL-2) for the treatment of patients with malignant brain tumor. Neurosurg 23: 725–732

Moseley R, Zalutsky MR, Coakham HB et al. (1987) Distribution of 131I 81C6 monoclonal antibody (Mab) administered via carotid artery in patients with glioma. J. Nucl Med 28: 603–604

Nagai M (1988) Clinical use of interferons in the treatment of malignant brain tumor. In: Revel M (ed) Clinical Aspects of Interferons. Kluwer Academic Publishers, Boston, pp 183–194

Neuwelt EA (1984) Therapeutic potential for blood brain barrier modification in malignant brain tumors. Prog Exp Tumor Res 28: 51–56

Rutka JT, Giblin JR, Berens ME et al. (1988) The effects of human recombinant tumor necrosis factor on glioma-derived cell lines: cellular proliferation, cytotoxicity, morphological and radioreceptor studies. Int J Cancer 41: 573–582

Saris SC, Rosenberg SA, Friedman RB et al. (1988) Penetration of recombinant interleukin-2 across the blood–cerebrospinal fluid barrier. J Neurosurg 69: 29–34

Sawamura Y, Hosokawa M, Kuppner MC et al. (1989) Antitumor activity and surface pheno-types of human glioma-infiltrating lymphocytes after in vitro expansion in the presence of interleukin-2. Cancer Res 49: 1843–1849

Shibata S, Mori K, Moriyama T et al. (1987) Randomized controlled study of the effect of adjuvant immunotherapy with Picibanil on 51 malignant gliomas. Surg Neurol 27: 259–263

Wikstrand CJ, McLendon RE, Carrel S et al. (1987) Comparative localization of glioma-reac-tive monoclonal antibodies in vivo in an athymic mouse human glioma xenograft model. J Neuroimmunol 15: 37–56

Yoshida S, Tanaka R, Takai N et al. (1988) Local administration of autologous lymphokine-activated killer cells and recombinant interleukin 2 to patients with malignant brain tumors. Cancer Res 48: 5011–5016

Zuber P, Accolla RS, Carrel S et al. (1988) Effects of recombinant human tumor necrosis factor-a on the surface phenotype and the growth of human malignant glioma cell lines. Int J Cancer 42: 780–786

Index